Biophysical Chemistry of Membrane Functions

Biophysical Chemistry of Membrane Functions

Arnošt Kotyk
Institute of Physiology
Karel Janáček
Institute of Microbiology
Jiří Koryta
J. Heyrovský Institute of Physical Chemistry and Electrochemistry
Czechoslovak Academy of Sciences, Prague

Wiley-Interscience

JOHN WILEY & SONS
Chichester · New York · Brisbane · Toronto · Singapore

Library of Congress Cataloging-in-Publication Data:

Kotyk, Arnošt.
Biophysical chemistry of membrane functions.

1. Cell membranes. 2. Cytochemistry. 3. Biophysics.
4. Biological transport. I. Janáček, Karel
II. Koryta, Jiři. III. Title. [DNLM: 1. Biological
Transport. 2. Cell Membrane—physiology. 3. Membrane
Proteins—metabolism. QH 601 K87b]
QH601.K67 1988 574.87′5 87-13325
ISBN 0 471 91657 9

British Library Cataloguing in Publication Data:

Kotyk, Arnošt
Biophysical chemistry of membrane functions.
1. Cell membranes
I. Title II. Janáček, Karel
III. Koryta, Jiří
574.87′5 QH601

ISBN 0 471 91657 9

Phototypesetting by Thomson Press (India) Ltd., New Delhi
Printed in Great Britain by Page Bros. (Norwich) Ltd.

Contents

Most of the sections of the book were written by the first author. The ones written by Karel Janáček are specified by KJ, the ones written by Jiří Koryta by JK.

List of Abbreviations ... xi
List of Symbols ... xiii
Preface ... xvii

1 The Membrane Principle of Cell Organization ... 1
1.1 Biomembranes past and present ... 1
1.2 Principal membrane functions ... 5
1.3 Membranes of cells and organelles ... 8
1.3.1 The plasma membrane ... 8
1.3.2 The nuclear membrane ... 20
1.3.3 The endoplasmic reticulum ... 22
1.3.4 The Golgi apparatus ... 23
1.3.5 The lysosomes ... 25
1.3.6 The vacuole ... 26
1.3.7 The microbodies ... 27
1.3.8 The mitochondrion ... 30
1.3.9 The chloroplast ... 34
1.3.10 Chromaffin granules ... 39
1.3.11 Other membranous organelles ... 39

2 Membrane Structure ... 41
2.1 Structure-forming elements of cells ... 41
2.2 Types of membrane lipids ... 43
2.2.1 Glycerolipids ... 43
2.2.2 Sphingolipids ... 45
2.2.3 Sterols ... 47
2.2.4 Other nonprotein membrane components ... 48
2.2.5 Distribution of lipids ... 49
2.2.6 Fatty acids in membrane lipids ... 51
2.3 Organization of membrane lipids ... 54
2.3.1 Different forms of lipid structure ... 54
2.3.2 Asymmetry of membrane lipids ... 57

2.4 Physical state of lipids in membranes 60
- 2.4.1 Membrane microviscosity 63
- 2.4.2 Membrane order 65
- 2.4.3 Phase transitions 66

2.5 Mobility within and of lipid membranes 73
- 2.5.1 Rotation of intramolecular bonds 73
- 2.5.2 Rotation of lipid molecules 73
- 2.5.3 Lateral diffusion 74
- 2.5.4 Transbilayer movement 75
- 2.5.5 Movements of the bilayer and membrane stability 76

2.6 Membrane proteins 80
- 2.6.1 Types of membrane proteins 81
  - Glycoproteins 86
  - Lipoproteins 87
- 2.6.2 Mobility of membrane proteins 88
- 2.6.3 Interaction of proteins with lipids 92

2.7 Electrostatic properties of membranes (JK) 96

2.8 Membrane biogenesis 104
- 2.8.1 Lipid synthesis and turnover 105
- 2.8.2 Protein synthesis and turnover 108
  - ER-dependent membrane protein synthesis 109
  - Membrane protein synthesis in bacteria 113
  - ER-independent membrane protein synthesis 113
  - Membrane protein turnover 115

3 Membranes as Mediators of Flow of Matter **117**

3.A FUNDAMENTAL CONCEPTS OF THERMODYNAMICS (KJ) 117
- Thermodynamic systems 117
- Zeroth law 118
- First law 118
- Second law 118
- Free energy 119
- Chemical potential 120
- Electrochemical potential 121
- Energetics of chemical reactions 121

3.B REACTION KINETICS (KJ + AK) 122
- Diffusion-controlled reactions 124
- Enzyme-catalyzed reactions 125

3.1 Membrane-bound enzymes 125

3.2 Membrane transport 129

3.3 Nonspecific permeation 131
- 3.3.1 General aspects 131

3.C PRINCIPLES OF TRANSPORT IN THE CONTINUOUS PHASE (JK) 131

3.3.2 Diffusion of nonelectrolytes across membranes (KJ) .. 138
3.D PARTITION EQUILIBRIUM (JK) 140
Particle–solvent interaction 142
Rate theory of membrane transport 143
Molecular dynamics 148
Effects of the so-called unstirred layers 149
3.3.3 Electrodiffusion and membrane potentials (KJ) 150
The rate theory of ion transport 152
Goldman's solution of electrodiffusion 154
Piek's extension of the Goldman equation 156
The Hodgkin–Horowicz equation and its extensions 158
Addition of an electrogenic pump 159
Ussing's flux ratio 160
3.3.4 Mechanisms of simple diffusion 161
Passage through nonpolar membrane parts 161
Passage through polar channels 162
3.3.5 Membrane currents (KJ + JK) 163
3.3.6 Specifics of water transport (KJ) 168
3.E THERMODYNAMICS OF IRREVERSIBLE PROCESSES (KJ) 168
Osmotic equilibrium 169
Flow of water and of solute 171
Electro-osmosis 177
3.4 Molecular basis of aqueous channels 181
3.4.1 Connexin channels 181
3.4.2 Porins 181
Prokaryotic porins 182
Organelle porins 183
3.4.3 Extraneous pore formers 184
The membrane-attack complex of complement 184
Polyene antibiotics and saponins 184
Bacteriocins 185
3.5 Protein-mediated transport 185
3.5.1 Introduction 185
3.5.2 Rate of transport through a specific channel 187
3.5.3 Rate of transport by a specific carrier 189
3.5.4 Kinetics of a two-site carrier 194
3.5.5 Kinetic identity of carriers with multiconformational channels 195
3.5.6 Active transport 197
Thermodynamic approach (KJ) 198
Kinetic approach 202
Primary active transport 202
Secondary active transport 203
Antiport systems 207

3.5.7 Properties of channel and carrier systems 208
Permeability.. 208
Dependence on temperature..................... 208
Dependence on pH............................... 209
Effect of surface potential 210
Effect of unstirred layers 210
3.5.8 Transport-associated chemical reactions............. 211
3.6 Specific channel and carrier proteins...................... 213
3.6.1 Band 3 protein.. 213
3.6.2 The ADP/ATP carrier of mitochondria 214
3.6.3 The lactose transport system of bacteria 215
3.6.4 Binding protein systems in bacteria................. 215
3.6.5 Sodium-glucose symport of the intestinal brush border 217
3.6.6 Extraneous carrier like molecules 217
3.6.7 Extraneous specific channel formers 218
3.7 Transport of macromolecules.............................. 221
3.7.1 Introduction.. 221
3.7.2 Mechanisms without vesiculation 221
3.7.3 Mechanisms with vesiculation 222
Fluid-phase endocytosis.......................... 223
Endocytosis via coated pits and vesicles........... 223
Receptor-mediated endocytosis.................... 225
Exocytosis 227
3.7.4 Intermediate mechanisms........................... 230
3.7.5 Cell fusion 231

4 Membranes as Convertors of Energy **234**
4.1 Flowsheet of bioenergetics (JK + AK).................... 234
4.2 Radiation energy conversion into electrochemical potential gradients.. 238
4.2.1 Photosynthesis without chlorophyll................. 239
4.2.2 Chlorophyll-associated photosynthesis 241
General organization of photosynthetic systems.... 242
Organization of the reaction centres 245
Transfer of excitation energy in photosynthetic membranes.. 250
Neutralization of charge in the PS II reaction centre 255
Electron transfer in the photosynthetic membrane.. 255
Sites of proton translocation.................... 260
Regulation of energy distribution between PS I and PS II 260
4.3 Free energy of oxidation converted into other forms of energy 262
4.3.1 General principles.................................. 262
4.A ELECTRON TRANSFER PROCESSES (JK)............ 263
4.3.2 Free energy of oxidation converted into electrochemical potential gradients 269

Terminal oxidation in mitochondria 269
Components of the electron transfer chain......... 269
Complex I 269
Complex II.................................. 271
Ubiquinone (Coenzyme Q) 271
Complex III 271
Cytochrome *c* 273
Complex IV 273
Summary of reactions 276
The path of electrons 276
The paths of protons 276
The protonmotive force 278
Plant mitochondria 279
Oxidations in bacteria 279
Electron transport under anaerobic conditions..... 279
Electron transport under aerobic conditions 281
Components of the bacterial electron-transport chain 282
Catabolism in 'Na-motive' bacteria 283
Redox reactions in secondary active transport 285
4.3.3 Free energy of oxidation converted into electromagnetic radiation 285
4.4 Electrochemical potential gradient of ions converted into other forms of energy 293
4.4.1 Ion gradients generating gradients of other ions or molecules 293
4.4.2 Ion gradient energy converted into chemical energy .. 294
4.4.3 Conversion of $\Delta\tilde{\mu}_{ion}$ into mechanical energy 300
4.4.4 Conversion of $\Delta\tilde{\mu}_{H^+}$ into heat 302
4.5 Hydrolysis of chemical bonds for driving transport systems.. 303
4.5.1 Membrane adenosinetriphosphatases 303
Na, K-ATPase 307
Ca-ATPase 307
K, H-ATPase 309
H-ATPase 310
V-type ATPases 310
4.5.2 The phosphotransferase system 311
4.5.3 Group translocation without phosphate 312

5 Processing of Information in Membranes **314**
5.A INFORMATION PRINCIPLES IN BIOLOGY (KJ)...... 314
5.1 Physical signals 316
5.1.1 Light as a signal 316
Photomovement 316
Visual rhodopsin 318
5.1.2 Heat as a signal 319
5.1.3 Electric potential as a signal 321

5.1.4 Magnetic field as a signal 321
5.1.5 Response to mechanical pressure.................. 322
Cell adhesion.................................. 322
Inner ear hair cells 323
5.2 Chemical attractants and repellents....................... 324
5.2.1 Methylation-dependent chemotaxis................ 325
5.2.2 Methylation-independent chemotaxis 327
5.3 Chemical signalling in multicellular organisms 328
5.3.1 Extrinsic chemical signals 328
5.3.2 Intrinsic chemical signals......................... 328
Hormones as first messengers.................... 329
Information carried by the nerve impulse 337
Neurotransmitters........................... 339
Action potential............................ 341
5.4 Signalling in the immune system 349

CODA **354**

General references **355**

Index **357**

List of Abbreviations

Ab	antibody
Ag	antigen
ADP	adenosine diphosphate
AMP	adenosine monophosphate
ATP	adenosine triphosphate
BChl	bacteriochlorophyll
BPh	bacteriopheophytin
bR	bacteriorhodopsin
cAMP	cyclic adenosine 3′, 5′-monophosphate
cGMP	cyclic guanosine 3′, 5′-monophosphate
Ch	cholesterol
Chl	chlorophyll
CL	cardiolipin
CoA	coenzyme A
CTP	cytidine triphosphate
cyt	cytochrome
DAP	dihydroxyacetone phosphate
dATP	deoxyadenosine triphosphate
DCCD	dicyclohexylcarbodiimide
DES	diethylstilbestrol
DIDS	4, 4′-diisothiocyano-2, 2′-stilbene disulphonate
DPG	diphosphatidylglycerol
DNP	2, 4-dinitrophenol
FA	fatty acid
FAD	flavin adenine dinucleotide
FCCP	carbonylcyanide fluorophenylhydrazone
Fd	ferredoxin
FMN	flavin mononucleotide (riboflavin 5′-phosphate)

Fuc	fucose
Gal	galactose
GalNAc	N-acetyl-D-galactosamine
Glc	glucose
GlcNAc	N-acetyl-D-glucosamine
GTP	guanosine triphosphate
hR	halorhodopsin
Ig	immunoglobulin
ITIES	interface of two immiscible electrolyte solutions
ITP	inosine triphosphate
LPC	lysophosphatidylcholine
Man	mannose
NAD	nicotinamide adenine dinucleotide
NADP	nicotinamide adenine dinucleotide phosphate
NEM	N-ethylmaleimide
PC	phosphatidylcholine; plastocyanin
PCMB	*p*-chloromercuribenzoate
PA	phosphatidic acid
PE	phosphatidylethanolamine
PG	phosphatidylglycerol
Ph	pheophytin
PI	phosphatidylinositol
PQ	plastoquinone
PS	phosphatidylserine; photosystem
Q	quinone
SA	sialic (neuraminic) acid
SITS	4-acetamido-4′-isothiocyano-2, 2′-stilbenedisulphonic acid
SP	sphingomyelin
sR	sensory rhodopsin
TG	triglyceride
TPAs	tetraphenylarsonium
TPB	tetraphenylborate
TPT	triphenyltin
UQ	ubiquinone
UTP	uridine triphosphate
TMT	trimethyltin

List of Symbols

$^{\circ}$(such as $\mu^{\circ}, G^{\circ}, E^{\circ}$)	standard quantities relevant when activities rather than concentrations are used
$^{\ominus}$(such as $\mu^{\ominus}, E^{\ominus}$)	formal quantities relevant when concentrations rather than activities are used and the pH is defined
$^{\neq}$(such as $K^{\neq}, G^{\neq}, S^{\neq}, H^{\neq}$)	quantities related to the activated state of reaction
A	Helmholtz free energy
A	surface area
$\mathscr{A}, A_r$	affinity of a chemical reaction
a_i	activity of substance i
C	capacity
$\bar{c}$	mean or equilibrium concentration
c_i	molar concentration of substance i (in mol dm$^{-3} \equiv$ M)
c_m	membrane capacity (per unit length)
D	diffusion coefficient
d	distance
E	elastic energy
E	redox potential
E_a	activation energy
E_i	Nernst–Donnan potential of substance i
e	elementary charge ($1.602 \cdot 10^{-19}$ C)
F	Faraday constant (96.49 kC mol^{-1})
G	Gibbs free energy
G_i	membrane conductivity for substance i
H	enthalpy (heat content)
$\mathscr{H}$	Hamiltonian
h	Planck constant ($6.626 \cdot 10^{-34}$ J s)
I	ionic strength
i	electric current

J	flux
$\vec{J}, \overleftarrow{J}$	unidirectional fluxes
J_{max}	maximum rate of transport
j_i	electric current carried by ion i
K	equilibrium constant
K	splay bending elasticity
$\bar{K}$	saddle-splay bending elasticity
K_m, K_T	Michaelis or half-saturation constant of an enzyme and transport reaction, respectively
k	rate constant
k_B	Boltzmann constant ($1.38 \cdot 10^{-23}$ J K^{-1})
L_D	Debye length
L_{ij}	phenomenological coefficients
l	membrane thickness
M	molar mass
M_r	relative molar mass (molecular weight)
m	molecular mass
N_A	Avogadro number ($6.023 \cdot 10^{23}$ mol^{-1})
P	permeability coefficient (constant)
p	momentum
p	pressure
q	generalized coordinate
q	heat
R	gas constant (8.314 J mol^{-1} K^{-1})
R_{ij}	phenomenological coefficients of resistance nature
r	resistance
S	entropy
S	order parameter
T	absolute temperature
T_i	transference number for substance i
t	time
U	difference of electric potential
U	internal energy
U_i	electrolytic mobility of substance i
u_i	mobility of substance i
V	volume
$\bar{V}_i$	partial molal volume of substance i

w	work
w'	useful work
W	probability
X_i (i = 1, 2. . . .)	generalized force in steady-state thermodynamics
X_i	mole fraction of substance *i*
Z	number of particle collisions
z	valency (charge number)
α	coefficient of thermal expansion
β	adsorption coefficient
β	coefficient of compressibility
β	partition coefficient
Γ_i	surface concentration of component *i*
γ	activity coefficient
γ	interfacial tension
δ, δ_N	thickness of the unstirred (Nernst) layer
δ_0	thickness of the Prandtl layer
ε	permittivity, dielectric constant
ζ	electrokinetic potential
η	viscosity
θ	surface free energy
κ	compressibility
κ	conductivity
λ	ionic conductivity
λ	lattice distance
μ_i	chemical potential of substance *i*
$\tilde{\mu}_i$	electrochemical potential of substance *i*
ν	frequency
ν	stoichiometric coefficient
ξ	pore edge energy
$d\xi$	extent of chemical reaction
π	osmotic pressure
ρ	density
σ	reflection coefficient
τ	lifetime
Φ	dissipation function
ϕ	electric potential
$\Delta\phi_m$	membrane potential
ψ, ψ_s	surface potential

Preface

Membranology is a highly interdisciplinary science and to write a book about it requires merging three major scientific areas, biology, chemistry and physics. This is to justify the tripartite title of the book. By sheer coincidence, the functions dealt with are also of three categories, flow of matter, flow of energy and flow of information. And, for better or worse, three authors cooperated in writing the volume, most parts having been written by a single author, but all of them discussed by all three.

We spent hours debating the scope, the level of sophistication and the extent of literature references to be cited. In the end, we decided to attempt to write a detailed textbook-like volume, with as much concrete information as possible without jeopardizing its readability, and of necessity to cut down the number of literature sources to a selected few.

We are indebted to several of our colleagues who read the parts that were farther off the area of our main interest, in particular to Dr I. Šetlík of the Institute of Microbiology, Czechoslovak Academy of Sciences, Třeboň. Our thanks are due to Dr Jiří Pečený who prepared about a half of the drawings appearing in the book.

Dr Tomáš Soukup of the Institute of Physiology, Dr Eva Streiblová of the Institute of Microbiology, both in Prague, and Professor Oldřich Nečas with his team at the Faculty of Medicine, J. E. Purkyně University, in Brno, provided us with some unique microphotographs and their assistance is gratefully acknowledged.

The first author is especially thankful to his friend Dr Jiří Luštinec who let him use for several weeks the facilities of his mountain retreat where, far from the irritating obligations of living and working in a large city but with a carload of books and reprints (and a grand piano for the moments or despair), he was able to complete the major part of the manuscript.

Prague, February 1987

ARNOŠT KOTYK
KAREL JANÁČEK
JIŘÍ KORYTA

1 *The Membrane Principle of Cell Organization*

1.1 BIOMEMBRANES PAST AND PRESENT

Periods of data accumulation alternate with periods of synthetic accomplishments. This holds perhaps more in biology than in other areas of science and membranology is no exception. After half a century of gathering information on membranes—and during the early years of seeking arguments to support their credibility—a new cell biology has emerged, one that has membranes in its emblem.

Membranes have not only been recognized as objective entities but they have been found to be endowed with an immense variety of capabilities indispensable for cell survival. Those who see farther ahead in the epistemology of biology and probe deeper into the abysses of subcellular life recently attempted to categorize the principles according to which life—represented materially by the cell—is constructed to function as it has for the nearly four billions of years.

One of the architectural principles put forth, and certainly the more ubiquitous of the two, is the membrane principle of cell organization, the companion principle being that of the cytoskeleton. This principle can be described by stating that living matter in its functional form is composed of self-contained compartments with different internal contents, some of the compartments being often enclosed within other, larger compartments. The boundaries of these compartments are of a uniform fundamental type and are called membranes, biomembranes or biological membranes. The principle is obviously independent of the material from which membranes are composed—it would be equally valid if membranes were made of pure protein, polysaccharide or, for that matter, polymethacrylate. However, although many types of membranes are easily constructed by polymer chemists, biomembranes are unique, being characterized as organized planar structures made of lipids, with an appreciable content of admixed proteins.

The origin of membranes is obscure and probably will remain so. The earliest indications of the existence of material that has some of the physicochemical properties of membranes take the origin back to the time when the solar system itself became differentiated: The interior of the Murchison carbonaceous chondrite (1–3 per cent organic carbon), a meteorite found immediately after

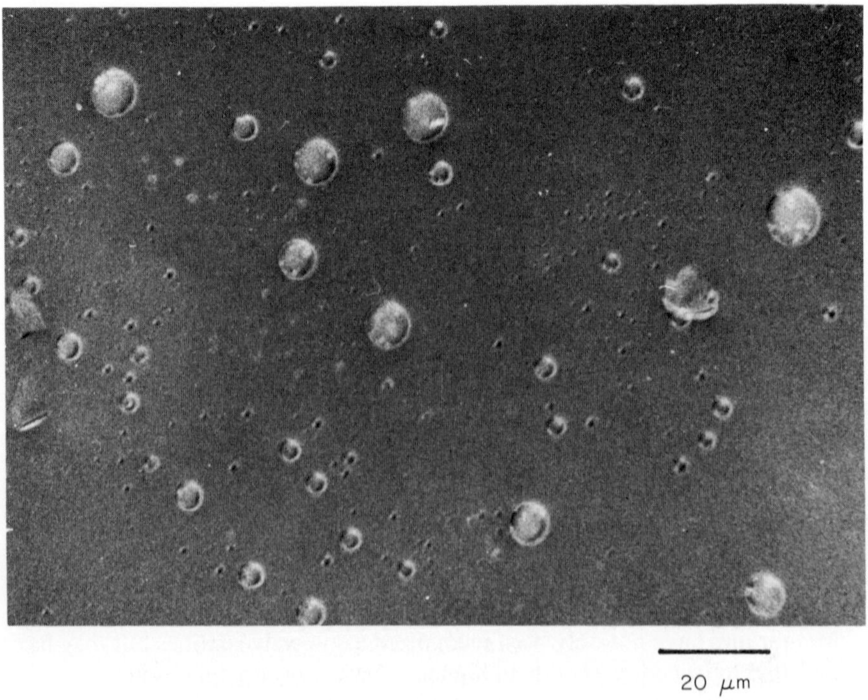

Fig. 1.1. 'Lipid' droplets isolated from the Murchison meteorite. (Courtesy of Dr D. W. Deamer, Davis, CF, USA)

its fall in Australia in 1969, was extracted to obtain a selection of amino acids plus carbon material which, when mixed with water, produced globular structures that, in many respects resembled lipid droplets (Fig. 1.1). This would date the first 'membrane material' at about 4.5 billion years old, certainly well before any life existed on Earth or in the rest of the solar system. It must have taken another billion years before true membranes appeared together with the earliest forms of life.

There are, on the one hand, the finds of bacteria- or yeast like forms in several different localities, viz. Isua in SW Greenland, Pilbara in Western Australia and Onverwacht in South Africa, dating back $3.5–3.8 \cdot 10^9$ years (Fig. 1.2). Not only are the fossilized 'cellular' forms clearly discernible (Fig. 1.3) and have been assigned systematic names (*Isuasphaera* from Greenland, *Ramsaysphaera* from South Africa) but their chemical analysis revealed the presence of isoprenoid hydrocarbons, such as phytane ($C_{20}H_{42}$) and pristane ($C_{19}H_{40}$), obviously degradation products of chlorophyll side chain alcohol phytol ($C_{20}H_{41}O$), as well as straight-chain alkanes, the carotenoid sporopellenin and remnants of porphyrins and an increased ratio of $^{12}C/^{13}C$ which is indicative of photo-assimilation of CO_2. As all the compounds mentioned occur in contemporary cells exclusively in membranes (the porphyrin-containing haemoglobin and

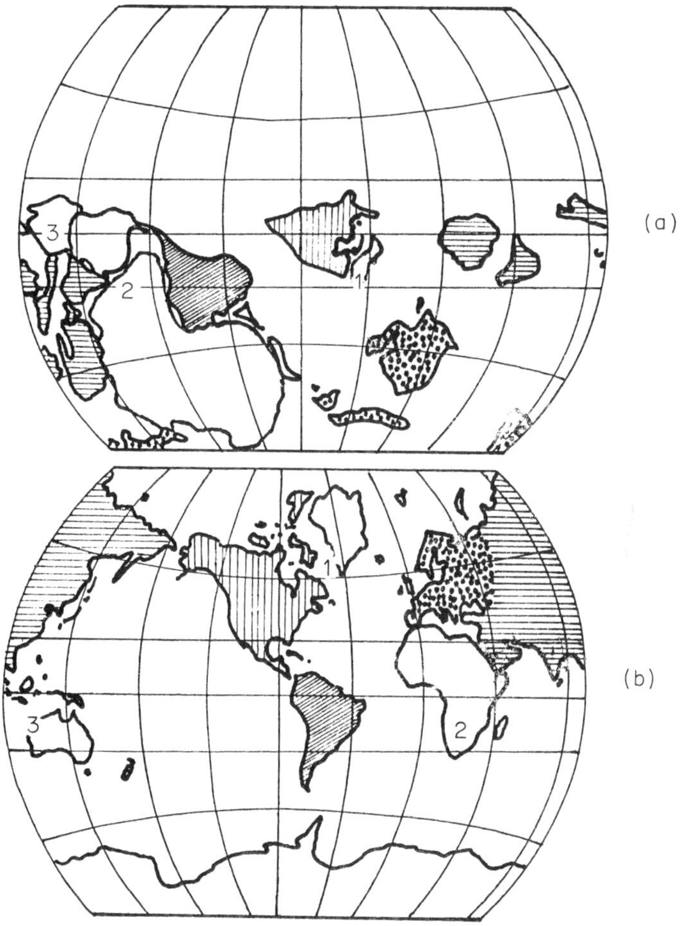

Fig. 1.2. Localities where primitive prokaryotic forms were found in pre-Cambrian shales. 1—Isua; 2—Onverwacht; 3—Pilbara. (a) situation 550 million years ago; (b) present topography

myoglobin and soluble enzymes in the cytosol are of much later derivation) it may be surmised that the primeval cells contained lipid membranes. Moreover, apparently even the earliest cells were capable of some type of photosynthesis, apparently anoxygenic, as massive oxygen release appeared on this planet only about $2.5 \cdot 10^9$ years ago.

The perseverance of sterol-like, isoprenoid compounds is documented by the presence in oil-bearing sedimentary strata of vast amounts of pentacyclic molecules called the hopanoids (after hopane, a kind of triterpene). It is claimed that the quantity of hopanoids in the fossil oils exceeds substantially the total mass of organic matter in the biosphere.

Whether the membranes enclosing the hypothetical precellular forms of life

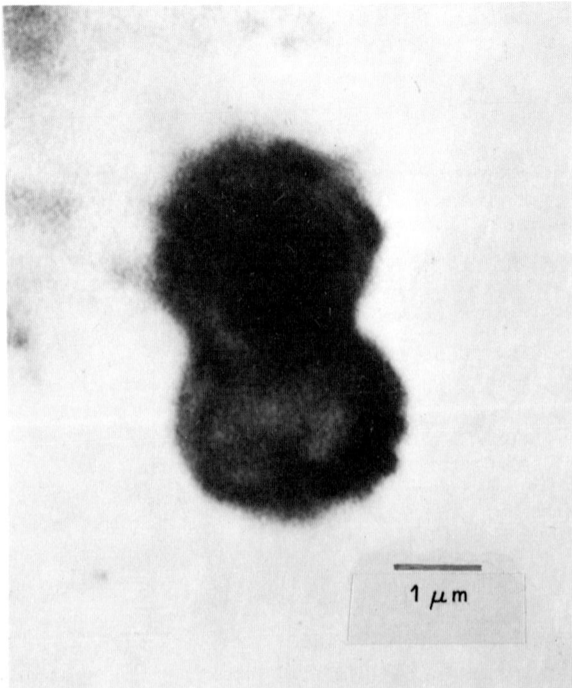

Fig. 1.3. A fossilized pre-Cambrian cell undergoing division; from South Africa. (Courtesy of Dr H. Reinbothe, Halle, GDR)

(precytes) were made of lipids other than hopanoids or whether they contained much protein is something that is not amenable to experimental verification within the bounds of the Earth.

For all one knows the primordial 'membranes' may have formed on a microcrystalline clay support (such as montmorillonite) from a completely different, abiotic, material and it was only after some basic biosynthetic process had become available that modern-type lipid membranes were formed. At any rate, biomembranes have apparently been around for the entire duration of cell-associated life. Their function as barriers between compartments of different composition, most typically intracellular vs. extracellular, is often considered to be primary, both from the point of view of evolution and in the context of the present-day versatile functions of membranes. Membranes, by definition separate different compartments and this is their universal property, which is implicit even in some of the more complicated membrane-bound events.

However, some functions other than the separatory one were apparently present already in the primordial membranes and endowed them with the ability to maintain the cell's continued existence. Likewise, some functions, particularly of cell organelles, such as the inner mitochondrial membrane or the thylakoid or the Golgi body, are physiologically equally or more important than the separation of two different compartments.

SYNOPSIS

Lipid-protein membranes are the fundamental structural elements of all cells and cell organelles. Such membranes probably surrounded the primordial cells on the Earth as early as $3.8 \cdot 10^9$ years ago.

FURTHER READING

Cairns-Smith, A. G. The first organisms, *Sci. Amer.*, **252**, 74–82 (1985).
Ourisson, G., Albrecht, P., and Rohmer, M. The microbial origin of fossil fuels, *Sci. Amer.*, **251**, 34–41 (1984).
Reinbothe, H., and Krauss, G.-J. *Entstehung und molekulare Evolution des Lebens*, G. Fischer-Verlag, Jena (1982).
Schwemmler, W. *Reconstruction of Cell Evolution: A Periodic System*, CRC Press, Boca Raton (1984).

1.2 PRINCIPAL MEMBRANE FUNCTIONS

Whatever the primeval function of biomembranes may have been the present-day membranes are instrumental in three categories of phenomena that would not occur as expeditiously without them.

1. Flow of matter—translocation of substances through membranes.
2. Flow of energy—transduction of one form of energy into another.
3. Flow of information—transformation of physical or chemical signals.

The *flow of matter* can be of two principally different types: (1) movement of material within membranes, including rotation, lateral diffusion, flip-flop movements, as well as shifts of molecules during cell fusion and exchange of membrane components; (2) movement of extraneous material from one side to the other, as in vectorial transport, as well as movements, vectorial or scalar, of substrates and products during reactions catalyzed by membrane-bound enzymes, including some types of protein synthesis.

Most of the *flow of energy* is restricted to special membranes, the bacterial, the inner mitochondrial and the thylakoid, where the free energy released by oxidation of substrates or the energy of electromagnetic radiation can be harnessed and stored in the form of an electrochemical potential gradient of H^+ and, exceptionally, of Na^+ or even of Cl^-, and ultimately in the structure of ATP or analogous, 'energy-rich' compounds. Some of the energy transduction (and in resting cells a sizable portion of it) has to do with active membrane transport, a more ubiquitous property of plasma, as well as intracellular, membranes.

Only some membranes specialize in facilitating the *flow of information*. They must be equipped with suitable receptors for tactical stimuli, for hormones, for

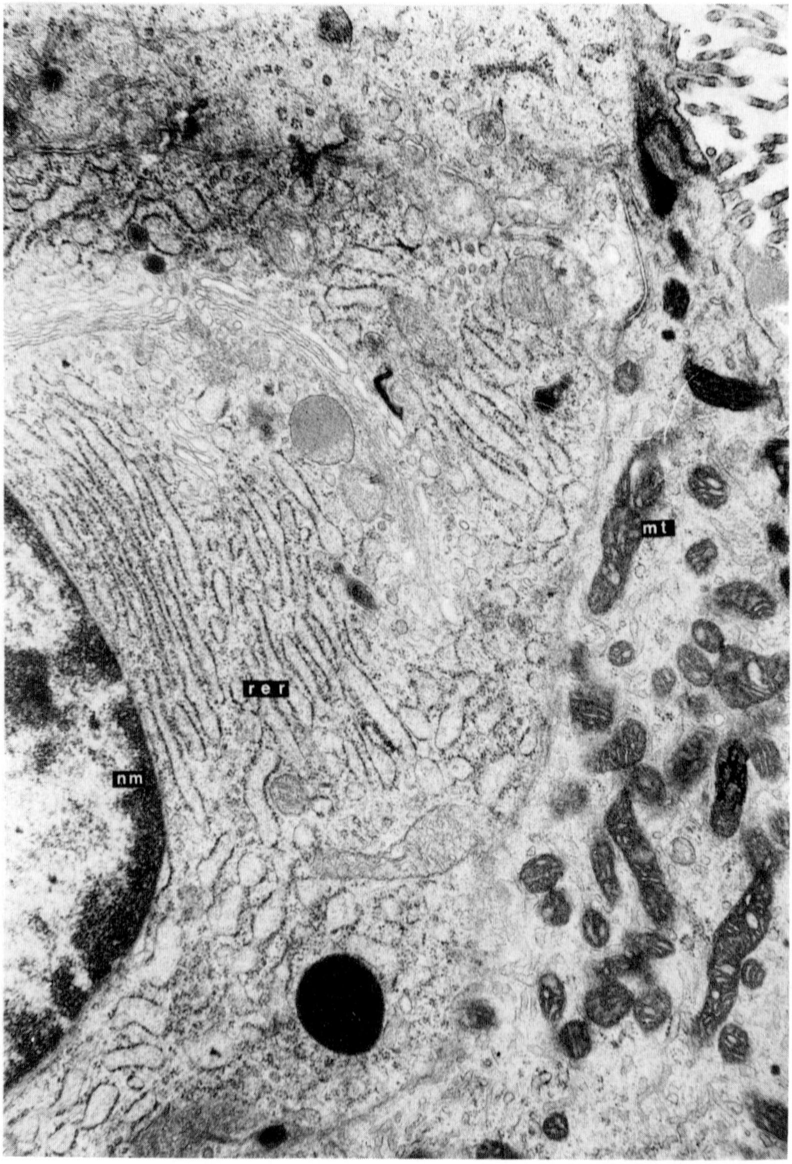

(a)

1 μm

Fig. 1.4. (a) An ultrathin section through rat respiratory epithelial cell. nm—nuclear membrane; mt—mitochondrion; rer—rough endoplasmic reticulum. (Courtesy of Dr T. Soukup, Institute of Physiology, Prague)

(b)

Fig. 1.4. (b) A freeze-etched preparation of the yeast *Saccharomyces cerevisiae* embedded in ethylene glycol. The lower smaller bud on a mother cell shows a nearly smooth vacuole, several storage vesicles, and a pitted nuclear membrane, part of a nucleus just passing through the neck between cells. The mother cell contains, in addition, several oval mitochondria to the right. The straight line nearly at the top is part of the endoplasmic reticulum. (Courtesy of Professor O. Nečas, J. E. Purkyně University, Brno) antigens or for whatever piece of information may arrive from the outside. Furthermore, they must contain the appropriate machinery—mainly of enzyme nature—to process the information received through the binding to a form better recognized by intracellular components where the final acceptor of the external stimulus is usually found.

SYNOPSIS

Biomembranes can perform a variety of functions. They are involved in the flow of material, either their own or extraneous, such as transmembrane movement of solute molecules and ions, or enzyme-catalyzed conversions of substrates; they are singularly active in the transductions of various forms of energy, especially where the electrochemical potential gradients of H^+ and Na^+ are involved; they mediate the transmission of physical and chemical signals between the environment and the cell or organelle interior.

FURTHER READING

Finean, J. B., Coleman, A., and Michell, R. H. *Membranes and Their Cellular Functions*, Blackwell Sci. Publ., Oxford (1984).

1.3 MEMBRANES OF CELLS AND ORGANELLES

The membrane principle of cell organization can be best appreciated by examining electron microphotographs prepared from ultrathin (50–100 nm thick) sections and from freeze-etched objects, the first-named preparations showing the two-dimensional layout of membranes in cells, the second then providing some idea of the three-dimensional disposition of the cell (Fig. 1.4). It may help the reader in his progress through the subsequent chapters if he realizes that the internal topology of all cells is such that every compartment is lined by membrane surface of the same type, the corollary of this being that across the lining membranes must lie compartments of a different type, themselves enclosed by a unique kind of membrane. This is indeed so and intracellular compartments can be divided into two classes, the plasmatic and the noncytoplasmatic ones, such as the examples of a schematized bacterial, plant and animal cell indicate (Fig. 1.5(a), (b), (c)).

Before proceeding to discuss the functions of membranes in specific detail (see Chapters 2–5), let us consider the properties of various cell membranes and point out which of the main categories of events take place in them.

1.3.1 The plasma membrane

This is the name reserved for the membrane surrounding the protoplast, i.e. the entire cell interior. In most animal cells, the plasma membrane is the outermost structure of the cell while in plants, fungi and prokaryotes, it is surrounded by

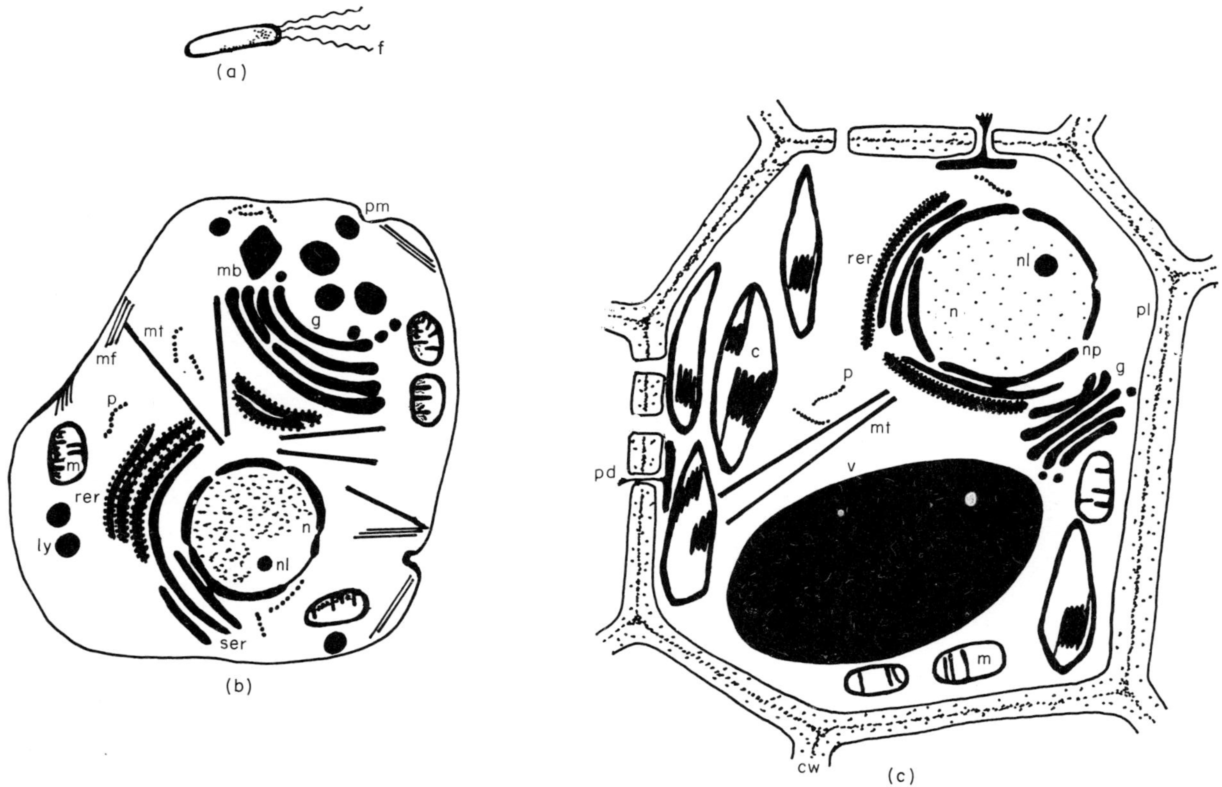

Fig. 1.5. Schematic drawings of (a) a bacterium, (b) an animal cell, and (c) a plant cell. The noncytoplasmic compartments are shown in black. c—Chloroplast, cw—cell wall, f—flagellum, g—Golgi complex, ly—lysosome, m—mitochondrion, mb—microbody, mt—microtubule, n—nucleus, nl—nucleolus, np—nuclear pore, p—polysome, pd—plasmodesma, pm—plasma membrane, rer—rough endoplasmic reticulum, ser—smooth endoplasmic reticulum, v—vacuole

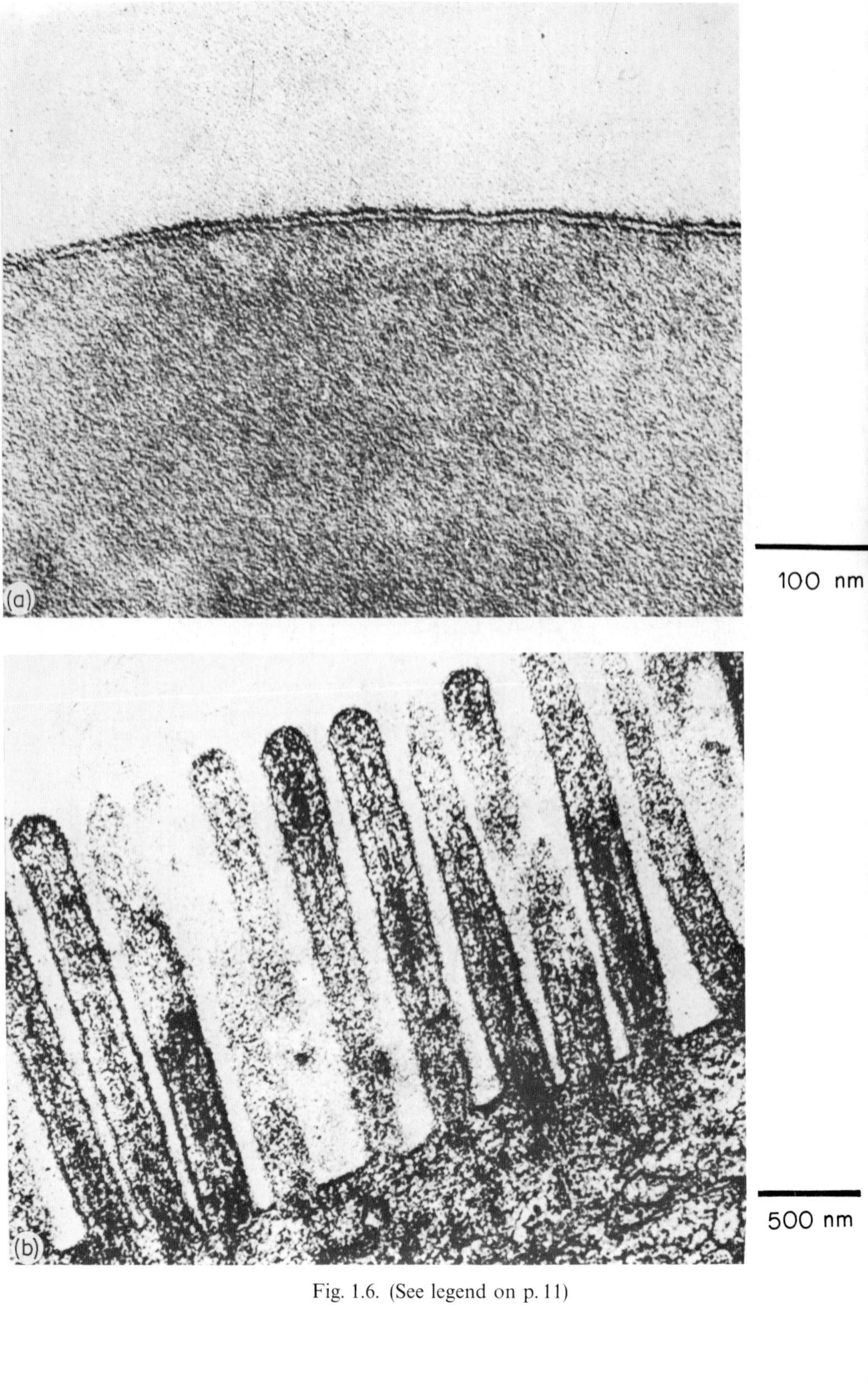

Fig. 1.6. (See legend on p. 11)

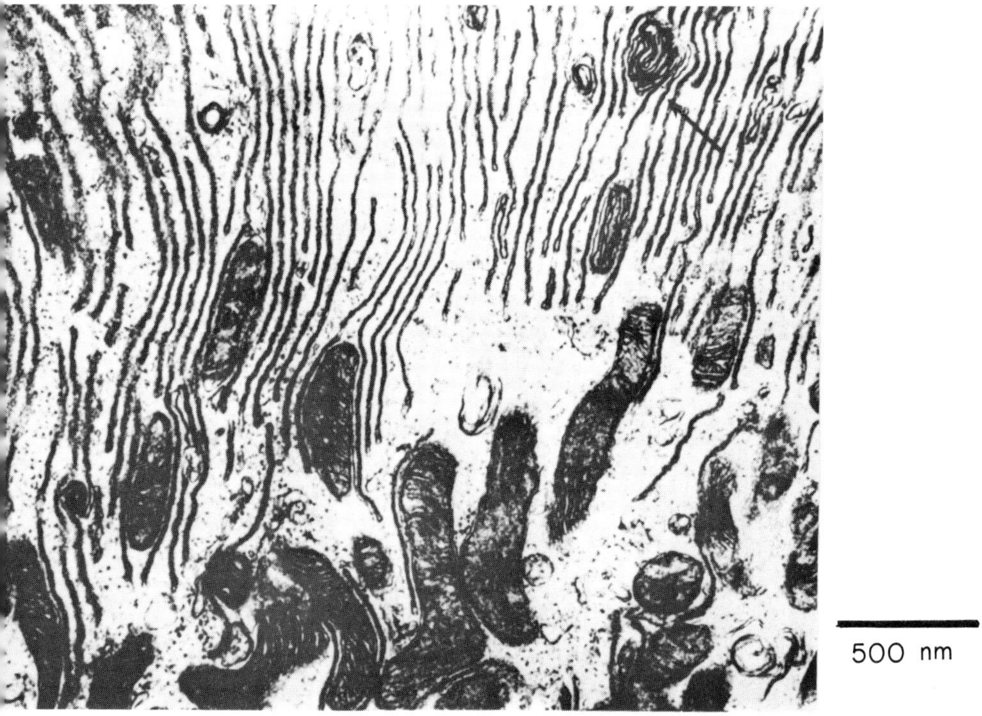

Fig. 1.6. Different modifications of the plasma membrane: (a) the smooth plasma membrane of a human erythrocyte; (b) intestinal brush border membrane with microvilli; (c) part of a cell of mosquito larva anal papilla; the thin structures are projections of the basal plasma membrane. (Taken with permission from A. Kotyk and K. Janáček: Membrane Transport, Academia, Prague, 1977)

a simple or complex polysaccharide wall. In plants in particular the plasma membrane is called plasmalemma.

It is usually thicker than other cell membranes (9–12 nm after OsO_4 fixation), this being apparently due to the frequent presence of glycoproteins on its external surface and of peripheral proteins on its internal surface.

Table 1.1. Increase of the surface area (by factor *F*) of various plasma membranes due to invaginations or projections

Membrane type	*F*
Erythrocyte (discocyte)	1.0
Bacillus subtilis	1.5–2.5
Candida utilis	3.0–3.5
Luminal part of enterocyte	25–200
Anal papilla of mosquito larva	1000–5000

The variety of forms of different plasma membranes is astounding. The flat type of the erythrocyte or of the basolateral membrane of enterocytes is the paradigm of the trilaminar (railroad-track) 'unit' membrane (Fig. 1.6(a)). Invaginations are observed in many membranes, with the apparent teleonomic objective of extending the membrane surface (Table 1.1). Plasma membranes of microorganisms, both prokaryotic (bacteria) and eukaryotic (yeasts and fungi) are grooved, the folds being more prominent in a hypertonic medium. The luminal membrane of absorptive cells of the intestine or kidney tubule lining is organized into fingerlike projections, the microvilli (which may be found all over the surface of such cells in tissue culture) (Fig. 1.6(b)). Insect intestinal cells contain plasma membranes forming extremely long centripetal projections reaching all across the cell (Fig. 1.6(c)).

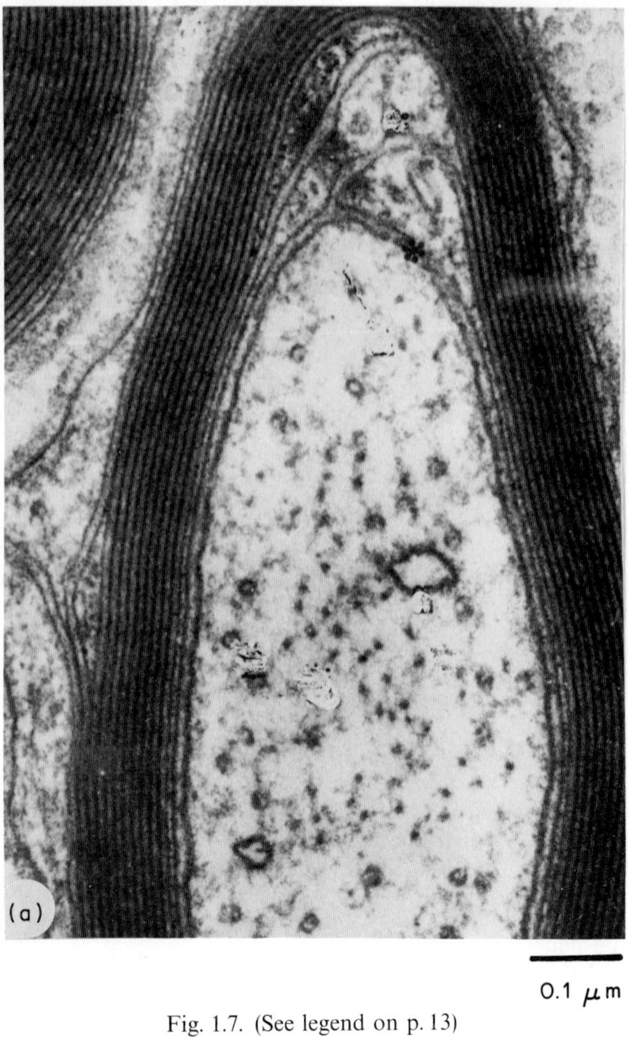

Fig. 1.7. (See legend on p. 13)

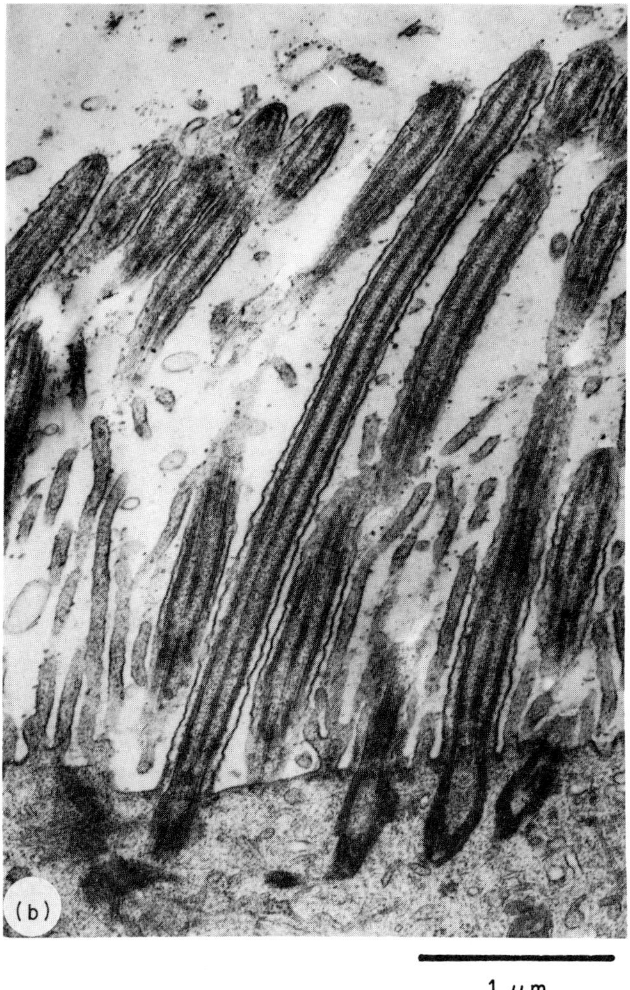

Fig. 1.7. Ultrathin sections of a myelin sheath of a crab nerve axon (a) and through the ciliated surface of rat respiratory epithelium (b). (Courtesy of Dr T. Soukup, Institute of Physiology, Czechoslovak Academy of Sciences, Prague)

Plasma membrane forms the sheath of eukaryotic flagella, which are constructed on a specific array of tubulin and dynein molecules (Fig. 1.7(a)). A somewhat special plasma membrane is that of Schwann cells forming the myelin sheath of thick nerve axons. These cells contain practically no cytoplasm but their plasma membrane is wrapped around several times, forming a regular pattern, amenable even to X-ray crystallographic analysis (Fig. 1.7(b)). Another special case of a double or triple plasma membrane is found in various blood flukes (*Schistosoma*, *Spirorchis*) where it arises from subtegumental vacuoles and may

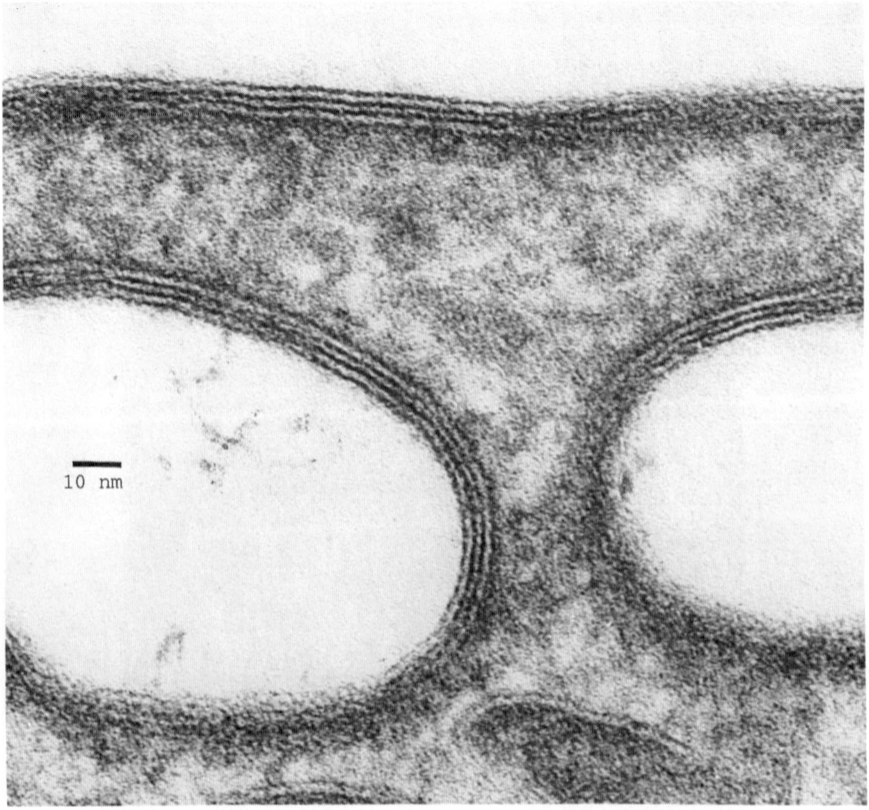

Fig. 1.8. The double outer membrane and parts of two vacuoles of the blood fluke *Schistosoma margrebowiei*. (Courtesy of Dr D. McLaren of Mill Hill, UK)

serve as a trap for insertion of host glycoproteins and glycolipids which render the blood fluke resistant toward the host's immune system (Fig. 1.8).

In bacteria, this plasma membrane is called the cytoplasmic membrane, to distinguish it from the phospholipid bilayer on the surface of the cell wall, usually called the outer membrane, of most gram-negative species (Fig. 1.9).

Plasma membranes are rarely in touch with any other membrane types but they do come in physical proximity of plasma membranes of neighbouring cells in animal organs. While for the most part such cells are separated by the interstitial space their plasma membranes can interact in several ways.

1. By direct adhesion in the so-called tight junctions (*zonulae occludentes*) where ribbons of integral membrane proteins are shared by the two apposed cells, the ribbons forming a perfect seal preventing interstitial fluid from passing along the cell boundaries (Fig. 1.10). In invertebrate tissues, tight junctions are replaced by characteristic septate junctions.

2. By forming the so-called gap junctions, a rather ubiquitous type of

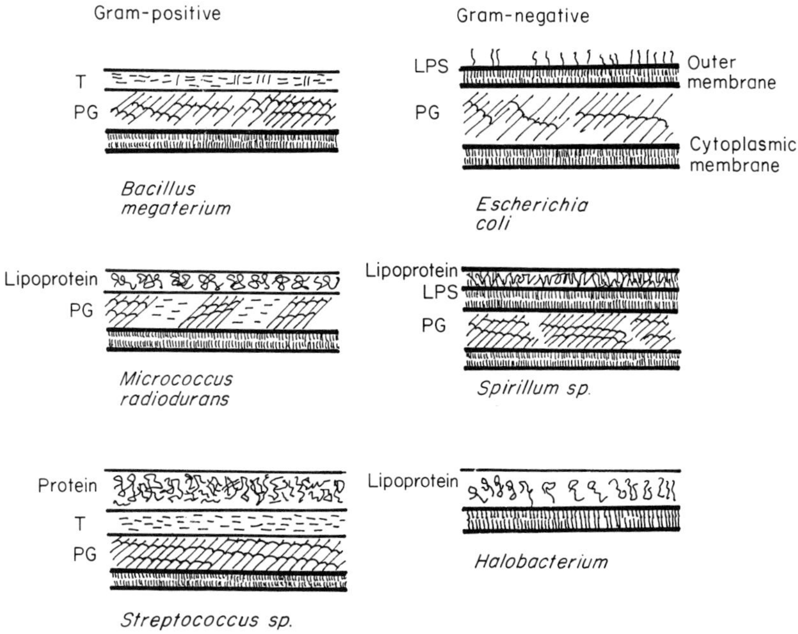

Fig. 1.9. Schematic cross-sections of bacterial envelopes showing one or two membranes and abundant polysaccharide wall material. T—teichoic acid, PS—polysaccharide, PG—peptidoglycan, LPS—lipopolysaccharide

intercellular contact. In these areas, the plasma membranes of two adjacent cells are about 3 nm apart (Fig. 1.11) but are interconnected by rather densely packed molecules of the protein connexin, to form a connexon. Owing to these gap junction patches the contents of small molecules and ions in tissue cells are apparently identical whereas the macromolecules of proteins and nucleic acids remain separated.

3. Via the so-called desmosomes (spot, belt, or hemi-), relatively complicated structures composed, in the case of spot desmosome (*macula adhaerens*), of round cytoplasmic plaques (Fig. 1.12) about 200–300 nm in diameter which are held in juxtaposition by a network of intermediate-type keratin filaments about 10 nm thick. Between the two plaques the two plasma membranes are somewhat farther apart than in the rest of the intracellular surface area and are held in position by thinner, actin-type filaments. A hemidesmosome is simply a half of a spot desmosome. A belt desmosome (*zonula adhaerens*) is not a round structure but a distended swath of actin-positioned plasma membranes.

Communication channels exist even in plant cells which are surrounded by a thick cellulose wall. The connections, called plasmodesmata, are approximately 20–40-nm-thick tubes lined by interconnecting plasmalemmas of two adjacent cells and housing a cytoplasmic cylinder which contains a desmotubule, a membranous structure contiguous with the endoplasmic reticulum of the two

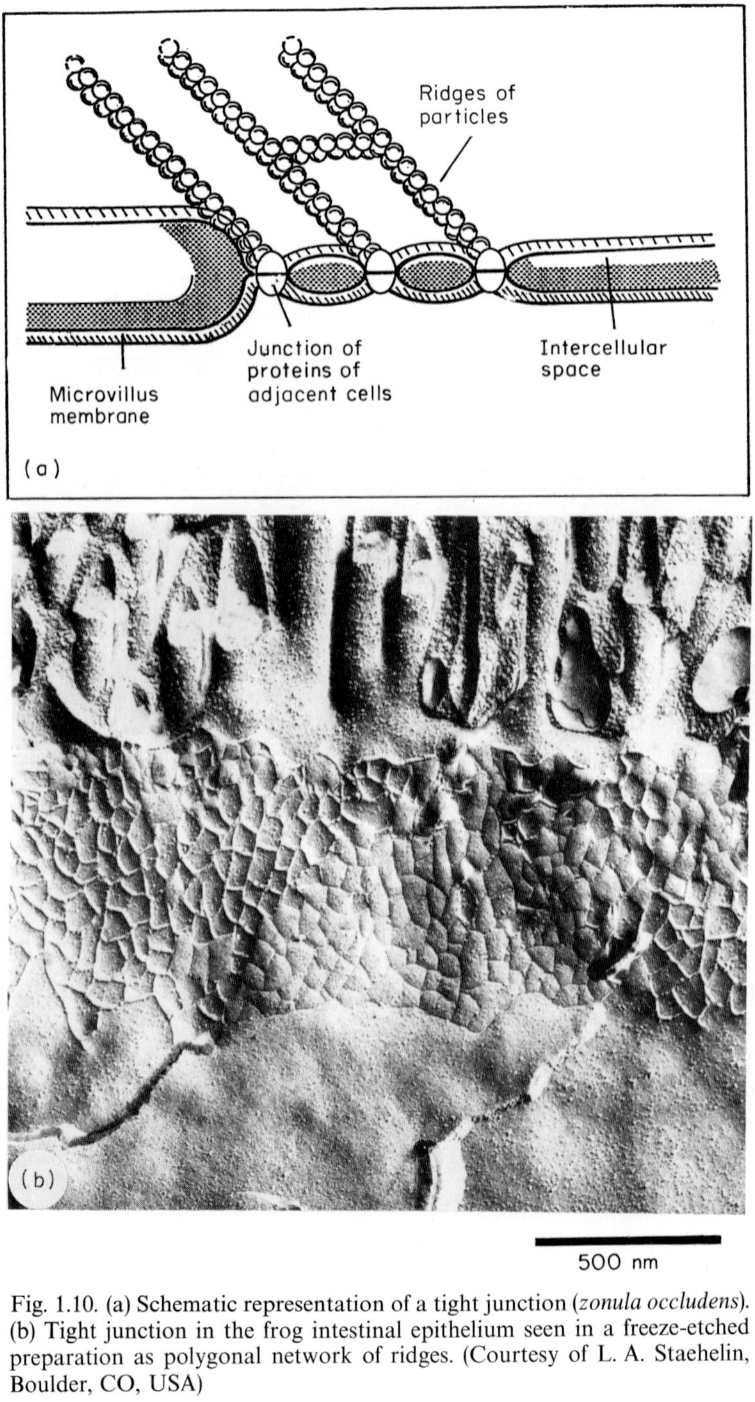

Fig. 1.10. (a) Schematic representation of a tight junction (*zonula occludens*). (b) Tight junction in the frog intestinal epithelium seen in a freeze-etched preparation as polygonal network of ridges. (Courtesy of L. A. Staehelin, Boulder, CO, USA)

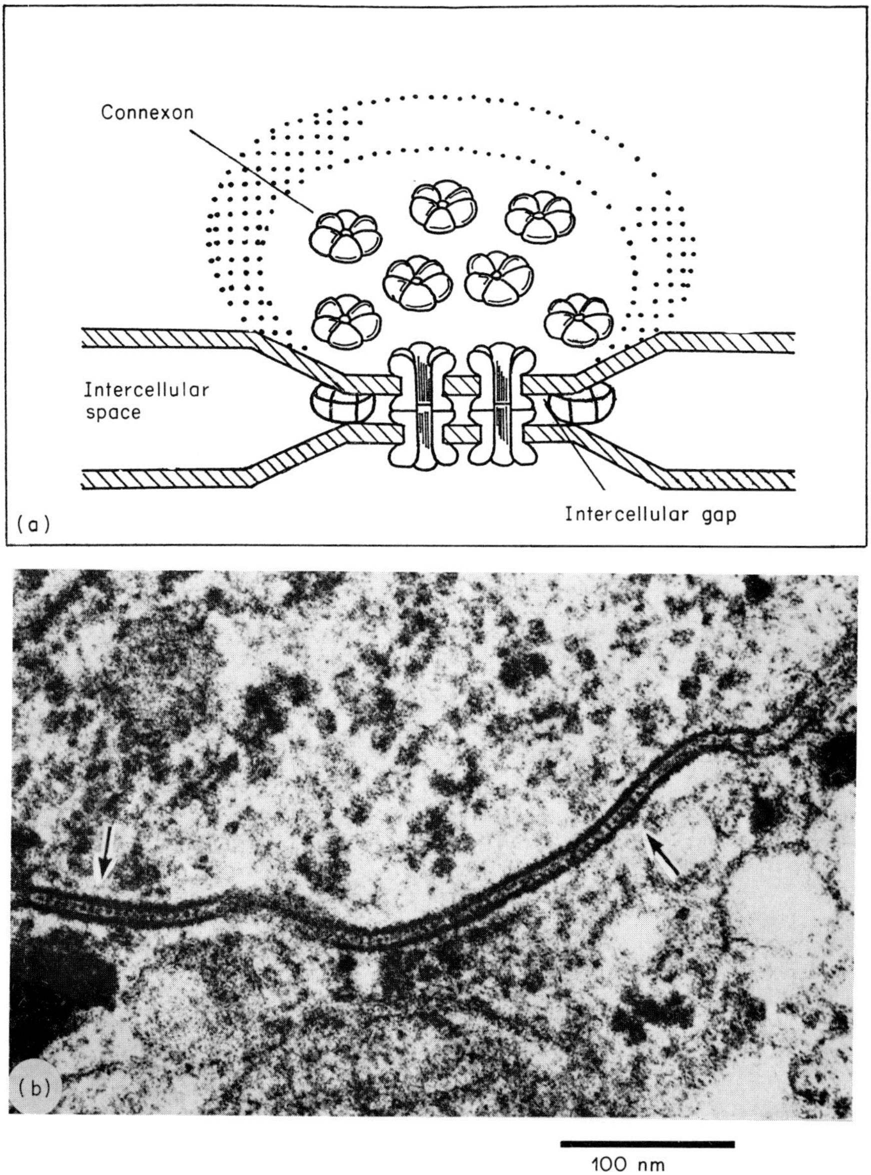

Fig. 1.11. (a) Schematic representation of a gap junction. (b) Cross-section of two adjoining cells of mouse connective tissue. The gap junctions are seen particularly clearly at the points shown by arrows where connexons are well distinguishable. (Courtesy of Prof. P. Sitte, Freiburg im Breisgau, FRG)

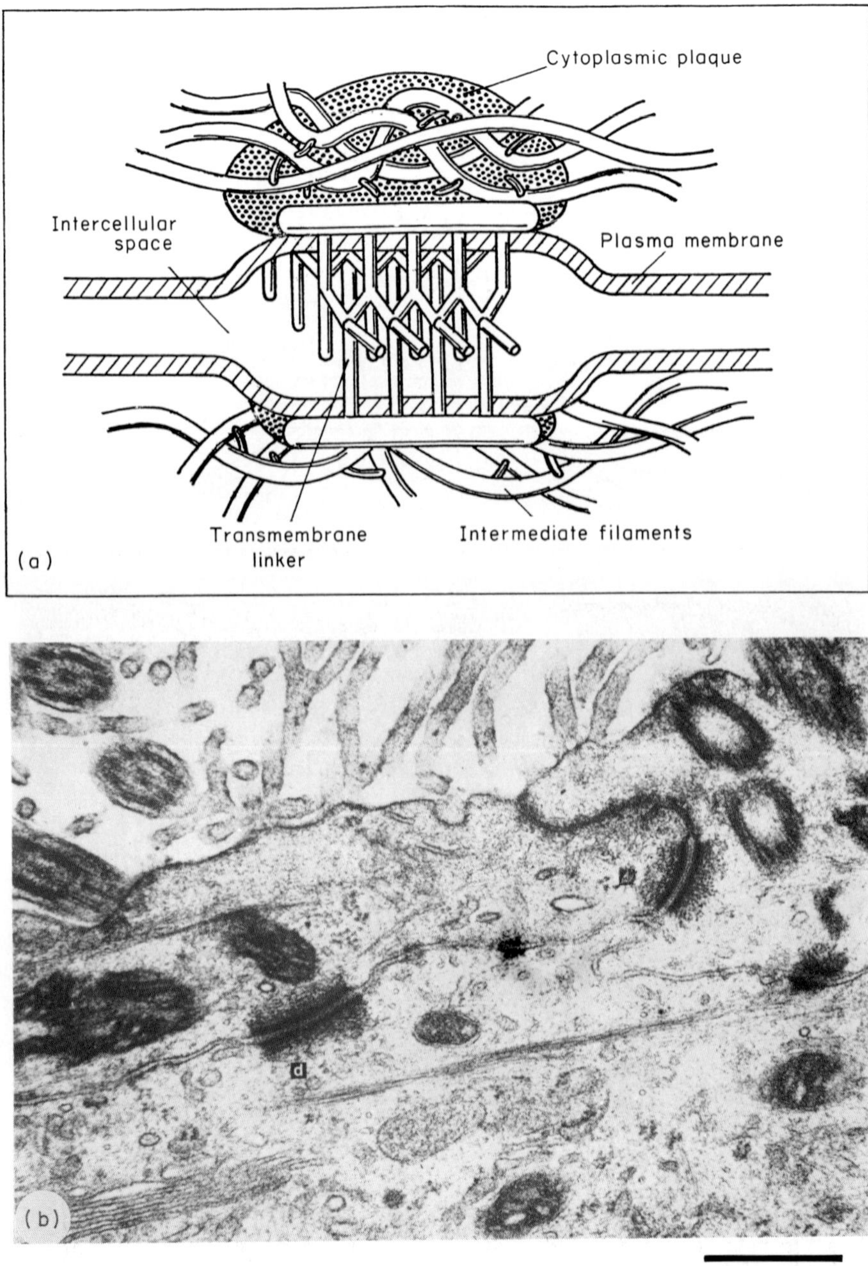

Fig. 1.12. (a) Schematic representation of a spot desmosome (*macula adhaerens*). (b) Ultrathin section of two neighbour cells of rat respiratory epithelium showing two desmosomes (d). (Courtesy of Dr T. Soukup, Institute of Physiology, Czechoslovak Academy of Sciences, Prague)

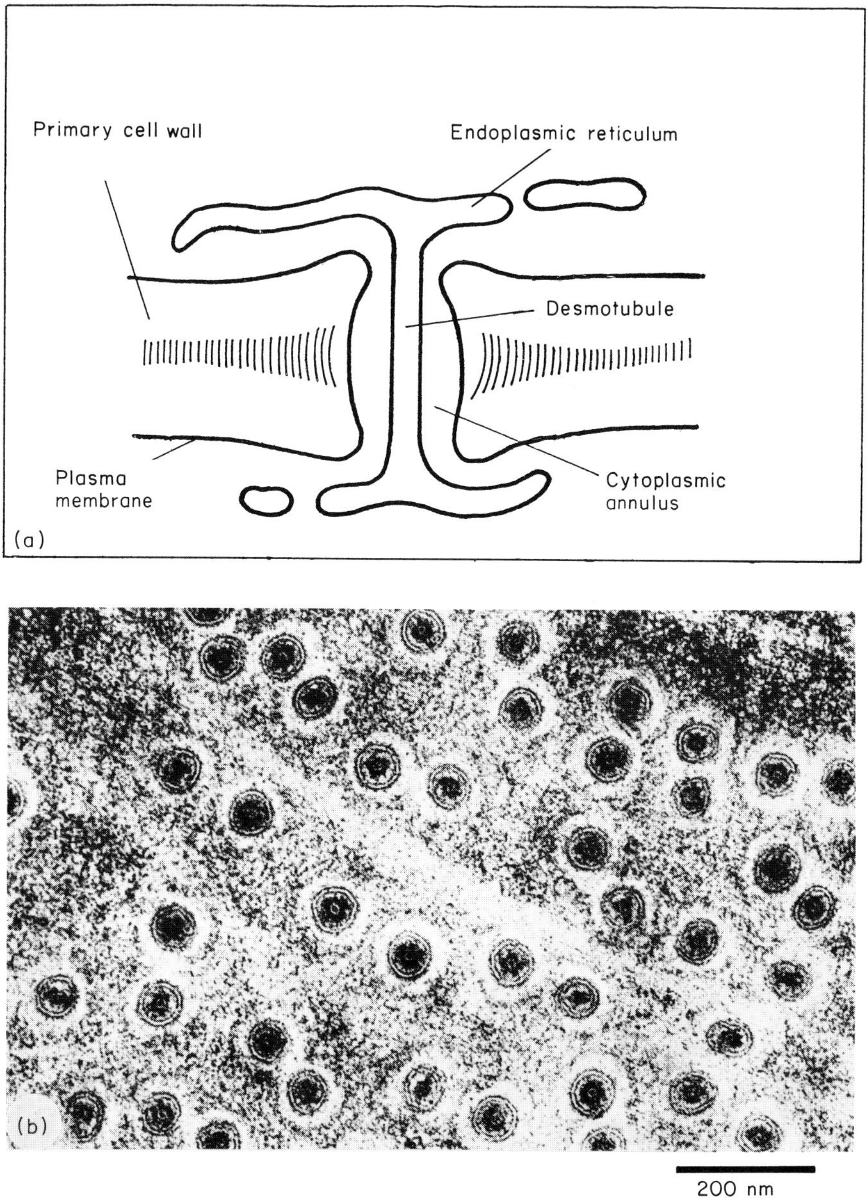

Fig. 1.13. (a) Schematic representation of a plasmodesma. (b) Cross-section through plasmodesmata in a needle of *Metasequoia glyptostroboides*. The plasma membrane lining the openings is clearly seen as a concentric double structure

Table 1.2. Functions performed by various plasma membranes

Type of function	Example of cells
Transport of small molecules and ions	All cells
Transport of large molecules and particles	Macrophages, amoebae
Fusion	Nearly all naked cells after a stimulus
Enzyme-reaction catalysis	Bacteria*
Antigenicity	Bacteria*
Conversion of free energy of oxidation into electrochemical potential gradients of ions	Bacteria*
Conversion of radiation energy into electrochemical potential gradients of ions	Photosynthetic bacteria,* halobacteria*
Perception of chemical and tactical stimuli	Bacteria, slime molds
Binding of hormones	Numerous animal and plant cells
Binding of antigens	Lymphocytes
Generation of electric current	Various nerve cells
Electrical insulation	Myelin sheath cells

* It might be argued that the prokaryotic plasma membrane is more closely related to the inner mitochondrial membrane than to the eukaryotic plasma membrane. However, from the point of view of the cell's survival it functions as the communication link between the cell and its surroundings, just as any 'higher' plasma membrane does.

cells (Fig. 1.13). The desmotubule, being more than 10 nm wide, permits the passage of many molecules and even virus particles from cell to cell. The abundance of plasmodesmata ranges from 1 to 12 per μm^2.

Of all the biomembranes the plasma membranes possess the greatest variety of capabilities, as listed in Table 1.2.

1.3.2 The nuclear membrane

The cell nucleus of eukaryotic cells is surrounded by a membranous envelope, a double-membrane structure enclosing cisternae about 10–30 nm wide. Each of the membranes has an average thickness of 7–8 nm; they carry some enzyme activities that they share with the smooth endoplasmic reticulum from which the nuclear envelope is generated as cells divide during mitosis. The envelope is perforated by openings about 50–70 nm in diameter around which the two membranes are fused (Fig. 1.14). The pores are lined by an octagonal system of subunits (10–18 nm in diameter) and often seem to be filled by electron-dense material. It is clear that all the nucleic acid and protein material associated with replication and transcription of nucleus-localized DNA must pass through the pores. There is recent evidence that tRNA passes through by a specifically

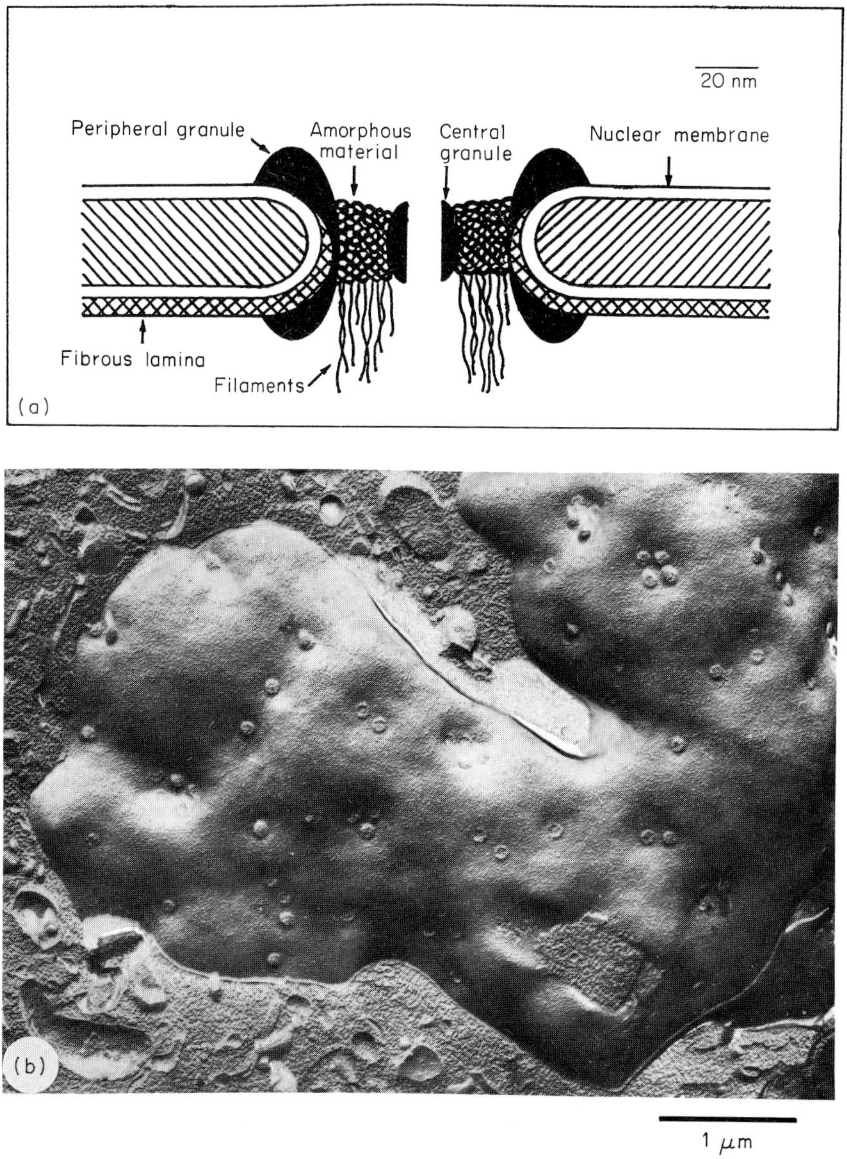

Fig. 1.14. (a) Schematic representation of a nuclear pore cross section. (b) Nuclear membrane of a guinea-pig macrophage with numerous nuclear pores. The freeze-etched preparation exposes the plasma face of the inner membrane of the nuclear envelope. (Courtesy of Dr R. Janisch of the J. E. Purkyně University, Brno)

mediated process which suggests that the pores are not freely open to diffusion. This view is further supported by observations made with highly sensitive electron probes which show a higher concentration of K^+ as well as Na^+ and Cl^- in the nucleus than in the cytoplasm.

Generally, active nuclei possess a higher number of pore complexes per unit area and so do larger nuclei (their volume to surface ratio is greater). A few examples demonstrate this trend (diameter of the nucleus in μm vs. the number of pore complexes per μm^2): chicken erythrocyte 1.7:3.5, baker's yeast 2.3:12.5, onion root tip 10:18, *Amoeba proteus* 41:60.

1.3.3 The endoplasmic reticulum

Frequently attached to the outer face of the nuclear envelope and extending into much of the cytoplasm, the endoplasmic reticulum forms a vast system of flat cisternae, about 20–30 nm inner diameter (Fig. 1.15).

The endoplasmic reticulum of most cells is of two types, smooth and rough. The *smooth* ER is important in two cell types: Those with a high detoxication ability, such as hepatocytes, and those where lipid metabolism is prominent,

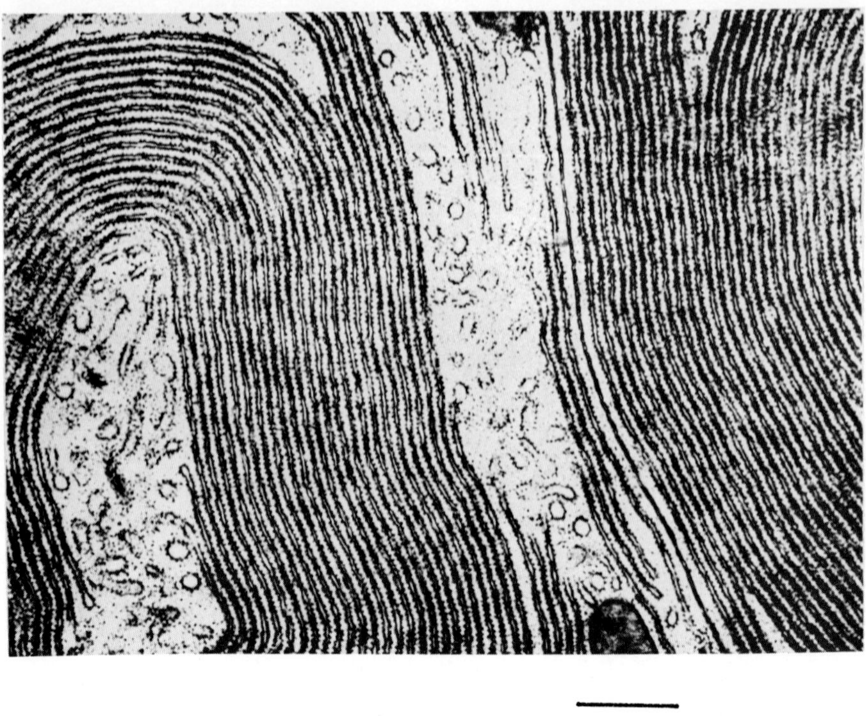

Fig. 1.15. Ultrathin section through a bat pancreatic acinar cell showing abundant rough endoplasmic reticulum. (Courtesy of Japan Society for Cell Biology)

such as the Leydig cell. The relevant enzymes are found in the smooth endoplasmic reticulum, thus (C indicates orientation toward cytoplasm, L toward the ER lumen):

Cytochrome b_5	C
Cytochrome-b_5 reductase (1.6.2.2)	C
NADPH-ferrihemoprotein reductase (1.6.2.4)	C
Cytochrome P-450	C, L
Cytochrome P-450 reductase (1.6.2.4)	L
Adenosinetriphosphatase (3.6.1.3)	C
5′-Nucleotidase (3.1.3.5)	C
Nucleoside diphosphatase (3.6.1.6)	L
GDP-mannosyl transferase (2.4.1.x)	C
Glucose-6-phosphatase (3.1.3.9)	L
β-Glucuronidase (3.2.1.31)	L
Hydroxymethylglutaryl-CoA reductase (1.1.1.88)	C
8–10 enzymes of steroid synthesis	C
10 enzymes of cholic acid synthesis	C

The *rough* ER is so called because its surface is studded with ribosomes—this is the locale of much of the protein synthesis (see section on membrane biogenesis) of secretory cells of both endocrine and exocrine glands, as well as of cells of the immune system.

The thickness of the ER membrane is 7–10 nm. On cell homogenization, this is the structure that gives rise to smooth or rough microsomes, vesicles about 100 nm in diameter.

The ER membrane apparently contains several specific transport systems, such as those for oligosaccharides used for glycosylation of ER-synthesized proteins and those for phospholipids synthesized on the plasmatic side of the membrane and destined for the nonplasmatic face of the ER bilayer. However, the ER membrane is virtually impermeable to all ions or uncharged molecules of M_r above 700.

1.3.4 The Golgi apparatus

The typical picture of several stacked cisternae, found in the vast majority of eukaryotic cells, is the Golgi apparatus, Golgi body, or dictyosome. The membrane is of the same type, both in thickness and in buoyant density, as that of the smooth ER. This is obvious if one considers the rapid exchange of vesicular material between the smooth (or transitional) ER and the Golgi apparatus (Fig. 1.16).

It has become clear in the past few years that this cisternal organelle is not a homogeneous structure but that it is polarized, with its *cis* side toward the ER and its *trans* side toward the cell periphery. This polarity has its corollary in the functions the Golgi apparatus performs, viz. glycosylation and trimming of membrane or secretory proteins, as they move centrifugally from the rough

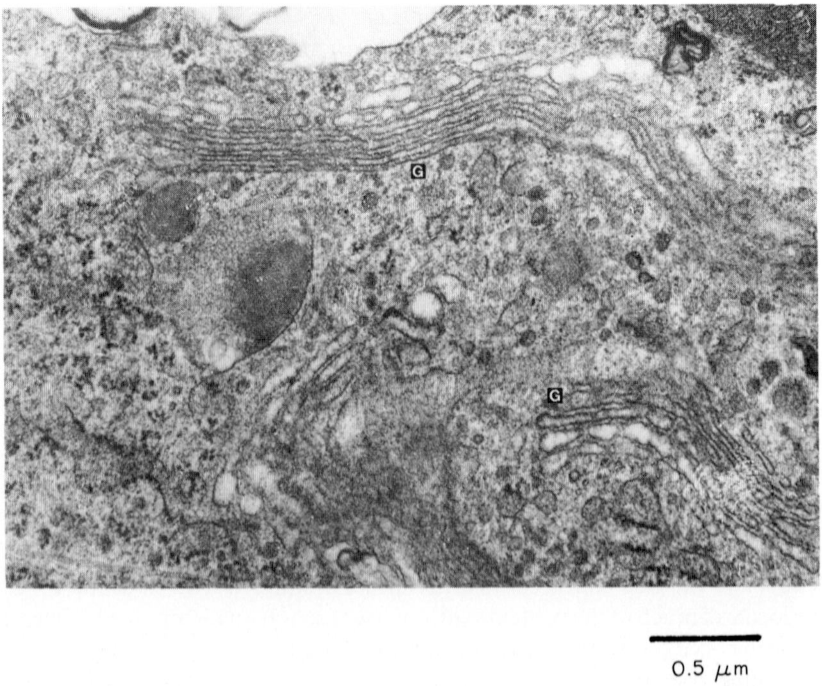

Fig. 1.16. The Golgi complex in a goblet secretory cell of rat nasal epithelium. (Courtesy of Dr T. Soukup, Institute of Physiology, Czechoslovak Academy of Sciences, Prague)

ER. This function is reflected in the complement of enzymes found in the Golgi: Besides 5′-nucleotidase, NADH-cytochrome *c* reductase, thiamine diphosphatase and NADPH-cytochrome *c* reductase, it contains UDP-galactose-N-acetylglucosamine galactosyl transferase and a number of other glycosylating enzymes.

Another function of the Golgi apparatus is apparently the recycling of membrane material through fusion with the so-called coated vesicles, an important vehicle in the transport of material between cell organelles and during pinocytosis.

In addition to its radial polarity the Golgi apparatus is heterogeneous as one goes from one sector to the next. Different types of enzymes with different final destinations are apparently sorted out in the Golgi apparatus and transmitted either to coated vesicles of various types or to secretory vesicles, to deliver their cargo at the right place.

Little is known about the transport systems of this complex organelle, except that it contains antiport systems for nucleotides and their sugar derivatives, thus: UMP out for UDP-galactose in; CMP out for CMP-sialic acid in; GMP out for GDP-mannose in. Phosphate is transported out by a separate system.

1.3.5 The lysosomes

There are several types of intracellular vesicles that contain a relatively acid (pH 5) complement of hydrolytic enzymes, including proteinases, glycosidases, lipases, phosphatases, nucleases and sulphatases. Their obvious function is to digest all kinds of material that is either ingested through the plasma membrane or that is destined to be degraded, originating from intracellular organelles (Fig. 1.17).

The primary lysosomes are small vesicles about 50 nm across that are pinched off the Golgi apparatus in the clathrin-coated form and are then stripped of the clathrin. As uncoated primary lysosomes they can fuse with other membranous structures to form secondary lysosomes, much larger bodies (300–500 nm across) where the hydrolytic process takes place.

The membranes of lysosomes are of the usual thickness but contain several unique proteins, some to recognize the site of their destination after formation, others to transport the products of digestion (nucleosides, monosaccharides including sialic acid, amino acids and dipeptides, cobalamin) from the lysosome to the cytoplasm and, in addition, a proton-transporting ATPase to keep the

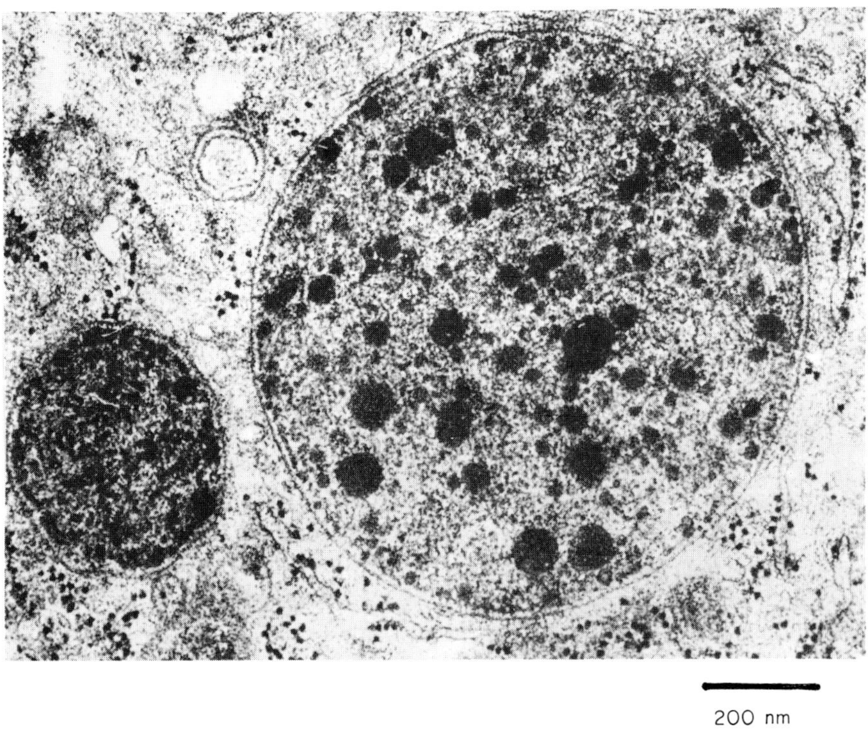

Fig. 1.17. A secondary lysosome in a rat Kupffer cell. (Courtesy of Prof. P. Sitte, Freiburg im Breisgau, FRG)

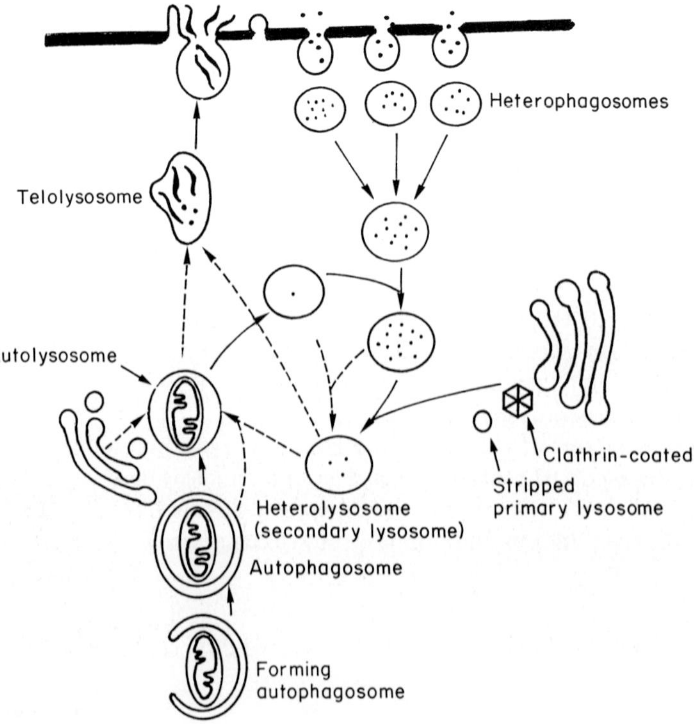

Fig. 1.18. The circulation of lysosomes and analogous organelles in an animal cell capable of pinocytosis. Hypothetical routes are shown in dashed lines

inside of the organelle at a low pH. A comprehensive view of the formation and degradation of lysosomes is given in Fig. 1.18.

1.3.6 The vacuole

The most distinctive feature of plant and fungal cells is the vacuole, in some plants often taking up as much as 95 per cent of the cell volume. The membrane surrounding it, the tonoplast, is a smooth flat membrane about 8 nm thick which apparently arises in young cells from the fusion of primary lysosomes or other vesicles pinched off the Golgi apparatus but may very well be formed directly from smooth endoplasmic reticulum. The vacuole is analogous to animal lysosomes in its content of proteinases and the relatively low pH (about 5.5). The tonoplast contains a number of transport systems, among others an H^+-translocating ATPase of a special type.

The vacuolar content usually reflects the storage needs and preferences of the cell. In yeast, it contains a remarkably high amount of inorganic polyphosphate, with P–O cross-links provided by Fe and Mn, plus various basic amino acids, presumably to compensate for the charge of the phosphate groups (Fig. 1.19).

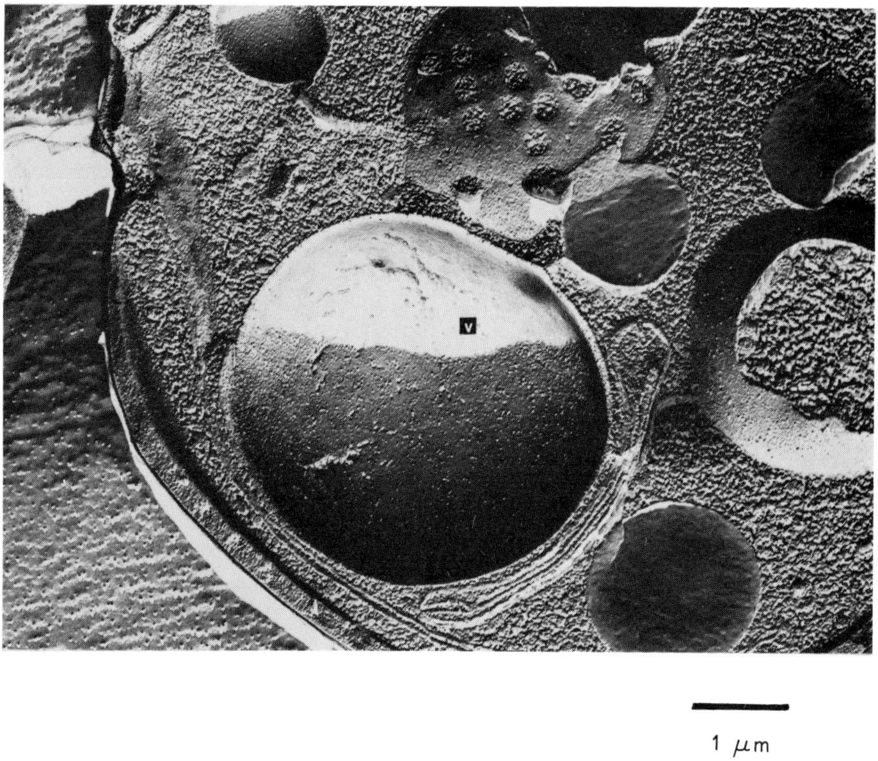

Fig. 1.19. A freeze-etched preparation of a sphacelial cell of the ergot *Claviceps purpurea* with a large vacuole (v) (Courtesy of Dr E. Streiblová, Institute of Microbiology, Czechoslovak Academy of Sciences, Prague)

1.3.7 The microbodies

Similar in size and appearance to secondary lysosomes, the various microbodies contain predominantly oxidative enzymes.

The *peroxisomes* are organelles specializing in the oxidation of a number of substances, such as amino acids, uric acid, ethanol, phenols and fatty acids (to acetyl-CoA), the key enzyme being catalase which removes the H_2O_2 arising through the action of the flavin-containing oxidases. The peroxisomes range in size from 0.2 to 1.5 μm, with a 5–8 nm membrane, and may be formed in the ER from which they bud. However, peroxisomes are now considered to be mainly self-reproducing organelles (in common with chloroplasts and mitochondria but minus their genetic equipment). Probably all of their enzymes are synthesized in the cytosol but the peroxisomes divide together with cell division and, if the cell divides by budding, pass into the bud along with the nucleus and other organelles.

The small peroxisomes are widely distributed in practically all eukaryotic cells while the large ones are mainly known from liver cells. Apparently, their membranes will contain special transport systems for the substrates to be oxidized.

Another kind of microbody is the plant peroxisome, more properly the *glyoxysome*, which contains enzymes of the glyoxylate pathway, involved either in photorespiration of glycollate or in the conversion of acetyl-CoA to succinate. Glyoxysomes are only known from plant cells (Fig. 1.20) and are especially abundant in cells where oleosomes, with their content of triacylglycerols, are present and apparently serve as the reserve of fat which is split by a membrane-bound lipase of the glyoxysome to fatty acids and glycerol, the former being converted eventually to succinate and this then oxidized in mitochondria.

It should be noted in this connection that the enzymes of the glyoxylate cycle, a frequent partner or alternative to the Krebs cycle, are found in the glyoxysomes of plants but in the mitochondria of animal cells.

The much studied yeast microbodies contain an array of 40 enzymes, as well as a membrane-bound H^+-ATPase.

Rather unique are organelles that have been found so far in parasitic trichomonads and infusorians where, both aerobically and anaerobically, they convert pyruvate to acetate, giving rise to ATP, a special bonus of the anaerobic pathway being the production of molecular hydrogen—hence their name

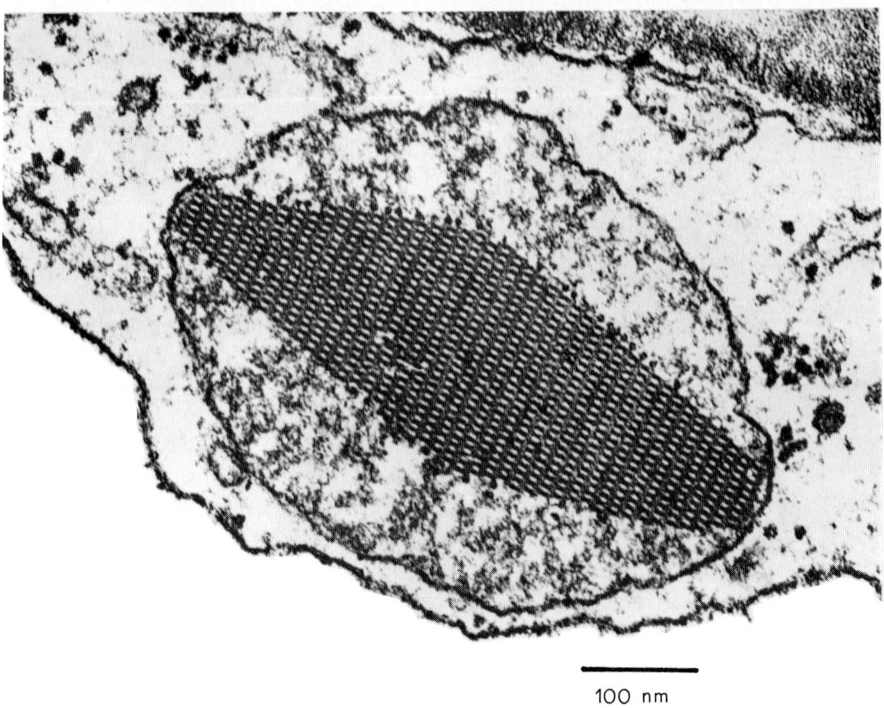

Fig. 1.20. Ultrathin section of a peroxisome in a parenchyma cell of oats. A crystalline enzyme inclusion dominates the microbody. (Courtesy by Prof. B. Gebhardt, Münster)

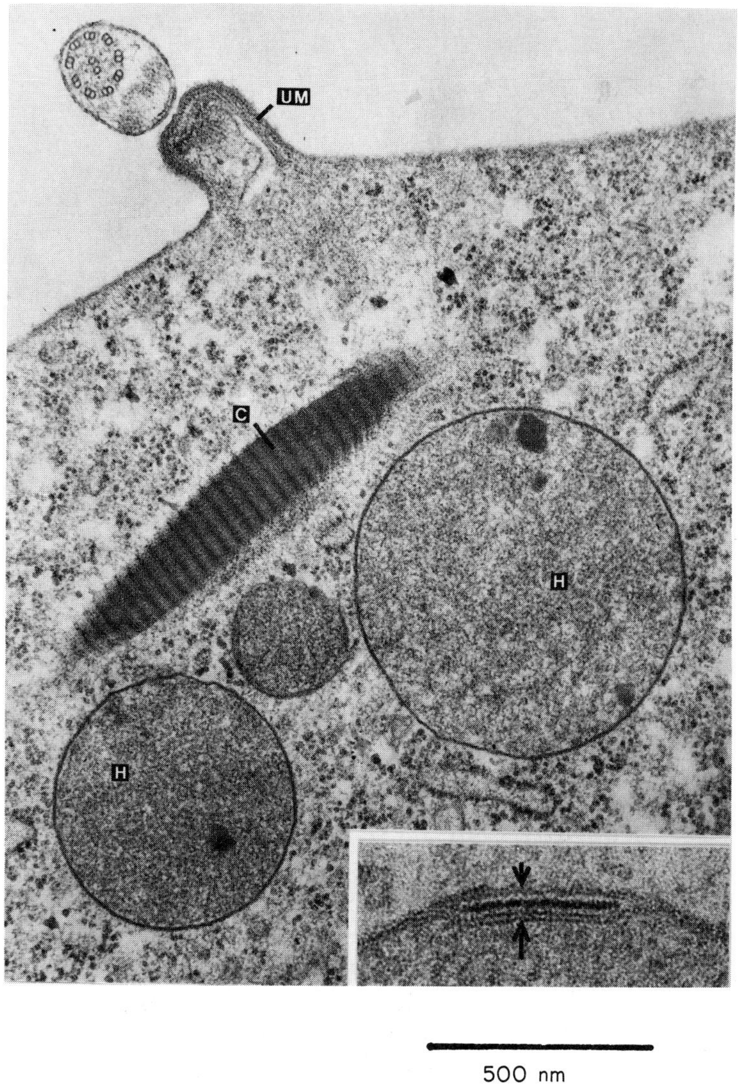

Fig. 1.21. Ultrathin section of a hydrogenosome-containing protozoan *Tritrichomonas foetus* which also shows the costa (c), a cytoskeletal support for the locomotor organelle, and the undulating membrane (UM). The two bilayers enclosing the hydrogenosome come apart only at the sites of membrane plaques (inset) containing calcium-rich amorphous material. (Courtesy of Dr J. Kulda, Faculty of Science, Charles University, Prague)

hydrogenosomes. The organelle has a double membrane with an assortment of transport systems for organic acids and presumably ATP, like in mitochondria (Fig. 1.21).

1.3.8 The mitochondrion

This double-membrane organelle about $0.5 \times 1.5\,\mu m$ in size is the principal power generator of practically all eukaryotic cells and contains, in addition, a variety of metabolic enzymes that serve a function preparatory to the terminal oxidation. The mitochondrial outer membrane is akin to the membranes of the ER while its inner membrane is unique and resembles in many respects the cytoplasmic membrane of bacteria (Figs 1.22, 1.23). The physical properties of the two membranes are summarized in Table 1.3.

Mitochondria are generally believed to have arisen through the invasion of a bacterium (the present-day *Paracoccus* is a candidate for that historic role) into a primitive eukaryotic cell and its gradual assimilation by the cell (the endosymbiont theory). Although other views exist, such as the compartment hypothesis, the plasmic hypothesis, the gene-duplication-segregation hypothesis,

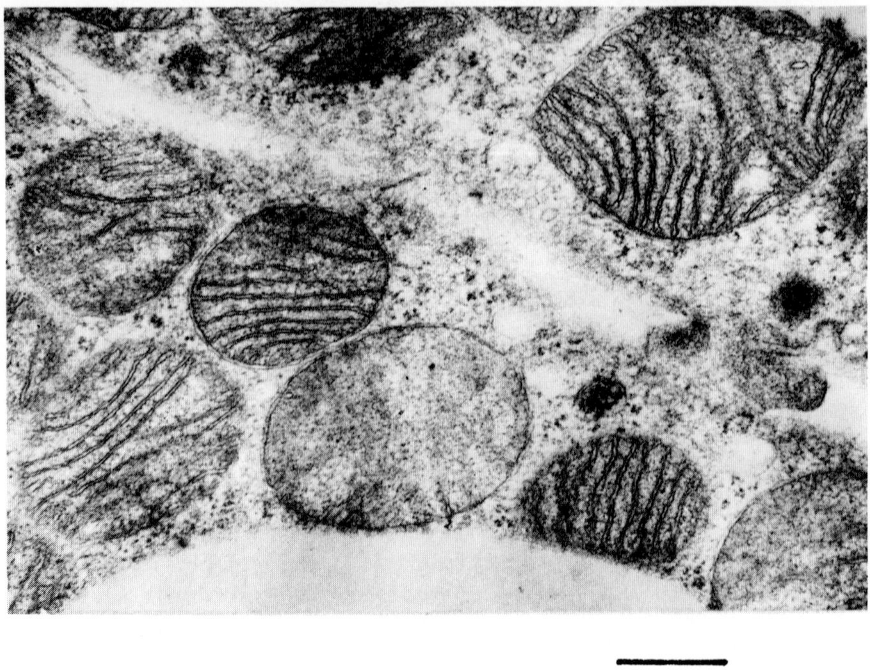

Fig. 1.22. Ultrathin section through an adipose cell of a newborn rat showing numerous mitochondria. (Courtesy of Dr T. Soukup, Institute of Physiology, Czechoslovak Academy of Sciences)

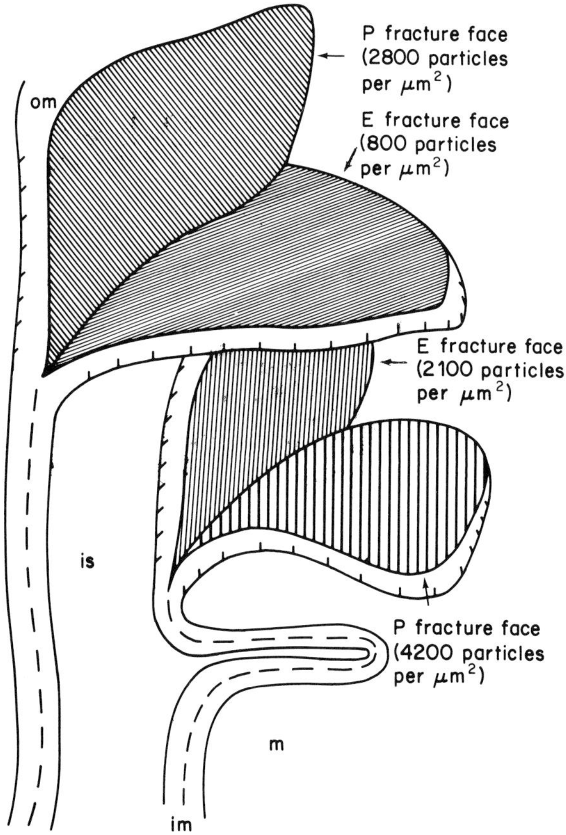

Fig. 1.23. Schematic split-up of mitochondrial outer (om) and inner (im) membrane showing the intermembrane space (is) and the matrix (m). The average number of particles on a freeze-fractured preparation is given for the endoplasmic (E) and for the protoplasmic (P) fracture faces

Table 1.3. Properties of the two mitochondrial membranes

Property	Inner membrane	Outer membrane
Thickness (nm)	7	7
Density (g/ml)	1.19–1.23	1.09–1.12
Particles on freeze-fractured faces (μm^{-2})		
E-face	2100	800
P-face	4200	2800
Phospholipid-protein ratio	0.32	0.89
Sterol-protein ratio	0.003	0.03
Limit of free permeation of nonelectrolytes	100	10 000

Table 1.4. Mitochondrial enzymes

Outer membrane	
Amine oxidase (flavin-containing)	1.4.3.4
Cytochrome b_5 reductase	1.6.2.2
NADH dehydrogenase	1.6.99.3
Kynurenine monooxygenase	1.14.13.9
Long-chain-fatty-acid-CoA ligase	6.2.1.3
Intermembrane space	
Intermembrane space	
Glycerol-phosphate dehydrogenase	1.1.99.5
Sulphite dehydrogenase	1.8.2.1
Adenylate kinase	2.7.4.3
Nucleoside diphosphokinase	2.7.4.6
Inner membrane	
3-Hydroxybutyrate dehydrogenase	1.1.1.30
3-Hydroxyacyl-CoA dehydrogenase	1.1.1.35
L-Lactate dehydrogenase (cytochrome)	1.1.2.3
Choline dehydrogenase	1.1.99.1
Glycerol-phosphate dehydrogenase	1.1.99.5
D-2-Hydroxy-acid dehydrogenase	1.1.99.6
Succinate dehydrogenase	1.3.99.1
Butyryl-CoA dehydrogenase	1.3.99.2
Acyl-CoA dehydrogenase	1.3.99.3
NAD(P) transhydrogenase	1.6.1.1
NADH transhydrogenase (ubiquinone)	1.6.5.3
Cytochrome-*c* oxidase	1.9.3.1
Ubiquinol–cyt-*c* reductase	1.10.2.2
Carnitine acetyltransferase	2.3.1.7
Acetyl-CoA acetyltransferase	2.3.1.9
H^+-transporting ATP synthase	3.6.1.34
Enoyl-CoA hydratase	4.2.1.17
3-Hydroxybutyryl-CoA epimerase	5.1.2.3
Dodecenoyl-CoA Δ-isomerase	5.3.3.8
Matrix	
Alcohol dehydrogenase (an isoenzyme)	1.1.1.1
Malate dehydrogenase	1.1.1.37
Malate dehydrogenase (decarboxylating)	1.1.1.39
Aldehyde dehydrogenase (NAD^+)	1.2.1.3
Pyruvate dehydrogenase complex	1.2.4.1, 1.8.4.1, 2.3.1.12
2-Oxoglutarate dehydrogenase complex	1.2.4.2, 2.3.1.61
4-Enoyl-CoA reductase	1.3.1.34
Superoxide dismutase	1.15.1.1
Pyruvate dehydrogenase kinase	2.7.1.99
Nucleoside-triphosphate adenylate kinase	2.7.4.10
3-Oxoacid CoA-transferase	2.8.3.5
Pyruvate dehydrogenase phosphatase	3.1.3.43
Phosphoenolpyruvate carboxylase	4.1.1.31
Citrate (*si*)-synthase	4.1.3.7
Fumarate hydratase	4.2.1.2
Aconitate hydratase	4.2.1.3
Acid-CoA ligase and specific ligases	6.2.1.1–6.2.1.5, 6.2.1.10

and the genome-invagination hypothesis, the fact remains that mitochondria possess their own genome and can multiply independently of the cell (but only within it!).

The four locales of the mitochondrion, the outer membrane, the intermembrane space, the inner membrane, and the matrix, contain assortments of enzymes (Table 1.4).

In addition to these enzymic activities, the inner membrane is endowed with transport systems for both ions and for nutrient molecules (Table 1.5).

Table 1.5. Transport systems of the mitochondrial inner membrane

Carrier for	Features
H^+	1. Through the respiratory chain 2. By transhydrogenase 3. By ATP synthase
Na^+, K^+	By antiport with H^+
Ca^{2+}	1. By an active efflux 2. By various electrophoretic movements
ATP/ADP	An electrogenic antiport
Phosphate anion	Antiport with OH^-
Pyruvate$^-$ (lactate$^-$, acetoacetate$^-$)	Antiport with OH^-
Malate^{2-}, succinate^{2-}, malonate^{2-}, sulphate^{2-}, phosphate^{2-}	A promiscuous antiport
Citrate^{3-}, isocitrate^{3-}, malate^{2-}, *cis*-aconitate^{3-}, phosphoenol-pyruvate^{3-}	A promiscuous antiport using H^+ to compensate one of the three charges
2-Oxoglutarate^{2-}, malate^{2-}, malonate^{2-}, succinate^{2-}, oxaloacetate^{2-}	A promiscuous antiport
Lower fatty acids ($n_C < 9$)	Antiport with OH^-?
Higher fatty acids ($n_C > 9$)	Carnitine cycle
Glutamate	1. With H^+ by antiport with aspartate 2. By antiport with OH^-
Amino acids	Various more or less specific systems

Table 1.6. Mammalian mitochondrial gene products

21S rRNA	Subunits I, II and III of cytochrome *c* oxidase
15S rRNA	Apocytochrome *b*
tRNAs for all amino acids (a total of thirty)	Subunits 6, 8 and 9 of ATP synthase
	The var-1 protein of the small ribosomal subunit

This is true for yeast and plants where subunit 9 is of mitochondrial origin but not in fungus *Neurospora* and in mammals

Although the mitochondrion contains an autonomous chromosome of the bacterial (circular) type and the entire machinery for protein synthesis the number of proteins coded and synthesized in the mitochondrion is very small indeed (Table 1.6). The great majority of mitochondrial enzymes and their subunits are synthesized in the cytoplasm and are brought into the mitochondrion through the cytosol (see the chapter on membrane biosynthesis). The number of mitochondria per cell ranges from a single one in parasitic trypanosomes to 10^5 in some frog oocytes.

1.3.9 The chloroplast

All plants contain a distinct type of organelle, generally called the plastid—if chlorophyll is present the plastid is a chloroplast (Fig. 1.24). Like mitochondria,

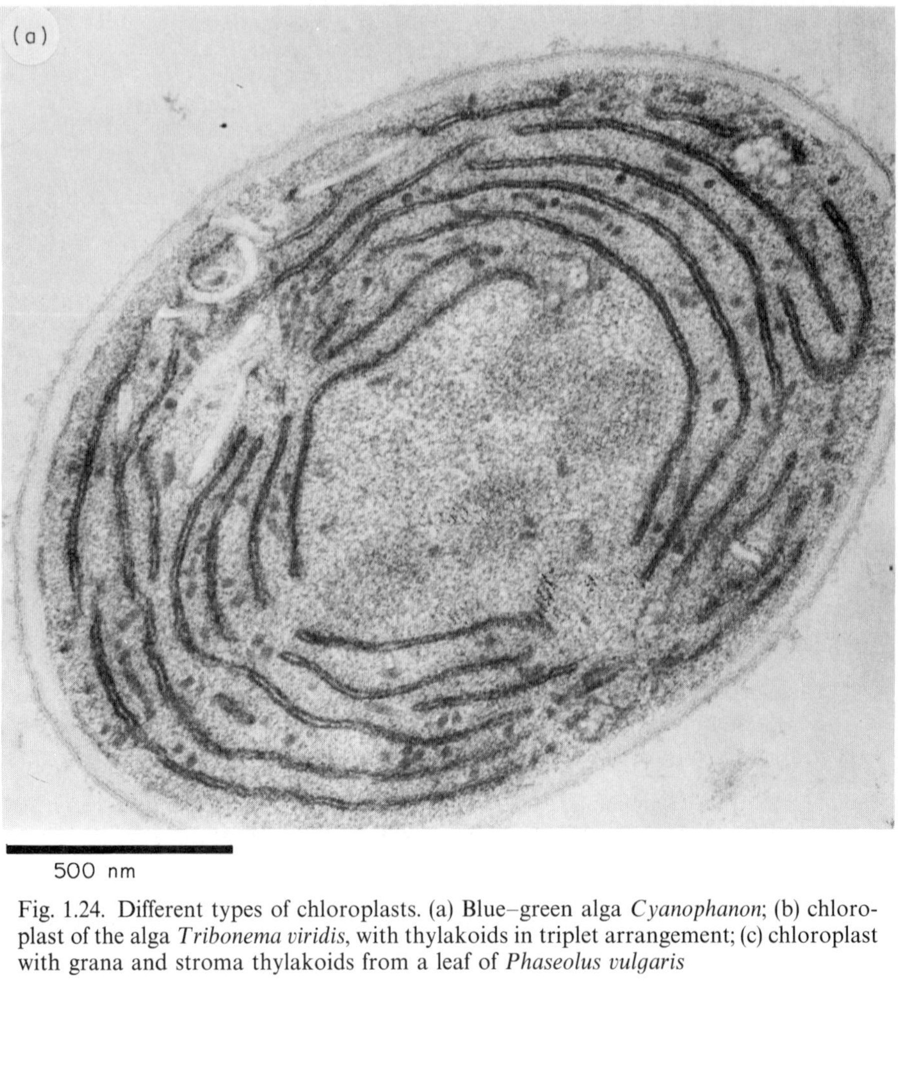

Fig. 1.24. Different types of chloroplasts. (a) Blue–green alga *Cyanophanon*; (b) chloroplast of the alga *Tribonema viridis*, with thylakoids in triplet arrangement; (c) chloroplast with grana and stroma thylakoids from a leaf of *Phaseolus vulgaris*

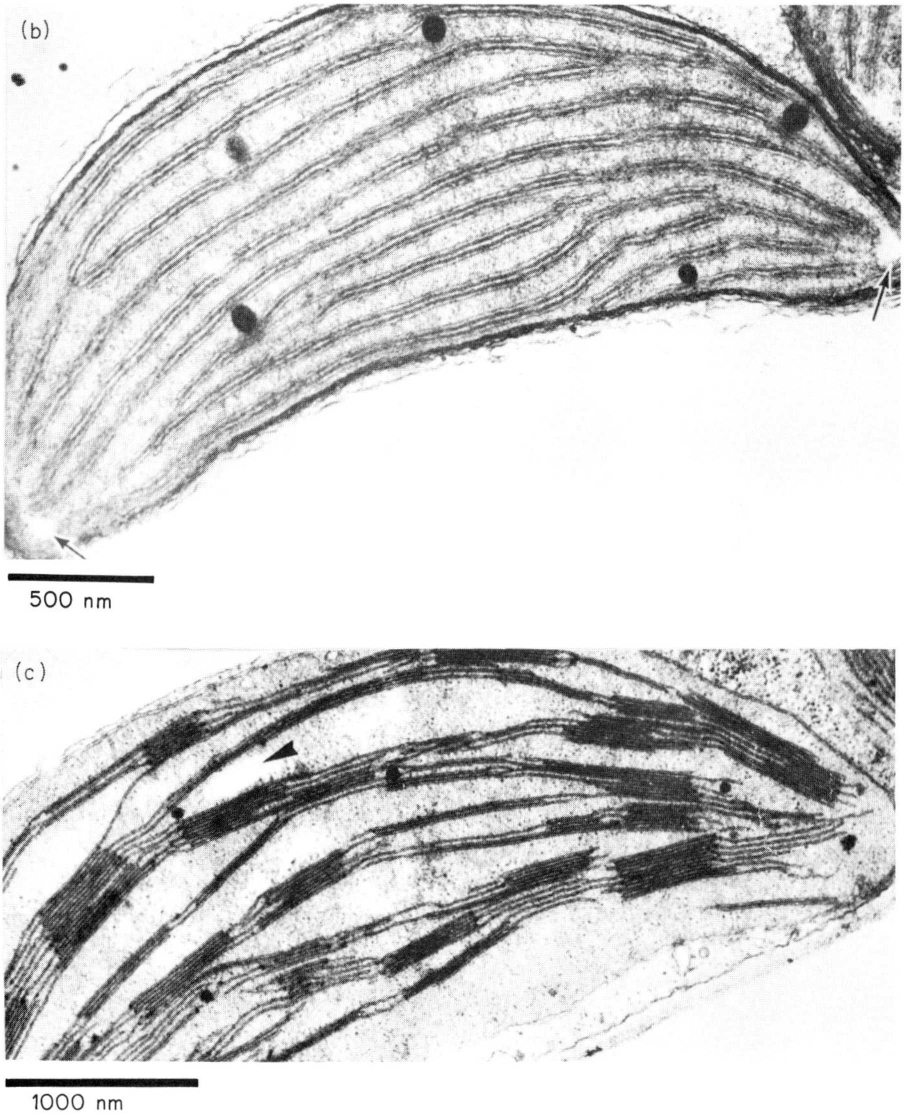

Fig. 1.24. (See legend on p. 34)

chloroplasts have their genomes and are capable of autonomous reproduction. Both the chloroplast size (typically $2 \times 5\ \mu m$) and the magnitude of its genome are greater than those of mitochondria. A typical list of chloroplast-synthesized proteins is included in Table 1.7.

Like mitochondria, chloroplasts have a double membrane, the outer one being of the eukaryotic type in containing sterols, the inner one being distinct from any other cell membrane (Fig. 1.25), particularly in that tiny self-enclosed

Table 1.7. Chloroplast gene products

23S rRNA 16S rRNA 5S rRNA 4.5S rRNA tRNA of 25–30 kinds mRNA for the large subunit of ribulose bisphosphate carboxylase mRNA for one of the PS II polypeptides	Ribosomal proteins (2–5) Soluble proteins (up to 90) including the large subunit of ribulose bisphosphate carboxylase, elongation factors G and Tu, fatty acid synthetase Thylakoid proteins (up to 20), including 5 subunits of ATP synthase, 5 subunits of PS I and II, cyt f, cyt b_6, cyt b_{559}

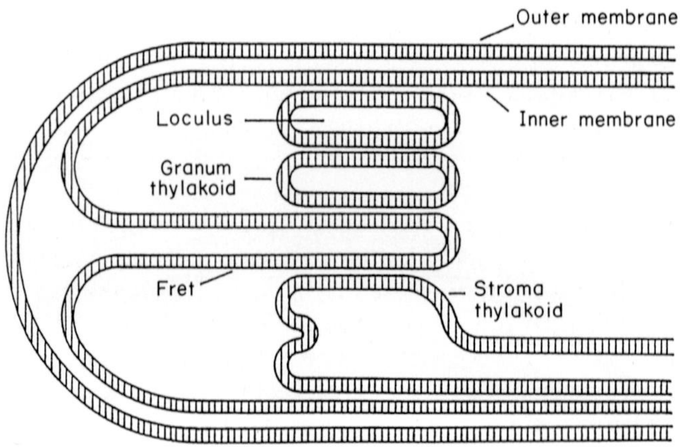

Fig. 1.25. Schematic cross-section of a chloroplast showing the various structural forms observed

vesicles, the thylakoids, are usually fully separated from the inner membrane, although clearly having originated from it.

The chloroplast contains a variety of enzymes and enzyme systems, either soluble in the stroma (analogy of the mitochondrial matrix) or embedded in its different membranes.

The outer membrane contains a porin-like protein permitting the passage of rather large molecules across the membrane, the inner membrane contains various specific transport systems for the following substances.

1. Dicarboxylates.
2. Glycollate.
3. Phosphoglycollate.
4. The malate shuttle, taking oxaloacetate out to be reduced by NADH to malate which is taken in to be oxidized to oxaloacetate by NADP.
5. The aspartate shuttle, taking out aspartate (and oxoglutarate in parallel) to be transaminated to oxaloacetate and glutamate which are taken in and

retransaminated to oxoglutarate and aspartate, apparently to provide oxaloacetate for the malate shuttle.

6. The triose phosphate shuttle, starting with dihydroxyacetone phosphate (DAP) being transported out in antiport with inorganic phosphate. DAP undergoes isomerization, oxidation (by NAD) and phosphorylation, dephosphorylation to phosphoglycerate which is transported inward and, in the stroma, is reduced eventually to DAP (by NADPH). The inner membrane is impermeable to sugars, phosphorylated sugars, diphosphate, NAD and NADP, and to neutral amino acids.

7. Pyruvate plus phosphate in antiport for phosphoenolpyruvate (in the C_4 mesophyll chloroplasts).

The thylakoid membrane contains the whole set of enzymes and factors for carrying out the photochemical part of photosynthesis, including the proton translocators and ATP synthase. Also present are carriers for K^+ and Mg^{2+}.

The lipid composition of the outer chloroplast membrane resembles that of the endoplasmic reticulum and Golgi body while the inner membrane (IM) and the thylakoid (TH) contain some unusual lipids and lipid-soluble components, e.g., in spinach chloroplasts, galactosylglycerides (69 per cent in IM and 37 per cent in TH), sulpholipids (1 per cent in IM and 7 per cent in TH), and chlorophyll (21 per cent in TH).

A great many enzymes are present in the aqueous stroma (Table 1.8). In C_4 plants, the principal enzyme of CO_2 fixation, phosphoenolpyruvate carboxylase (4.1.1.31), as well as the transaminase that in some cells converts oxaloacetate to aspartate, are actually in the cytosol of mesophyll cells, while NADP-malate dehydrogenase (1.1.1.82), pyruvate orthophosphate dikinase (2.7.9.1), the NADP-linked malic enzyme (1.1.1.40) and the whole Calvin cycle are located in the chloroplast stroma of the bundle-sheath cells.

Table 1.8. Selection of enzymes present in the chloroplast stroma of C_3 plants

Phosphogluconate dehydrogenase	1.1.1.44
Glucose-6-phosphate dehydrogenase	1.1.1.49
Glyceraldehyde-phosphate dehydrogenase (NADP)	1.2.1.13
Transketolase	2.2.1.1
1,2-Diacylglycerol 3-β-galactosyltransferase	2.4.1.46
3-β-D-galactosyl-1,2-diacylglycerol 1,6 α-galactosyltransferase	2.4.1...
Phosphoribulokinase	2.7.1.19
Phosphoglycerate kinase	2.7.2.3
Fructosebisphosphatase	3.1.3.11
Sedoheptulose bisphosphatase	3.1.3.37
Ribulose bisphosphate carboxylase	4.1.1.39
Fructose bisphosphate aldolase	4.1.2.13
Ribulose-phosphate-4-epimerase	5.1.3.4
Triosephosphate isomerase	5.3.1.1
Ribosephosphate isomerase	5.3.1.6

In addition to the enzymes listed in Table 1.8, the stroma contains the whole set of enzymes for fatty acid synthesis, for isoprenoid synthesis and for porphyrin synthesis. Besides these, the chloroplast contains enzymes involved in nitrate reduction and glutamate synthesis, in sulphate reduction and sulpholipid synthesis, and a superoxide dismutase.

The autonomy of the chloroplast is as obvious as that of the mitochondrion. In fact, there are present-day analogies of an endosymbiosis of a chlorophyll-containing cell in a larger host cell. The most spectacular of these endosymbionts are the cyanelles, derived from the blue–green algae which, unlike similar endosymbionts of animals (ranging from protozoans to molluscs) cannot be cultivated outside the host cell and may thus form a transition from a free-living cyanobacterium to a cell-bound chloroplast.

It is now generally believed that both mitochondria and chloroplasts invaded primitive eukaryotes approximately $2 \cdot 10^9$ years after the first cells had appeared on this planet, according to an evolutionary tree recently designed to include the archaebacteria (Fig. 1.26).

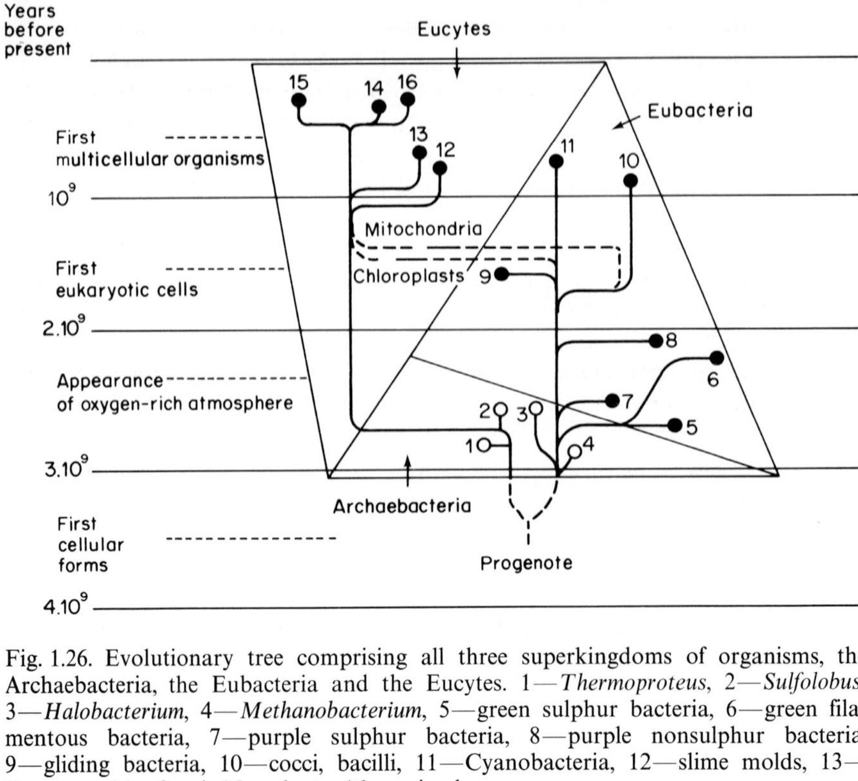

Fig. 1.26. Evolutionary tree comprising all three superkingdoms of organisms, the Archaebacteria, the Eubacteria and the Eucytes. 1—*Thermoproteus*, 2—*Sulfolobus*, 3—*Halobacterium*, 4—*Methanobacterium*, 5—green sulphur bacteria, 6—green filamentous bacteria, 7—purple sulphur bacteria, 8—purple nonsulphur bacteria, 9—gliding bacteria, 10—cocci, bacilli, 11—Cyanobacteria, 12—slime molds, 13—Protozoa, 14—fungi, 15—plants, 16—animals

1.3.10 Chromaffin granules

These tiny vesicles (200–400 nm across) literally fill the cells of some specific organs, in particular those of adrenal medulla where they store adrenaline and noradrenaline, together with various enzymes and other proteins. Much like lysosomes, they are formed from the Golgi complex and, while adrenaline is being synthesized in them, they move toward the cell periphery where, upon suitable stimulus, they fuse with the plasma membrane and empty their contents into the external medium. The membranes of chromaffin granules contain an H^+-ATPase, the transport protein for dopamine (a noradrenaline precursor), and cytochrome b_{561} which acts as a redox carrier, shuttling electrons from ascorbic acid in the cytosol to semi-hydroascorbate in the granule.

1.3.11 Other membranous organelles

Besides the various membranes described in detail in the preceding sections there exist perhaps a hundred membrane-bounded organelles in different cells, some being simply modified versions of an organelle from a 'typical' cell, others, especially in protozoans, being quite unique. A sampling of these follows.

A number of organelles are implicated in the lysosomal cycle (see Fig. 1.18), thus an *autolysosome* is formed from a *phagosome* by fusion with cytoplasmic particles containing acid hydrolases, itself turning into a *telolysosome.* To this category belong also *heterophagosomes* and *heterolysosomes*, as well as acid-phosphatase-containing *neolysosomes*, soluble-material-containing *pinosomes*, *receptosomes* or *endosomes* which are pinched off the plasma membrane into the cell and contain substances absorbed originally on the surface on appropriate receptors.

Some ciliates contain *acidosomes* as sources of acid for phagocytotic vacuoles, trypanosomes keep their oxidative enzymes in a *cercidosome* and their glycolytic enzymes in a *glycosome* (0.27 μm across, about 230 per cell), while in coelenterates the locale of bioluminescence are the *luminosomes.*

Red algae and cyanobacteria keep their antenna pigment-proteins in *phycobilisomes*, while green sulphur bacteria contain their bacteriochlorophyll in *chlorosomes.*

In the retinal cones of some fish the cytochrome pigments are present in *ellipsosomes*, melanogenesis takes place in the *melanosomes* of animal melanoma tissues. In sperm cells, the anterior region is capped by the *acrosome. Synaptosomes* are vesicles located near the presynaptic nerve endings.

Lipid substances, in particular oils, are stored in plant cells within *spherosomes*, a special kind of which are the *oleosomes* of seeds and fruits. Their membranes, although derived from endoplasmic reticulum, may not be of the bilayer type but, on account of their lipid content, be composed of a single amphipathic lipid layer.

Extrusomes and *prostasomes* are organelles that are extruded from some unicellular parasites and from the prostate gland, respectively, their contents being unknown (and possibly unimportant) but their membranes carrying various enzymes. Their common feature is that they are extruded from the cell as such through an opening of the outer membrane, not by pinching off the external membrane, by a process called *diacytosis*.

Some annelid worms contain *photosomes*, derivatives of the endoplasmic reticulum, made up of membrane tubules about 20 nm in diameter which are instrumental in the bioluminescence of these animals.

SYNOPSIS

All cells are surrounded by the plasma membrane and the eukaryotic ones contain, in addition, several types of membranous organelles. The principal ones are the nucleus in which DNA is replicated and transcribed, the endoplasmic reticulum on which much of the protein synthesis takes place when ribosomes are attached to it, the Golgi apparatus where proteins are glycosylated and sorted out according to their destination, the mitochondria which serve as the power station of most cells, and the chloroplasts which, in blue–green algae, higher algae and plants, carry out photosynthesis. In addition to these, many special and often unusual membrane-bounded organelles are known from both plant and animal kingdoms.

FURTHER READING

Bendall, D. S. (ed.). *Evolution from Molecules to Man*, Cambridge University Press, Cambridge (1983).

Fewson, C. A. *Archaebacteria. Biochem. Education*, **14**, 103–15 (1986).

Hilling, B., and Amelunxen F. On the development of plant vacuole. Further evidence for endoplasmic reticulum origin, *Eur. J. Cell Biol.*, **38**, 195–200 (1985).

Kleinig, H., and Sitte, P. *Zellbiologie*, Gustav Fischer Verlag, Jena (1984).

Kuehn, G. D. Some one-hundred 'somes', *Trends Biochem. Sci.*, **10**, 227–30 (1985).

Matile, P. *The Lytic Compartment of Plant Cells*, Springer-Verlag, Vienna (1975).

McLaren, D. J., and Hockley D. J. Blood flukes have a double outer membrane, *Nature*, **269**, 147–9 (1977).

Prebble, J. N. *Mitochondria, Chloroplasts and Bacterial Membranes*, Longman, London (1981).

Tolbert, N. E., and Essner, E. Microbodies: peroxisomes and glyoxysomes, *J. Cell Biol.*, **91**, 271s–83s (1981).

Tzagoloff, A. *Mitochondria*, Plenum Press, New York (1982).

Ude, J., and Koch, M. *Die Zelle, Atlas der Ultrastruktur*, Gustav Fischer Verlag, Jena (1982).

van den Bossche, H. (ed.). *Biochemistry of Parasites and Host-Parasite Relationship*, North Holland, Amsterdam (1976).

2 Membrane Structure

2.1 STRUCTURE-FORMING ELEMENTS OF CELLS

The three classes of cellular structures, the cell walls, the cytoskeleton, and the biomembranes, can be defined by their content of a prevalent molecular species.

The *cell walls* of yeasts, fungi, and plants, as well as the slime or cuticular coats of some animal cells, especially the free-living protozoans, are composed of polysaccharides, such as cellulose, glucans, mannans, xylans, pectins and chitin. Even the peptidoglycan envelopes of bacteria contain a sizable portion of alternating molecules of N-acetylmuramic acid and N-acetylglucosamine. (It should be observed that some of the plant cell-wall polysaccharides represent the most abundant organic molecules on Earth—there may be 10^{12} tons of cellulose and about half that amount of xylans present in the biosphere.) Particularly in bacteria, the peptidoglycan cell walls are in intimate contact with outer membranes.

The *cytoskeleton* of eukaryotic cells is exclusively protein in composition—there are now tens of different proteins known to be either integral or associated components of the three cytoskeletal structural types.

The *microtubules* are hollow cylinders 20–25 nm in diameter with a lumen of about 10 nm, and are composed of α- and β-tubulin subunits (dimer $M_r = 110\,000$). They interact with dynein (600 000 + 700 000) possessing an ATPase activity, especially in eukaryotic flagella, and nexin, further with clathrin, the main component of coated vesicles, with microtubule-associated proteins (MAP) of M_r of 340 000 and 270 000, as well as with ankyrin (200 000) and with tau factors (55 000–62 000) which promote the elongation of microtubules. Finally, they are associated with the recently discovered kinesin (2 × 110 000 + 65 000/70 000) which is essential for microtubule movement.

The *intermediate filaments* (10 nm in diameter) are composed of a number of different proteins, depending on cell type and function, thus glial filament proteins (51 000), cytokeratins (40 000–68 000), vimentin (58 000), titin, plectin, synemin, neurofilamins (68 000), desmin (55 000) and filaggrin.

The *microfilaments* are principally made of actin (42 000) and myosin (200 000) plus α-actinin (2 × 100 000), β-actinin (37 000) and tropomyosin (40 000). The total amount of actin represents as much as 5–10 per cent of the total cell protein while that of myosin amounts to 0.3–1.5 per cent of the total proteins. The microfilaments associate with perhaps twenty different proteins, some of

Table 2.1. Composition of various membranes expressed as their protein-lipid weight (W) and area (A) ratios

Membrane	W	A
Bovine myelin	0.28	0.43
Yeast tonoplast	0.67	—
Human erythrocyte	1.11	2.5
Outer mitochondrial membrane	1.22	—
Amoeba plasma membrane	1.34	—
Retinal rod outer segment	1.44	—
Hepatocyte plasma membrane	1.50	—
Escherichia coli cytoplasmic membrane	1.50	—
Pseudomonas aeruginosa	1.71	—
Micrococcus lysodeikticus	1.91	4.3
Brain synaptic vesicles	1.94	—
Saccharomyces cerevisiae plasma membrane	1.98	—
Leukovirus	2.10	—
Chloroplast lamella	2.22	—
Thermoplasma	2.28	—
Acholeplasma	2.33	1.1
Halophilic bacteria	2.86	—
Bacillus megaterium	3.10	5.4
Inner mitochondrial membrane	3.14	—
Myxoviruses	3.65	—

them involved in Ca^{2+} traffic (and some of them Ca^{2+}-dependent), such as actinogelin (115 000), caldesmin (150 000), calmodulin (16 700), gelsolin (91 000), villin (95 000), brevin (90 000), fragmin (43 000). Ankyrin (200 000), spectrin ($2 \times 220\,000 + 2 \times 240\,000$) and fodrin (240 000) are supporting elements of the erythrocyte membrane; fascin (58 000), acumentin (30 000), filamin ($2 \times 250\,000$), fimbrin (68 000) and vinculin (130 000) bundle acting filaments; profilin (16 000) stabilizes G-actin.

Most of the cytoskeletal elements, particularly intermediate filaments in gap junctions and microfilaments in the intracellular support are in close contact with plasma membranes.

The *biomembranes* are the only class of architectural elements where lipids form the basic structure in which or on which other components may be anchored. However, the overall average weight composition of biomembranes is only 40 per cent lipid and 60 per cent protein, there being a considerable variation in their proportion (Table 2.1).

FURTHER READING

Birchmeier, W. Cytoskeleton structure and function, *Trends Biochem. Sci.*, **9**, 192–5 (1984).

Haest, C. W. M. Interactions between membrane skeleton proteins and the intrinsic domain of the erythrocyte membrane, *Biochim. Biophys. Acta*, **694**, 331–52 (1982).

Seltmann, G. *Die bakterielle Zellwand*, Gustav Fischer-Verlag, Jena (1982).

2.2 TYPES OF MEMBRANE LIPIDS

There are several ways one can divide and classify the lipid components of membranes, e.g. into phospholipids (containing a phosphoryl group), glycolipids (containing sugar moieties), and others; into neutral and charged lipids; into esters and nonesters, etc.

We shall use here a structure-based classification which divides membrane lipids into derivatives of glycerol (glycerolipids), derivatives of 2-aminoethanol (sphingolipids), and sterols.

2.2.1 Glycerolipids

The general formula here is

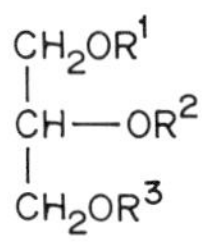

Only rarely are the hydroxyls esterified simply by one, two or three fatty acids, giving rise to neutral mono-, di-, and triglycerides. A more frequent situation is the one where R^1 and R^2 are acyl groups while R^3 is a phosphoric acid group. Without further esterification, this is then phosphatidic acid (PA). However, the phosphatidic acid is almost always further substituted (Table 2.2), giving rise to a family of important phospholipids.

Occasionally, one of the acyl groups of a substituted phospholipid, mostly that at C_1 of glycerol, is replaced with an alk-l-enyl chain, giving rise to a *plasmalogen*. If the substituent is a saturated alkyl chain the compound formed is simply a glycerol ether or an alkyl phospholipid. Several types of such mono- and diethers are known, some of them containing an acylcholine substituent at the phosphoryl group, thus generating a species with three long alkyl chains (see formula). Some of the compounds may be of high therapeutic value but it is not clear in what amounts they actually occur in biomembranes.

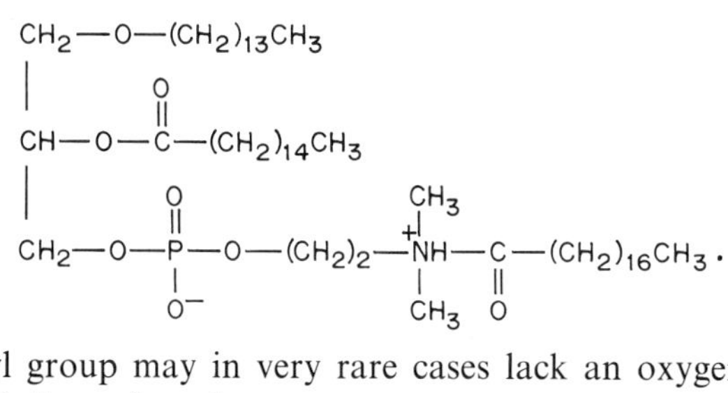

The phosphoryl group may in very rare cases lack an oxygen, whereupon a *phosphonolipid* is formed, such as

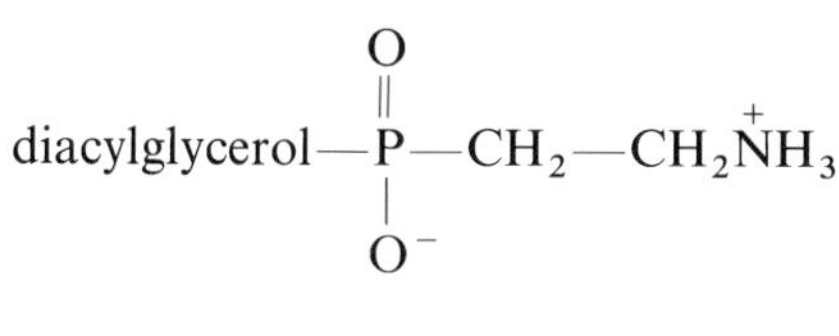

Table 2.2. Structural formulas of principal glycerolipids

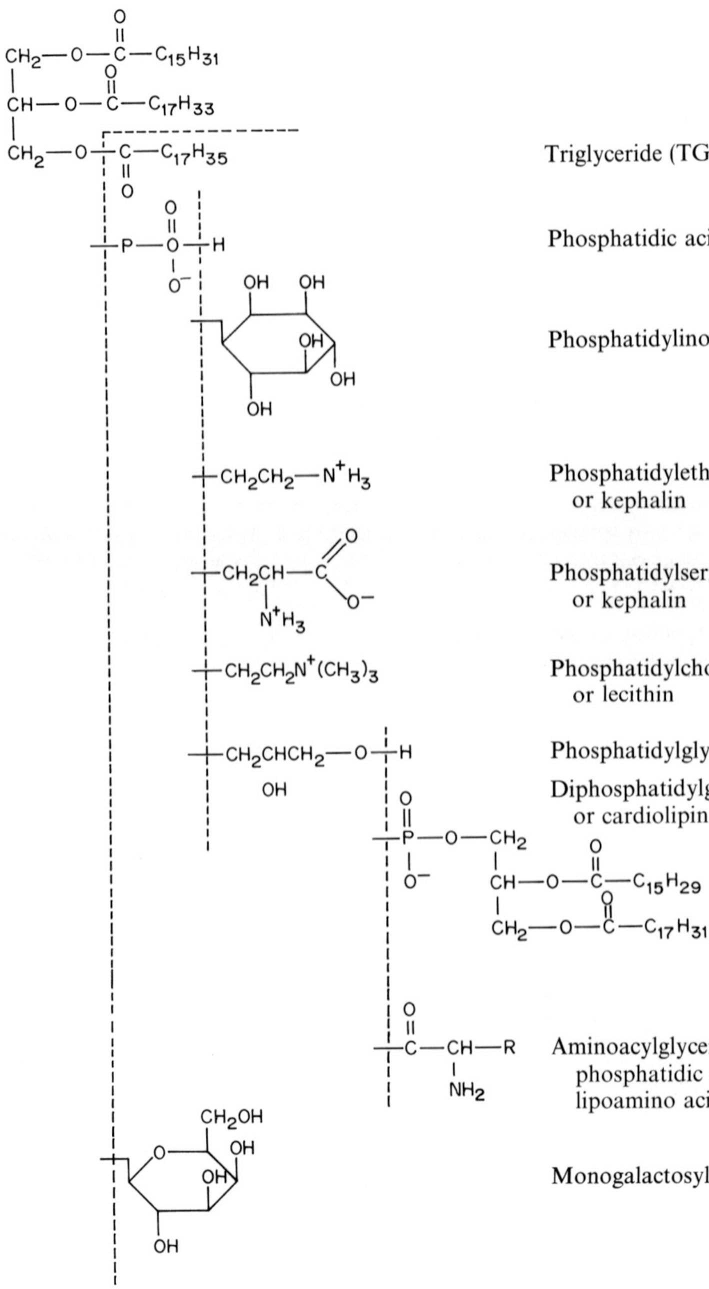

- Triglyceride (TG)
- Phosphatidic acid (PA)
- Phosphatidylinositol (pI)
- Phosphatidylethanolamine (PE) or kephalin
- Phosphatidylserine (PS) or kephalin
- Phosphatidylcholine (PC) or lecithin
- Phosphatidylglycerol (PG)
- Diphosphatidylglycerol (DPG) or cardiolipin (CL)
- Aminoacylglycerol-phosphatidic acid or lipoamino acid (LAA)*
- Monogalactosyldiacylglycerol[†]

*R may be Arg, Phe, Thr, Leu, Orn, Ala, Lys, Glu, Asp or His.
[†]Also glucosyl, diglucosyl and sulphodeoxyglucosyl diacylglycerols are known.

The fatty acid substituent R^2 may sometimes be missing in a phosphatidyl derivative; in that case we have to do with a *lysophospholipid* which plays a special role in membrane integrity.

It is important to note in the context of membrane surface potentials and of cell–cell interactions that at neutral pH both PC and PE are neutral, PS and PI carry one negative charge while CL and PA carry two negative charges.

A highly unusual kind of glycerol-based phospholipid is found in some archaebacteria, such as *Thermoplasma* or *Sulfolobus*. Instead of acyl chains at glycerol carbon atoms 1 and 2 there are diphytanyl (reduced isoprenoid) bridges spanning covalently the whole membrane:

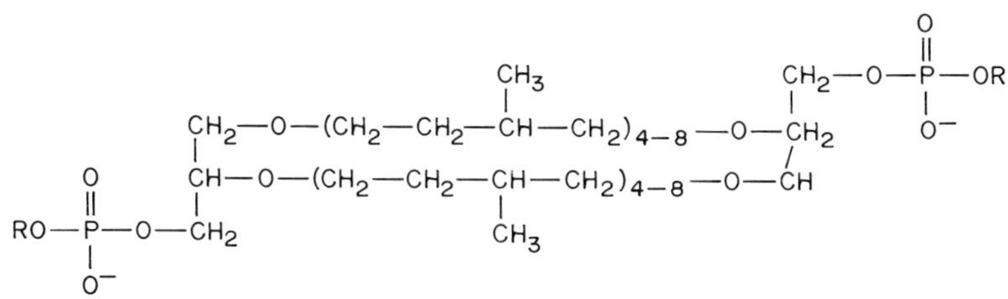

R can be glycerol, glycerol phosphate or glycerol sulphate. Occasionally, 1–4 cyclopentane rings replace the same number or isoprene units in the chain. It is noteworthy that the chirality of glycerol carbon 2 is opposite to that in other biological kingdoms, viz. L or *sn*-2, 3, rather than D or *sn*-1,2.

Occasionally, 'half' molecules (with C_{20} or C_{25} terpenoid chains) are found which resemble in size and shape the usual phosphoglycerolipids.

2.2.2 Sphingolipids

Their core structure is 1-amino-2-hydroxyethane, substituted at three positions as follows:

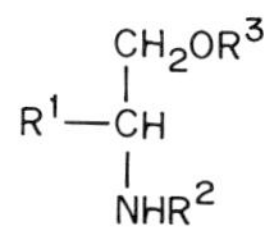

The nomenclature of the derivatives can be based either on the type of substituent at R^1 or R^3 (Table 2.3). The R^2 is usually an acyl or an α-hydroxyacyl group with 16–26 carbon atoms. The R^3 in its fundamental form is H, the compound then being called ceramide. It is this very substituent which is highly variable. The most abundant of all sphingolipids is the phosphoryl choline derivative, called *sphingomyelin* (SP). It carries no electrical charge, like the corresponding glycerol derivative. However, the phosphoryl inositol derivative carries one negative charge, like the corresponding glycerol-based phospholipid.

The R^3 substituent is often a simple of complex sugar; thus, with glucosyl or galactosyl a *cerebroside* is formed, with an oligosaccharide a *ganglioside* is generated. The various gangliosides, including the blood-group substances, occur abundantly in nervous tissue and in erythrocytes. There is obviously a

Table 2.3. Structural formulas of sphingolipids

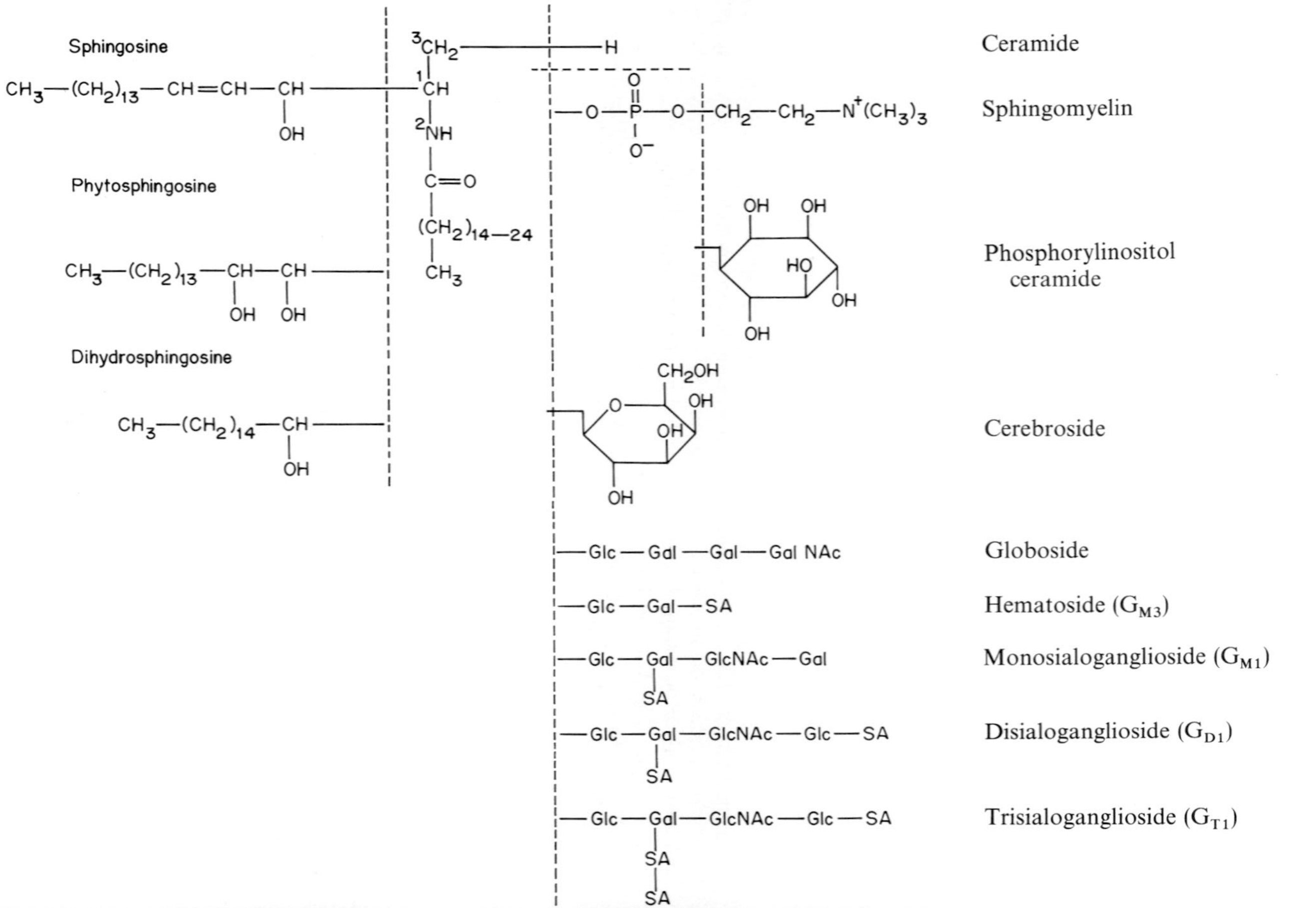

vast range of possibilities of structural isomers if only four or five different sugars are arranged in different sequence, the variety being further enhanced by differences in bonds (α, β, $1 \rightarrow 2$, $1 \rightarrow 3$, $1 \rightarrow 4$, etc.). To give an example, the ABO antigens of the various blood groups are shown. Two determinant glycosyl chains are involved in each group on two different sphingolipids. While one is always galactose linked $\beta 1 \rightarrow 4$, the other chain is specific, thus:

group A contains
$$\text{GalNAc} \xrightarrow{\alpha 1\rightarrow 3} \underset{\substack{|\,\alpha 1\rightarrow 2\\ \text{Fuc}}}{\text{Gal}} \xrightarrow{\beta 1\rightarrow 3} \text{GalNAc}$$

group B contains
$$\text{Gal} \xrightarrow{\alpha 1\rightarrow 3} \underset{\substack{|\,\alpha 1\rightarrow 2\\ \text{Fuc}}}{\text{Gal}} \xrightarrow{\beta 1\rightarrow 3} \text{GlcNAc}$$

group O contains
$$\underset{\substack{|\,\alpha 1\rightarrow 2\\ \text{Fuc}}}{\text{Gal}} \xrightarrow{\beta 1\rightarrow 3} \text{GlcNAc.}$$

(It should be noted that the determinants of MN blood-group specificity depend on the substitution by glycosyl units of glycoproteins, especially glycophorin.)

The sugars contained in gangliosides are D-glucose, D-galactose, N-acetyl-D-glucosamine, N-acetyl-D-galactosamine, D-fucose, and sialic or neuraminic acid which is

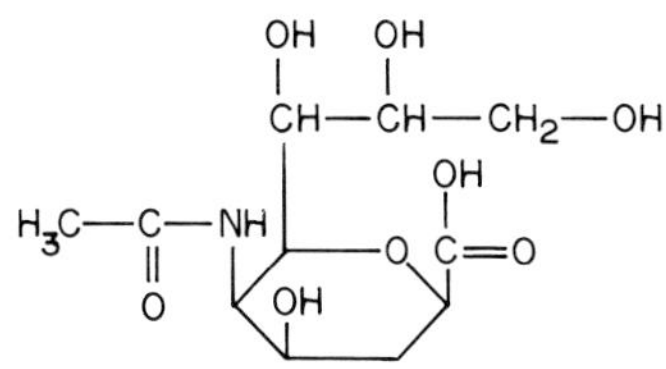

If the R^3 of a sphingolipid is a sugar sulphate the compound is one of the sulphatides.

2.2.3 Sterols

Different both in structure and in molecular shape from the first two groups, the sterols are an important and ancient class of compounds, occurring predominantly in eukaryotic cell membranes and representing a major portion of the lipid complement of some membranes (e.g., up to 40 per cent in mammalian brain cells).

The best-known representative is cholesterol of animal membranes while ergosterol occurs in lower eukaryotes and β-sitosterol in plants.

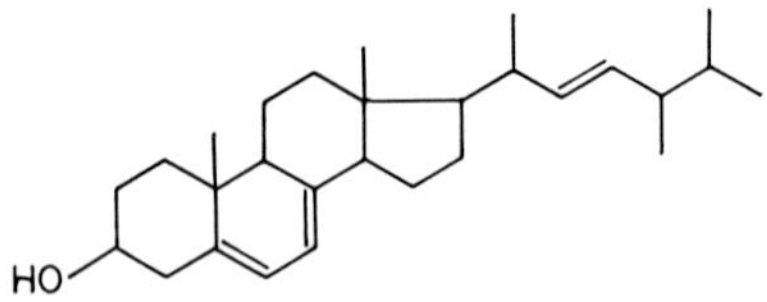

Ergosterol

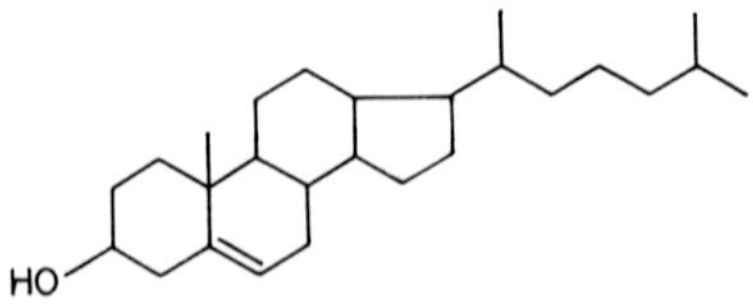

Cholesterol

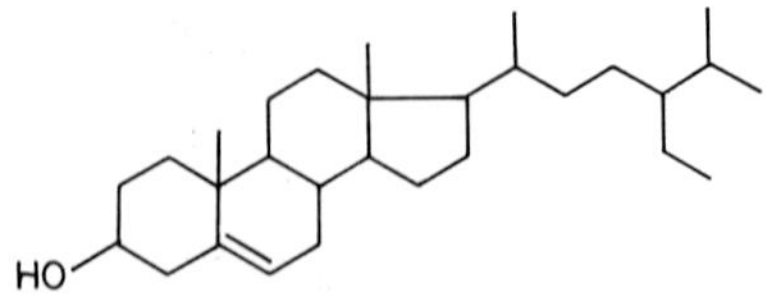

β-sitosterol

Now these are only a few of the several hundred sterols known to be present in biomembranes. There is a variety of sterols in lower plants and particularly in marine animals of various phyla. These sterols may contain some esoteric side chains at $C_{(17)}$, including cyclopropane rings and terminal methylene groups. The variety of the sterols is somewhat difficult to understand but apparently the lack of selection pressure at the level of the basic sterol structure permitted the evolution into the many compounds that now exist side by side.

Sterols are akin to the hopanoids mentioned in the introductory chapter on membrane evolution. The similarity between a hopanoid and lanosterol, the important precursor of most sterols of modern-cell membranes, is clearly apparent. While most abundant in fossil deposits the hopanoids occur in present-day bacteria and some eukaryotic cells, such as *Tetrahymena*, where they are found either free or bound to adenosine (via a hydroxyl group of D-ribose). More than 150 different hopanoids have been described so far.

Hopanoid

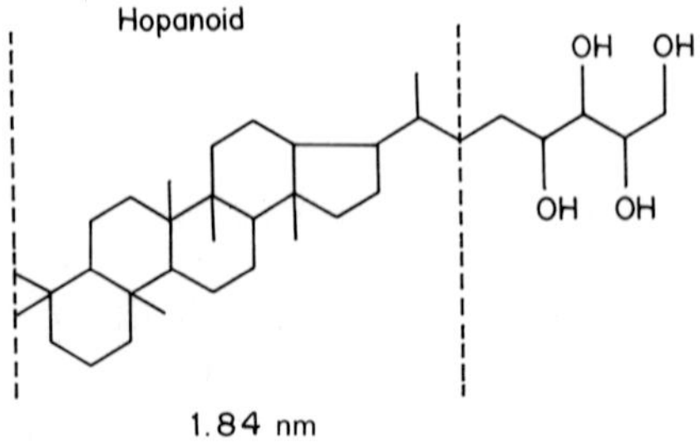

Lanosterol

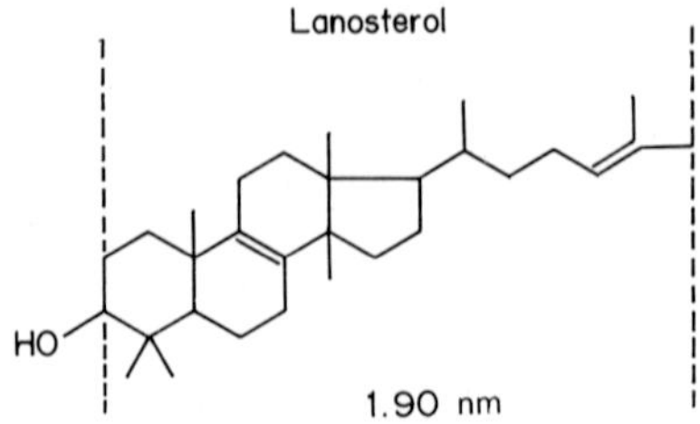

All the above classes of lipids, although structurally quite different, share one property in common. They are *amphiphilic* (= *amphipathic*) which indicates that one part of their molecule is relatively more hydrophilic than the other. This has then the obvious consequences for lipid organization in the bilayer, as discussed below.

2.2.4 Other nonprotein membrane components

Some special membranes may contain molecules that are not lipids in the true sense of the word but still can be 'dissolved' in the lipid domains of the membrane.

They include:

(a) terpenoid compounds, such as vitamin K, carotenes, ubiquinone and plastoquinone, squalene, branched alkylbenzenes,
(b) wax esters,
(c) sugar esters with fatty acids,
(d) diol esters,
(e) sulpholipids, mainly alkyl sulphates,
(f) halosulpholipids, such as chloroalkyl sulphates,
(g) lipopolysaccharides, present in the outer layer of gram-negative bacteria,
(h) chlorophyll and bacteriochlorophyll,
(i) nonprotein redox compounds, especially in bacterial membranes.

2.2.5 Distribution of lipids

Ubiquitous as membranes are in living matter there are important differences in their composition between one group of cells and another (Table 2.4). Although there are no absolute rules about the occurrence of the different lipid classes in various organisms some general conclusions can be made.

1. Membrane-enclosed viruses contain most lipid classes as they derive from the lipids of their hosts. The same holds for the parasitic mycoplasmas and acholeplasmas (primitive subbacterial types of organisms).

2. True bacteria contain no sphingolipids, very few glycolipids, and only traces of sterols. The major components of bacterial membranes are PG, PE and CL, with PC being notably absent.

3. The inner mitochondrial membrane as well as some bacterial membranes contain appreciable amounts of CL.

4. Glycolipids are mainly restricted to plasma membranes of eukaryotes.

5. Nerve cell membranes contain large amounts of glycosphingolipids.

6. Uncharged lipids are found in animal as well as plant plasma membranes while acidic ones are common in mitochondria.

7. There are some striking interspecies differences in the content of various phospholipids. Thus, cow and sheep erythrocytes contain no PC while in dog erythrocytes it represents nearly one half of all phospholipids; PI makes up 0.3 per cent of horse erythrocyte phospholipids but 7.4 per cent in the cat; SP is low in the dog erythrocyte (10.8 per cent) and very high in sheep (51 per cent). In comparison, there is less variety between various organs of the same animal.

It is interesting to observe the changing composition of a membrane undergoing recycling in a rat liver cell. The smooth endoplasmic reticulum contains 48 per cent PC, 5 per cent SP, 19 per cent PE and 6 per cent cholesterol. By the time the membrane becomes part of the plasma membrane it contains 19 per cent PC, 12 per cent SP, 12 per cent PE and 20 per cent cholesterol.

One can envisage the importance for certain reactive groups of lipids not to be exposed on a given membrane surface (PC, SP) or the usefulness of charged groups in contributing to the surface potential (e.g., CL on the internal, matrix, face of the inner mitochondrial membrane).

Table 2.4. Lipid composition of selected membranes (in molar %)

Type of membrane	PC	PE	PS	PI	PG	CL	Glyco-glycero-lipids	Sphingo-lipids	Sterols	Others
Sarcina lutea	—	—	—	8	90	1	—	—	—	1
Micrococcus roseus	—	—	—	—	66	32	—	—	—	2
Salmonella typhimurium										
Cytoplasmic	—	60	—	—	32	7	—	—	—	1
Outer	—	81	—	—	15	2	—	—	—	2
Bacillus subtilis	—	—	—	—	78	3	—	—	—	19
Anacystis nidulans	—	—	—	—	14	—	70	—	—	16
Saccharomyces cerevisiae										
Plasma membrane	27	10	14		—	—	—	—	40	9
Tonoplast	33	15	43							
Rat hepatocyte										
Plasma membrane	28	17	6	6	—	1	—	12	27	3
Smooth endoplasmic reticulum	41	16	5	10	—	1	—	7	17	3
Erythrocyte										
Rat	30	13	7	2	—	—	2	16	25	5
Pig	15	18	11	1	—	—	10	18	27	—
Lobster nerve plasma membrane	29	18	8	—	—	—	—	9	36	—
Rabbit muscle plasma membrane	23	9	1	—	—	—	—	4	57	6
Human myelin sheath	11	14	5	1	—	—	—	32	25	12
Inner mitochondrial membrane										
Rat liver	40	35	1	5	2	17	—	1	1	1
Cauliflower	37	34	—	4	3	13	—	—	5	1
Baker's yeast	36	29	2	8	1	19	—	—	1	4
Outer mitochondrial membrane										
Rat liver	46	21	1	10	2	5	—	2	10	3
Cauliflower	36	21	—	17	8	2	—	—	12	4
Baker's yeast	43	22	3	20	1	6	—	—	1	4
Spinach chloroplast										
Outer membrane	22	—	—	1	9	—	68	—	—	—
Inner membrane	6	—	—	1	8	—	79	—	—	6
[illegible]	[illegible]	[illegible]	[illegible]	[illegible]	[illegible]	[illegible]	[illegible]	[illegible]	[illegible]	7

2.2.6 Fatty acids in membrane lipids

Two major groups of membrane lipids, glycerolipids and sphingolipids, as well as some of the minor groups of esters, contain one, two or three acyl groups derived from a medium or long-chain fatty acid. The variety of these fatty acids adds to the varied pattern of lipid classes so that hundreds upon hundreds of

Table 2.5. Selection of fatty acids occurring in biomembranes

Straight-chain, saturated			
12:0	Lauric	22:0	Behenic
14:0	Myristic	23:0	—
16:0	Palmitic	24:0	Lignoceric
17:0	Margaric	26:0	Cerotic
18:0	Stearic	28:0	Montanic
19:0	—	30:0	Melissic
20:0	Arachic		
Straight-chain, monounsaturated			
12:1 (9 *cis*)	Lauroleic	18:1 (11 *cis*)	*cis*-Vaccenic
14:1 (9 *cis*)	Myristoleic	18:1 (11 *trans*)	*trans*-Vaccenic
16:1 (7 *cis*)	Hypogaeic	20:1 (9 *cis*)	Gadolenic
16:1 (9 *cis*)	Palmitoleic	22:1 (11 *cis*)	Getolenic
18:1 (9 *cis*)	Oleic	22:1 (13 *cis*)	Erucic
18:1 (9 *trans*)	Elaidic	24:1 (15 *cis*)	Nervonic
Straight-chain, oligounsaturated			
16:2	—	20:4 (8, 11, 14, 17 all-*cis*)	—
16:3	—	20:5	—
18:2 (9, 12 all-*cis*)	Linolic	22:4	—
18:3 (9, 12, 15 all-*cis*)	Linolenic	22:5	—
18:3 (9 *cis*, 11, 13 *trans*)	α-Eleostearic	22:6	—
18:4 (all-*cis*)	—		
20:3 (8, 11, 14 all-*cis*)	—		
20:4 (5, 8, 11, 14 all-*cis*)	Arachidonic		
Branched-chain			
12:0 (10-Me)	Isolauric	16:0 (14-Me)	Isopalmitic
14:0 (12-Me)	Isomyristic (sarcinic)	17:0 (14-Me)	—
		17:0 (15-Me)	—
15:0 (12-Me)	—	19:0 (10-Me)	—
15:0 (13-Me)	—		
Hydroxy			
14:0 (3-OH)	β-Oxymyristic	18:0 (9, 10-OH)	Dioxystearic
16:0 (3-OH)	β-Oxypalmitic	24:0 (2-OH)	Cerebronic
18:0 (3-OH)	β-Oxystearic	24:1 (15 *cis*, 2-OH)	α-Oxynervonic
18:1 (9 *cis*, 12-OH)	Ricinoleic	26:0 (2-OH)	α-Cerotic

Table 2.5 *(Contd.)*

Containing a cyclopropane ring	
17:0 (9, 10 *cis*, methylene)	—
18:1 (8 *cis*, 8, 9 *cis*-methylene)	Malvalic
19:0 (11, 12 *cis*-methylene)	Lactobacillic
19:0 (9, 10 *cis*-methylene)	Dihydrosterculic
19:0 (9 *cis*, 9, 10 *cis*-methylene)	Sterculic
Others	
2, 4, 6-Trimethylheptacos-2-enoic	Phthionic
2-Tetradecyl-3-hydroxyoctadec-11-enoic	Corynomycolenic
3, 5-Dihydroxy-3-methylpentanoic	Mevalonic

different lipid molcules may be encountered in biomembranes, the usefulness of this complexity being little understood (Table 2.5 and 2.6). The tables are far from listing all fatty acids found in cell membranes. Marine sponges contain an assortment of C_{30} acids (occasionally 40 per cent of the total), in protozoans and algae C_{36} acids are fairly common. The bacterium *Mycobacterium phlei* contains no less than 55 different fatty acids in its outer envelope, among them such polyunsaturated rarities as 40:6, and some with as many as 56 carbon atoms.

In lipids that contain two fatty acyl residues, such as glycerophospholipids, very rarely are the two fatty acids identical. The double bonds encountered in many of the fatty acids lie beyond C_9. If one of the two fatty acids is unsaturated it is the one nearer the phosphoryl group.

There is only one general rule that may be applied in this context, viz. that with increasing number of double bonds in the fatty acid the lipid becomes more liquid which is reflected in a lower melting point. This subject will be dealt with in more detail in the section on lipid transitions.

SYNOPSIS

The principal classes of membrane lipids are the glycerol-based fatty acyl esters and the ethanolamine-based fatty acyl esters and long-chain aliphatic ethers, often containing a phosphoryl group and a nitrogenous base or a mono- or oligosaccharide. The third main group are the sterols. Other lipid substances, such as waxes, isoprenoids, and lipopolysaccharides, are known to occur in membranes. The great variety of structural forms of lipids is further increased by the assortment of fatty acids present in the lipids which range from lauric (12:0) to melissic (30:0), with frequent double bonds (1–6) in the chain.

Table 2.6. Fatty acid composition of selected membranes (in molar %)

Organism or membrane	12:0	13:0	14:0	15:0	16:0	16:0	16:1	17:0	18:0	18:0	18:1	18:2	18:3	20:0	20:3	20:4	20:5	22:0	22:4	22:5	22:6	23:0	24:0	24:1
Acholeplasma laidlawii	5.0	3.4	30.8	6.6	40.1	–	–	3.2	7.7	–	1.8	1.2	–	–	–	–	–	–	–	–	–	–	–	–
Acholeplasma laidlawii grown on 18:1	2.2	–	6.6	–	22.1	–	–	–	4.5	–	64.4	–	–	–	–	–	–	–	–	–	–	–	–	–
Bacillus licheniformis	–	–	4.2	–	3.9	50.4	12.0	–	0.2	28.2	11.1	–	–	–	–	–	–	–	–	–	–	–	–	–
Escherichia coli B	–	–	6.8	–	42.0	–	2.3	–	1.7	–	25.7	–	–	–	–	–	–	–	–	–	–	–	–	–
Saccharomyces cerevisiae	0.4	–	1.4	0.5	11.2	–	47.6	–	6.0	–	30.1	1.0	0.1	1.4	–	–	–	0.7	–	–	–	–	0.1	–
Chlorella pyrenoidosa	–	–	+	–	+	–	3	+	+	–	14		43	–	–	–	–	–	–	–	–	–	–	–
Apple fruit	–	–	–	–	25	–	–	–	4	–	7		65	–	–	–	–	–	–	–	–	–	–	–
Hypnum moss	–	–	+	–	14	–	5	–	2	–	7		44	–	–	12	4.7	5	–	–	–	–	–	–
Human erythrocyte PC	–	–	–	–	33.0	–	1.0	–	11.7	–	20.6	18.2	–	–	–	5.0	+	+	+	5.4	1.1	+	+	–
Human erythrocyte SP	–	–	–	+	41.3	–	0.1	–	9.1	–	5.2	3.7	–	+	–	0.1	–	8.0	–	–	–	+	15.0	15.5
Bovine erythrocyte SP + PC	–	–	–	2.1	17.1	–	–	1.9	4.2	–	2.2	3.7	1.8	+	1.8	+	–	10.7	–	–	–	3.2	42.2	10.5
Neural synaptic plasma membrane PE	–	–	–	–	7.5	–	–	0.4	23.7	–	6.6	0.2	0.4	–	0.3	18.0	–	–	8.4	1.5	32.9	–	–	–
PC	–	–	–	–	50.7	–	–	0.2	12.6	–	24.2	0.6	1.0	–	0.1	5.6	–	–	0.8	–	32.9	–	–	–
PS	–	–	–	–	0.5	–	–	–	48.6	–	7.6	–	–	–	–	2.0	–	–	3.9	3.2	34.1	–	–	–
PI	–	–	–	–	10.0	–	–	–	38.4	–	8.5	0.3	–	–	–	36.9	–	–	–	–	4.1	–	–	–
SP	–	–	–	–	5.2	–	–	–	87.5	–	0.1	–	–	1.9	–	–	–	–	–	–	–	–	1.3	–
Mitochondria																								
Pig heart	–	–	–	–	15	–	4	–	14	–	17	34	–	–	–	11	–	–	–	–	–	–	–	–
Sycamore	–	–	–	–	26	–	–	–	2	–	4	48	–	–	–	19	–	–	–	–	–	–	–	–
Tetrahymena	–	–	–	–	5	–	3	–	2	–	6	30	48	–	–	–	–	–	–	–	–	–	–	–

[a]Also some 19:0, 26:0, 26:0(OH); [b]also some 18:4, 16:2, 16:3 and 36:1. The figures shown here are not always of the same degree of reliability, having been obtained with different methods, and can only serve for rough comparisons

FURTHER READING

Ansell, G. B., Hawthorne, J. N., and Dawson, R. M. C. (eds). *Form and Function of Phospholipids*; Elsevier, Amsterdam (1973).
Cronan, J. E. Jr., and Vagelos, P. R. Metabolism and function of the membrane phospholipids of *Escherichia coli*, *Biochim. Biophys. Acta*, **265**, 25–60 (1972).
Daum, G. Lipids of mitochondria, *Biochim. Biophys. Acta*, **822**, 1–42 (1985).
Kramer, R., Kopp, F., Niedermeyer, W., and Fuhrmann, G. F. Comparative studies of the structure and composition of the plasmalemma and the tonoplast in *Saccharomyces cerevisiae*, *Biochim. Biochim. Acta*, **507**, 369–80 (1978).

2.3 ORGANIZATION OF MEMBRANE LIPIDS

2.3.1 Different forms of lipid structure

From the very beginnings of membranology, which may be identified with the proposal made by Gorter and Grendel in 1925 that cells can be enclosed in lipid-type envelopes, it has been generally assumed that the arrangement of lipids in a membrane is that of a molecular bilayer, with polar heads of the molecules exposed to the outside, the hydrophobic parts interacting within the bilayer. Various physical methods, among them low-angle X-ray diffraction, nuclear magnetic resonance (NMR) and electron-spin resonance (ESR), as well as electron-microscopic observations and model experiments with mixtures of lipids and salt solutions, bore out the principal correctness of this view (Fig. 2.1). Even if some membranes contain extraordinarily high proportions of proteins, such as the inner mitochondrial or thylakoid membranes do, it is believed that these proteins are connected, or the vacancies between them filled, by a flat sheet of two molecular layers of lipids.

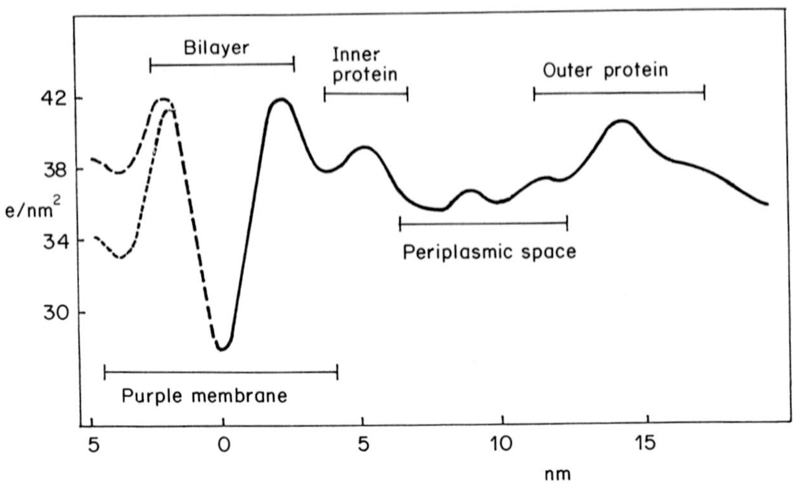

Fig. 2.1. Profile of the *Halobacterium halobium* envelope, showing the typical bilayer membrane on the left. The absolute density of electrons is plotted on the ordinate

It is now understood that there are situations in which the bilayer temporarily collapses or when its properties are pronouncedly altered by the presence of integral proteins but much useful information on the behaviour of biomembranes can be obtained from considerations of relatively simple, protein-free systems of lipids organized in bilayers.

The bilayer is one of the thermodynamically most stable configurations (see Section 3.D) and, functionally speaking, the most economic one, considering the amount of material required for the formation of a given surface. The lipid molecules are laterally packed in each monolayer to a density which depends both on pressure and on temperature. While lateral (or surface) pressure will not vary over more than one order of magnitude in biomembranes under different physiological conditions (being estimated to be 1–20 mN m^{-1}), temperature can change or be changed drastically in such a way that the overall state of the lipids is in a liquid crystalline or in a gel state.

Above the phase-transition temperature the individual lipid molecules have a relatively greater freedom of alkyl chain rotation and may take up 0.30–0.40 nm^2 per chain and roughly twice that area per molecule of a common, doubly acylated lipid. This is a situation that corresponds to the liquid expanded state of a monolayer spread over an aqueous surface. As temperature drops

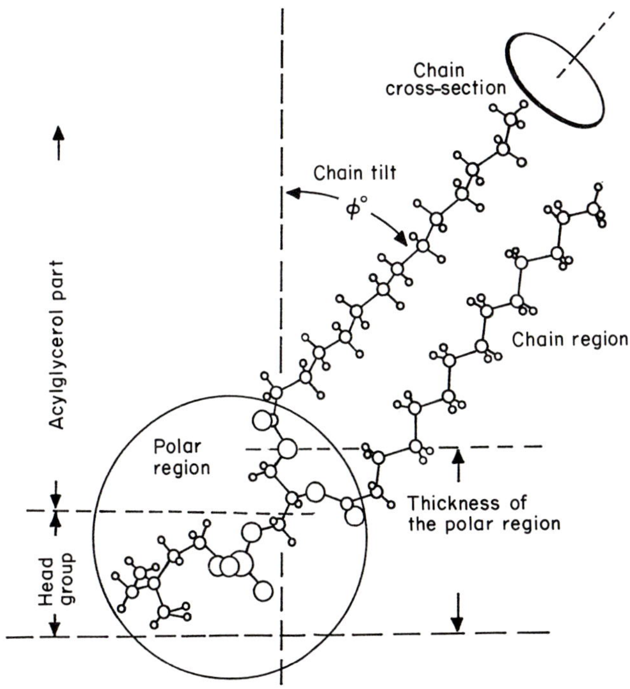

Fig. 2.2. Location of a lecithin molecule in the membrane when in the $L_{\beta'}$ phase

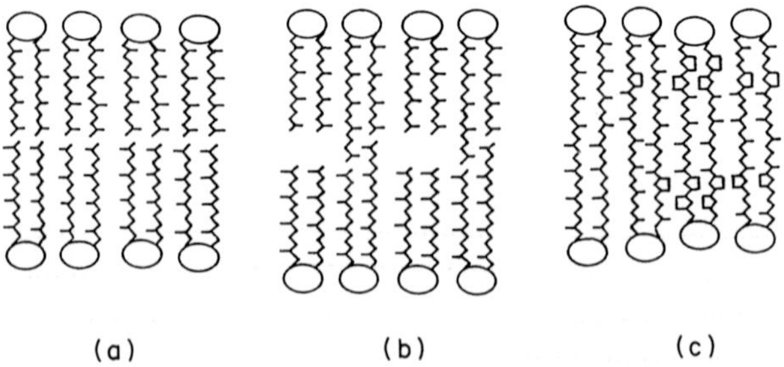

Fig. 2.3. Possible types of organization of archaebacterial lipids. (a) bilayer of $C_{20}C_{20}$ diethers; (b) zip membrane composed of $C_{20}C_{20}$ and $C_{25}C_{20}$ diethers; (c) monolayer of $C_{40}C_{40}$ tetraethers

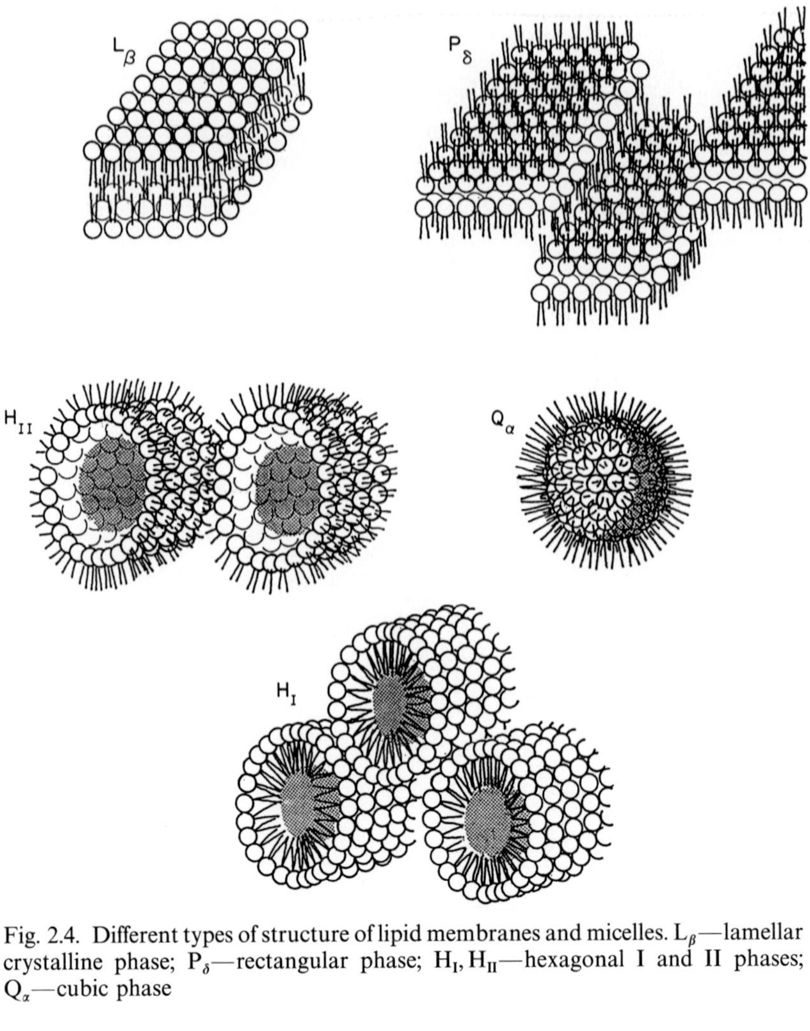

Fig. 2.4. Different types of structure of lipid membranes and micelles. L_β—lamellar crystalline phase; P_δ—rectangular phase; H_I, H_{II}—hexagonal I and II phases; Q_α—cubic phase

below the phase-transition point, the lipid molecules assume a liquid condensed state and each alkyl chain takes up about 0.20–0.23 nm^2. If lateral pressure should now increase a solid condensed state might form where all alkyl chains are *trans* and where the cross-section area per chain is less than 0.20 nm^2. Details of a PC molecule in this gel state are shown in Fig. 2.2.

The thickness of a pure phospholipid bilayer with no proteins involved is 4–5 nm and one can calculate that 1 mole of a common phospholipid organized in a bilayer would take up an area of about 0.1 km^2 (roughly 25 acres) with a total volume of less than 1 dm^3.

Unusual structures are found in the membranes of some archaebacteria, particularly those living at high temperatures and/or high acidities of the medium. There a monolayer of tetraether lipids or a hybrid lipid monolayer–bilayer of diether and tetraether molecules forms the resistant outer membrane (Fig. 2.3).

Although lipids are generally considered to be in the liquid crystalline lamellar phase (L or $L_{\beta'}$ bilayer), at temperatures above 100 °C, especially in rather dehydrated lipids, and transiently during fusion or domain separation even at physiological temperatures, other types of lipid organization are known to occur. Typical are the hexagonal phases H_I and H_{II}, the cubic phase Q and the banded or rectangular phase P (Fig. 2.4). Most of what is known about these phases derives from observations on artificial lipid-water-salt mixtures and relevance of these data to natural membranes is still under discussion.

2.3.2 Asymmetry of membrane lipids

All lipids are inserted into membranes from the cytoplasmic face or are directly synthesized at this face. It is only through the action of special enzymes that lipids can appear at the noncytoplasmic face. These enzymes either catalyze the

Table 2.7. Shapes of representative lipid molecules in membranes

Shape	R*	Lipid type
Cone	—	Lysophospholipids
Truncated cone, rapidly converging toward the small base	1.3/0.6	Nonionic lipids, lysolecithin
Truncated cone, slowly converging toward the small base	0.7/0.4	PC, PS, PG, PA, SP, disugar diglycerides, single-chain lipids with small uncharged head groups
Cylinder	0.5/0.7	PE, PS (with Ca^{2+})
Inverted truncated cone	0.4/1.2	Nonionic lipids, with poly-unsaturated chain in *cis* configuration

* A typical value of diameter at membrane surface over diameter in membrane interior (in nm).

Table 2.8. Distribution of lipids in the two monolayers of various membranes. C/N = ratio of lipid exposed on the cytoplasmic face to that on the noncytoplasmic face

Membrane type	PC	PE	PS	PI	SP	Ch	CL
Human erythrocyte	0.32	4.15	100	—	0.17	1	—
Rat erythrocyte	0.53	4.42	9.6	—	0.01	1	—
Hepatocyte							
Plasma membrane	0.22	100.00	12.9	—	5.38	—	—
Golgi body	1.28	11.80	4.5	0.28	0.23	—	—
Lysosomes	1.03	3.88	—	0.26	0.23	—	—
Mitochondrial inner membrane	1.41	1.35	—	0.18	—	—	0.26
Pig platelet plasma membrane	1.02	2.19	—	4.27	—	—	—
Influenza A virus	1.5	2.50	3.0	—	3.20	—	—

passage of a complete molecule or pull it through the membrane while modifying it chemically. The relative specificity and restricted extent of this movement probably account for the fact that all existing membranes are highly asymmetric, not only in their protein and sugar content but also in their composition of lipids.

Membranes tend to assume a thermodynamically favourable configuration, with minimum strain arising from the curvature of the monolayers and lack of alignment of hydrophilic ends of molecules in the approximately two-dimensional membrane sheet. This leads to preferences in the detailed distribution of lipid molecules according to their size and shape. Table 2.7 shows the different general shapes of lipid molecules that occur in membranes. It is clear that lecithin combined with cholesterol will pack more densely and take up less space than in a membrane composed of either lecithin or cholesterol alone.

The degree of lipid asymmetry may range from a complete unilaterality (SP in rat erythrocytes) to complete symmetry (PC in blood platelet membranes), the reason for this, if any, being ill understood (Table 2.8).

The steric or thermodynamic nature of some of the asymmetric distribution of lipids in biomembranes can be documented in preparations of mixed phospholipid vesicles, e.g. PC + PS, PC + PE, and the like, where there was always a preference for PC to be in the outer face and for PS or PE in the inner face of the vesicle membrane, in much the same way as in a living erythrocyte or hepatocyte.

The lipid asymmetry strongly depends on the degree of nonideality of lateral mixing of membrane components. The nonideality (deviation from complete homogeneity) may be expressed as excess Gibbs energy

$$\rho = Z[E_{nc} - 0.5(E_{nn} + E_{cc})] \tag{2.1}$$

Z being the number of nearest neighbours of a membrane component, E_{nc}, E_{nn} and E_{cc} the energies of interaction between pairs of neutral–charged, neutral–neutral, and charged–charged partners. Fig. 2.5 shows the transbilayer distribution of component A

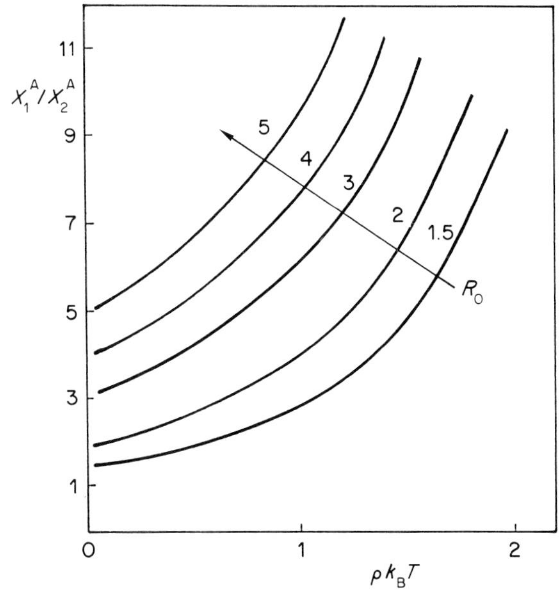

Fig. 2.5. Dependence of transbilayer distribution of component A on the nonideality parameter ρ in a membrane containing 35 per cent of A and 65 per cent of B. R_0 stands for lipid asymmetry in the case of ideal mixing

in a mixture of A:B = 35:65, for several levels of lipid asymmetry R_0 in case of ideal mixing. The actual asymmetric distribution of component A (X_1/X_2) is seen to rise the more steeply the higher the initial lipid asymmetry.

On the other hand, there is good evidence that both ATP and cytoskeleton are required for maintaining the asymmetry of lipids in erythrocytes so that the whole problem is still open to further discoveries, that should also shed more light on the functionality of this asymmetry.

SYNOPSIS

The most common type of lipid organization in membranes is the lamellar, bilayer, form where the polar head groups are on the surfaces and the hydrophobic chains in the membrane core. The shape and possibly the electric charge of the lipids favour their localization either in the outward- or inward-facing monolayer so that all natural lipid bilayers are asymmetric.

FURTHER READING

Blaurock, A. E. Evidence of bilayer structure and of membrane interactions from X-ray diffraction analysis, *Biochim. Biophys. Acta*, **650**, 167–207 (1982).

Nicolau, C., and Paref, A. (eds). *Structural and Kinetic Approach to Plasma Membrane Functions*, Springer-Verlag, Berlin (1977).
Robertson, R. N. *The Lively Membranes*, Cambridge University Press, Cambridge (1983).
Singer, S. J., and Nicolson, G. L. The fluid mosaic model of the structure of cell membranes, *Science*, **175**, 720–31 (1972).

2.4 PHYSICAL STATE OF LIPIDS IN MEMBRANES

The static organization of membrane lipids in a bilayer depends on temperature in a conspicuous way. Its mesomorphism finds the general expression in the existence of two distinct lamellar phases, the L_α form at temperatures above the transition point, the L_β form at temperatures below it (Fig. 2.6). The transition temperature for a single-lipid membrane is a well-defined quantity which, due to the cooperative nature of the melting, may be narrowed down to less than a degree. In actual membranes, however, the passage from one phase to the other may extend over several degrees because of the heterogeneity of membrane composition.

The L_β form corresponds to a solid crystalline phase in which there is very little lateral movement of lipid molecules and where the acyl chains are in the *trans* configuration. Depending on the character of the lipid, the acyl chains may extend perpendicular to the membrane plane, e.g. with PE or in the presence of very low water contents (the L_β phase), or be tilted by several degrees, e.g. with PC or PG (the $L_{\beta'}$ phase). This is apparently due to the large area taken up by PC polar heads, compared to the cross-section of the two acyl chains. Viewed along the carbon–carbon bonds of the molecule this configuration appears as in Fig. 2.7(a). If there is a double bond in *trans* position involved, the longitudinal view is the same; however, a *cis*-double bond will project as

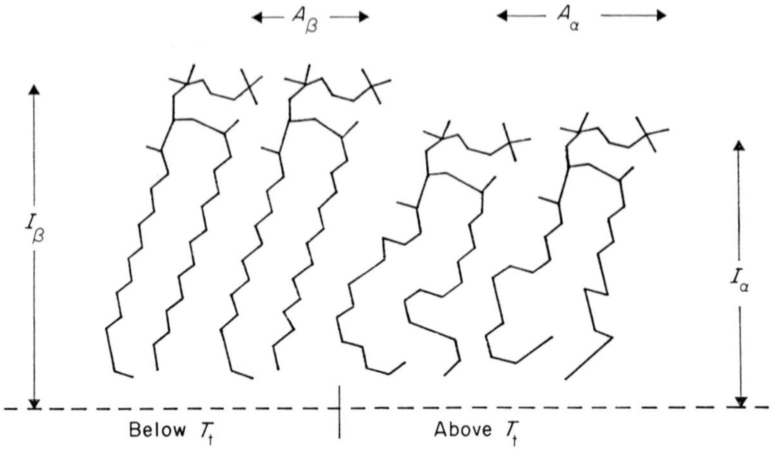

Fig. 2.6. Lipid configurations in the ordered $L_{\beta'}$ state and in the disordered liquid-crystalline L_α state

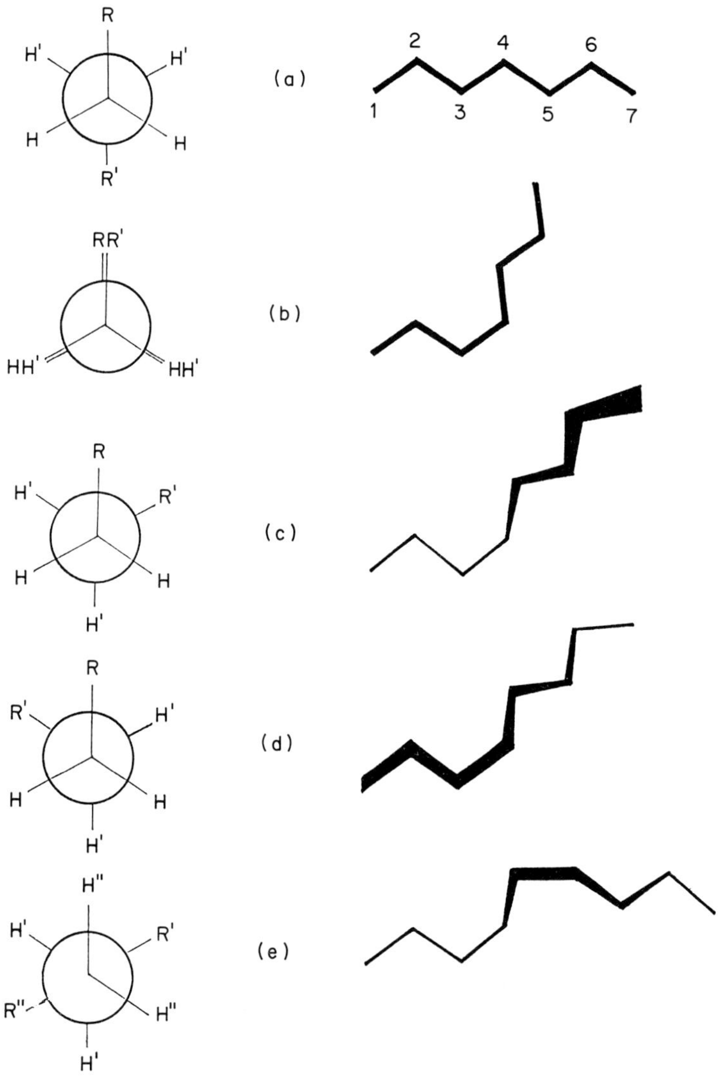

Fig. 2.7. Different configurations of the alkyl chain in a lipid molecule. Left: schematic view along the C_3–C_4 bond. Right: idealized side view of the chain. (a) *trans*, (b) *cis*, (c) *gauche* at C_4 to the right, (d) *gauche* at C_4 to the left, (e) *gauche* at C_4 to the right followed by *gauche* to the left at C_6

shown in Fig. 2.7(b) and the molecule will occupy more space than in the all-*trans* configuration.

Now in the L_α form the liquid-crystalline organization prevails which permits free lateral diffusion, as well as rotation around the axis normal to the membrane plane. This results in a frequent formation of rotational isomers (rotamers) of the acyl chains, based on a 120-degree turn of a C–C bond in either one or the other direction, the resulting configuration being called *gauche* (Fig. 2.7(c), (d)).

The *trans-gauche* transition is generally accompanied by a *gauche-trans* reversal one or two carbon atoms further along the chain so that the direction of the alkyl chain is restored but a vacancy is formed between adjacent chains, called the *kink* (Fig. 2.7(e)).

At temperatures well below the transition temperature the frequency of kinks (n_k/n) per CH_2 is less than 0.01, at the transition point there may be an average of one kink per acyl chain ($n_k/n \simeq 0.07$) but at higher temperatures two or more kinks may occur ($n_k/n > 0.4$), all this resulting in a rather loose organization of the lipid molecules and considerable freedom of several types of motion, as well as increased possibilities of a certain type of transmembrane diffusion.

It may be seen from Fig. 2.8 how, owing to greater disorder of the lipid molecules above the phase transition, the membrane area increases while its thickness decreases. The overall effect is one of greater volume, between 1 and 8 per cent, depending mainly on the length of the acyl chains involved.

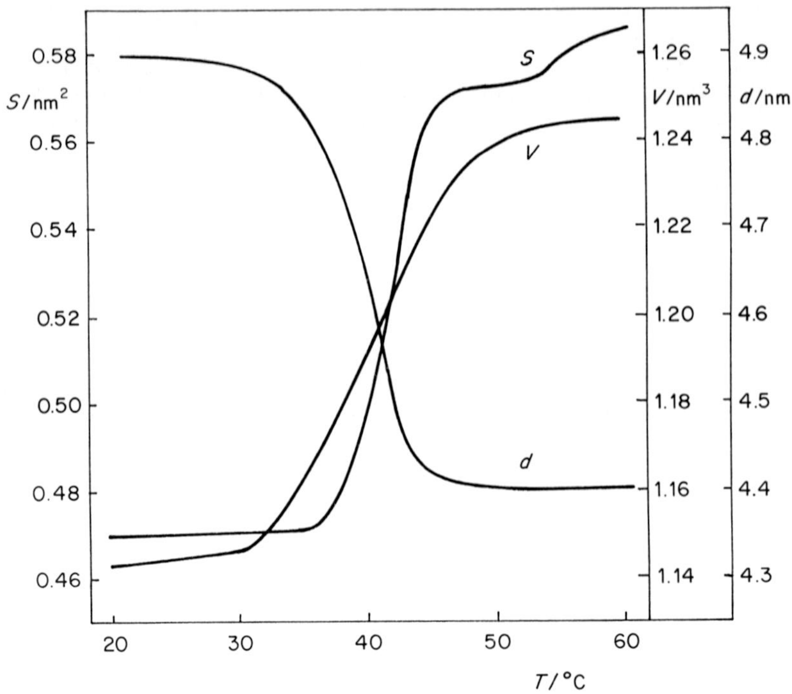

Fig. 2.8. Temperature dependence of cross-section area (S), volume (V) and double height ($=$ bilayer thickness, d) of a molecule of dipalmitoyl-PC. The incomplete agreement between $S \cdot d/2$ and V is caused by differences in the 'free space' between molecules

2.4.1 Membrane microviscosity

It is in the context of molecular organization of membranes that the concepts of *viscosity* (= 1/fluidity) and *order* (= 1/disorder) emerge. It is incorrect to speak of true membrane viscosity since the term as is used in macroscopic dimensions implies an isotropic medium which the membrane certainly is not. Therefore, the term *microviscosity* is often used, implying an averaging out of the anisotropic features of the membrane. Although still ambiguous, microviscosity, or rather its reverse, microfluidity, can be taken as a measure of the ease with which the individual molecules in a membrane move with respect to one another. In fact, the correlation time of a spherical particle, inversely proportional to the rate of rotational movement, is defined as

$$\tau_c = \tfrac{4}{3}\pi r^3 \frac{\eta}{k_B T}. \tag{2.2}$$

For a molecule 0.5 nm in diameter ($2r$) and membrane viscosity η of 0.6 Pa.s (a value found for a 1:1 mixture of phospholipid and cholesterol at 25 °C) the correlation time would be about 75 ns. In water, the same molecule would rotate with a correlation time of 0.12 ns. Clearly, these values are merely an indicator of the orders of magnitude of actual motion in membranes as both the size and the shape of the molecule will have a pronounced influence on the correlation time. For example, fatty acids in lipid bilayers (using an electron-spin label as probe) have correlation times of the order of 1 ns. However, this is true for the C atoms near the glycerol or ethanolamine backbone. The terminal carbons, near the membrane centre, will have τ_c of 12 ps.

Membrane microviscosity may be affected by a variety of factors, both physical and chemical in nature. Viscosity as such depends on temperature even without a phase transition, according to

$$\eta = \eta_0 \exp(E_a/RT) \tag{2.3}$$

where E_a is the flow activation energy (the energy required for free volume formation, i.e. for moving a molecule out of the bulk liquid). Its values in a typical animal membrane are about 30 kJ/mol while in plant membranes they are lower at about 15 kJ/mol. In an animal phospholipid membrane containing cholesterol (about 25 per cent) η would be 0.7 Pa.s at 15 °C and 0.13 Pa.s at 40 °C.

Another factor affecting membrane microviscosity is pressure applied to the membrane. Its effect is reflected in changes of membrane volume and that depends on membrane *compressibility* which, under isothermal conditions, is defined as

$$\kappa = -\frac{1}{V}(\partial V/\partial P)_T \tag{2.4}$$

and, for mixtures of phospholipids with neutral lipids, is of the order of $10^{-9}\,m^2\,N^{-1}$ (compared with $4.7 \cdot 10^{-10}\,m^2\,N^{-1}$ for water). From measurements of dielectric breakdown, however, higher values of $10^{-7}\,m^2\,N^{-1}$ were obtained.

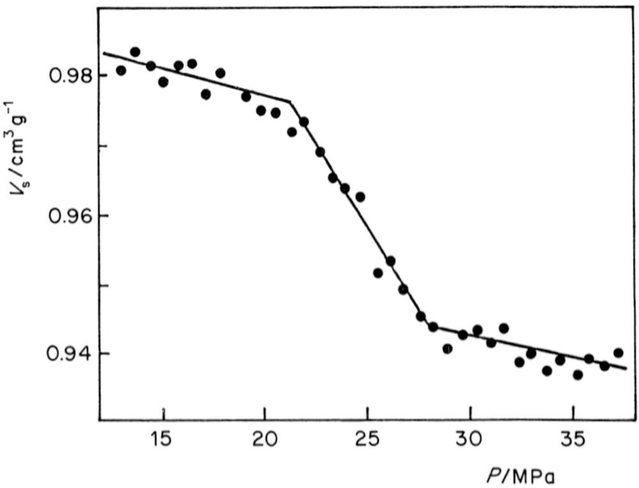

Fig. 2.9. Dependence of specific volume (V_s) of dipalmitoyl-PC on isotropically applied pressure P at 320 K

The compressibility is considerably increased during the main phase transition, attaining values greater by nearly one order of magnitude (Fig. 2.9). It will be seen in this connection that pressure will bring about phase transition just as temperature can do, the pressure-temperature equivalents being known for some lipids, as well as natural membranes, to be about 0.2 K/kPa.

Outside the phase transition, either in the L_α or L_β phase, membrane volume is related to viscosity by the following empirical (Baltschinski) equation:

$$\eta = k/(1/\rho - 1/\rho_\infty) = k/\Delta\bar{V} \tag{2.5}$$

where ρ is the density at experimental conditions (reciprocal of specific volume) and ρ_∞ the density at maximum (infinite) viscosity (820 kg m^{-3} for lipids in a fluid bilayer) so that $\Delta\bar{V}$ can be considered as the free specific volume of the membrane. The value of k for fluid lipid bilayers was found to be about $5.9 \cdot 10^{-5}\,m^2\,s^{-1}$, its dimensions being those of a diffusion coefficient.

The effects of pressure on microviscosity (mediated by changes in volume) are different depending on whether the pressure is applied isotropically (like hydrostatic pressure in a mixture of lipid bilayers in water) or anisotropically (like osmotic pressure acting normal to the membrane plane) because the lateral and transverse compressibilities of a membrane can be very different. The equation applicable here is

$$(\eta_\infty - 1)/\eta_p = (\eta_\infty/\eta_0 - 1)\exp(-p\Delta V^{\neq}/RT) \tag{2.6}$$

where η_0, η_p, and η_∞ are viscosities at zero, experimental, and limiting pressure, respectively, and $\Delta V^{\neq}$ is the change in molar volume during flow activation (as the activation energy of a molecule is reached, i.e. a constant for a given system) so that, in an approximation,

$$\ln(\eta_\infty/\eta_p - 1) = \text{constant} - k'p. \tag{2.7}$$

For egg lecithin liposomes the value of $\eta_\infty/\eta_0 - 1$ was 0.878 while that of k' was $1.96 \cdot 10^{-5}\,\text{Pa}^{-1}$.

By virtue of its exerting a pressure on the membrane the membrane potential difference can cause a volume decrease which may be reflected in changes of membrane viscosity. Using a modified Coulomb's law

$$p = C\Delta\phi_m/2\delta \tag{2.8}$$

where C is the membrane capacitance (about $10\,\text{mF}\,\text{m}^{-2}$), $\Delta\phi_m$ the membrane potential (commonly 0.1 V) and δ the membrane thickness (say 7 nm), one can derive a pressure of 7 kPa. Now volume is related to pressure by the above compressibility factor. Using the value of $10^{-9}\,\text{m}^2\,\text{N}^{-1}$, the volume change brought about by the potential would be only $7 \cdot 10^{-6}$ of the total. Since the actual effect of membrane potential on membrane viscosity is much greater one must assume that it acts by imposing an energy barrier on lipid flow within the membrane and its effect can be calculated from a Boltzmann-type equation relating η to $\eta_0 \exp(\Delta\phi_m\Delta_Q^{\neq}/RT)$ where $\Delta_Q^{\neq}$ is the difference in charge distribution in a flow unit between the activated and the normal state (it was found to be about $3\,\text{kC}\,\text{mol}^{-1}$).

Membrane fluidity can also be affected by interactions with extraneous proteins, the cytoskeleton, by various metabolic processes and by a number of artificially applied chemical reagents. It is increased by binding of acetylcholine, nicotine, complement, adrenocorticotropic hormone, and decreased by insulin, serotonin, lymphocyte-activating factor, thyrotropin and fibrinogen. More information on these effects will be presented after membrane proteins have been treated in subsequent sections.

2.4.2 Membrane order

The other term often employed in connection with the level of membrane organization is *order*. The concept of order or disorder refers to rather local, submicroscopic deviations from the crystal-like arrangement of a lipid, such as is assumed to occur in the subtransition range of temperatures. In a good approximation the deviation from complete order with all acyl chains in the *trans* configuration (the order parameter $S = 1$) corresponds to the fraction of acyl chains that are *gauche*. When complete order prevails, i.e. when the membrane is isotropic, the order parameter S is zero. The minimum value of S is -0.5 when all the chains are randomly distributed in the plane of the membrane at exactly 90 degree to the normal.

The molecular order parameter at a given position of the alkyl chain is defined as

$$S = 0.5(\overline{3\cos^2\beta_i - 1}) \tag{2.9}$$

where $\overline{\cos^2/\beta_i}$ is the time-averaged value of angular fluctuations of the chain segment with respect to the axis perpendicular to the membrane plane.

The order parameter can be estimated both by ESR and by NMR methods at different depths of the bilayer and is seen to be rather high near the polar heads, decreasing toward practically zero at the depth of the terminal methyl group. There is thus a gradient of mobility from the membrane surface toward the interior. As the order parameter reflects the flexibility of a part of a molecule in a way that averages the deviations from anisotropy over space rather than over time, it does not necessarily yield true information on fluidity which, however, usually changes qualitatively in the same direction as disorder does.

The order parameter S is related to both the area per lipid molecule and to the chain length of the lipid, S decreasing with increasing area. A useful relation is that of $Z = l/l_0$, where l is the effective length of the acyl chain and l_0 its maximum length when fully extended (all-*trans*). Empirically it may be shown that

$$Z^3 < S < Z^2.$$

As an example, dipalmitoyl-PC in 5 per cent water at 35 °C has a Z value of 0.91 and S equal to 0.81; at 49 °C (L_α phase) the Z value is 0.61 and S is equal to 0.29.

2.4.3 Phase transitions

The transition from the crystalline (gel, or L_β) phase to the liquid crystalline (fluid, or L_α) phase depends outstandingly on the composition and mutual organization of membrane components.

The following properties have a direct bearing on the transition temperature T_t of a phospholipid (Table 2.9).

1. Length of the acyl chains; with longer chain T_t will increase.
2. Degree of unsaturation of the acyl chains: with higher unsaturation T_t will decrease.
3. Position of the double bond in acyl chains: the lowest T_t is obtained if the double bond is in the middle of the chain.
4. Character of the two different acyl chains in a lipid molecule; a longer or a more saturated chain near the phosphate group will decrease the T_t.
5. Position of branching of acyl chains: branching nearer the polar head will decrease the T_t.
6. Nature of the head group: PE will cause the highest T_t of all common phospholipids, thus: PE > PS > PC $\geqslant$ PG.
7. The pH of the medium: introduction of a charge on a phospholipid lowers the T_t (PS has no net charge at pH 1, one negative charge at pH 7 and two negative charges at pH 13).

The enthalpy of phase transition measured readily by a differential scanning calorimeter will generally increase with the T_t in a homologous series of phospholipids but will not necessarily follow the trend on going from one group

Table 2.9. Transition temperatures (in °C) of different phospholipids

Fatty acid in position 1	Fatty acid in position 2	Head group in position 3	T_t
12:0		PC	−2
14:0		PC	24
16:0		PC	41
18:0		PC	55
22:0		PC	72
18:1 (9 *trans*)		PC	5
18:1 (2 *cis*)		PC	41
18:1 (4 *cis*)		PC	23
18:1 (6 *cis*)		PC	1
18:1 (9 *cis*)		PC	−22
18:1 (12 *cis*)		PC	−8
18:1 (16 *cis*)		PC	34
14:0	16:0	PC	35
16:0	14:0	PC	27
16:0	18:0	PC	47
18:0	16:0	PC	44
18:0	18:1 (9 *cis*)	PC	3
18:1 (9 *cis*)	18:0	PC	15
12:0		PE	31
14:0		PE	50
16:0		PE	64
18:0		PE	82
14:0		PG	23
16:0		PG	41
18:0		PG	54
16:0		PS (pH 13)	32
16:0		PS (pH 7)	55
16:0		PS (pH 1)	65
At all three positions:	16:0		36
	18:1 (9 *cis*)		25
	18:2 (9, 12 *cis*)		16

of phospholipids to another. Thus dimyristoyl-PC has a ΔH_{trans} equal to 22.6 kJ/mol, dipalmitoyl-PC 36.4 kJ/mol and distearoyl PC 44.3 kJ/mol. However, although the ΔH_{trans} of dimyristoyl PE is similar (24.2 kJ/mol) to that of dimyristoyl-PC its T_t is 50 °C as compared to 24 °C of the PC derivative.

So far, the phase transition has been treated as a well defined process occurring at a given temperature. However, this is not true even for single-lipid membranes and much less for membrane containing different lipids. The broadening of the phase transition has an entropic character. The gel or crystalline phase existing below the transition temperature contains defects in the regular lattice, the number of which increases with temperature (so does entropy during melting) and hence areas with a greater concentration of defects will melt faster, this giving rise to a transient formation of domains of the other phase. If the transition temperature is approached from below (melting) islands of fluid lipid will spread

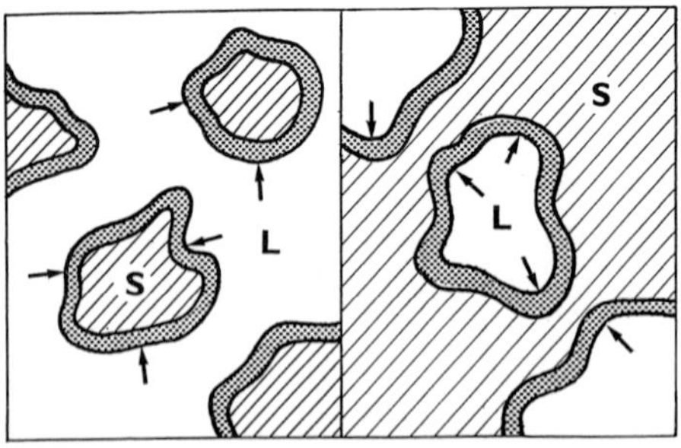

Fig. 2.10. The concept of interfacial lipid occurring during phase transition of a bilayer. Left: process of solidification where the liquid lipid L is under tension; right: process of melting where the crystalline lipid is under compression. Interfacial lipid is shown by the stippled rings around islets of S and L phases

in a 'sea' of crystalline phase; if the transition is approached from above (freezing) islands of solid will float in a 'sea' of fluid. In both cases the solid will experience pressure from the liquid as it takes up less space than the liquid does (Fig. 2.10). The domains may be 10–20 nm in diameter, containing thus a hundred or more molecules of lipid.

The islands of either the L_α phase in the bulk of L_β or islands of L_β in the bulk of L_α are surrounded by rims of interfacial lipid, roughly two lipid molecules thick, where neither the highly anisotropic ($S \simeq 1$) nor the more irregular ($S < 1$) organization of lipid molecules prevails. These are areas of molecules with a relatively high free energy content where both enthalpy and entropy terms contribute to its size. They reflect the ill matching of two fundamental phases and may represent areas through which passive permeation of small molecules and ions occurs. Indeed, in artificial phospholipid membranes, the highest passive permeability to both charged and uncharged solute species is found at the transition temperature when the amount of interfacial lipid is highest. In natural membranes, the total phase transition is a much broader event so that at any given temperature the amount of interfacial lipid will be relatively small and will not perceptibly contribute to passive permeability.

The different spatial interaction between solid and liquid phases as the transition temperature is approached from below or from above results in the possibility of hysteresis (Fig. 2.11) which is particularly pronounced with charged lipids where the repulsion of identical charges favours membrane area expansion and hence melting to the L_α phase.

The various structural features of fatty acids bearing on membrane phase transition are shown in Table 2.10. A few comments should be made here on the mutual proportion of membrane lipids with particular emphasis on the role of membrane sterols.

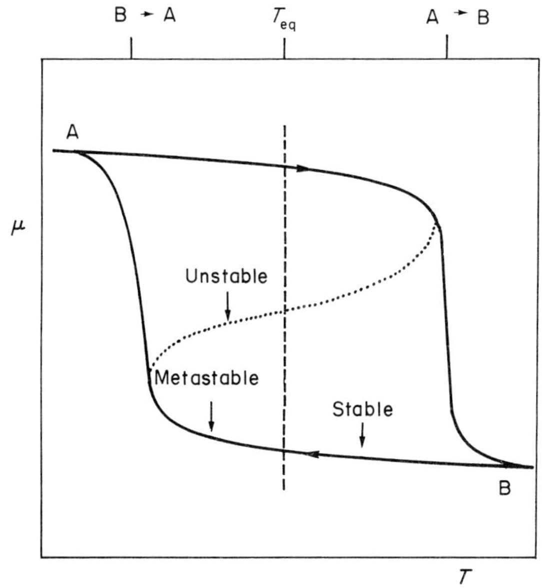

Fig. 2.11. Hysteretic behaviour of melting (upper curve) and solidification (lower curve) of a lipid system shown by changes in the chemical potential μ of the species. B → A—temperature of transition from liquid to crystalline phase; A → B—temperature of transition from crystalline to liquid phase; T_{eq}—temperature of thermodynamic equilibrium transition

Sterols, especially the most abundant and best investigated cholesterol, are shorter than either glycero- or sphingolipids carrying two fatty acyl chains. The total length of an extended cholesterol molecule is less than 2 nm, as compared with 2.2–2.4 nm of a doubly acylated lipid (Fig. 2.12). The important feature of the sterol molecule is its rigid four-ring core which prevents the adjacent acyl chains from moving freely to assume the *gauche* configuration typical of the L_α phase. Thus, even above the assumed transition temperature, there is more rigid order in the membrane than without cholesterol. On the other hand, when the membrane lipids freeze, the cholesterol molecules remain *in situ* without forming separate domains which prevents the surrounding phospholipids from forming a regular crystal lattice. These two effects combined gradually abolish the transition from L_α to L_β or back, without actually shifting the transition temperature.

In a mixture of PC with cholesterol, the enthalpy change accompanying the phase transition disappears when 22 per cent or more cholesterol is present. This tallies nicely with the computed structural arrangement of cholesterol within the membrane where below 22 molar per cent proportion each cholesterol molecule is surrounded by seven acyl chains (= 3.5 phospholipid molecules) while above this proportion this is no more possible. Then cholesterol dimers

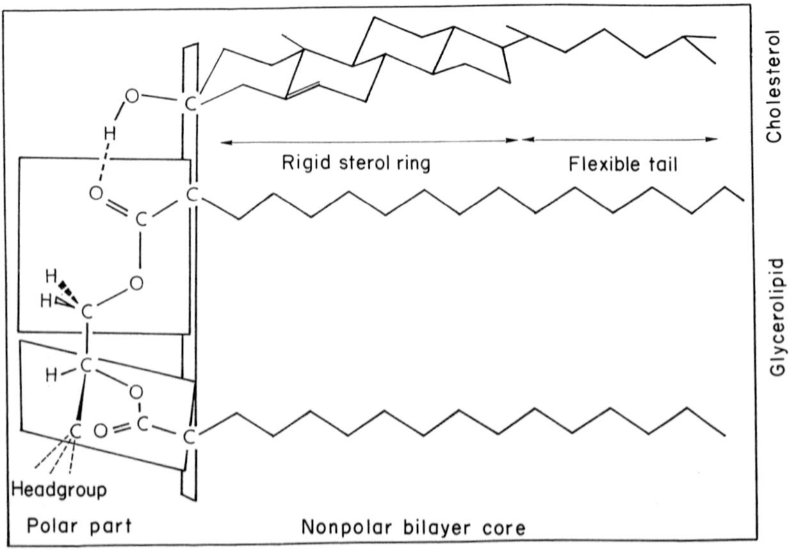

Fig. 2.12. Interaction of cholesterol with a phospholipid. The regions in boxes are planar

begin to be formed, each surrounded by nine acyl chains. Above 30 molar per cent cholesterol, even this structure becomes untenable and the system becomes rather disorganized. The described scenario is nicely depicted in freeze-etched preparations of PC-cholesterol liposomes where the regular pattern of ripples disappears above 20 molar per cent cholesterol.

It should be noted that only sterols with a β-hydroxyl group at C_3 and phospholipids capable of hydrogen bonding with it through the C=O group of the first acyl chain show the abolition of enthalpy change described above. Thus, the 3α-OH analogue (epicholesterol), on the one hand, and plasmalogens, on the other, will not show the effect.

It appears that in prokaryotic cells the role of sterols is taken over by

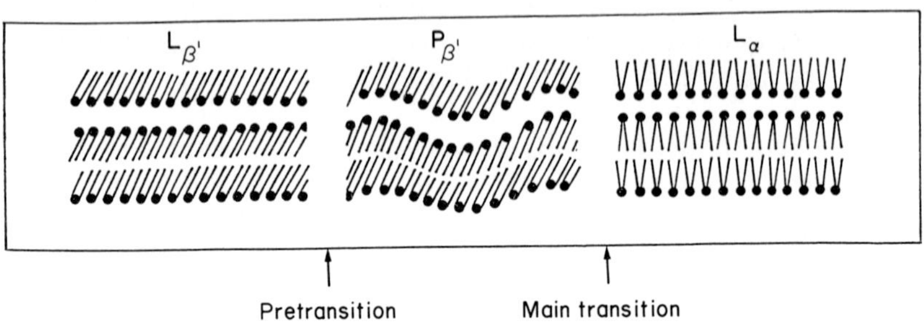

Fig. 2.13. Pretransition and transition changes in the structural organization of a bilayer

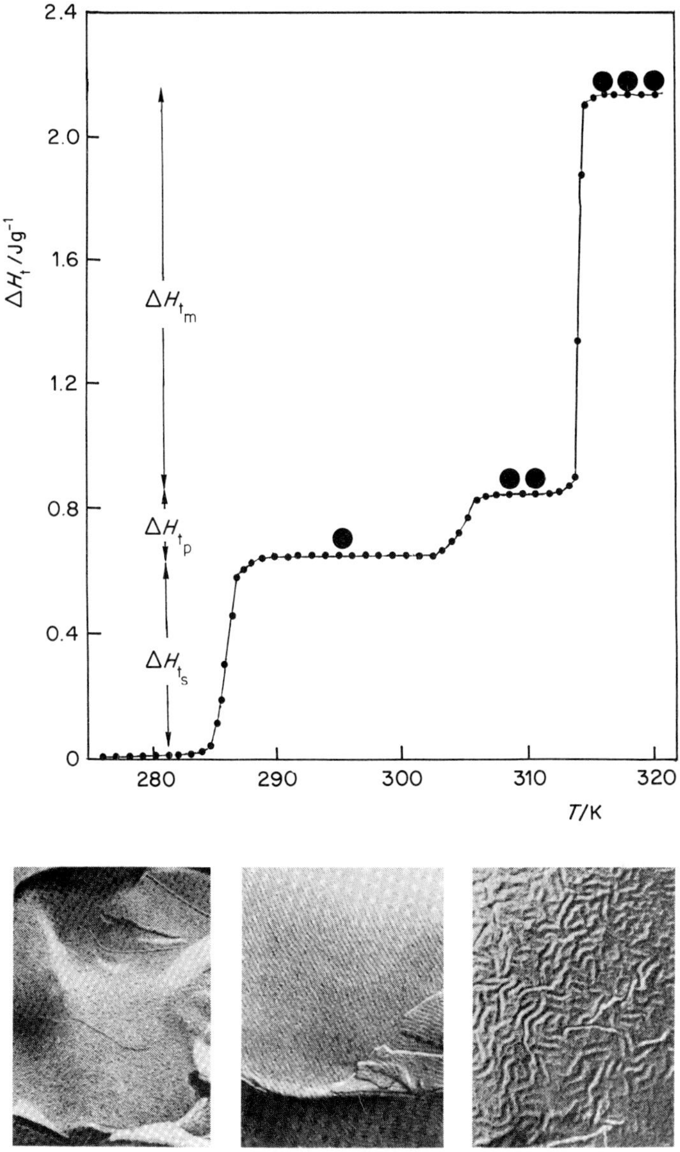

Fig. 2.14. Transition enthalpies ΔH_t of a subtransition (at about 285 K), of a pretransition (at about 305 K) and of the main transition (at about 315 K) of dipalmitoyl-PC, together with the freeze-fracture appearance of a 4:1 mixture of dipalmitoyl-PC and dimyristoyl-PC in the various temperature ranges. (Compiled from various sources)

hopanoids which have the same type of amphipathy and the same order of molecular dimensions.

The transition temperature discussed so far has been associated with the $L_\alpha \rightarrow L_\beta$ phase change as a single event describing the melting or freezing of a membrane. However, both in single-lipid membranes and in mixed-lipid ones, a slight endothermic reaction is observed below the main transition temperature, the so-called pretransition. It is now believed that this is associated with a shift from the L_β or $L_{\beta'}$ phase to a phase designated as $P_{\beta'}$ and characterized by a ripple distortion of the bilayer plane (Fig. 2.13). This phase then passes to L_α at the main transition temperature.

As the $L_{\beta'} \rightarrow P_{\beta'}$ change incurs substantial displacement of molecules it is a relatively slow event with half-times as long as 10 min. It is this very change which underlies the hysteresis phenomenon discussed above.

To make the story more complicated, recently a subtransition enthalpy change was observed with lipids that had been stored at below 0 °C for some time. This occurs well below the pretransition and apparently indicates another, still poorly described configuration of phospholipids at very low (physiologically) temperatures, called L_c. The enthalpy change of palmitoyl PC as temperature is raised, together with the freeze-fracture appearance of the phase, is shown in Fig. 2.14.

Even in a membrane containing solely a lamellar organization of lipids, domains of a single lipid species may be formed in the fluid state if the composition of lipids is favourable. This is an extremely rapid process and is not to be confounded with different phase domains at or near the transition temperature. The time required for such a domain formation is given by

$$\tau \simeq AN/4\pi Dx \tag{2.10}$$

where A is the cross-section area of the lipid molecule (typically 0.8 nm^2), N the number of molecules that will be comprised in a domain, D its lateral diffusion coefficient (typically 10^{-12} m^2 s^{-1}) and x the mole fraction of the lipid examined. Thus, for $x = 0.25$ and $N = 100$, $\tau = 64\ \mu$s. For domains with greater numbers of molecules the τ will linearly increase.

SYNOPSIS

Membrane lipids occur in several mesomorphic states. In the L_α phase, the liquid-crystalline phase, the bilayer takes up more area and slightly more volume due to many *gauche* configurations of the alkyl chains. In the L_β or $L_{\beta'}$ phase it resembles a two-dimensional crystal with most alkyl chains all-*trans*. In the $P_{\beta'}$ phase which precedes melting of β into α, the bilayer is distorted by a periodic ripple. The melting phase transition is accompanied by an abrupt change in microfluidity (decrease in microviscosity) and a decrease of order which characterizes the fraction of acyl chains that are *gauche*. The phase transition is pronouncedly influenced by the structure, particularly the number of double bonds, of the fatty acyl chains.

FURTHER READING

Corda, D., Pasternak, C., and Shinitzky, M. Increase in lipid microviscosity of unilamellar vesicles upon the creation of transmembrane potential, *J. Membr. Biol.*, **65**, 235–42 (1982).

Edidin, M. Rotational and translational diffusion in membranes, *Annu. Rev. Biophys. Bioengin.*, **3**, 179–201 (1974).

Lenaz, G. Lipid properties and lipid-protein interactions. In: *Membrane Proteins and Their Interactions with Lipids* (R. A. Capaldi, ed.), pp. 47–149, Marcel Dekker, New York (1977).

Poste, G., and Nickolson, G. L. (eds). *Membrane Fusion*, Elsevier-North Holland, Amsterdam (1978).

Shinitzky, M., and Yuli, I. Lipid fluidity at the submacroscopic level: Determination by fluorescence polarization, *Chem. Phys. Lipids*, **30**, 261–82 (1982).

Wallach, D. F. H., and Fischer H. (eds). *The Dynamic Structure of Cell Membranes*, Springer-Verlag, New York (1971).

Tosh, R. E., and Collings, P. J. High pressure volumetric measurements in dipalmitoyl-phosphatidylcholine bilayers, *Biochim. Biophys. Acta*, **859**, 10–14 (1986).

2.5 MOBILITY WITHIN AND OF LIPID MEMBRANES

Many of the phenomena described so far in the context of bilayer organization and phase transitions are based on different types of movement of the lipid molecules in membranes.

2.5.1 Rotation of intramolecular bonds

The movement that is common to all lipid molecules is rotation about the C–C bonds of the alkyl chains, the correlation times being of the order of 0.1 ns. Reference to this type of movement was made in connection with the formation of 'kinks' in the liquid crystalline phase. The kinks thus formed have very short lifetimes of approximately 1 ns.

The various rotational and bending movements of the chains as well as of the polar heads are also extremely rapid. Unlike the formation of kinks, they have apparently no relevance to any particular membrane function. The intramolecular mobility of the various segments of a glycerolipid increases in the direction away from the glycerol backbone which is rather rigid. The mobility of the individual methylene groups is best expressed by the longitudinal spin-relaxation time τ_1 obtained in NMR studies (the time required for a 0.63 change toward a new equilibrium, i.e. $1 - 1/e$). While the τ_1 of glycerol C atoms in dipalmitoyl-PC in D_2O at 325 K is 0.1 s, the C atoms of the alkyl chain up to about $C_{(10)}$ have a τ_1 of 0.6 s and the terminal C atom one of 3.3 s.

2.5.2 Rotation of lipid molecules

The whole lipid molecules including sterols can undergo fast rotation about their long axis, spinning with correlation times of the order of 10 ns. This type of movement is occasionally described as rotational diffusion and is then expressed in terms of a rotational diffusion coefficient

$$D_{\mathrm{R}} = k_{\mathrm{B}}T/4\pi r^2 h\eta \tag{2.11}$$

where r is the radius of the rotating molecule, h its height and η the viscosity of the surrounding medium. Substituting values for lecithin in a lecithin membrane (0.5 nm, 2.2 nm and 0.5 Pa.s, respectively), one obtains at 300 K a value of $1.2 \cdot 10^6\,\mathrm{s}^{-1}$. In analogy to lateral diffusion, the angular distance (in radians) is related to the diffusion coefficient by

$$\overline{\theta^2} = 2D_{\mathrm{R}}t. \tag{2.12}$$

2.5.3 Lateral diffusion

Lateral diffusion of lipids in the membrane is a spectacular process that can be even seen under a microscope using a photobleaching recovery technique.

At the molecular level one can envisage lateral diffusion as a series of jumps from one position to the next, involving a vacancy in the liquid-crystal arrangement of the fluid membrane. These jumps occur with a frequency of 10^7–$10^8\,\mathrm{s}^{-1}$ but their efficiency in terms of translational movement of a given molecule is by several orders of magnitude lower in a membrane below the transition temperature (the L_β phase).

The theoretical derivation of a lateral diffusion coefficient D_{L} for a lipid in a lipid membrane proceeds from Einsteins's theory of Brownian movement. The collision frequency ν_{col} determined by spin-labelled or fluorescent reporter molecule interactions is defined as

$$\nu_{\mathrm{col}} = n_{\mathrm{s}}^{-1}\nu_{\mathrm{j}} \tag{2.13}$$

where ν_{j} is the jumping frequency of the molecule and n_{s} is the average number of jumps the diffusing molecule can make before it interacts with another of its kind:

$$n_{\mathrm{s}} = (2/\pi x)\ln(2/x) \tag{2.14}$$

where x is the mole fraction of the lipid under investigation referred to the total membrane lipid.

The lateral diffusion coefficient is obtained as

$$D_{\mathrm{L}} = \tfrac{1}{4}\nu_{\mathrm{j}}d^2 \tag{2.15}$$

where d is the average distance between membrane lipid molecules. The meaning of D_{L} can be better grasped from considering it as a value depending on the free area available for diffusion, thus

$$D_{\mathrm{L}} = A\exp(a_0/a_{\mathrm{T}}) \tag{2.16}$$

where a_0 and a_{T} are the packed area per molecule and the mean free area per molecule at temperature T, respectively; a_{T} is equal to the increment of free area during melting (main transition) plus the increment above phase transition (cf.

Table 2.10. Coefficients of lateral diffusion of lipids in membranes

Membrane	Diffusing molecule	$T/°C$	$D_L/m^2\,s^{-1}$	Method
Erythrocyte	PC	30	$3\cdot10^{-13}$	Photobleaching
			10^{-11}	ESR
Dimyristoyl-PC (L_α-phase)	PC	30	$7\cdot10^{-12}$	Both methods
Dimyristoyl-PC ($L_{\beta'}$-phase)	PE	15	$3\cdot10^{-15}$	Photobleaching
Egg PC	Fatty acid	20	10^{-13}	ESR
Egg PC	Fatty acid	25	10^{-12}	ESR
Sarcoplasmic reticulum vesicles	Phospholipid	40	$6\cdot10^{-12}$	ESR
Rat liver microsomes	Fatty acid	40	10^{-11}	ESR
Escherichia coli	Fatty acid	40	10^{-12}	ESR

Fig. 2.8). The pre-exponential factor (with dimensions of a diffusion coefficient, i.e. $m^2\,s^{-1}$) is interpreted either as $k_B T/f$ with f being the viscous frictional coefficient due to interaction of the diffusing molecule with the aqueous medium, as well as with the apposed monolayer (in fact, $6\pi\eta r$) or as $D_0 \exp(-E_a/RT)$, where E_a is the energy required to break the transient bonds with neighbouring lipids as a vacancy appears in the liquid-crystalline lattice. Unfortunately, the a_0/a_T dependence on temperature obscures the temperature dependence of A and it is thus not possible at present to decide between the first and the second interpretations.

The actual values of D_L for membrane lipids obtained by various techniques are shown in Table 2.10.

It should be observed that the photobleaching recovery technique which examines long-range paths of molecules gives lower D_L values in natural membranes than the excimer techniques studying immediately neighbouring molecules. This is apparently due to the tortuosity of the paths caused by the presence of proteins and immobilized lipid around proteins in membranes as will be discussed in the next section.

The distance x travelled by a lipid molecule within time t can be derived from Einstein's diffusion equation which, for a two-dimensional system, is (see p. 134)

$$\overline{x^2} = 4D_L t. \tag{2.17}$$

It may be easily seen that in an artificial lipid vesicle a lecithin molecule will travel an average of 5.3 μm per second while in the erythrocyte the average distance will be only 1.1 μm.

2.5.4 Transbilayer movement

Flip-flop movement of lipids is the term used for lipid molecules flipping over from one membrane monolayer to the other. This is an extremely slow process apparently because in its course the polar head of the lipid would have to pass through the hydrophobic membrane interior. The slow rate is found even with cholesterol with its unassuming polar hydroxyl group but this is apparently because this very group interacts by hydrogen bonds with neighbouring molecules of phospholipids and it is thus kept in position.

The half-times of flip-flop equilibration range from 10^4 to 10^6 s in all-lipid membranes but can be reduced in the vicinity of some proteins and especially in growing membranes. The fastest flip-flop rate reported was for PC vesicles of the electric organ of the electric eel *Electrophorus electricus*, viz. several minutes.

2.5.5 Movements of the bilayer and membrane stability

The whole lipid bilayer undergoes thermally energized ($k_B T$) undulating movement with an amplitude that may exceed the membrane thickness about ten times and with a frequency of 0.2–30 Hz, depending on temperature and composition and hence elasticity of the membrane. It is becoming clear that this movement can facilitate the approach of two membranes or two cells and lead to recognition and eventual fusion of cells.

The vibration modes can be of two types: the squeezing mode which may lead to membrane rupture, and the stretching mode which results in bending or buckling of the membrane (Fig. 2.15).

(a) The squeezing (or symmetric) fluctuations leading to breaking of membrane continuity can be assisted by the action of chemicals (e.g., detergents, bee-venom melittin), increase in temperature and by electric fields. Cytolysis is a typical rupture event which, in a human erythrocyte or in cancer cells, occurs as the membrane tension exceeds about 20 mN m^{-1} if caused by mechanical stretching, or 5 mN m^{-1} if caused by osmotic swelling.

(b) The stretching (antisymmetric) fluctuation mode is observed as buckling of membranes due to mechanical stress, chemical reagents (amphipathic solvents, caffeine, tetracaine) or osmotic pressure.

Both types of instability greatly depend on shear and bending elasticity of the membrane. A useful approach to defining the regions of membrane stability under different stresses is the introduction of dimensionless terms, $M = \pi \gamma l^2 / A$

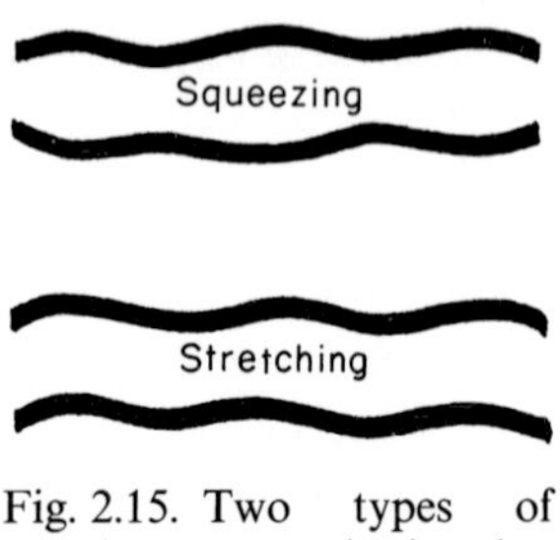

Fig. 2.15. Two types of membrane perturbation due to thermal movement

and $N = \pi l^3 G/A$ where l is the membrane thickness, γ is the interfacial tension (e.g., 0.1 mN m^{-1}), G is the shear elastic modulus, averaged for membrane anisotropy (1 mN m^{-2}) and A is the Hamaker constant, expressing the energy of van der Waals interaction (10^{-21}–10^{-23} J). The M term is thus a measure of the surface tension forces with respect to intermolecular interactions, the N term is a measure of the elastic restoring forces with respect to intermolecular interactions.

A bilayer membrane with solute molecules and ions present on both sides but without any proteins in the membrane can be considered as tangentially immobile. An analysis of the appropriate stability equation shows that (a) biomembranes are stable in the stretching mode, (b) in the squeezing mode, for a fully fluid membrane, the membrane is unstable but it becomes stable at situations found in biological membranes, with proteins present.

If the membrane is anisotropic, as all membranes are, another restriction becomes important. The squeezing mode can never lead to membrane rupture as the anisotropy prevents local thinning of the membrane.

Membrane instability is greatly increased by the presence of surface and/or membrane electric potential difference at the two membrane faces. The membrane will break down if

$$M - 1 + N_2/2 - P < 0 \tag{2.18}$$

where N_2 corresponds to the formula containing the transverse shear modulus G_2 and

$$P = \frac{0.5C\phi^2 + (4k_B T/e)^2(\varepsilon_b \kappa/8\pi)\psi^2/(1-\psi_s^2)}{A/8\pi l^2} \tag{2.19}$$

with C the specific membrane capacity in Fm^{-2}, $\Delta\phi_m$ the transmembrane potential in V, ψ_s the surface potential in V, e the elementary charge ($1.602 \cdot 10^{-19}$ C), κ the reciprocal of Debye length and ε_b the dielectric constant of the bulk fluid (about 70 nF m^{-1}). Thus, for $\phi_m = 0$ and $G_2 = 10$ kN m^{-2}, the membrane becomes unstable at $\psi_s = 20$ mV for interfacial tension $\gamma = 10\ \mu$N m^{-1}, but at $\psi_m = 80$ mV for $\gamma = 1$ mN m^{-1}.

However, in cell membranes with a high protein content and interactions with a supportive cytoskeleton, the instability boundary can be shifted to substantially higher potentials.

Now even before breaking down completely, a membrane responds to forces, such as perpendicular electric field or differences in surface tension at the two membrane faces, under more or less normal conditions, by generating discontinuities in its two-dimensional lipid plane.

Particularly in highly curved membranes both the electric and the 'elastic' force become prominent. The polarization defined as

$$P = f(1/R_1 + 1/R_2) \tag{2.20}$$

includes radii of curvature of the polar head of the lipid molecule and of its entire length and yields a flexoelectric coefficient of the order of 0.03 aC.

The elastic energy of a curved bilayer per unit area is defined as

$$E = 0.5K(1/R_1 + 1/R_2 - 1/R_0)^2 + \bar{K}/R_1R_2 \tag{2.21}$$

where R_0 is the radius of curvature of the whole membrane arising from the chemical difference of the two membrane faces, K is the so-called splay bending elasticity and $\bar{K}$ is the saddle-splay bending elasticity important when pore edges are considered. Both K and $\bar{K}$ were found to be of the order of 10^{-19}–10^{-20} J so that the energy E, with $R_1 = 0.3$ nm, $R_2 = 1$ nm and $R_0 = 1$ μm, becomes equal to about 0.5 J m^{-2}.

Now the total energy of pore formation is taken as

$$\Delta E = 2\pi\zeta r - \pi\theta r^2 \tag{2.22}$$

with ζ the pore edge energy (about $2 \cdot 10^{-11}$ J m^{-1}) and θ the surface free energy of a planar bilayer (about 10^{-3} J m^{-2}), and r the pore radius in m.

Thus, for a 0.6 nm pore ($r = 3 \cdot 10^{-10}$ m) $\Delta E = 3.7 \cdot 10^{-20}$ J. By a somewhat simplified reasoning then at least 10^{19} such pores could be formed on 1 m^2.

On the basis of such considerations it is now assumed that membranes above the temperature transition (in the L_α phase) contain a large number of transient (statistical) pores which are viewed as cylindrical structural defects in the planar (or gently curved) bilayer membrane. Apparently a membrane becomes saturated with pores at about 40 000 pores per μm^2.

The pores are assumed to display a certain distribution of radii from virtually zero, given by the molecular mechanical properties of the membrane, to some 0.8–0.9 nm. The pore radius continuously changes, with a mobility given by (from statistical mechanics)

$$u_p = \bar{v} V_w / 8\pi l^2 k_B T \tag{2.23}$$

where $\bar{v}$ is the molecular velocity of water (about 17 m s^{-1}), V_w is the volume of a water molecule (0.03 nm^3), l the thickness of the membrane. From this, at $T = 300$ K, $u_p = 1.5 \cdot 10^{10}$ ms^{-1} N^{-1}. The lifetime of such a pore is less than 1 μs.

The continuous existence of these pores provides a rationale for explaining two important instability phenomena related to the application of external voltage to membranes and, by extension, to electrofusion of membranes. The first is the so-called reversible electric breakdown, the second irreversible membrane rupture.

It follows from considerations of the permittivities of water and of the lipid matrix of the membrane (80 times and 2–3 times ε_0, respectively) that it is unfavourable for electric charges to be in the vicinity of the low-permittivity medium if water is available and hence that, up to a certain size, these pores are nonconductive and, even in the presence of ionic solutions, are considered to contain only water.

A quantitative estimate of the effect of externally applied voltage is provided by Eq. (2.22) in which θ is replaced with

$$\theta + \frac{\varepsilon_0 \varepsilon_l}{2l}\left(\frac{\varepsilon_w}{\varepsilon_l} - 1\right)U^2 \tag{2.24}$$

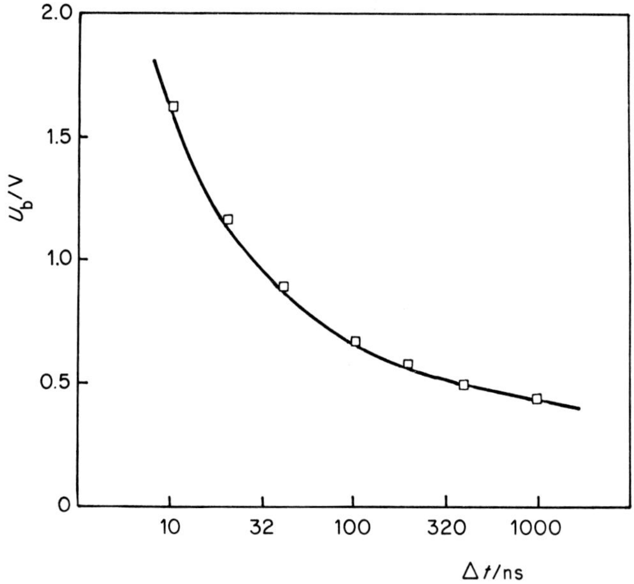

Fig. 2.16. Dependence of the potential at which electric breakdown occurs on the duration of applied pulse Δt. A 0.5 V pulse applied for 1 μs will cause the same membrane damage as a 1.5 V pulse applied for 10 ns

where ε_0 is the permittivity of the vacuum, ε_l the relative permittivity of the membrane interior (2.1) and ε_w of water (80) and U is the applied potential in V.

As the external voltage is raised to 0.5–1.0 V for a short period (100 ns–100 μs) the membrane conductance rises dramatically and so does the permeability to large molecules which, normally, would not cross the membrane (Fig. 2.16).

This electric breakdown is reversible as the enlarged pores permit a quick attainment of electroneutrality as soon as the voltage pulse is terminated. However, as the pulse length increases above 100 μs, even a potential of less than 1 V will rupture the membrane, the mechanism apparently being that of a few subcritical-size pores expanding rapidly to cause a complete tearing up of the membrane.

The mean lifetime of a membrane, i.e. the time required for the membrane to reach its critical rupture energy $E_c = \pi\zeta^2/\theta$ may be derived on the basis of rather complicated, and often unwarranted, assumptions, to be

$$\tau = (k_B^{3/2}T^{3/2}/4\pi N_t A_m D_p \theta' l)(\theta + 0.5CU^2)^{-1/2} \exp[(\theta'^2 l^2/k_B T)/(\theta + 0.5CU^2)]. \quad (2.25)$$

Now for $k_B = 1.38 \cdot 10^{-23}\,\mathrm{J\,K^{-1}}$, $T = 300\,\mathrm{K}$, $N_t = 10^{14}$ pores per m^2, A_m (the membrane area) $= 10^{-10}\,\mathrm{m^{-2}}$, $D_p = 8 \cdot 10^{-13}\,\mathrm{m^2\,s^{-1}}$, θ (the surface tension in the plane) $= 50\,\mathrm{mN\,m^{-1}}$, θ' (the surface tension at the pore edge) $= 30\,\mathrm{mN\,m^{-1}}$, U (the applied voltage) $= 0.5\,\mathrm{V}$, and C (the capacity of the membrane, equal to $\varepsilon_w/\varepsilon_1 - 1$, times the specific capacity)$=0.4\,\mathrm{F\,m^{-2}}$, we obtain a τ of 2.10^{11} s. If, however, the pore edge tension is decreased to $20\,\mathrm{mN\,m^{-1}}$, τ is equal to 0.01 s! Hence, accurate computations of membrane lifetimes depend enormously on a quantity that is particularly hard to determine.

By controlling the reversible breakdown of membranes cells can be made to fuse in the relatively small electric field of an electrophoretic apparatus as an electric pulse is applied for 5–50 μs at a voltage intensity of $\mathrm{kV\,cm^{-1}}$.

SYNOPSIS

The components of a lipid bilayer membrane exhibit a variety of movements: (1) rotation and bending of C–C bonds, (2) rotation of lipids about an axis perpendicular to the membrane, (3) lateral diffusion in the membrane plane, (4) flip-flop movement between the monolayers, and (5) undulation of the whole bilayer which affects the stability of the membrane as a whole.

FURTHER READING

Bivas, I., and Petrov, A. G. Flexoelectric and steric interactions between two bilayer lipid membranes resulting from their curvature fluctuations, *J. Theoret. Biol.*, **88**, 459–83 (1981).

Chernomordik, L. V., Sukharev, S. I., Abidor, I. G., and Chizmadzhev, Yu. A. Reversible breakdown of lipid bilayer membranes in an electric field, *Biologicheskie Membrany*, **1**, 229–43 (1984).

Dimitrov, D. S., and Jain, R. K. Membrane stability, *Biochim. Biophys. Acta*, **773**, 437–468 (1984).

Edholm, O., Nilsson, L., Berg, O., Ehrenberg, M., Claesens, F., Gräslund, A., Jönsson, B., and Teleman, O. Biomolecular dynamics, *Quart. Rev. Biophys.*, **17**, 125–52 (1984).

Galla, H. J., Hartmann, W., Theilen, U., and Sackmann, E. On two-dimensional passive random walk in lipid bilayers and fluid pathways in biomembranes, *J. Membr. Biol.*, **48**, 215–36 (1979).

Powell, K. T., Derrick, E. D., and Weaver, J. C. A quantitative theory of reversible electrical breakdown in bilayer membranes. *Bioelectrochem. Bioenerget.*, **15**, 243–55 (1986).

Vaz, W. L. C., Goodsaid-Zalduondo, F., and Jacobson, K. Lateral diffusion of lipids and proteins in bilayer membranes, *FEBS Lett.*, **174**, 199–207 (1984).

Zimmermann, U. Electric-field-mediated fusion and related electrical phenomena, *Biochim. Biophys. Acta*, **694**, 227–77 (1982).

2.6 MEMBRANE PROTEINS

It should be clear from the preceding sections that membranes composed of lipids are thermodynamically stable structures and that the arrangement of lipids into a lamellar bilayer has been the form of choice during the evolution of primitive cells which has persisted up to the present. However, it should be equally clear that natural membranes contain on the average more than one-half their weight of proteins and that it is this membrane component which endows them with a vast array of functional capabilities.

2.6.1 Types of membrane proteins

Two classes of membrane proteins are distinguished. 1. Those that interact noncovalently with the polar groups of lipids, often through electrostatic attraction, or with parts of peptide chains protruding from the membrane, are called peripheral or extrinsic. 2. Those that interact noncovalently at least in part with the hydrophobic core are called integral or intrinsic (Fig. 2.17).

The integral proteins contain hydrophobic stretches while the peripheral ones resemble soluble enzymes and are, indeed, readily washed off the membrane. The integral proteins are synthesized somewhat differently, with signal or address sequences, depending on their final location. It is the integral proteins that are instrumental in all the three types of catalytic functions treated in this book, viz. flow of matter (transmembrane transport), flow of energy (conversion of one type into another) and flow of information (transformation of external signals).

The peripheral proteins are either accessory to the integral protein function (e.g., in bacterial 'binding protein' transport) or contribute to the contact between membranes and cytoskeletal components (e.g., ankyrin and spectrin of erythrocytes).

Of the types of integral protein insertion into membranes the two spanning the membrane are abundant in the field of transport and energy transduction, the two submerged partly in the membrane may form parts of signal transformation systems, and the complex proteins are particularly common in mitochondrial, chloroplast and bacterial membranes where transport and energy transduction processes are combined. It is the fully membrane-surrounded

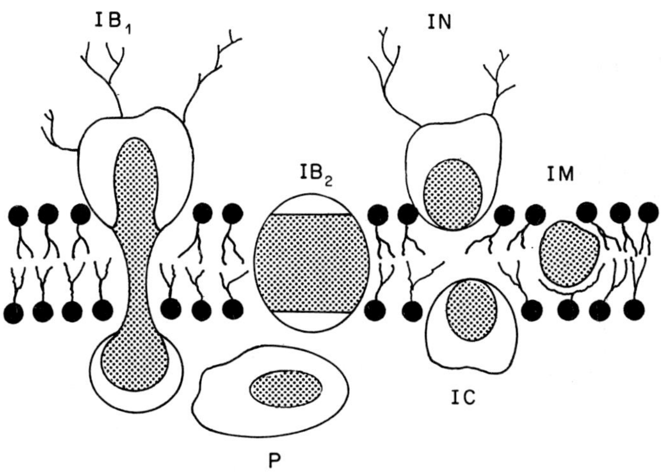

Fig. 2.17. Different types of disposition of membrane proteins. IB_1—integral, membrane-bridging protein, with a single polypeptide span; IB_2—the same with several spans; IN and IC—integral noncytoplasmic and cytoplasmic proteins; IM—integral, buried proteins; P—peripheral protein

protein that is a rare event. It is known only from myelin where the major (55 per cent) of the three membrane-associated proteins, the Folch–Lees proteolipid, is so hydrophobic that it prefers olive oil to water as solvent. Its small size (M_r 5000 or, as a hexamer, 30 000) would permit its complete submersion in the hydrophobic lipid core. Some components of complex proteins, e.g., those of the F_0 sector of the mitochondrial ATP synthase, are probably similarly buried in the membrane.

The degree and type of protein interaction with a given lipid membrane are unequivocally determined by its primary and secondary structure. Hydrophobicity of the individual amino acids and their arrangement in α-helices favours an intramembrane location so that the presently known amino acid sequences of many membrane proteins as well as the probability (determined both theoretically and empirically) of their submersion in membrane lipids make it possible to predict the actual localization of an amphipathic protein in a membrane.

Originally, the information used for such predictions derived from the hydrophobicity indexes of individual amino acids, based on Gibbs free energy of transfer from water to an apolar solvent; a more accurate measure was that based on transfer of an amino acid residue in an α-helix from water to an apolar

Table 2.11. Properties of amino acids related to their insertion into the hydrophobic membrane core

Amino acid	A	B1	B2	B3	B4	C
Alanine	−1.75	−1.0	—	—	−1.0	1.56
Arginine	1.96	−3.7	7.5	7.5	11.3	0.45
Asparagine	2.97	−2.1	5.0	—	2.9	0.27
Aspartic acid	3.02	−1.9	5.0	4.3	7.4	0.14
Cysteine	−2.49	−1.5	—	—	−1.5	1.23
Glutamic acid	2.12	−2.9	5.0	3.8	5.9	0.23
Glutamine	2.49	−2.6	5.0	—	2.4	0.51
Glycine	0.00	0.0	—	—	0.0	0.62
Histidine	1.80	−3.0	5.0	1.4	3.4	0.29
Isoleucine	−6.68	−2.5	—	—	−2.5	1.67
Leucine	−7.00	−2.4	—	—	−2.4	2.93
Lysine	1.06	−3.1	2.5	4.8	4.2	0.15
Methionine	−5.99	−2.7	—	—	−2.7	2.96
Phenylalanine	−8.80	−3.4	—	—	−3.4	2.03
Proline	−2.28	−1.7	5.0	—	3.3	0.76
Serine	2.70	−1.0	2.5	—	1.5	0.81
Threonine	0.69	−1.6	2.5	—	0.9	0.91
Tryptophan	−6.31	−4.5	2.5	—	−2.0	1.08
Tyrosine	−5.57	−3.9	5.0	—	1.1	0.68
Valine	−4.77	−2.0	—	—	−2.0	1.14

A—free energy of transfer from water to *n*-octanol (kJ mol^{-1}); B—free energy of transfer of a residue in an α-helix from water to an apolar solvent (kJ mol^{-1}); B1—hydrophobic contribution, B2—H-bond contribution, B3—charge contribution, B4—total; C—optimized preference parameter for submersion in membrane (in a linear relative scale).

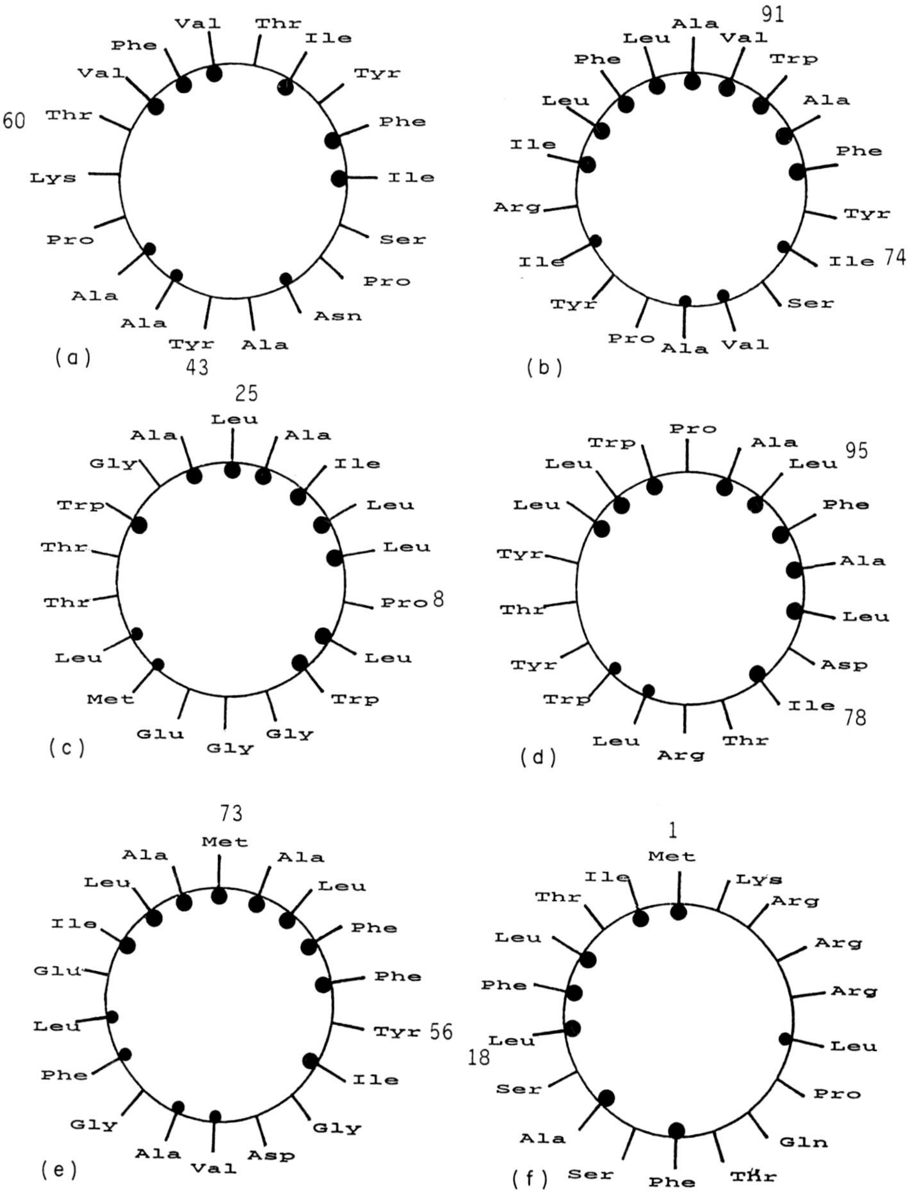

Fig. 2.18. Membrane-spanning polypeptide stretches plotted as helical wheels. (a) Bovine rhodopsin segment between Tyr 43 and Thr 60; (b) bovine rhodopsin segment between Ile 74 and Val 91; (c) bacteriorhodopsin helix A between Pro 8 and Leu 25; (d) bacteriorhodopsin helix C between Ile 78 and Leu 95; (e) proteolipid of *Neurospora crassa* ATP-synthase, between Tyr 56 and Met 73; (f) presequence of yeast cytochrome *c* oxidase subunit IV, between Met 1 and Leu 18. The individual positions of amino acids follow by rotating the 'wheel' by 100 degrees, in (a) and (b) counterclockwise, in (c)–(f) clockwise. The black dots mark hydrophobic amino acids

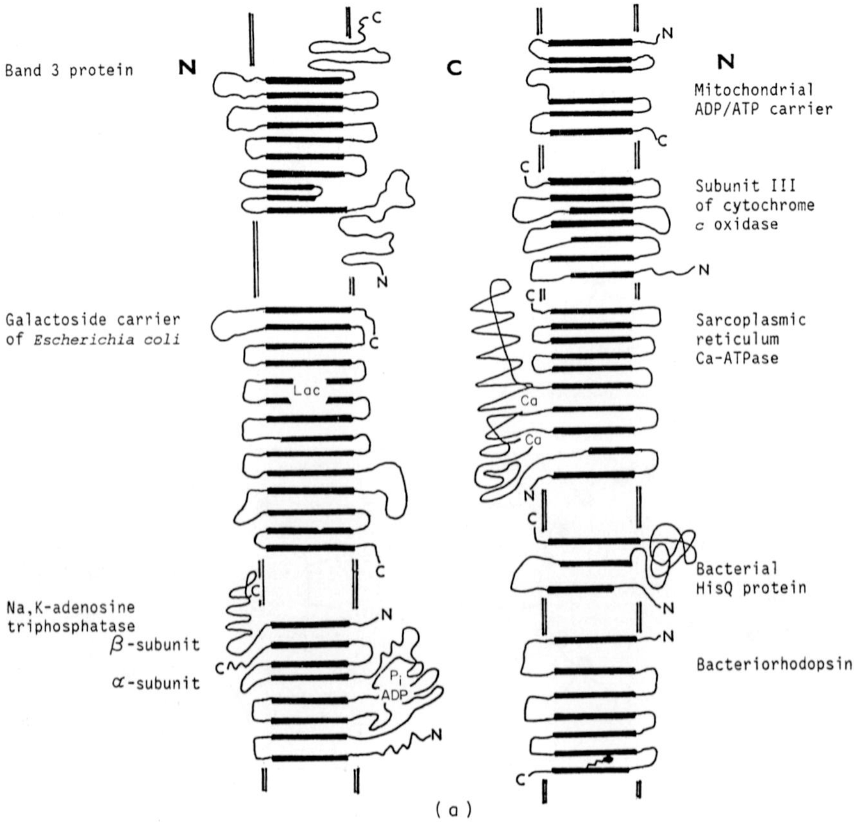

Fig. 2.19. (See legend on p. 85)

solvent; finally, a combination of hydrophobicity data with the actual occurrence of amino acids in the membrane-spanning segments of known proteins is used and an optimally weighted membrane-associative propensity of individual amino acids is reached (Table 2.11).

The α-helical secondary structure is always preferred to any other type in the intramembrane protein segments as it offers the possibility of the highest number of hydrogen bonds to be formed within the protein. The nonpolar membrane medium does not yield itself to hydrogen bonding at all.

If the integral protein spans the membrane several times (as most, perhaps all, transport proteins do) the peptide columns are generally organized in a 'cell' with a hydrophilic channel inside (cf. with data on bacteriorhodopsin, cytochrome *c* oxidase, connexins, porins, etc.). Clearly, such an arrangement would favour a biased distribution of hydrophobic amino acids on those places in the α-helix that are adjacent to the lipid, leaving the rest for the more hydrophilic ones. Indeed, this appears to be the case with the proteins studied so far (Fig. 2.18).

The anchor provided by the hydrophobic segment within the membrane can be surprisingly strong. It may be calculated that to pull out an ideal α-helix,

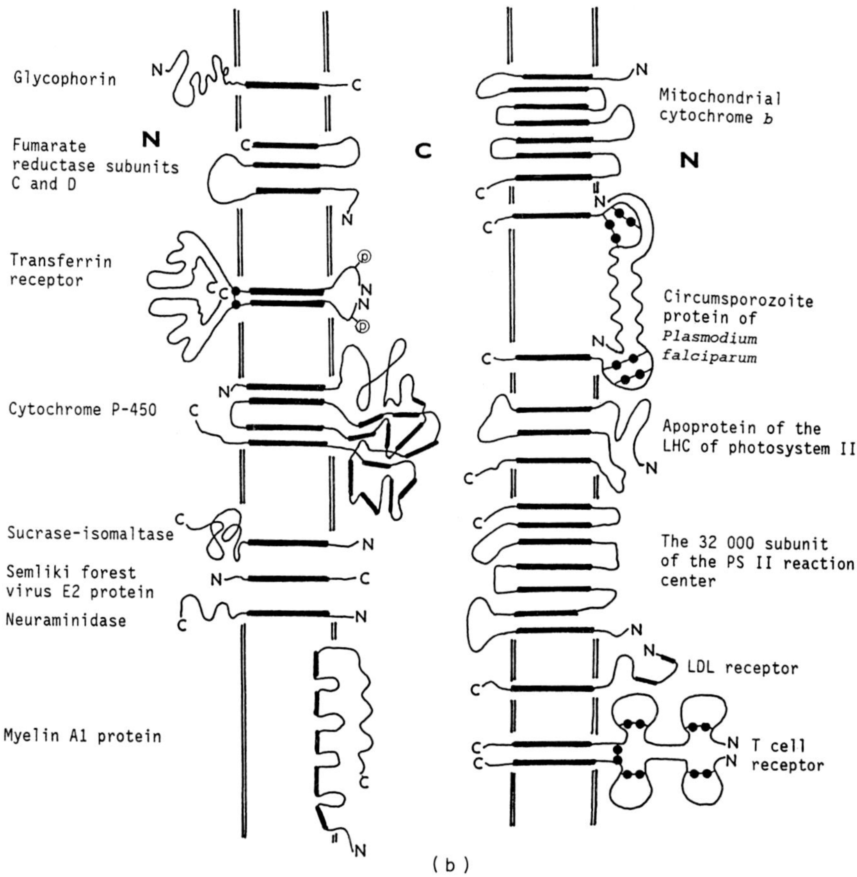

Fig. 2.19. A 'rolled-out' view of various membrane proteins. (a) Transport proteins; (b) proteins with other functions. The helical spans are shown in heavy lines. The actual conformation of the proteins is obviously more compact but is not known with certainty in all cases. The hydrophilic stretches are shown with a great degree of arbitrariness for similar reasons. C—Cytoplasmic face; N—noncytoplasmic face of the membrane

interacting with all its 20–25 amino acids of the membrane-spanning segment with the surrounding lipid, from the membrane, would require 100–120 kJ mol^{-1}, an energy level practically never available from the thermal motion of the protein molecule.

The depth of penetration and the number of membrane-spanning segments may vary over a considerable range, a selection of integral proteins being shown in Fig. 2.19.

Membrane proteins range in size from M_r 5000 of the myelin proteolipid to M_r about 700 000 of the mitochondrial and bacterial complex I (NADH-ubiquinone reductase). The largest single polypeptide chains are those of the E_1E_2 ATPase subunits with M_r about 100 000. (If peripheral proteins are counted then the larger subunit of laminin with M_r of 400 000 would probably take the lead.)

In some membranes a single protein species may represent a substantial portion of the total, e.g. the ADP/ATP carrier of the inner mitochondrial membrane (12 per cent), the band 3 protein of the erythrocyte membrane (30 per cent), the Ca-ATPase of sarcoplasmic reticulum (50–60 per cent), the lipoprotein of the outer membrane of gram-negative bacteria (30 per cent by weight, 75 molar per cent).

While the bulk of integral proteins is formed by amino acid residues, two important complex protein types are known to occur in membranes, the glycoproteins and the lipoproteins.

Glycoproteins

Practically all animal and some plant and fungal cells contain sugar residues on their surface which play an important role in membrane recognition, in antigenic properties and in hormone reception. Some of the sugar residues are attached to lipids, mainly sphingolipids, but a great number are covalently linked to the parts of proteins that protrude into the noncytoplasmic spaces of cells, principally the external medium. In some animal cells this is the substantial part of the glycocalyx.

The sugars are either O-linked or N-linked. The first type bind via serine or threonine (and, exceptionally, hydroxylysine), the second type do so via asparagine. The O-linked glycoproteins contain obligatorily N-acetyl-D-galactosamine (GalNAc) as the first, and often the second, sugar, the N-linked ones bind always to N-acetyl-D-glucosamine (GlcNAc) and always contain the sequence Man $\xrightarrow{1,4}$ GlcNAc $\xrightarrow{1,4}$ GlcNAc → Asn, e.g.

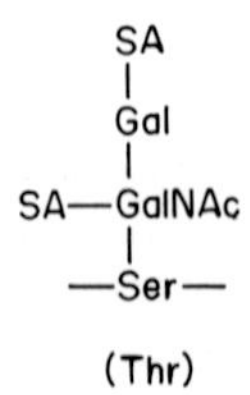

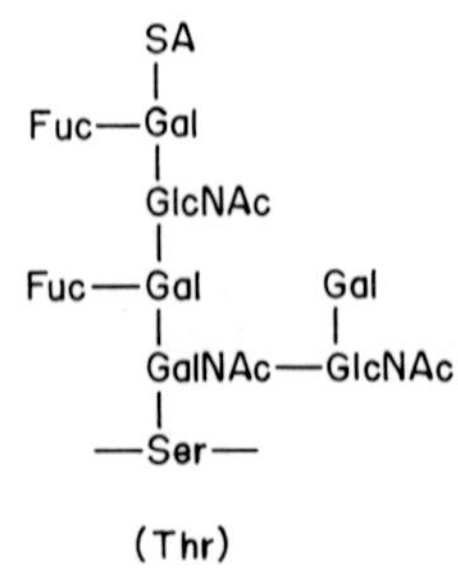

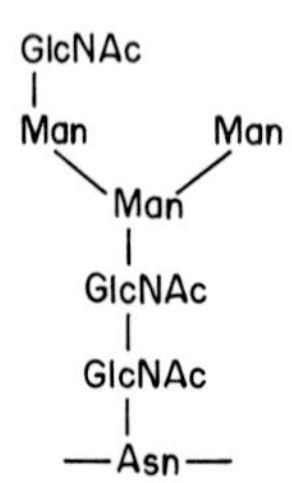

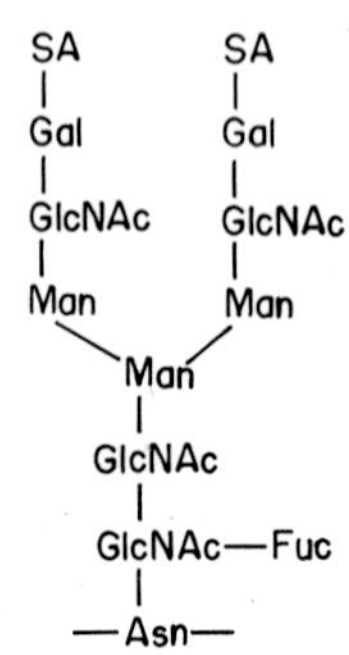

(Man = D-mannose, Fuc = D-fucose, Gal = D-galactose, SA = sialic or N-acetylneuraminic acid.)

A typical representative of glycoproteins is the erythrocyte glycophorin with 131 amino acid residues plus a full 60 per cent of total weight in sugar residues, 15 of the oligosaccharide chains O-linked, one N-linked. Other glycoproteins include immunoglobulins, fibronectin, transferrin, interferon, HLA antigens, yeast invertase and acid phosphatase, viral coat proteins, most eukaryotic transport proteins and a host of others.

Lipoproteins

Covalently linked peptide chain and lipid is not a common occurrence in membranes. Some examples are known from bacteria, specifically the outer membrane lipoprotein from *Escherichia coli* and the secreted penicillinase of *Bacillus licheniformis*. The attachment of the lipid part takes place at the terminal cysteine as follows

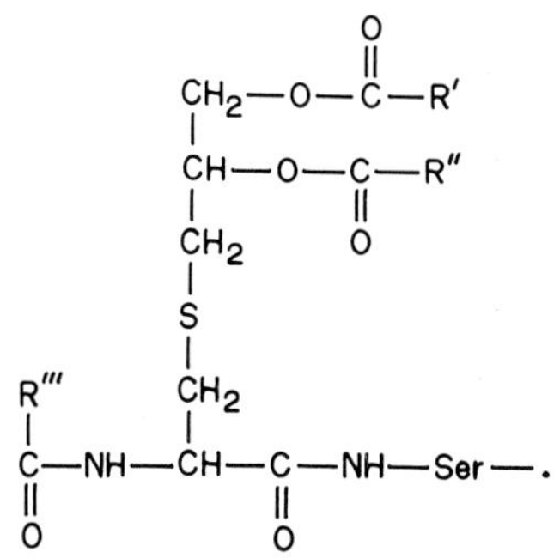

Many more proteins are known in which one or more fatty acids esterify directly an amino acid residue. Although there is no certain proof yet available it seems that serine and threonine hydroxyl groups are the most likely candidates as acylation sites. However, in one case at least the terminal amino group was esterified to an acyl amide.

Many of the fatty-acid-containing proteins are also glycoproteins, e.g., the virus envelope glycoproteins, human transferrin receptors, xanthine oxidase of milk fat globules and the human histocompatibility antigen.

Of the nonglycosylated proteins, the typical examples of a covalently linked lipoprotein are the proteolipids of myelin and of the sarcoplasmic reticulum ATPase.

There is a special class of lipoproteins that are linked to phosphatidylinositol, the two fatty acids of which are anchored in the membrane (Fig. 2.20). An ethanolamine-glycan bridge forms the covalent link between the protein and the lipid. Structures of this type were found in alkaline phophatase, 5′-nucleotidase, acetylcholinesterase, and a number of other proteins ranging from trypanosomes to human erythrocytes.

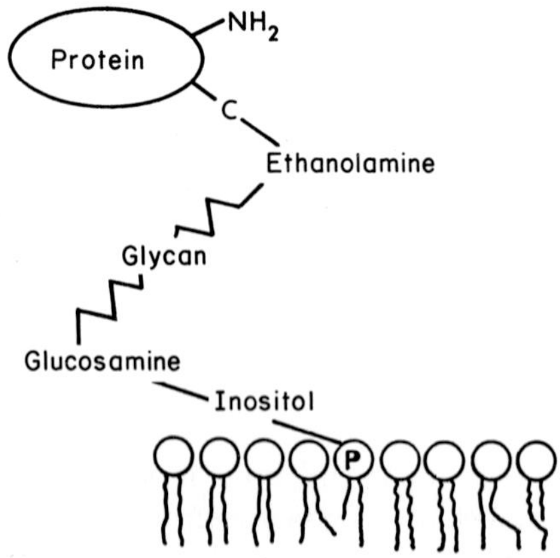

Fig. 2.20. A 'peripheral' lipoprotein anchored in a membrane

2.6.2 Mobility of membrane proteins

In analogy to membrane lipids, but even more so because of the considerably greater bulk and complexity of their structure, proteins exhibit a variety of movements.

1. The aliphatic side chains of amino acids undergo rotational oscillations (much like the *trans-gauche* configuration change of fatty acyl chains of lipids), the methyl groups and the Tyr, Phe and Trp residues rotate with correlation times between 0.1 and 10 ns. Even the rigid peptide chain vibrates. It would be premature to speculate on the functional significance of these movements but it appears that some of them may play a role in overcoming the activation energy barrier in conformational changes.

2. Another movement that occurs in all proteins is a vibration or flipping of segments of the protein. In general, only some protein segments are so rigid as to be virtually immobile (such as would require about 100 kJ mol^{-1} activation energy to undergo movement). All the rest and obviously especially the exposed peptide chains with little internal hydrogen bonding or steric hindrance will experience thermally energized movement. The frequencies of flipping of individual amino acid residues (measured especially for the aromatic amino acids Phe and Tyr) vary over several orders of magnitude (10^1–$10^7\,s^{-1}$) while whole segments of the protein molecule move at frequencies of 10^0–$10^6\,s^{-1}$. These movements are apparently of significance in finding and binding ligands (e.g., the Y-shaped immunoglobulin molecule anchored in the membrane can flip its Fab ends over an angle of some 150 degrees) and in promoting the attainment of an activated state both in enzyme catalysis and in carrier transport. Membrane proteins in particular should display this type of mobility if they are to act as energy or information transducers. NMR studies now permit to classify proteins according to this sort of mobility. Thus, cytochrome *c* is a relatively

highly ordered protein with low mobility while calmodulin is a medium-mobile one and glycophorin and immunoglobulins are very mobile membrane proteins.

3. Proteins in membranes rotate about the axis normal to the membrane plane, the correlation times being listed in Table 2.12.

4. Proteins undergo lateral diffusion as lipids do but in this case it is not self-diffusion and an expression taking account of the difference in protein and lipid size is usually employed. It is based on hydrodynamic considerations and bears the name of Saffman and Delbrück:

$$D_L = (k_B T/4\pi\eta h)[\log(\eta h/\eta_w r) - \gamma] \tag{2.26}$$

with η the membrane viscosity (say, 0.2 Pa s), η_w the aqueous phase viscosity (1.0 mPa s), h the height of the (cylindrical) protein molecule (say 5 nm) and r its radius (say 2 nm); γ is Euler's constant equal to 0.5722. The diffusion coefficient thus calculated would be $7 \cdot 10^{-13}\,\mathrm{m^2\,s^{-1}}$, in agreement with experimental results obtained mainly by the photobleaching-recovery method (Table 2.13).

It should be observed that, in keeping with data on the transition from the L_α to the L_β or $L_{\beta'}$ phase, the lateral diffusion coefficients of proteins drop very substantially, e.g. for the M-13 coat protein by a factor of 100, for bacteriorhodopsin by a factor of 200, for glycophorin by a factor of 1000.

The data of Tables 2.12 and 2.13 indicate that both the rotational correlation times and the lateral diffusion coefficients are in some cases well below the value calculated on the basis of the molecular parameters of the protein molecules. Likewise, it is well known that many proteins of enzyme or carrier nature remain localized in one part of the cell plasma membrane throughout the cell's life (e.g., the intestinal brush-border cells) which is incompatible with what one would calculate from Eq. (2.26). The lower mobility of proteins may be due to several phenomena.

Table 2.12. Correlation times of rotation of membrane proteins

Protein	Source	$\tau_c/\mu s$
Rhodopsin	Retinal rod outer segment	20
Cytochrome a_3	Mitochondrial inner membrane	500
Cytochrome P-450	Microsomal membrane	270
Cytochrome b_5	Microsomal membrane	0.4
Band 3 protein	Erythrocyte	4000
Ca^{2+}-ATPase	Sarcoplasmic reticulum	70, 130, 200 (different methods, below T_t)
Acetylcholine receptor	*Torpedo* electroplax	0.7
Glycophorin	Erythrocyte	1–2 (monomer) 10–20 (dimer)
Epidermal growth factor receptor	Human epidermoid carcinoma cell	350
Bacteriorhodopsin	*Halobacterium* purple membrane	20 000

Table 2.13. Lateral diffusion coefficients of membrane proteins

Protein	Cell type	$D_L/m^2\,s^{-1}$
Rhodopsin	Frog retinal rod outer segment	$3.5 \cdot 10^{-13}$
Acetylcholine receptor	Rat myotubes	10^{-16}
Epidermal growth factor receptor	Epidermoid carcinoma cells	$7 \cdot 10^{-14}$
Epidermal growth factor receptor	Fibroblast	$3 \cdot 10^{-14}$
Coat protein (in liposomes)	M-13 virus	$7 \cdot 10^{-13}$
Concanavalin A receptor	Rat myotubes	$3 \cdot 10^{-15}$
Agglutinin receptor	Wheat germ	$4–8 \cdot 10^{-15}$
Immunoglobulin E	Various cells	$2–7 \cdot 10^{-14}$
Band 3 protein	Erythrocyte	$4 \cdot 10^{-15}$
Bacteriorhodopsin		
(in liposomes)		10^{-12}
(in purple membranes)		10^{-16}
ADP/ATP carrier	Mitochondria	$2 \cdot 10^{-13}$
Cytochrome *c*	Mitochondria	10^{-12}

All the measurements were done at a supratransition temperature (L_α-phase), mostly at 20–25 °C, but the acetylcholine receptor diffusion coefficient was estimated at 35 °C.

(a) Domain formation of lipids where islands of high affinity for proteins occur together with those from which proteins are excluded. This leads both to a certain tortuosity in protein movement in membranes and to the possibility of formation of protein aggregates.

(b) Clustering of proteins which may occur on the basis of electrostatic interactions or hydrophobic effects. This would generate macromolecules with effectively very low lateral diffusion coefficients. Such cases are indeed fairly common in the presynaptic membrane (a gridlike structure of membrane proteins) as well as in the postsynaptic membrane (acetylcholine receptors are oligomers that may form a pseudocrystal in the membrane). Likewise, even in the living cell, bacteriorhodopsin exists as a two-dimensional crystal with an apparent molecular weight 10^3–10^4 times greater than the bacteriorhodopsin monomer.

(c) Interaction of integral proteins with peripheral elements and especially with the intracellular protein skeleton. The situation has been most extensively explored in human erythrocytes where a complicated structural organization connecting the membrane, the cytoskeleton and individual enzyme proteins appears to exist (Fig. 2.21). Here the dissociation constants of the membrane-cytoskeleton complexes are known, e.g., band 3 protein–ankyrin $5 \cdot 10^{-8}$ M, ankyrin–spectrin up to $2 \cdot 10^{-8}$ M, band 4.1–spectrin 10^{-7} M, etc. The importance of the binding between membrane proteins and cytoskeletal elements for

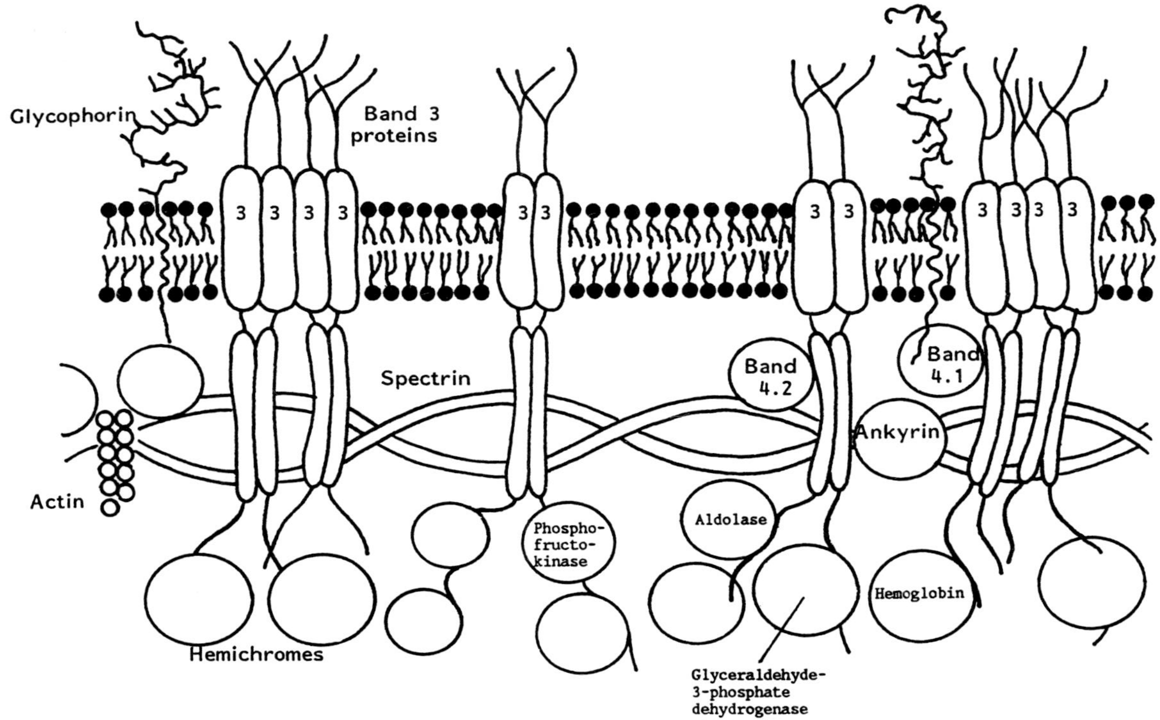

Fig. 2.21. A model representation of links between membrane and cytoskeletal components

immobilizing the former follows from data obtained in spectrin-deficient erythrocytes where the D_L for band 3 protein was $2.5 \cdot 10^{-13}\,m^2\,s^{-1}$, two orders of magnitude greater than in a normal cell; the correlation time of rotation was 200 μs, again 200 times shorter than with cytoskeletal interaction.

2.6.3 Interaction of proteins with lipids

Both the peripheral and the integral proteins are affected by their lipid neighbours and, conversely, may exert some influence on the lipids.

With peripheral proteins this interaction is not altogether clear and, while for instance the myelin Al protein certainly interacts with the polar head groups of bilayer lipids, the true peripheral proteins both on the cytoplasmic and the noncytoplasmic sides of the membrane probably bind to integral proteins if high specificity is involved. Interaction with the polar head groups is not excluded, however, as many enzymes become lipid-dependent as they are associated with membranes although they do not require lipids when soluble (brain hexokinase, pyruvate dehydrogenase, α-oxoglutarate dehydrogenase). Some specificity of peripheral proteins for phospholipids is indeed known. Cytochrome *c* prefers CL, spectrin prefers PS, myelin Al protein binds PG and PS, melittin and band 4.1 protein of human erythrocytes bind PS, etc.

Still, the area of major interest is the interaction between integral proteins and membrane lipids. To begin with, membrane proteins are more or less evenly distributed in membranes only above the transition temperature. Below this temperature they tend to cluster and are surrounded by a poorly organized ring of lipids (cf. with the annular region discussed below). Inevitably this leads to a decrease of their activity. On the other hand, it should be observed that, because of this noncrystalline lipid ring, the proteins retain some of their activity (membrane carriers function with a higher activation energy but still at appreciable rates even after the membrane has virtually solidified). However, the highly laterally and transversally mobile valinomycin (see p. 217) introduced into a lipid membrane does cease to function abruptly as the membrane shifts to the L_β phase—apparently because it is too small to form appreciable disorder in the surrounding lipids (Fig. 2.22). (Valinomycin behaves as a mobile carrier even in the covalently bound monolayer membranes of archaebacteria.)

Even with the highest protein contents, such as are found in the inner mitochondrial membrane, the molar percentage is very small (less than 4 per cent). However, due to the very bulk of the protein molecules they will affect the transition temperature of the bilayer in a remarkable way, often to the extent of completely smoothing out the temperature dependence (Fig. 2.23).

The microviscosity of the membrane is another quantity that is increased by the presence of proteins and, consequently, the lateral diffusion coefficients decrease. These are diminished in fact more than one would expect from the microviscosity change based merely on decreased free volume of lipids. The additional decrease is apparently caused by the proteins forming steric obstacles to the movement of lipids so that their free path is considerably reduced (cf. Table 2.10).

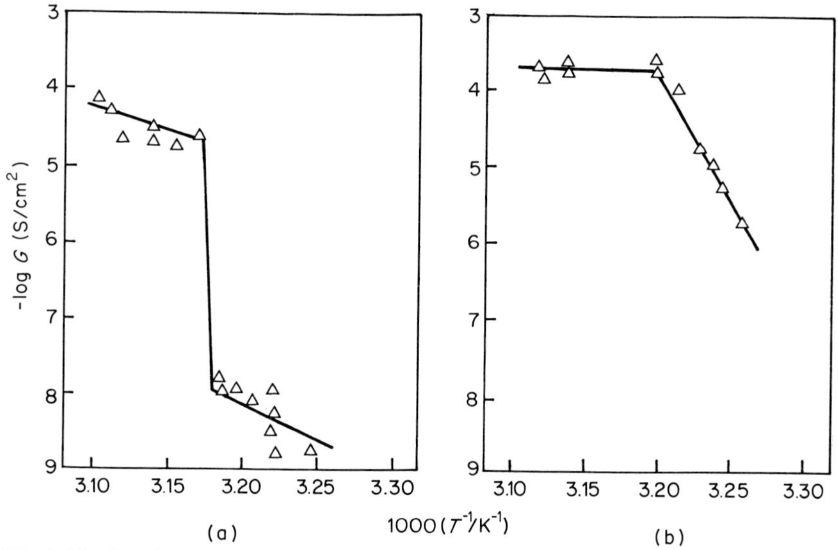

Fig. 2.22. Conductance (G) of dipalmitate-PC bilayer membranes in the presence of 0.1 μM nonactin, a mobile ion carrier (a), and of 1 nM gramicidin A, a channel former (b)

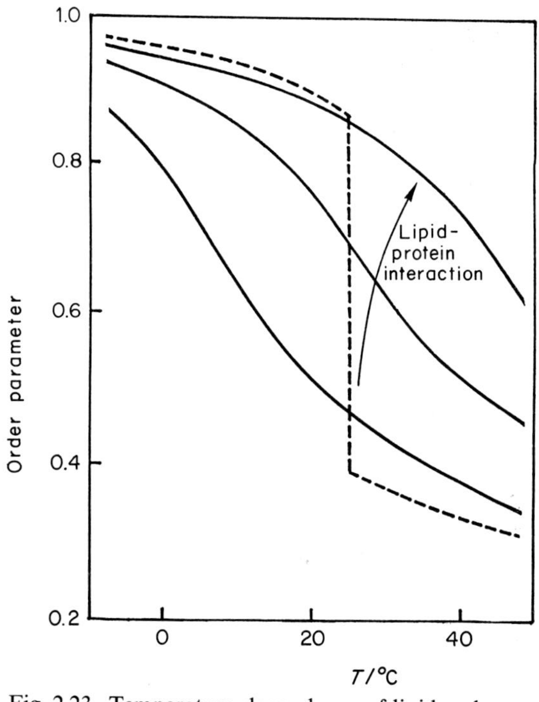

Fig. 2.23. Temperature dependence of lipid order on the annular lipid. Lipid protein interaction increases in the direction of the arrow. The dashed line shows the behaviour of bulk lipids

All these measurable parameters refer to the membrane as a whole. However, it should be understood that there is a special region of interaction between the protein and the bulk lipid which is called annular region or boundary layer. In this region the lipid molecules behave quite differently from those in the bulk.

1. The degree of order pertaining to the acyl chains is greater but apparently does not change as the rest of the membrane crystallizes.

2. The lateral diffusion of the annular region lipids is lower by perhaps two orders of magnitude than in the bulk. This has led to describing the lipid molecules involved as immobilized. The number of molecules thus immobilized was at one time considered to be rather high (as many as 450 around an average-size protein) but now the numbers are known to be substantially lower (Table 2.14). A general observation is that cholesterol is excluded by most integral proteins from their annular domains.

The role of the annular regions in enzyme activity is seen in Arrhenius plots of log activity vs. $1/T$ (Fig. 2.24). An enhanced restriction of lipid movement by membrane proteins appears when protein molecules are clustered and the lipids between them (at a thickness of 3–5 molecules) are effectively trapped and virtually immobile.

It may be remembered from the introductory chapter that some membranes appear thicker than others after negative staining and contrasting in the electron microscope. This is most probably due to their protein content and protein disposition. While the proteins may increase the membrane thickness by their being located on the surface or by protruding from the membrane, there is a secondary effect which is connected with protein-lipid interaction. If a protein interacts optimally with a membrane bilayer that is not completely planar it will tend to bend or buckle the membrane and thus increase its local thickness (Fig. 2.25). The depth to which a protein will be submerged in a membrane depends on the interfacial energy components of the protein with the surround-

Table 2.14. Number of lipid molecules (n) forming the annular region around integral membrane proteins.

Protein	n	Preferred lipid
Glycophorin dimer	30	PS
Rhodopsin	24	PS
Na, K-ATPase	58	PS, PA, CL
Ca-ATPase (sarcoplasmic reticulum)	30	All phospholipids but no sterols
β-Hydroxybutyrate dehydrogenase	30	PC
Cytochrome *c* oxidase	55	CL
Myelin proteolipid	10	PA, fatty acids
Glucose carrier of erythrocytes	—	PS, PA, PG, PC

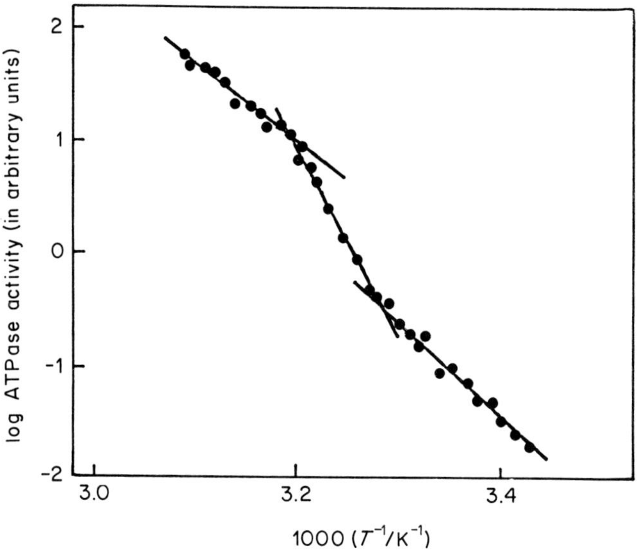

Fig. 2.24. An Arrhenius plot of Ca-ATPase activity in a dipalmitoyl-PC bilayer

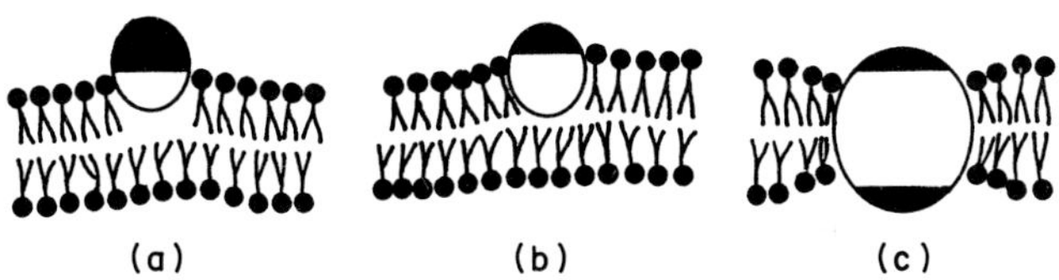

Fig. 2.25. (a), (b) Distortion of a planar bilayer caused by insertion of a protein with hydrophilic (black) and hydrophobic (white) domains. A membrane-spanning integral protein can increase or decrease (c) the bilayer thickness

ing lipid and water. Thus

$$\Delta G = \Delta G^{\circ} - 2(\gamma_{mw} - \gamma_{pw} + \gamma_{pm})hr + \gamma_{mw}h^2 \tag{2.27}$$

ΔG° being the free energy in the case of full submersion of the protein in the membrane, h the distance the protein protrudes from the membrane, r its radius (taking a globular protein molecule), γ's are the interaction energies between protein$_p$, water$_w$, and membrane$_m$. For an amphipathic protein, the minimum free-energy state will be reached when

$$h/r = 1 + (\gamma_{pm} - \gamma_{pw})/\gamma_{mw}. \tag{2.28}$$

Thus, the greater the interaction energy of protein with membrane the farther it will stick out from the membrane.

SYNOPSIS

Biomembrane proteins are of two principal types. The integral or intrinsic proteins are submerged in the lipid bilayer or span it with their peptide chains. The peripheral or extrinsic proteins are loosely attached to the polar head groups of lipids or to parts of integral proteins. They may form links to the cytoskeletal system. The integral proteins invariably contain one or more α-helical stretches which represent the parts bridging the two membrane surfaces. Almost all integral proteins of surface membranes carry simple or complex oligosaccharides, O-linked or N-linked, in the part that protrudes out of the cell, and are generically called glycoproteins. Rather infrequently a covalently linked lipoprotein is found.

Membrane proteins undergo many types of movement. The individual amino acid residues rotate, the peptide chain vibrates. Segments of protein molecules can flip about a considerable angle. They rotate about the axis normal to the membrane plane and they display a slow lateral diffusion.

Some membrane proteins require the presence of general or, occasionally, specific lipids for proper function.

FURTHER READING

Argos, P., Rao, J. K. M., and Hargrave, P. A. Structural prediction of membrane-bound proteins, *Eur. J. Biochem.*, **128**, 565–75 (1982).

Chapman, D., Gomez-Fernandez, J. C., and Goñi, F. M. Intrinsic protein-lipid interactions: Physical and biochemical evidence, *FEBS Lett.*, **98**, 211–23 (1979).

Garland, P. B., and Williams, R. J. P. (eds). *Mobility and Migration of Biological Molecules*, The Biochemical Society, London (1981).

Geiger, B. Microfilament-membrane interaction, *Trends Biochem. Sci.*, **10**, 456–61 (1985).

Gennis, R. B., and Jonas, A. Protein-lipid interactions, *Annu. Rev. Biophys. Bioengin.*, **6**, 195–238 (1977).

Hoffmann, W., and Restall, C. J. Rotation and lateral diffusion of membrane proteins as determined by laser techniques. In: *Biomembrane Structure and Function* (D. Chapman, ed.), pp. 257–318, MacMillan Press, London (1983).

Vanderkooi, G., and Bendler, J. T. Dynamics and thermodynamics of lipid-protein interactions. In: *Structure of Biological Membranes* (S. Abrahamsson, and I. Pascher, eds), pp. 551–70, Plenum Press, New York (1977).

2.7 ELECTROSTATIC PROPERTIES OF MEMBRANES

All biomembranes carry on their surfaces a substantial charge which ranges from 1 to 100 mC m^{-2}. The charge is mainly negative, representing about 10^7–10^8 anionic groups per average animal cell, compared with 10^6–10^7 cationic groups. This charge has a profound influence on a number of membrane phenomena, particularly on the kinetics of ion transport as it causes accumulation of ions of different sign in layers around the cell. The physical chemistry of the phenomenon follows.

The membrane/aqueous solution interface resembles other interfaces between electrically charged phases by the presence of an *electric double layer*. This interface is not analogous to the metal/electrolyte solution interface as the electronic charge present in the metal results in specific phenomena such as the electron cloud overlapping with the electrolyte phase that are not encountered in systems consisting of electrolytes as the only charged species.

The simplest 'model' interface containing ions as the only charged particles is the interface of two immiscible electrolyte solutions (ITIES). If both the solvents have a rather high permittivity (for example, water and nitrobenzene), ion adsorption at the interface due to other effects than electrostatic is absent (with the exception of phospholipids). Let us suppose that the phases contain an electrolyte with one ionic species more hydrophilic (or less hydrophobic) than the other. A rational approach to these concepts, hydrophobicity and hydrophilicity, will be attempted in Section 3.D. In the distribution equilibrium of this electrolyte between the two phases in contact an electric double layer is formed as the more hydrophobic ion has a tendency to cross the interface into the organic phase while the more hydrophilic ion has the opposite tendency. This is, however, possible only in the range of electrostatic attraction forces between ions of opposite charge. Thus a *space-charge region* or the *diffuse electric layer* is formed at both sides of the surface. In the Verwey–Niessen model of such an interface there is no other electrical potential present in ITIES and the space-charge regions are described by the Gouy–Chapman approximate approach. This is based on the solution of the Poisson–Boltzmann differential equation (details can be found in the references listed under 'Further reading'). The surface charge density in this case is defined for the phase w as

$$\sigma_w = (8\varepsilon_w RTc_w)^{1/2} \sinh(zF\psi_{s,w}/2RT) \tag{2.29}$$

where ε_w means the permittivity of water, c_w is the concentration of the z-z-valent electrolyte and $\psi_{s,w}$ the surface potential in phase w. As no dipolar layer is probably present at the interface the surface potential ψ_s and its contribution due to the ionic space charge region, $\psi_{s,ion}$ (the electric potential difference in the diffuse electric layer) are identical. It is defined as the difference of the electric potential in the bulk phase w and of the electric potential close to the interface ϕ_0

$$\psi_{s,w} = \phi_w - \phi_0. \tag{2.30}$$

The surroundings of the interface encompassing both space regions is the electric double layer. The electric double layer as a whole is electroneutral,

$$\sigma_w + \sigma_o = 0 \tag{2.31}$$

where the symbols w and o denote water and the organic 'oil' phase, respectively. When applying the fundamental relationship for the hyperbolic sine the equation for the surface potential $\psi_{s,w}$ results:

$$\psi_{s,w} = (RT/zF)\ln[\sigma_w/(8\varepsilon_w RTc_w)^{1/2} + (\sigma_w^2/8\varepsilon_w RTc_w + 1)^{1/2}]. \tag{2.32}$$

The total electric potential difference between the phases is

$$\Delta_o^w\phi = \psi_{s,w} - \psi_{s,o} \tag{2.33}$$

(for a discussion of this quantity, see Section 3.3.3).

When the bulk concentration of an ion with a charge number z_i is c_i then, due to electrostatic attraction or repulsion in the diffuse electric layer, its concentration at the interface follows the Boltzmann law,

$$c_{i,s,w} = c_{i,w} \exp(z_i F \psi_{s,w}/RT). \tag{2.34}$$

The situation at the interface of water and a low-permittivity organic solvent (e.g. a liquid hydrocarbon) is different. No space charge can be formed in the latter medium which is, in fact, a liquid dielectric only. When hydrophobic ions penetrate into such an organic phase electroneutral particles, ion pairs, are instantly formed. Thus, the only process resulting in electric double layer formation is the adsorption of amphiphilic (amphipathic) ions at the interface. An amphiphilic species is characterized by the presence of a hydrophilic and a hydrophobic group or region in its molecule. The amphiphilic substances adsorbing at interfaces are called *surfactants.*

Adsorption at an interface is, roughly speaking, equivalent to accumulation of one component of the phases in contact at the interface. The quantity characterizing adsorption is the surface concentration, denoted Γ_i (mol m^{-2}). The interface can accommodate only a limited amount of the adsorbate. Thus, the term of relative surface coverage is introduced as

$$\Theta = \Gamma_i/\Gamma_{i,m} \tag{2.35}$$

where $\Gamma_{i,m}$ is the maximum surface concentration accessible.

The relationship between the amount adsorbed, usually expressed as surface concentration or the coverage and the concentration (molar concentration, molar fraction, etc.), in the bulk phase (in the present case, in water) is termed the adsorption isotherm. A considerable number of adsorption isotherms has been suggested, many suitable for uncharged adsorbates, some for adsorbed ions.

For low coverages the linear (or Henry law) isotherm is usually valid,

$$\Gamma_i = \beta_i c_i \tag{2.36}$$

where β_i is the adsorption coefficient (an equilibrium quantity, analogous to an equilibrium constant). If interaction between adsorbed particles is absent and the coverage is appreciable the situation is described by the Langmuir isotherm

$$\Gamma_i = \Gamma_{i,m} \beta_i c_i/(1 + \beta_i c_i). \tag{2.37}$$

The attraction or repulsion of the adsorbed particles is taken into account in the Frumkin isotherm,

$$\beta_i c_i = \frac{\Gamma_i}{\Gamma_{i,m} - \Gamma_i} \exp(-a\Gamma_i) \tag{2.38}$$

where a is the interaction coefficient ($a > 0$ for attraction, $a < 0$ for repulsion).

The surface concentration data are usually obtained from the measurement of the interfacial tension γ as a function of concentration of the adsorbing species by an appropriate method (drop volume method, Wilhelmy plate techniques, etc.) according to the Gibbs equation

$$\Gamma_i = RT\partial\gamma/\partial \ln c_i. \qquad (2.39)$$

If the only adsorbing species is the ionic surfactant with a charge number z_i the surface charge σ is given by

$$\sigma = z_i F\Gamma_i. \qquad (2.40)$$

If σ is known the surface potential is given by Eq. (2.32). In this equation, however, the concentration term $c_{i,w}$ can denote the concentration of an excess indifferent electrolyte which is much larger than the surfactant concentration. The success of this procedure depends on a number of assumptions, particularly that (i) no interfacial ion-pair formation occurs that would decrease the surface charge; (ii) the Gouy–Chapman theory is a satisfactory description of the system (for a discussion see 'Further reading'). In a good approximation, the Gouy–Chapman theory is applicable to low-concentration systems while at higher concentrations the more advanced treatments respecting the discreteness of charge should be used, which is, however, often impeded by the fact that the form of the theoretical results is not suitable for direct application.

The system of an adsorbed surfactant at the interface of an aqueous electrolyte solution and a liquid hydrocarbon strongly resembles the bilayer lipid membrane or even a biological membrane. These systems are complicated by the fact that we have to do here with two surface potentials, each on one surface of the membrane. Furthermore, the interfacial region contains not only ions but also dipoles. These are either already present in the polar heads of the phospholipids (for example, in lecithin the negative pole of the dipole is situated in the phosphate unit while the quaternary nitrogen group bears the positive charge) or are formed in the interaction of ions from the aqueous solution with charged polar head groups. Formation of such ion pairs is favoured by the low permittivity of the interfacial region as water dipoles are aligned with the strong electric field of the electric double layer. Thus, the surface potential consists, in analogy with classical electrochemistry and solid-state physics, of two parts $\psi_{s,ion}$ and ψ_{dip}

$$\psi_s = \psi_{s,ion} + \psi_{dip}. \qquad (2.41)$$

In contrast to the ionic component the dipole component of the surface potential $\psi_{s,dip}$ strongly depends on the distance x of the point where it is to be determined, from the plane of the centres of the dipoles

$$\psi_{s,dip} = -\Gamma_{dip}\mu/\pi\varepsilon(4x^2 - l^2) \qquad (2.42)$$

where Γ_{dip} is the surface concentration of the dipoles, μ the dipole moment and l the distance between the poles of the dipole. For $x \gg 1$

$$\psi_{s,dip} = -\Gamma_{dip}\mu/4\pi\varepsilon x^2. \qquad (2.43)$$

When in Eq. (2.36) c_i is substituted with $c_{i,s}$ from the Boltzmann equation the Stern equation is obtained in the form

$$\Gamma_i = \beta_i c_i \exp(z_i F\psi_{s,ion}/RT). \qquad (2.44)$$

The membrane potential $\Delta\phi_m$, is defined as the difference of the electric potential of the two aqueous phases, 2 and 1, in contact with the membrane

$$\Delta\phi_m = \phi_2 - \phi_1. \tag{2.45}$$

When no external field is imposed between phases 1 and 2, no internal field is formed in the inner (hydrocarbon) region of the bilayer, so that

$$\Delta\phi_m = \psi_{s,2} - \psi_{s,1} \tag{2.46}$$

(see Fig. 2.26(a)). However, in the presence of an external potential difference, an internal potential difference ϕ_{int}, appears in the internal region. Thus, for example, when the phases 1 and 2 are short circuited $\Delta\phi_m = 0$, and

$$\phi_{int} = \psi_{s,1} - \psi_{s,2} \tag{2.47}$$

(Fig. 2.26(b)).

The membrane capacity C_m (F m^{-2}) consists of three components in parallel, the capacities of the two interfacial regions $C_{s,1}$ and $C_{s,2}$ and the capacity of the internal region C_i,

$$1/C_m = 1/C_{s,1} + 1/C_{s,2} + 1/C_i. \tag{2.48}$$

The surface charge of the membrane is determined by measuring the electrophoretic mobility of liposomes or individual cells, by assessment of the conductivity of bilayer lipid membranes in the presence of mobile ionophores or hydrophobic ions, or by changing the transmembrane field strength and by measuring the transport rate of hydrophobic ions.

The most important electrokinetic effect in membranology is the electrophoresis of cells, organelles and liposomes. It is characterized by the electrophoretic mobility u_{em} (m s^{-1} V^{-1}), defined by the Smoluchowski equation

$$u_{em} = 2\varepsilon\zeta/3\eta \tag{2.49}$$

where η is the viscosity of the medium and ζ is the electrokinetic potential. This quantity is the electric potential of the surface of shear (this is a rather vaguely defined position in the neighbourhood of the interface where a distinct motion of the aqueous phase is observed—definitely closer to the interface than the outer edge of the Prandtl layer, see p. 136) with respect to the bulk of the aqueous phase. Thus

$$|\zeta| \lessapprox |\psi_{s,ion,w}|. \tag{2.50}$$

As shown above, the membrane surface potential depends on the ionic composition of the solutions in contact with the membrane. If the solution composition is changed a different surface potential is generated and a corresponding change of the field strength will occur even if the externally applied voltage is kept constant. All field-dependent membrane parameters are then shifted along the potential axis. The shift is equal to the change of the surface potential and, in this way, to the change of the surface charge.

The determination of the membrane interfacial tension (in fact, the bilayer

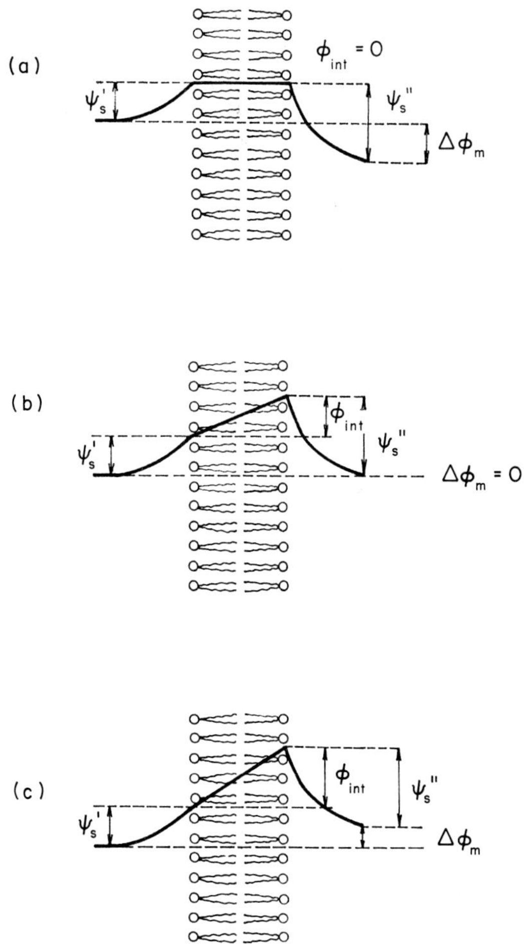

Fig. 2.26. Potential profiles across a membrane for a membrane without an internal potential (a) when the transmembrane potential difference is given by the difference in surface potentials; for a short-circuited membrane (b); for a realistic 'biological' membrane (c) where both the surface potentials ψ_s and the internal membrane potential ϕ_{int} contribute to the overall potential difference across the membrane $\Delta\phi_m$

tension) is achieved by contact angle measurement at the boundary of the bilayer, of the Plateau–Gibbs border and of the aqueous solution.

Particular attention has been paid to the interaction of divalent cations, in the first place Ca^{2+}, with phospholipid bilayer membranes. The substitution of monovalent cations by divalent ones results in shrinking of the diffuse electric layer and in recharging of the surface to more positive values. The latter phenomenon is due partly to electrostatic, partly to chemical interaction with the phosphate groups present on the membrane surface. Usually one divalent cation is bound to two phosphate groups, which resembles the situation in a liquid membrane of the calcium ion-selective electrode based on dialkyl-phosphate as liquid ion-exchanger. The interaction of the cations with the membrane can be described by the Langmuir isotherm by means of which the Stern equation is modified. This simple approach is restricted to small divalent cations while the large ones with separated positive charges, such as hexamethonium ion $(CH_3)_3N^+{-}N^+(CH_3)_3$ do not fit into this picture.

When in an electrolyte solution two bilayer membranes mutually approach in a parallel position the pressure exerted on each of them is

$$P = \frac{2\varepsilon}{L_D^2}\psi_{s,ion}^2 \exp(-d_w/L_D) \tag{2.51}$$

where d_w is the interbilayer distance, which is sufficiently larger than the Debye length L_D given by the equation

$$L_D = (\varepsilon k_B T/2e^2 N_A I)^{1/2} \tag{2.52}$$

here k_B is the Boltzmann constant, e the elementary charge, N_A the Avogadro constant and I the ionic strength of the medium

$$I = \tfrac{1}{2}\sum c_i z_i \tag{2.53}$$

where c_i are the concentrations of ionic species present in the solution and z_i their charge numbers.

Under certain conditions membranes behave as passive components of electrical circuits. This is true for simple unmodified bilayer membranes prepared of pure solvent-free phospholipids by the Montal–Mueller method while the solvent-containing and modified bilayer membranes and, particularly, biological membranes show a more complicated performance with frequency-dependent resistivity and conductivity.

The unit area resistivity of a simple unmodified bilayer membrane is rather high, reaching with pure preparations values of $10^{15}\,\Omega\,cm^2$ which means that such a membrane is an excellent insulator. The resistivity of solvent-filled membranes is lower. Thus, with oxidized cholesterol in a 1:1 decane/tetradecane mixture the initial resistivity of about $8\cdot10^6\,\Omega\,cm^2$-$3\cdot10^7\,\Omega\,cm^2$ decreases after a month to $0.5\cdot10^6\,\Omega\,cm^2$–$1.4\cdot10^7\,\Omega\,cm^2$.

The capacity of a bilayer membrane, which is mainly due to the low-permittivity ($\varepsilon_r \approx 2.1$) hydrocarbon region is usually less than $1\,\mu F\,cm^{-2}$. Figure 2.27 shows the temperature dependence of the capacity of a lecithin (stearoylmyristoyl-phosphatidylcholine) membrane on temperature. At temperatures higher than the transition (Kraft) point the capacity is $0.83 \pm 0.02\,\mu F\,cm^{-2}$ and below this point $0.74 \pm 0.04\,\mu F\,cm^{-2}$. For similar membranes containing a solvent (a hexane–chloroform mixture) lower values of capacity were found. The already mentioned oxidized cholesterol solvent-filled membranes show a gradual decrease in capacity with time.

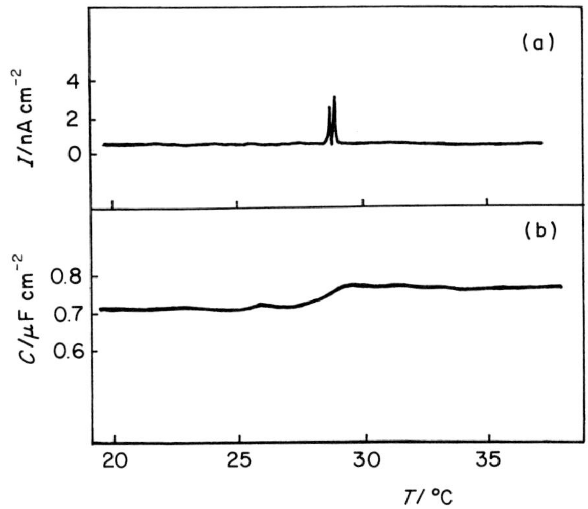

Fig. 2.27. Behaviour of current flowing across a bilayer PC membrane (A) and of membrane capacity (B) during the cooling process

The decrease in capacity between the liquid-crystalline and gel aggregation of the bilayer is due to the increase of the thickness of its hydrocarbon region (cf. Fig. 2.8). Similarly, an increase of capacity appears when an increasing potential difference is applied to the membrane because of decrease of the membrane thickness due to the increasing electric field.

SYNOPSIS

The surface of a membrane is characterized by electric double-layer formation. As the simplest model of this surface the interface of two immiscible electrolyte solutions ITIES is considered. Here, the electric double layer only consists of two diffuse electric layers, in a reasonable approximation described by the Gouy–Chapman equation. The electric potential differences in these diffuse layers, termed surface potentials, also appear at genuine membrane surfaces. In the case of ion adsorption at this surface the surface charge due to the adsorbed ions and the surface potential are linked by the Stern equation. Divalent cations display a profound influence on electric properties of the membrane. Important electric characteristics of the membrane are its resistance and the capacity of the inner region of the bilayer.

FURTHER READING

Alvarez, O., and Latorre, R. Voltage-dependent capacitance in lipid bilayers made from monolayers, *Biophys. J.*, **21**, 1–17 (1978).

Boheim, G., Hanke, W., and Eibl, H. Lipid phase transition in planar bilayer membranes and its effect on carrier and pore-mediated ion transport, *Proc. Natl. Acad. Sci. USA*, **77**, 3403–07 (1980).

Cherny, V. V., Sokolov, V. S., and Abidor I. G. Determination of surface charge of bilayer lipid membranes, *Bioelectrochem. Bioenerget.*, **7**, 413–20 (1980).

Haydon, D. A., and Hladky, S. B. Ion transport across thin lipid membranes: A critical discussion of mechanisms in selected systems, *Quart. Rev. Biophys.*, **5**, 187–282 (1972).

Henderson, D., Blum, L., and Lozada-Cassou, M. The statistical mechanics of the electrical double layer, *J. Electroanal. Chem.*, **150**, 291–303 (1983).

Ketterer, B., Neumcke, B., and Läuger, P. Transport mechanisms of hydrophobic ions through lipid bilayer membranes, *J. Membr. Biol.*, **5**, 225–45 (1971).

Lau, A. L. Y., McLaughlin, A. C., and McLaughlin, S. G. A. The adsorption of divalent cations to phosphatidylglycerol bilayer membranes, *Biochim. Biophys. Acta*, **645**, 279–92 (1981).

Läuger, P., Benz, R., Stark, G., Bamberg, E., Jordan, P. C., Fahr, A., and Brock, W. Relaxation studies of ion transport systems in lipid bilayer membranes, *Quart. Rev. Biophys.*, **14**, 513–98 (1981).

Loosley-Millman, M. E., Rand, R. P., and Parsegian, V. A. Effects of monovalent ion binding and screening on measured electro-static forces between charged phospholipid bilayers *Biophys. J.*, **40**, 221–32 (1982).

McLaughlin, S. Electrostatic potentials at membrane-solution interface, *Curr. Topics Membranes and Transport*, **9**, 71–144 (1977).

Needham, D., and Haydon, D. A. Tensions and free energies of formation in solventless lipid bilayers, *Measurement for high contact angles. Biophys. J.*, **41**, 251–8 (1983).

Ohki, S. Membrane potential and ion permeability of lipid bilayer membranes, *Bioelectrochem. Bioenerget.*, **7**, 487–501 (1980).

Robinson, R. L., and Strickholm, A. Oxidized cholesterol bilayers. Dependence of electrical properties on degree of oxidation and aging, *Biochim. Biophys. Acta*, **509**, 9–20 (1978).

Spach, G. (ed.), *Physical Chemistry of Transmembrane Ion Motions*, Elsevier, Amsterdam (1983).

Stark, G. Hydrophobic ions, carriers and pore formers in lipid bilayers as studied by fast kinetic methods. In: *Biomembranes, Dynamics and Biology* (R. M. Burton, F. C. Guerra, eds), pp. 193–224. Plenum Press, New York (1984).

Szabo, G., and Waldbillig, R. C. Lipid model membranes. In: *Methods of Experimental Physics*, Vol. 20, pp. 513–43. Academic Press, Orlando (1982).

Ti Tien, H. *Bilayer Lipid Membranes.* Marcel Dekker, New York (1974).

Trasatti, S., and Parsons, R. Interphases in systems of conducting phases, *Pure Appl. Chem.*, **58**, 437–54 (1983).

White, S. H., and Thomson, T. E. Capacitance, area and thickness variations in thin lipid films, *Biochim. Biophys. Acta*, **323**, 7–22 (1973).

2.8 MEMBRANE BIOGENESIS

There is no definitive information available that membranes can be formed *de novo*, from their lipid and protein constituents. The scant morphological evidence that membrane fragments associate into vesicles and organelles is inconclusive and hence it is likely that all membranes grow by insertion of their components into existing membranes and all organelles are formed by division or budding of existing 'mother' organelles. In other words, like *omnis cellula ex cellula*, we have *omnis membrana ex membrana.*

There are two principal ways how a membrane can grow: (1) by insertion of individual lipids and proteins, (2) by fusion with membrane vesicles circulating in the cell.

2.8.1 Lipid synthesis and turnover

In *prokaryotic organisms*, phospholipids (there is virtually no other kind of lipid present) are synthesized by enzymes attached to the cytoplasmic face of the protoplast membrane. The sequence of reactions is as follows:

$$sn\text{-}1,2\text{-glycerol-3-phosphate} + 2\,\text{acyl-CoA} \longrightarrow$$

$$\text{phosphatidic acid} \xrightarrow{\text{CTP} \quad \text{P-P}_i} \text{CDP-glyceride}$$

$$\text{(A)} \xrightarrow{\text{serine} \quad \text{CMP}} \text{phosphatidylserine} \xrightarrow{\text{CO}_2} \text{phosphatidylethanolamine}$$

$$\text{(B)} \xrightarrow{sn\text{-glycerol-3-phosphate} \quad \text{CMP}} \text{phosphatidylglycerol phosphate}$$

$$\xrightarrow{\text{P}_i} \text{phosphatidylglycerol} \xrightarrow{\text{CDP-diglyceride} \quad \text{CMP}} \text{diphosphatidylglycerol.}$$

As the enzymes, with the somewhat strange exception of phosphatidylserine synthetase which is attached to ribosomes, are all associated with the membrane there is no obstacle for the incorporation of the phospholipids into it. However, to reach the outer monolayer and eventually the outer membrane, they must cross the membrane. This does not take place by a flip-flop mechanism–this is much too slow, but apparently there is a specific protein present at least in growing cells which catalyzes the shift across the bilayer. The rate of incorporation of cytoplasm-synthesized phospholipids into the outer layer of the cytoplasmic membrane may be 10^5 faster than expected from a flip-flop mechanism.

In *archaebacteria*, where phytanylglycerol ethers predominate in membranes the reaction sequence for synthesis appears to be

$$sn\text{-}2,3\text{-glycerol-3-phosphate} + 2\,\text{phytane} \longrightarrow$$
$$\text{diphytanylglycerolphosphate diether} \longrightarrow$$
$$\text{bis-diphytanyldiglycerolphosphate tetraether.}$$

In *eukaryotic cells*, all lipids, i.e. glycerolipids, sphingolipids and sterols, are synthesized on the membranes of smooth endoplasmic reticulum. Some lipids characteristic for certain organelles may, however, be synthesized in the organelle membranes, such as cardiolipin in mitochondria and the highly glycosylated lipids in chloroplasts.

The pathway for glycerolipid synthesis is different from that in bacteria:

1. $\text{ethanolamine} \xrightarrow{\text{ATP} \quad \text{ADP}} \text{phosphorylethanolamine} \xrightarrow{\text{CTP} \quad \text{P P}_i}$

CDP-ethanolamine $\xrightarrow{\text{diacylglycerol} \quad \text{CMP}}$ phosphatidylethanolamine or, in mammals,

CDP-ethanolamine $\xrightarrow{\text{phosphatidic acid} \quad \text{CDP}}$ phosphatidylethanolamine

$\xrightarrow{\text{serine} \quad \text{ethanolamine}}$ phosphatidylserine.

2. choline $\xrightarrow{\text{ATP} \quad \text{ADP}}$ phosphorylcholine $\xrightarrow{\text{CDP} \quad \text{P P}_i}$

CDP-choline $\xrightarrow{\text{diacylglycerol} \quad \text{CMP}}$ phosphatidylcholine

or

phosphatidylserine $\xrightarrow{\text{H}^+ \quad \text{CO}_2}$ phosphatidylethanolamine

$\xrightarrow{3[\text{CH}_3]\,3\text{H}^+}$ phosphatidylcholine

(the methyl groups are supplied here by S-adenosylmethionine).

3. The 'prokaryotic' pathway to CDP-diglyceride exists also in eukaryotes and CDP-diglyceride can serve as the starting point for the synthesis of phosphatidylinositol, phosphatidylglycerol and diphosphatidylglycerol.

Sphingolipids are synthesized by the following sequence:

Acyl-CoA + serine $\longrightarrow$ dehydrosphingosine $\xrightarrow[\text{FAD} \quad \text{FADH}_2]{\text{NADPH} \quad \text{NADP}}$ sphingosine

(A) $\xrightarrow{\text{CDP-choline} \quad \text{CMP}}$ sphingomyelin

acyl-CoA CoA

(B) $\xrightarrow{\text{acyl-CoA} \quad \text{CoA}}$ ceramide $\xrightarrow{\text{UDP-galactose} \quad \text{UDP}}$ cerebroside

(C) $\xrightarrow{\text{acyl-CoA} \quad \text{CoA}}$ ceramide $\xrightarrow{\text{UDP-glucose} \quad \text{UDP}}$ glucosylceramide

$\xrightarrow{\text{UDP-galactose} \quad \text{UDP}}$ galactosylglucosylceramide

$\xrightarrow{\text{CMP-SA} \quad \text{CMP}}$ sialylgalactosylglucosylceramide, etc.

Sulphatides are formed from cerebrosides under the action of phosphoadenosine phosphosulphate.

It is a general rule that sugar moieties are attached to glycolipids in the *trans* zone of the Golgi apparatus, only a few minutes before they appear at their final destination in the plasma membrane.

Sterols, like other lipids, are synthesized on the smooth endoplasmic reticulum by a highly complex set of reactions which have a start common with the synthesis of isoprenoid compounds, phytane, hopanoids, etc.

$$2\ \text{Acetyl-CoA} \xrightarrow{H_2O\ \ CoA} \text{3-hydroxy-3-methylglutaryl-CoA}$$

$$\xrightarrow[\to CoA]{2NADPH\ \ 2NADP} \text{mevalonate} \xrightarrow[CO_2\,H_2O]{2ATP\ \ 2ADP} \text{isopentenyl diphosphate}$$

$$\xrightarrow{\text{isomerization}} \text{dimethylallyl diphosphate (isoprenyl diphosphate)}$$

(A) 6 steps to diterpenes,
(B) 9 steps to carotene,
(C) 30–32 steps to cholesterol (via lanosterol in animals).
(D) about 30 steps to stigmasterol (via cycloartenol in plants).

Although some of the lipids synthesized on the smooth endoplasmic reticulum remain anchored in that membrane most of them must be transported to a different target membrane. This can be achieved in three different ways: 1. By pinching vesicles off the ER membrane and by their fusion with the target membrane; 2. by their movement through the cytoplasm in the form of individual molecules or in complex with special carrier proteins; 3. by lateral diffusion along membranes and their links to other membranes.

1. The first mechanism certainly exists and it is the major pathway of membrane recycling. This process is observed particularly in cells with intense endo- and exocytosis but apparently exists in all eukaryotic cells. It has some features in common with vesicle-mediated protein transport (p. 222) but, apparently in the Golgi apparatus, vesicles carrying only membrane material without any (important) intravesicular content are sorted to their final destination. Whether there are any specific address structures on these vesicles remains to be seen.

2. Movement of lipids along membrane planes, if it takes place, can hardly provide for specific sorting toward the inner membranes of the semiautonomous organelles, such as mitochondria and chloroplasts, and even toward peroxisomes, as they are not connected by any obvious bridges to the ER.

3. An important possibility for phospholipid traffic is offered by the so-called exchange proteins isolated from many eukaryotic cells (Table 2.15). These proteins are particularly active in a more or less specific replacement of an 'old' phospholipid with a 'new' one but they can also insert newly synthesized phospholipids into growing membranes.

Studies of lipid turnover indicate that the picture may be even more complex as, even for a single phospholipid species, the half-life is different when it is labelled with ^{32}P or with ^{14}C. Moreover, the half-lives are considerably affected by such subtle differences as one or two double bonds in the acyl chains forming the given phospholipid. Thus, obviously, the interplay of various phospholipases in the case of glycerophospholipids and some sphingophospholipids, or sterol oxidases in the cases of sterols, with the exchange catalyzed by the lipid-exchange

Table 2.15. Lipid-exchange proteins

Transported lipids	Molar mass of protein	Isoelectric point	Source
PC	25 000	5.8	Bovine liver
PC, SP	21 000	4.7	Bovine brain
PC > SP	26 000	5.9	Bovine heart
PI > PC	29 500	5.2	Bovine brain
PI > PC	33 000	5.3	Rat liver
PC	24 500	8.4	Rat liver
PE, PC, PI, SP, sterols	14 500	9.5	Rat liver mitochondria
PC, PI, PE	22 000	?	Potato tubers

Table 2.16. Half-lives of membrane lipids

Lipid species	Membrane	Half-life (d)
Cholesterol	*Mycoplasma gallisepticum*	0.15
Total phospholipids	Rat liver	5–10
PC	Myelin sheath	40–90
PI	Rat liver	12.5
SP	Rat liver	40
PG	Rat liver	0.04
CL	Rat liver	0.08
PE	Rat liver	Extremely long
PE	*Escherichia coli*	10–20
PG, CL	*Escherichia coli*	0.03

proteins must be taken into account. Some representative apparent half-lives of lipids are listed in Table 2.16.

The half-lives of some phospholipids may be specifically altered if they are involved in metabolic or regulatory phenomena. A typical case is the phosphatidylinositol which plays a role in the hormone-triggered regulation of metabolism. Clearly, upon stimulation, when phosphatidylinositol diphosphate is decomposed to acyl diglyceride and triphosphoinositol, the overall turnover of phosphoinositol increases tremendously.

2.8.2 Protein synthesis and turnover

Membrane proteins are synthesized by the same sequence of reactions on ribosomes, through translation of information carried by messenger RNA, as are soluble proteins. However, virtually all nascent membrane proteins carry a sequence of amino acids at the N terminus which has been called signal sequence in the case of proteins synthesized on the endoplasmic reticulum (and on bacterial membranes) and signal or address or presequence in the case of proteins translocated through the cytosol to their target membrane. These sequences are practically always cleaved by a particular peptidase once the protein has been properly targeted.

As there are some distinctive features in the synthesis of proteins as they cross the ER membrane (the so-called cotranslational insertion) and that of proteins wandering through the cell to their final destination (the so-called posttranslational insertion) the processes will be described separately.

ER-*dependent membrane protein synthesis*

Proteins that will form part of the plasma membrane of eukaryotic cells or become inserted in the nuclear membrane or organelle membranes (with the exception of mitochondria, chloroplasts, and various microbodies) and finally those that are secreted from cells, all share several initial steps of their synthesis.

The messenger RNA to be translated into one of the above membrane proteins attaches to a cytosolic ribosome and translation begins with a sequence coding for the signal sequence of amino acids. This signal sequence is characterized by continuous stretches of nonpolar amino acids and it has been possible, on the basis of a bank of several hundred eukaryotic signal sequences, to design a 'minimum' signal sequence which reads (NP standing for nonpolar amino acid and X for any amino acid)

$$\text{Met-Lys-NP-NP-NP-NP-NP-NP-NP-Gly-Pro-X-Ala-X}\overset{\downarrow}{\text{-}}\text{Ala-}$$

(The arrow indicates the bond cleaved later in the scenario by the signal peptidase.)

Peptide synthesis continues until some 70 amino acids have been linked, whereupon translation stops while a cytosolic factor, called the signal recognition particle (SRP), attaches to the ribosome at a site on the large ribosomal subunit near where the nascent peptide chain emerges. SRP contains six polypeptide chains (M_r equal to 72 000, 68 000, 54 000, 19 000, 14 000 and 9000) and a 7S RNA (some 300 bases) and interacts with the signal peptide sequence deeply embedded in the membrane. This complex then binds via the SRP to an ER membrane receptor protein called the SRP receptor or docking protein, the binding being presumably stabilized by one or two ribophorins (glycoproteins of M_r 63 000 and 65 000), the interaction energy being about -200 kJ per 'mole', i.e. $6 \cdot 10^{23}$ ribosomes.

It is only now that the N-terminal part of the nascent peptide enters the membrane (Fig. 2.28) and immediately forms a loop in which the N terminus emerges back on the cytoplasmic face of the ER. The signal sequence is then cleaved by luminal signal peptidase and the newly formed N terminus, now free in the lumen, continues to be pushed through the membrane by further protein synthesis in the ER membrane. The SRP is released during membrane penetration by the nascent peptide chain. Protein synthesis continues, the fate of the synthesized protein being two-fold.

It can be either completely liberated in the lumen to be eventually secreted from the cell, or it can become anchored (possibly by a stretch of apolar followed

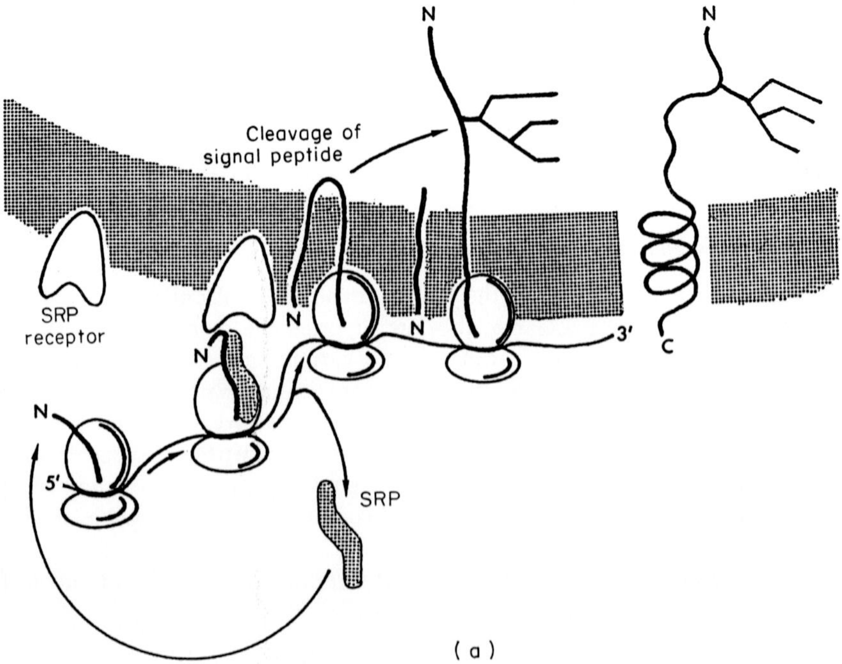

Fig. 2.28. (See legend on p. 111)

by polar amino acids at the C terminus) in the membrane and become its permanent part.

This mechanism is apparently valid both for proteins spanning the membrane once and for those crossing it several times, as most of the transport-engaged proteins do. In the latter case, the α-helices within the membrane are probably preformed before entering the membrane and they do so spontaneously, following the path toward minimum free energy of the system.

Once a substantial part of the protein has been synthesized and dangles freely in the ER luminal space post-translational modifications give it its final conformation. First, proteins with cysteine residues suitably located, such as immunoglobulins, are oxidized while still being extruded into the lumen, to form stable disulphide bonds, others, such as proinsulin, are so connected only after complete release into the lumen. Second, most proteins are synthesized in a proform, with extra stretches of amino acids, either at the N terminus or at both C and N termini, which are cleaved by endoproteases or carboxypeptidases. Examples of such proteins are proinsulin, proparathyroid hormone, proalbumin (which is a notable exception of a secretory protein without a signal sequence), proglucagon and pro-opiomelanocortin (this gives rise to a number of active hormones by cleavage at different sites, among them α-, β- and γ-melanocyte-stimulating hormone, corticotropin, β-endorphin, β-lipoprotein, enkephalin).

This processing takes place mainly after the proteins have moved to the Golgi

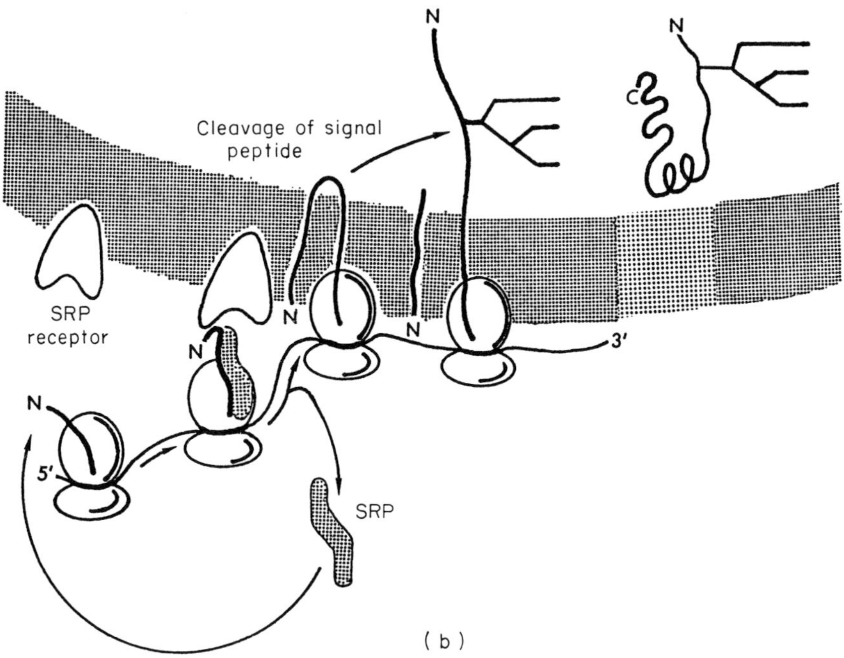

Fig. 2.28. Sequence of events taking place during protein synthesis on the rough endoplasmic reticulum. (a) The synthesized protein remains anchored in the membrane; (b) the synthesized protein is released into the RER lumen and is destined to be secreted from the cell

apparatus by way of fusion of the ER sections with the Golgi, following a centrifugal path involving innumerable fusions and fissions of vesicles during which the sorting process takes place. It is here that the vesicles are tagged in some not yet understood fashion to move either to the plasma membrane or toward the vacuole or simply to remain in the cytosol in the form of lysosomes.

The third, and the most characteristic of post-translational modifications, is protein glycosylation. Virtually all secreted and outward-facing membrane proteins are glycoproteins and they acquire their sugar groups either in the ER or in the Golgi apparatus. With the O-linked glycoproteins (via Ser, Thr or Hyl) the monosaccharide moieties are attached one by one from activated nucleotide-sugars, viz. CMP-sialic acid, UDP-N-acetylglucosamine, UDP-galactose, UDP-N-acetylgalactosamine. The nucleotide is liberated during the attachment.

The N-linked glycoproteins are synthesized differently. To begin with, there are two types of N-linked oligosaccharides, complex and high-mannose. The complex type contains mannose, fucose, galactose, N-acetylglucosamine (GlcNAc) and sialic acid and may contain 2–4 branches. The high-mannose type contains only two residues of GlcNAc and up to nine residues of mannose. All N-linked oligosaccharides are synthesized from a common precursor containing dolichol diphosphate, GlcNAc, and mannose (Man) and glucose (Glc).

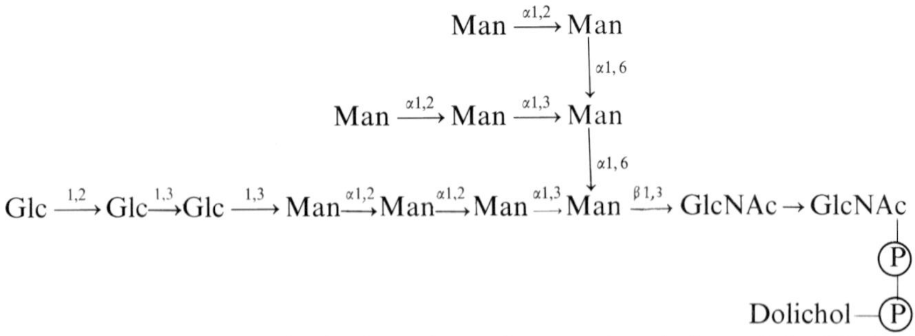

This precursor is synthesized partly at the cytoplasmic face of the ER from dolichol diphosphate (Dol-PP) to Dol-PP-(GlcNAc)$_2$-(Man)$_5$. This molecule then crosses the ER membrane to be modified in the lumen to its complete form Dol-PP-(GlnNAc)$_2$-(Man)$_9$-(Glc)$_3$. This is then transferred by an oligoglucan transferase to the nascent polypeptide at sequences of -Asn-X-Ser- or -Asn-X-Thr- (X being any amino acid). As the synthesis of the glycoprotein is completed and as it moves to the Golgi apparatus it is trimmed by a dozen different enzymes as shown in Fig. 2.29.

It is in the Golgi apparatus that it is decided where the (glyco) protein will be targeted, the possibilities being the plasma membrane (or secretion), the lysosome or vacuole, the Golgi apparatus or the ER. Phosphorylation of mannose on a glycoprotein targets the protein (enclosed in a vesicle) toward the lysosome, certain unknown signals (e.g., in galactosyl transferase) predestine it to remain in the Golgi, etc.

The diversification of the Golgi apparatus is astounding as it can also distinguish between proteins destined for different parts of the plasma membrane,

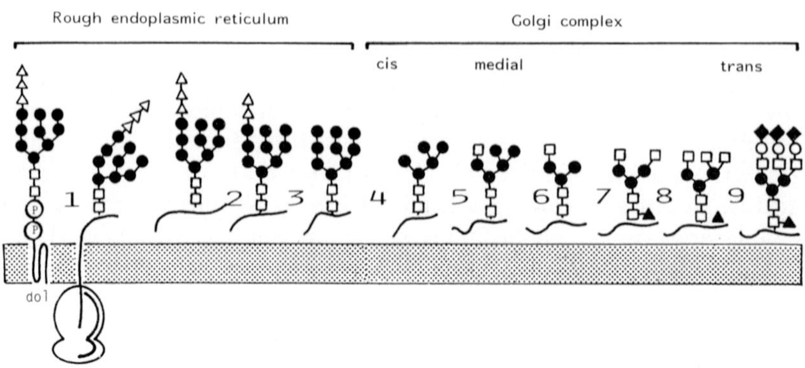

Fig. 2.29. Process of maturation of glycoproteins on their way toward cell periphery. □—N-acetylglucosamine, ●—mannose, △—glucose, ○—galactose, ▲—fucose, ◆—sialic acid. Step 1—transfer of oligosaccharide to nascent polypeptide; steps 2 and 3—removal of glucose residues; step 4—removal of mannose residues; step 5—attachment of N-acetylglucosamine residue; step 6—removal of further mannose residues; step 7—attachment of N-acetylglucosamine and of fucose; step 8—attachment of N-acetylglucosamine; step 9—attachment of galactose and sialic acid residues

such as in the highly polarized intestinal or renal tubule cells. It remains to be seen whether sliding along different types of cytoskeletal structures, such as actin microfilaments in yeast, or microtubules in nerve cells, could play a role here.

Membrane protein synthesis in bacteria

As ribosomes are often seen to be attached to the cytoplasmic membrane of bacteria it might be conjectured that membrane and secretory protein synthesis takes place on these ribosomes, much as it does in eukaryotic ER. Indeed, many such proteins are apparently synthesized in this way and extruded into the periplasm or into the outer membrane of gram-negative bacteria cotranslationally. As examples, one can cite the maltose-binding protein, alkaline phosphatase and β-lactamase in the periplasm, and the λ phage receptor, the OmpA protein and the major lipoprotein in the outer membrane. The signal sequences are similar to those of eukaryotic cells and the 'minimum' sequence deduced from roughly 200 prokaryotic enzymes appears to be

Met-Lys-NP-NP-NP-NP-NP-NP-NP-NP-Gly-Pro-X-X-Ala-X↓Ala-

(cf. with the eukaryote presequence on p. 109). While all the proteins of the outer membrane and of the periplasmic space and many proteins of the inner, cytoplasmic, membrane, such as the penicillin-binding proteins, two lipoproteins, and the serine chemotactic receptor Tsr, all contain a signal sequence that is cleaved off posttranslationally, some inner membrane proteins, such as the lactose carrier, histidine carrier and the b and c subunits of the F_0F_1 ATPase, do not contain such a cleavable sequence.

At the same time, some inner membrane proteins, in particular the phage M13 major coat protein, are assumed to be exported posttranslationally, following what has been termed the trigger mechanism of membrane insertion in which the protein is folded to a soluble tertiary structure and only on exposure to a hydrophobic environment does it assume the final 'membrane' conformation.

A similar post-translational translocation appears to apply to yeast prepro-α-factor, a pheromone protein of $M_r = 18\,600$.

ER-independent membrane protein synthesis

Mitochondria, plastids (especially chloroplasts) and microbodies (especially peroxisomes) are organelles without the full complement of enzyme-synthesizing machinery so that most of their proteins must be synthesized on cytosolic ribosomes and brought to the place of their destination by a diffusion-type movement through the aqueous intracellular medium.

The import of such proteins is best understood in the case of mitochondria where some 240 different enzymes or their subunits must be imported from the cytosol. Similarly to the ER-dependent membrane-protein synthesis, the mitochondrial proteins are synthesized on free polysomes and carry an N-terminal presequence which serves as the address sequence for finding the corresponding membrane receptor. Unlike the signal sequences of ER-synthesized proteins, these are rich in cationic amino acids Arg and Lys, and in hydroxylated amino acids Ser and Thr, and they contain no Asp and Glu.

Depending on the final destination of the protein, viz. outer membrane, intermembrane space, inner membrane, and matrix, the N-terminal presequence contains different information domains. All presequences contain a somewhat similar stretch of about ten amino acids which suffices to target the precursor protein to mitochondria. The mitochondrial surface contains protein receptors that may be of a very broad specificity so as to bind any short N-terminal sequence with, say, four cationic acids in it. On the other hand, the presequence may enter a membrane spontaneously as it does in liposomes–this, of course, is a process that would not permit specific targetting to mitochondria. Hence, a protein may be involved but only to permit access to rather than direct insertion into mitochondrial membrane.

It is now generally believed that the points of entry into mitochondria are the sites of close contact between the inner and outer mitochondrial membranes which display a lower surface charge and which may also serve for ADP/ATP translocation by the specific carrier protein.

1. If the protein is to stay in the outer membrane its presequence contains a long hydrophobic stretch which remains anchored in the outer membrane and no part of the peptide chain is cleaved (Fig. 2.30).

2. If the protein is destined for the intermembrane space it is anchored by a similar, more distant hydrophobic stretch, in the inner membrane and, after shifting along the membrane, the protein is cleaved from the presequence which remains in the inner membrane.

3. If the protein is targetted to the inner membrane its presequence contains no stretch of nonpolar amino acids so that it passes through both membranes and

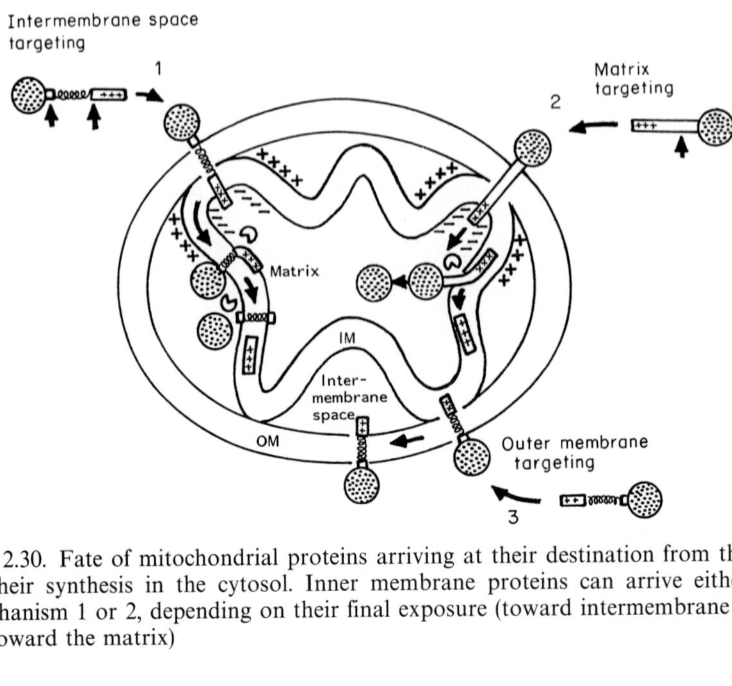

Fig. 2.30. Fate of mitochondrial proteins arriving at their destination from the site of their synthesis in the cytosol. Inner membrane proteins can arrive either via mechanism 1 or 2, depending on their final exposure (toward intermembrane space or toward the matrix)

remains anchored in the inner membrane because it contains nonpolar amino acids at its C-terminus.

4. If the protein is to go all the way into the matrix it follows the path as under (3) but cleavage of the presequence liberates it from the membrane.

For insertion into the membrane contact site the inner membrane must bear an appreciable electric potential but not necessarily a distinct pH gradient.

Import of proteins into chloroplasts is as massive as into mitochondria and its extent may be assessed from the fact that immature chloroplasts (proplastids) lack most of their thylakoid proteins. Like in mitochondria, the outer chloroplast membrane appears to contain several protein receptors of broad specificity. In one respect, import of proteins into thylakoids may be even more complicated than anything in the mitochondria–unless the protein moves laterally along the inner chloroplast membrane and the fret (see Fig. 1.26) into the thylakoid it must cross the aqueous stroma to reach its final goal.

Peroxisomal proteins, such as catalase, are also synthesized on cytosolic ribosomes together with an address sequence and reach their destination via a receptor or recognition site on their surface. In fact, some smooth ER proteins, such as cyt b_5 or cyt P-450, must cross the cytosolic aqueous medium to reach their membrane after being synthesized on free cytosolic ribosomes. They apparently insert their C-terminal and N-terminal sequence, respectively, into the membrane quite spontaneously.

Membrane protein turnover

Like with membrane lipids, two different mechanisms of turnover can be envisaged: the first involves more or less nonspecific membrane recycling during which the components are not released but simply change their membrane of residence (cf. lysosomal circulation or endocytotically active cells); the second has to do with specific protein degradation by intracellular proteinases. So far no specific membrane proteinase has been described although there is ample evidence available that proteins integrally located in membranes need not travel to the universal degradative compartments of cells, lysosomes and vacuoles, to be broken down. It is also not known whether the ubiquitin system is involved in labelling these membrane proteins for degradation. The half-lives of membrane proteins are of the order of days (Table 2.17).

SYNOPSIS

Membrane lipids are synthesized in prokaryotic organisms directly at the inner face of the membrane envelope and are inserted into the membrane spontaneously. To reach the outer membrane or the external monolayer of the cytoplasmic membrane they cross the membrane with the aid of special proteins.

In eukaryotic cells the synthesis takes place in the smooth endoplasmic reticulum and the lipids arrive at their destination either by moving within or with a membrane, or with the aid of special lipid-exchange proteins.

The half-lives of membrane lipids range from several hours to several months.

Table 2.17. Half-lives of some membrane proteins (in days)

Membrane or protein	Half-life
Nucleus*	2–3
Rough endoplasmic reticulum*	2–6
Smooth endoplasmic reticulum*	2–5
NAD nucleosidase	16
NADH dehydrogenase	3–4
Cytochrome b_5 reductase	5
Golgi apparatus*	1–2
Plasma membrane*	2–3
Mitochondrial inner membrane*	7–9
Peroxisomes*	3–4
Lysosomes*	7–8

*Bulk membrane protein of rat hepatocyte organelles.

Proteins destined for the endoplasmic reticulum, for the nuclear membrane, for the vacuolar membrane and for the plasma membrane are synthesized on rough endoplasmic reticulum and are characterized by a hydrophobic signal sequence which assists them in cotranslational penetration into the ER lumen. There they may undergo trimming by hydrolysis and attachment of sugar residues and, via the Golgi apparatus, arrive as part of vesicles to their destination. Secretory proteins are also synthesized in this way but these are not anchored in the vesicle membrane. Most proteins of mitochondria, chloroplasts and the microbodies are synthesized on free ribosomes in a precursor form and find their destination with the aid of an address presequence. They enter the autonomous organelles at places where their two membranes are in close contact.

Membrane proteins have half-lives of days.

FURTHER READING

Dunphy, W., and Rothman, J. E. Compartmental organization of the Golgi stack, *Cell*, **42**, 13–21 (1985).

Engelman, D. M., and Steitz, T. A. The spontaneous insertion of proteins into and across membranes, the helical hairpin hypothesis, *Cell*, **23**, 411–22 (1981).

Geuze, H. J., Slot, J. W., Strous, G. J. A. M., Hasilik, A., and von Figura, K. Possible pathways for lysosomal enzyme delivery, *J. Cell Biol.*, **101**, 2253–62 (1985).

Hurt, E. C., and van Loon, A. P. G. M. How proteins find mitochondria and intramitochondrial compartments, *Trends Biochem. Sci.*, **11**, 204–07 (1986).

Kelly, R. B. Pathways of protein secretion in eukaryotes, *Science*, **230**, 25–32 (1985).

Kornfeld, R., and Kornfeld, S. Assembly of asparagine-linked oligosaccharides, *Annu. Rev. Biochem.*, **45**, 631–64 (1985).

Pagano, R., and Sleight, R. Defining lipid transport pathways in animal cells, *Science*, **229**, 1051–57 (1985).

Schekman, R. Protein localization and membrane traffic in yeast, *Annu. Rev. Cell Biol.*, **1**, 115–44 (1985).

Silhavy, T. J., Benson, S. A., and Emr, S. D. Mechanisms of protein localization, *Microbiol, Rev.*, **47**, 313–44 (1983).

Walter, P., Gilmore, R., and Blobel, G. Protein translocation across the endoplasmic reticulum, *Cell*, **38**, 5–8 (1984).

3 Membranes as Mediators of Flow of Matter

Movement of intrinsic membrane components, various lipid and protein molecules, has been discussed in previous sections. However, movement of extraneous material, including neutral solute molecules and ions, macromolecules and even droplets of solution, is the function that is inseparable from membranes and that has, from the earliest periods of membrane research, attracted primary attention.

Membrane-associated flow of substances represents two categories of events, enzymic catalysis of chemical reactions (macroscopically mostly scalar process, even if the enzyme is embedded in an anisotropic membrane), and transmembrane movement of solvent and solute (always a vectorial process).

As the underlying concepts of both transport and enzymic reactions are those of kinetics and thermodynamics both these areas of physical chemistry will be now briefly presented.

3.A FUNDAMENTAL CONCEPTS OF THERMODYNAMICS

Thermodynamics plays a prominent role in understanding the phenomena of membrane transport and energy transduction. Whereas kinetic and molecular descriptions of these processes are in principle always amenable to further improvement following new experimental findings, phenomenological thermodynamical descriptions are likely to remain permanently valid. The fundamental laws of thermodynamics were discovered in the first half of the last century when little was known about the microstructure of matter and their validity is very general and independent of molecular mechanisms.

Thermodynamic systems

When considering various biological systems it is useful to define in what contact the given system is with its surroundings. Thus, an *isolated system* is one where neither energy nor matter can be exchanged with the exterior. Such systems may occur in Gedankenexperimente but are quite unrealistic. Clearly, a system may begin to resemble an isolated one as it increases in size since its volume will increase with the third power of linear dimension, the surface across which any exchanges with the surroundings take place, will increase only with the second power. A *closed system* is one where matter is retained within its boundaries but energy can pass through. A dead body or cell, and particularly an inanimate object can absorb energy either as work or as heat from the surroundings but will not exchange matter with them. Some such systems that are thermally isolated are called *adiabatic closed systems*.

Most objects of interest in biology are *open systems* where both matter and energy flow in and out. What is conserved here is the sum of energy and the sum of matter of the system studied *and* its surroundings.

Zeroth Law

The laws of thermodynamics postulate the existence of certain functions of state*—once a sufficient number of measurable variables of state, such as volume and pressure, are specified, these functions possess unique values independent of the history of the system. The first of these functions is *temperature*; its existence is based on the notion of thermal equilibrium, now recognized as the *zeroth law of thermodynamics*. Two bodies in thermal contact reach the same degree of subjectively perceived 'hotness' or empirical temperature, somewhere between their original temperatures. This state is called thermal equilibrium and whenever each of two bodies is in thermal equilibrium with a third one, they are also in thermal equilibrium with each other. If they are brought into contact, no change in their variables of state will result. Hence a function of state, temperature, had the same value for them already before they were placed together.

First Law

The *First Law of Thermodynamics* is the famous principle of conservation of energy or the impossibility to construct a perpetual machine of the first kind. For closed systems it may be stated as follows: To each transition of a system between two specified states corresponds the same sum of work performed on the system and of heat absorbed by it. Both the work and the heat may be negative; then the work is performed by the system and the heat given off by the system to its surroundings. The sum of the work and heat is independent of the path of the transition and there is an infinity of such paths. In some of them much work is performed and little heat absorbed, in others the opposite is true. Still, the algebraic sum of the two remains the same, corresponding thus to a change of the function of state called *internal energy* U

$$\Delta U = w + q \tag{3.A1}$$

where w is the work performed on the system and q the heat absorbed by it. (Sometimes the work performed by the system is considered as a positive quantity; then it appears in the above mathematical formulation of the first law with a minus sign.)

Second Law

The *Second Law of Thermodynamics* may be formulated as the impossibility to construct a perpetual motion machine of the second kind which would perform work in cycles (returning always to the same state) and at the same time cool only a single heat reservoir. From any of a number of equivalent statements of this law it may be deduced that there is a function of state called *entropy* S.

The change of entropy in a closed system resulting from any process at temperature T is given by

$$\Delta S = q/T. \tag{3.A2}$$

In a nonisothermal process an infinitesimal change would have to be integrated. The

* We prefer this somewhat old-fashioned designation to the more modern 'state function' to avoid any possible allusion to governmental affairs.

equality sign corresponds to reversible processes, proceeding so close to equilibria that they may be reversed without leaving behind any changes in other systems. In isolated systems entropy can either increase (in natural spontaneous processes) or remain constant (in idealized reversible processes). As soon as the entropy of an isolated system reaches its maximum value, no further spontaneous processes are possible and the system is in equilibrium.

Free energy

The whole framework of classical thermodynamics might be constructed with the above three functions of state: temperature, energy and entropy. However, it is convenient to introduce further functions of state, combining those above among themselves or with variables of state. The *Gibbs free energy* G is the most important of all in biology. Spontaneous processes in isolated systems are those in which entropy increases and reaches its maximum at equilibrium. In closed systems that are kept at constant temperature and pressure the Gibbs free energy is the criterion of the spontaneity of natural processes; it decreases in spontaneous processes and reaches its minimum in equilibrium. Constant pressure of the surrounding atmosphere and constant temperature secured by the thermal conductivity of surrounding tissues or media are conditions typical of living systems.

Changes of the Gibbs free energy are equal to the maximum useful work performed at constant temperature and pressure. The first law can be written as

$$\Delta U = q + w = q - p\,\Delta V + w' \qquad (3.A3)$$

where $p\,\Delta V$ is the pressure work, necessary to push away the surrounding atmosphere, but not useful for carrying out vital processes. On the other hand,

$$-w' = q - p\,\Delta V - \Delta U \qquad (3.A4)$$

is the useful work performed by the system. Maximum useful work is performed under reversible conditions, where $q = T\,\Delta S$ (maximum heat is absorbed and transformed into work). Thus

$$-w'_{\text{max}} = T\,\Delta S - p\,\Delta V - \Delta U = -\Delta G \qquad (3.A5)$$

the maximum useful work performed by the system is equal to the decrease of its Gibbs free energy.

In a system of constant volume the term $p\,\Delta V$ is equal to zero; all work performed by such a system is useful and its maximum value

$$-w_{\text{max}} = T\,\Delta S - \Delta U = -\Delta A \qquad (3.A6)$$

is equal to the decrease of A, the Helmoltz free energy or *thermodynamic potential.* In systems of constant temperature and volume A plays the same role as G in systems at constant pressure and temperature, i.e., it decreases in spontaneous processes and at thermodynamic equilibrium reaches a minimum. In condensed phase (liquid and solid state) the differences between the two functions are usually negligible.

Finally, let us observe that in the definition of the Gibbs free energy change

$$\Delta G = \Delta U + p\Delta V - T\,\Delta S \qquad (3.A7)$$

one can consider the sum $\Delta U + p\,\Delta V$ as a change, under constant temperature and pressure, in another function H, called *enthalpy* or heat content

$$\Delta H = \Delta U + p\,\Delta V. \qquad (3.A8)$$

According to the first law of thermodynamics the enthalpy change is equal to the heat absorbed by a system at constant pressure. Now

$$\Delta G = \Delta H - T\,\Delta S \tag{3.A9}$$

whereas

$$\Delta A = \Delta U - T\,\Delta S \tag{3.A10}$$

It is easy to show that the two criteria of the spontaneity of natural processes, that of entropy increase in an isolated system and that of free energy decrease in a closed system are, in fact, equivalent. A closed system with sufficiently large surroundings may be considered as an isolated system, in which the total entropy cannot but increase

$$\Delta S_{\text{total}} = \Delta S_{\text{system}} + \Delta S_{\text{surroundings}} > 0. \tag{3.A11}$$

On the other hand, in a closed system, e.g. one at constant pressure and temperature, the Gibbs free energy can only decrease

$$\Delta G_{\text{system}} = \Delta H_{\text{system}} - T\,\Delta S_{\text{system}} < 0. \tag{3.A12}$$

Multiplying the last equation by $-1/T$ one obtains

$$-\Delta G_{\text{system}}/T = -\Delta H_{\text{system}}/T + \Delta S_{\text{system}} > 0. \tag{3.A13}$$

The heat which left the system at constant pressure was absorbed reversibly by the surroundings, so that

$$-\Delta H_{\text{system}}/T = \Delta H_{\text{surroundings}}/T = \Delta S_{\text{surroundings}} \tag{3.A14}$$

and

$$-\Delta G_{\text{system}}/T = \Delta S_{\text{surroundings}} + \Delta S_{\text{system}} > 0. \tag{3.A15}$$

Since $T > 0$, if the total entropy increases the free energy of a closed system must decrease.

Chemical potential

What happens with, e.g., the Gibbs free energy of a system in which the amount of some substance i, given as the number of moles n_i, changes? Either the system is open and the substance i is transported across its boundaries, or a chemical reaction involving the substance takes place there. The change in the free energy G owing to the change in the content of i is a partial molal quantity (partial derivative with respect to the number of moles) and is called the chemical potential of i, μ_i:

$$\mu_i = (\partial G/\partial n_i)_{p,T,j\neq i} \tag{3.A16}$$

When taking the partial derivative it is understood that pressure, temperature and the amounts of other components in the system remain constant. It is convenient to express the chemical potential as a sum of the standard chemical potential μ, depending on pressure, temperature and various interactions of the particles of i with their surroundings, and a logarithmic function of the activity a_i:

$$\mu_i = \mu_i^\circ(p, T) + RT\ln a_i \tag{3.A17}$$

Sometimes, especially with water, it is useful to express the dependence of the chemical potential on pressure explicitly

$$\mu_i = \mu_i^\circ(T) + p\overline{V}_i + RT\ln a_i \tag{3.A18}$$

where $\overline{V}_i$, the partial molal volume of the substance i is another of the important partial molal quantities, equal to the volume corresponding to one mole of i.

Electrochemical Potential

With ions, the additional dependence on the electrostatic potential ϕ will be expressed explicitly in the so-called electrochemical potential $\tilde{\mu}_i$:

$$\tilde{\mu}_i = \mu_i^\circ(p, T) + RT \ln a_i + z_i F\phi \tag{3.A19}$$

where z is the valency of ion i and F the Faraday constant.

Let us use the concept of the electrochemical potential to derive the well-known condition for the Nernst–Donnan equilibrium potential between two phases $'$ and $''$. At equilibrium the electrochemical potentials of ion i must be the same in both phases, since then no decreases in free energy can result from the transport of the ion,

$$\mu_i^{\circ\prime}(p, T) + RT \ln a_i' + z_i F\phi' = \mu_i^{\circ\prime\prime}(p, T) + RT \ln a_i'' + z_i F\phi''. \tag{3.A20}$$

The standard potentials μ° are the same in the two phases if they represent aqueous solutions at the same pressure and temperature. Hence the equilibrium Nernst–Donnan potential will be

$$\phi'' - \phi' = (RT/zF) \ln (a_i'/a_i''). \tag{3.A21}$$

If more ionic species take part in Donnan equilibrium it is easy to see from this formula that, e.g.

$$a'_{K^+}/a''_{K^+} = a''_{Cl^-}/a'_{Cl^-} = (a'_{Ca^{2+}} a''_{Ca^{2+}})^{1/2}. \tag{3.A22}$$

Energetics of chemical reactions

Let us now consider changes in free energy related to chemical reactions. In the reaction

$$\nu_A A + \nu_B B = \nu_C C + \nu_D D \tag{3.A23}$$

where ν_i's are stoichiometric coefficients one can express the extent of the reaction $d\xi$ (a measure of its progress) as

$$d\xi = -dn_A/\nu_A = -dn_B/\nu_B = dn_C/\nu_C = dn_D/\nu_D. \tag{3.A24}$$

The overall change in the Gibbs free energy will be given by

$$(dG)_{p,T} = \mu_A\, dn_A + \mu_B\, dn_B + \mu_C\, dn_C + \mu_D\, dn_D. \tag{3.A25}$$

Setting $dn_A = -\nu_A\, d\xi, dn_B = -\nu_B\, d\xi$ etc., and dividing the last equation by $-d\xi$ one obtains

$$-(\partial G/\partial \xi)_{p,T} = \nu_A \mu_A + \nu_B \mu_B - \nu_C \mu_C - \nu_D \mu_D = \mathscr{A} \tag{3.A26}$$

where $\mathscr{A}$ is the affinity of the reaction. At equilibrium the changes in the free energy vanish, so that

$$\nu_A \mu_A + \nu_B \mu_B - \nu_C \mu_C - \nu_D \mu_D = 0 \tag{3.A27}$$

or, in general

$$\sum \pm \nu_i \mu_i = 0. \tag{3.A28}$$

When ν_A moles of substance A react with ν_B moles of B, giving rise to ν_C moles of C and ν_D of D, $\int d\xi = 1$ and

$$-\Delta G = \nu_A \mu_A + \nu_B \mu_B - \nu_C \mu_C - \nu_D \mu_D. \tag{3.A29}$$

Introduction of expressions for the chemical potential μ

$$\Delta G = -\nu_A \mu_A^\circ - \nu_B \mu_B^\circ + \nu_C \mu_C^\circ + \nu_D \mu_D^\circ + RT \ln \frac{a_C^{\nu C}\, a_D^{\nu D}}{a_A^{\nu A}\, a_B^{\nu B}} \tag{3.A30}$$

leads to

$$\Delta G = \Delta G^{\circ} + RT \ln \frac{a_{C}^{\nu C} a_{D}^{\nu D}}{a_{A}^{\nu A} a_{B}^{\nu B}} \tag{3.A31}$$

where ΔG^{0} is the standard free energy change. When all activities are equal to unity

$$\Delta G = \Delta G^{\circ}. \tag{3.A32}$$

At equilibrium $\Delta G = 0$ and

$$a_{C}^{\nu C} a_{D}^{\nu D} / a_{A}^{\nu A} a_{B}^{\nu B} = K \tag{3.A33}$$

where K is the *equilibrium constant*. It is thus obvious that

$$\Delta G^{\circ} = -RT \ln K. \tag{3.A34}$$

3.B REACTION KINETICS

Using the reaction discussed above (Eq. (3.A23))

$$\nu_{A}A + \nu_{B}B = \nu_{C}C + \nu_{D}D$$

its rate can be defined as

$$v = -\frac{1}{\nu_{A}}\frac{dc_{A}}{dt} = -\frac{1}{\nu_{B}}\frac{dc_{B}}{dt} = \frac{1}{\nu_{C}}\frac{dc_{C}}{dt} = \frac{1}{\nu_{D}}\frac{dc_{D}}{dt}. \tag{3.B1}$$

According to the experimentally established law of mass action the reaction rate depends on the active concentrations of reactants; in homogeneous systems on their actual concentrations. The rate of the forward reaction v_{12} is

$$v_{12} = k_{12} c_{A}^{\nu A} \cdot c_{B}^{\nu B} \tag{3.B2}$$

that of the back reaction

$$v_{21} = k_{21} c_{C}^{\nu C} \cdot c_{D}^{\nu D} \tag{3.B3}$$

where k's are rate constants which are temperature-dependent but independent of concentration. The overall reaction rate is $v = v_{12} - v_{21}$ and, at equilibrium, $v_{12} = v_{21}$, so that

$$c_{C}^{\nu} \cdot c_{D}^{\nu} / c_{A}^{\nu} \cdot c_{B}^{\nu} = k_{12}/k_{21} = K' \tag{3.B4}$$

where K' is the equilibrium constant defined in terms of concentrations.

The dimensions of the rate constants depend on the reaction order which is the sum of the exponents of the reactant concentrations. Thus, if $\nu_{A} = 1$ but $\nu_{B} = 0$ the reaction is first order and k_{12} is expressed in s^{-1} If $\nu_{A} = \nu_{B} = 1$, the reaction is second order and k_{12} is expressed in $M^{-1} s^{-1}$.

Such reactions, often termed bimolecular, can be quantitatively described by several approaches. The earliest is the one based on the empirical Arrhenius equation

$$k = A \exp(-E_{a}/RT) \tag{3.B5}$$

which is generally used to calculate the activation energy E_{a} of a reaction.

An equivalent but physically more sound expression can be derived either on the basis of the collision theory or of the theory of the transition or activated state.

The collision theory proceeds from an estimate of the number of collisions of the reacting partners and of the efficiency of the collisions that lead to a productive reaction.

In a gaseous phase the number of collisions Z_{AB} between reactants A and B can be shown to be

$$Z_{AB} = (r_{A} + r_{B})^{2}[8\pi k_{B} T(1/m_{A} + 1/m_{B})^{1/2}] c_{A} c_{B} \tag{3.B6}$$

where r_A and r_B are the radii of the reacting molecules, m_A and m_B their molecular masses, i.e. molecular weights divided by the Avogadro number (caution: if all units are in the SI framework molecular weights are also to be expressed in kg), and c_A and c_B are numbers of reactant particles per molar volume, i.e. $N_A/0.0224\,\mathrm{m}^3$ at normal pressure.

However, the efficiency of the collisions will be far from 100 per cent. The first reason is that the particles must possess a certain energy e_0 to react (the energy is much more effective if it is vibrational than translational), expressed by a Boltzmann distribution factor $\exp(-e_0/k_B T)$. The second reason is that they must possess proper orientation at the time of collision to make it productive. In the collision theory this is taken care of by a steric factor P so that the number of productive collisions Z^*_{AB} can be written as

$$Z^*_{AB} = PZ_{AB}\exp(-e_0/k_B T) \qquad (3.B7)$$

The reaction rate constant (related to molarities) is then

$$k_r = BT^{1/2}\exp(-E_0/RT) \qquad (3.B8)$$

where

$$B = N_A P(r_A + r_B)^2[8\pi R(1/M_A + 1/M_B)]^{1/2} \quad \text{with } M = mN_A.$$

The steric factor P attains values in the gaseous phase between 10^{-6} and 1.

A superior approach to the rate constant of bimolecular reactions proceeds from the theory of absolute reaction rates, according to which there is an equilibrium between the reactants A and B and the activated complex or transition state $AB^{\neq}$ at the top of the potential barrier. The bimolecular reaction is then

$$A + B \rightleftharpoons AB^{\neq} \rightarrow \text{products} \qquad (3.BI)$$

and one of the vibrational degrees of freedom of the activated complex is supposed to be equivalent to a translation along the reaction coordinate, i.e., formation of the products. The frequency of this degree of freedom is given by

$$\nu = k_B T/h \qquad (3.B9)$$

which, at normal temperatures, is of the order of $10^{13}\,\mathrm{s}^{-1}$! The forward reaction rate v_{12} is equal to the frequency multiplied by the concentration of the activated complex,

$$v_{12} = \frac{k_B T}{h} c_{AB} = k_{12} c_A c_B \qquad (3.B10)$$

so that

$$k_r = k_{12} = \frac{k_B T}{h}\frac{c_{AB}}{c_A c_B} = \frac{k_B T}{h} K^{\neq} \qquad (3.B11)$$

where $K^{\neq}$ is the equilibrium constant for the formation of the activated complex from the reactants. The last expression has a general validity; it is not limited to bimolecular reactions. Since each equilibrium constant is related to the standard free energy change in the process we can write

$$RT \ln K = -\Delta G^{\neq} \qquad (3.B12a)$$

where ΔG is the free energy of activation. Thus

$$K^{\neq} = \exp(-\Delta G^{\neq}/RT) \qquad (3.B12b)$$

and

$$k_r = \frac{k_B T}{h}\exp(-\Delta G^{\neq}/RT) = \frac{k_B T}{h}\exp(\Delta S^{\neq}/R)\exp(-\Delta H^{\neq}/RT) \qquad (3.B13)$$

which is the so-called Eyring equation.

Comparison of the three approaches shows that

$$k_r = A\exp(-E_a/RT) \qquad \text{Arrhenius (Eq. (3.B5))}$$
$$= BT^{1/2}\exp(-E_0/RT) \qquad \text{collision theory (Eq. (3.B8))}$$
$$= \frac{k_B T}{h}\exp(\Delta S^{\neq}/R)\exp(-\Delta H^{\neq}/RT) \qquad \text{transition-state theory (Eq. (3.B11)).}$$

Taking logarithms and differentiating with respect to T one obtains

$$E_a/RT^2 = \tfrac{1}{2}T + E_0/RT^2 = 1/T + \Delta H^{\neq}/RT^2 \tag{3.B14}$$

so that $E_a = \Delta H^{\neq} + RT$, while $E_0 = \Delta H^{\neq} + RT/2$ and, necessarily, $E_a > E_0 > \Delta H^{\neq}$. However, as the difference due to RT ($\approx 2.5\,\mathrm{kJ\,mol^{-1}}$) is small in relation to the usual values of 'activation' energy the three values can be generally taken as identical.

The pre-exponential factor in the Arrhenius equation is now equal to

$$\frac{ek_B T}{h}\exp(\Delta S/R)$$

and hence, its magnitude is related to entropy as one would intuitively expect of a quantity introduced to take account of the orientation and alignment of the reaction particles.

Diffusion-controlled reactions

The various theories of reaction rates discussed above were shown to provide satisfactory explanations of macroscopically measured rates in gases. The situation in a solution is quite different from that in the gaseous phase as the particles do not follow straight paths to randomly reach their reaction partner. They wind their way around the solvent molecules and their collision frequency is much reduced. However, there is a compensation for this in terms of the time the reaction partners stay near each other so that the probability of their reaction is greater.

To describe the situation it is useful to take an example from a single bimolecular reaction

$$\mathrm{A + B \underset{k'_d}{\overset{k_d}{\rightleftharpoons}} AB \xrightarrow{k_r} P} \tag{3.BII}$$

where AB may have the character of an activated complex. The rate of change of concentration of AB will be given by

$$\mathrm{d}c_{AB}/\mathrm{d}t = k_d c_A c_B - k'_d c_{AB} - k_r c_{AB}. \tag{3.B15}$$

In a stationary state of the reaction when c_{AB} is more or less constant then

$$c_{AB} = \frac{k_d}{k'_d + k_r} c_A c_B. \tag{3.B16}$$

The rate of product formation then

$$\mathrm{d}c_P/\mathrm{d}t = k_r c_{AB} = \frac{k_r k_d}{k'_d + k_r} c_A c_B. \tag{3.B17}$$

If the degradation of AB is much slower than the rate of product formation (i.e. $k'_d \ll k_r$) the complex coefficient reduces to k_d and the reaction is said to be *diffusion-controlled.* Diffusion control is probably present if $k_r \geqslant 10^9\,\mathrm{M^{-1}\,s^{-1}}$.

It should be mentioned here that the diffusion-controlled reaction described here are understood in a continuous phase or, at the best, from a continuous phase toward a membrane, but do not represent diffusion across membranes to be described in Section 3.3.

If, on the other hand k_r is small because of high activation energy involved (i.e. $k'_d \gg k_r$) the complex coefficient reduces to $k_r k_d / k'_d$ where the accumulation of energy required to overcome the activation barrier depends greatly on the surrounding solvent. These are the *activation-controlled* reactions.

The rate of a diffusion-controlled reaction can be derived in a fashion similar to that of the collision theory in gases and shown to be

$$v = \mathrm{d}c_{AB}/\mathrm{d}t = 4\pi(r_A + r_B)(D_A + D_B)N_A c_A c_B \tag{3.B18}$$

where

$$v/k_d = c_A c_B$$

and r_A and r_B are the radii of the reacting particles, D_A and D_B their diffusion coefficients and c_A and c_B their concentrations (caution: If all quantities are expressed in the SI framework, concentrations are to be expressed as mol m^{-3}).

Now the diffusion coefficient is approximately expressed by the Stokes–Einstein relation as

$$D = k_B T / 6\pi\eta r$$

with η the viscosity. If now the radii of interacting molecules are similar such that $r_A + r_B = 2r$ an extremely simple relation for k_d results

$$k_d = 8N_A k_B T/3\eta = 8RT/3\eta. \tag{3.B19}$$

Enzyme-catalyzed reactions

The reaction used above to describe the encounter of diffusing partners is the simplest reaction that is catalyzed by an enzyme. Enzymes are usually defined as (microheterogeneous) catalysts which emerge unchanged from the reaction but from the kinetic point of view the reaction of a substrate with an enzyme is a bimolecular reaction where the substrate (A) and enzyme (E) complex can be likened to an activated complex described in connection with Eyring's theory. In fact the AE complex usually undergoes several conformational changes, one of which does represent the activated (or transition) complex.

$$\mathrm{A} + \mathrm{E} \underset{k'_a}{\overset{k_a}{\rightleftharpoons}} \mathrm{AE} \xrightarrow{k_r} \mathrm{P} \tag{3.BIII}$$

Proceeding from the stationary-state assumption where $\mathrm{d}c_{AE}/\mathrm{d}t = 0$ and taking into account all the enzyme forms in a conservation equation ($c_{E_0} = c_E + c_{AE}$), the rate of product formation emerges as

$$v_A = \mathrm{d}P/\mathrm{d}t = c_{E_0}\frac{c_A}{c_A + K_m} \tag{3.B20}$$

where K_m, the Michaelis or half-saturation constant, is equal to $k_r + k'_a/k_a$.

Clearly, the rate depends linearly on enzyme concentration but hyperbolically on the concentration of substrate, such that at $c_A \ll K_m$ the dependence is linear (first-order) but at $c_A \gg K_m$ the rate is constant (maximum) and the reaction is zeroth order.

It will be seen subsequently that both membrane-bound enzymes and specific transports obey Michaelis-type kinetics, unless two or more substrate molecules are bound in a cooperative (either positive or negative) fashion (cf. p. 194).

3.1 MEMBRANE-BOUND ENZYMES

Of the roughly 2500 enzymes now recognized by biochemists, perhaps one-half are believed to be associated with intracellular structures, both of membrane and of cytoskeletal nature. Most of these enzymes would be classified as peripheral proteins as they are adsorbed on the structural support, apparently interacting mostly via hydrogen bonds between the $-NH_2$ and $-NH-$ groups of the peptide chain and the carbonyl groups of membrane fatty acyl esters, or electrostatically, between $-NH_3^+$ groups of proteins and phosphate residues of lipids, but some are distinctly integral components of the membrane. There may be several advantages derived from the membrane association.

1. The enzyme tertiary structure can be stabilized by this interaction and such enzymes, particularly those of thermophilic organisms, are less easily heat-denaturated than those free in solution.

2. One may assume that a certain orientation of the absorbed enzyme will facilitate access of substrate to its active site, this increasing the probability of a successful encounter.

Table 3.1. Enzymes activated by the presence of membrane lipids

Enzyme (source)	E.C. number	Activating lipid
3-Hydroxybutyrate dehydrogenase (beef-heart mitochondria)	1.1.1.30	PC
Malate oxidase (*Mycobacterium avium*)	1.1.3.3	CL
Pyruvate oxidase (*Escherichia coli*)	1.2.3.3	PC, FA, TG
Succinate dehydrogenase (beef-heart mitochondria)	1.3.99.1	CL, PE, PC
Acyl-CoA dehydrogenase (mitochondria)	1.3.99.3	TG, FA
NADH dehydrogenase (mitochondria)	1.6.99.3	PC, LPC
Cytochrome *c* oxidase (beef-heart mitochondria)	1.9.3.1	PE, CL, PC
Glucuronosyltransferase (liver microsomes)	2.4.1.17	Mixed lipids
Lipopolysaccharide galactosyltransferase (*Salmonella typhimurium*)	2.4.1.44	PE, PG, PS
Lipopolysaccharide glucosyltransferase I (*Salmonella typhimurium*)	2.4.1.58	PE, PA, PG
Undecaprenol kinase (*Staphylococcus aureus*)	2.7.1.66	PG, CL, PC, PE
Glucose-6-phosphatase (microsomes)	3.1.3.9	PE
Adenylate cyclase (liver cell plasma membrane)	4.6.1.1	PS, PC
Long-chain-fatty-acid-CoA ligase (mitochondria)	6.2.1.3	PC

Abbreviations: FA fatty acid, TG triglyceride, LPC lysophosphatidylcholine. Other abbreviations are defined in Table 2.2.

3. A distinct advantage over soluble enzymes is afforded by membrane localization of such enzymes as form parts of a reaction sequence in a metabolic pathway. Here the transit time that will be the rate-limiting factor in diffusion-controlled reactions can be reduced substantially, being virtually zero if the two enzymes in sequence are organized in such a way that the product of the first is immediately bound by the second as its substrate.

This third advantage really pertains more to enzymes that form an integral part of the membrane and it is these enzymes which are of particular interest in the membrane-mediated catalysis of reactions. A whole application field has recently emerged where enzymes are enclosed in artificial membrane or micellar structures and often function with greater efficiency and stability than in aqueous solution.

While some of these enzymes have an obvious justification for spanning the membrane, such as those involved in electron transport and ATP synthesis, or sterol and phospholipid synthesis, there are some where their integral localization is not readily understood, such as cytochrome P-450 in the endoplasmic reticulum.

Whatever the profit an enzyme may gain from being localized in a membrane some of them have adapted to the membrane environment during evolution and are in fact activated by the presence of lipids (Table 3.1).

The kinetics of membrane-bound enzymes is an area well explored in biotechnology where the (usually covalent) attachment of an enzyme to an artificial polymer support can bring about appreciable distortions of the tertiary structure and thus changes of both the half-saturation constant and the maximum rate of reaction are to be expected and, indeed, generally found (Table 3.2).

In native membranes no systematic studies have been done to compare the kinetic parameters of bound and free forms of the same enzyme but here again various possibilities exist and one will be able to predict the effect of binding to a support only when the actual atom–atom interactions can be estimated as is being done for a variety of proteins by advanced NMR techniques.

One property that distinguishes membrane-bound enzyme activities from the soluble ones is the dependence of their rate on temperature, specifically the discontinuity in the Arrhenius plot (Table 3.3). The discontinuities are often coincident with the main lipid transition of the membrane. A causal relationship

Table 3.2. Effect of membrane solubilization on the half-saturation constant K_m of some enzymes

K_m increased	Hexokinase	2.7.1.1
	Pyruvate oxidase	1.2.3.3
K_m decreased	Glucose-6-phosphatase	3.1.3.9
	Acetylcholinesterase	3.1.1.7
	Cytochrome-*c* oxidase	1.9.3.1
	ATP synthase	3.6.1.34
	Na, K-ATPase	3.6.1.37

Table 3.3. Discontinuities in Arrhenius plots of activities of membrane-bound enzymes

Enzyme	Transition temperature (°C)
Mitochondrial succinate dehydrogenase	16, 29
Mitochondrial ATP synthase	16
Mitochondrial cytochrome-*c* oxidase	26
Erythrocyte Na, K-ATPase	20
ATP synthase from *Acholeplasma* grown on stearic acid	18 (no cholesterol) 11 (normal membrane)
Sarcoplasmic reticulum Ca-ATPase	20
Erythrocyte acetylcholinesterase	20

is provided by experiments where culturing a cell in a particular mixture of fatty acids will shift the discontinuity according to the degree of saturation of the fatty acid. (However, not all breaks in Arrhenius plots are attributable to the membrane environment of a given enzyme—Na, K-ATPase is an example of many enzymes that show such breaks even when isolated and purified in solution.)

One general property of membranes does affect the kinetics of membrane-bound enzymes, viz. the surface charge reflected in a membrane surface potential. If the substrate carries an electric charge it will distribute itself between the aqueous medium and the membrane enzyme milieu according to $\exp(ze\phi/k_BT)$ where e is the elementary charge in C, ϕ the surface potential in V, and z the number of positive charges on the substrate. As the membrane surface potential is always negative it may be shown that the half-saturation constant measured with a membrane-bound enzyme K'_m is related to the 'true' Michaelis constant K_m by

$$K'_m = K_m \exp(ze\phi/k_BT) \tag{3.1}$$

Similarly, the pH dependence of the rate of a membrane-bound enzyme reaction will be different because again $c_{H^+(surface)}/c_{H^+(solution)} = \exp(-e\phi/k_BT)$ so that the optimum pH (for the maximum rate of enzyme reaction) will be shifted toward higher pH values.

SYNOPSIS

Many intracellular enzymes are associated with membranes, either as peripheral proteins or as deeply embedded integral proteins. Their association with membranes may render the enzymes more stable and give them a kinetic advantage if they are part of a metabolic sequence. The kinetics of individual enzymes are affected by surface potential and by local pH if charged substrates are involved.

FURTHER READING

Martinek, K., Levashov, A. V., Klyachko, N., Khmelnitski, Yu. L., and Berezin, I. V. Micellar enzymology, *Eur. J. Biochem.*, **155**, 453–68 (1986).

Sandermann, H. Regulation of membrane enzymes by lipids, *Biochim. Biophys. Acta*, **515**, 209–37 (1978).
Webb, E. C. (ed.). *Enzyme Nomenclature*, Academic Press, Orlando (1984).

3.2 MEMBRANE TRANSPORT

Together with chemical transformations, the translocation of matter, energy and electric charge is a fundamental physico-chemical characteristic of processes in living systems. There are two types of translocation phenomena, one represented by stochastic processes of large ensembles of uncharged molecules, ions and electrons, the other linked to the structure and conformational changes of individual (macro)molecules. The first group, including diffusion, migration and convection of matter, conduction of electric current and heat transfer, belong to the realm of classical physical and engineering sciences while the second group is specific for biology (as well as for polymer chemistry and physics).

Transport across membranes comprises both types of movement and they may occur either in a more or less pure form—transmembrane diffusion as a stochastic process and uptake of macromolecules as a structure-altering process, or in a mixed form—specific transport of large numbers of molecules where an intramolecular shift of the transporter is involved. In keeping with this classificaton, it is useful to divide membrane transport processes into three main categories: nonspecific permeation, specific transport, and translocation of macromolecules. Each group can be further subdivided, either on mechanistic grounds or on kinetic/thermodynamic grounds or on the basis of biosynthetic and genetic aspects (Table 3.4).

It will be seen in the following chapters that there exist a number of borderline cases, where the degree of specificity of a transport system is such that it can be considered either as nonspecific diffusion-type process or a specific protein-involving translocation. It is in these cases that the universal similarity of all transports of small molecules or ions becomes prominent.

In all sections on transport only movement in condensed, mainly liquid or liquid-crystalline phases will be considered as the gaseous reactants in biology ($O_2, CO_2, N_2, CO, H_2, CH_4, C_2H_4$) are not directly transported to the membrane surface because this is always covered with a layer of liquid.

SYNOPSIS

Processes of transport across membranes can be divided into several groups. The first is nonspecific permeation, often called simple diffusion, wherein molecules and ions cross either the lipid domains or flow through aqueous channels in a nonspecific way and are driven solely by the difference of their chemical or electrochemical potential at the two membrane sides. The second group is specific, saturable transport mediated by specific channels or specific carriers in the membrane. The third group are movements of macromolecules which require a local reorganization of the membrane to proceed.

Table 3.4. Classification of transport mechanisms

Mechanism	Example
A. Nonspecific permeation	
1. By direct passage through membrane core	Medium- and low-polarity solutes
2. By passage through aqueous pores	Small water-soluble molecules and ions
3. By flow through statistical pores	Small water-soluble molecules and ions
B. Specific transport	
1. Via specific channel-like proteins	
(a) Nongated	Anions in porins
(b) Gated	
(ba) By ligands	K^+ in some nerves
(bb) By voltage	Na^+ in nerves
2. Via carrier-like proteins	
(a) By mediated diffusion	
(aa) Of uniport type	Monosaccharides in erythrocytes
(ab) Of antiport type	ADP/ATP in mitochondria
(b) By an energy-linked process	
(ba) Primary active	
(baa) With ATP hydrolysis	Ions through ATPases
(bab) With use of oxidation free energy	H^+ in mitochondria
(bac) With capture of photons	H^+ in bacteriorhodopsin
(bad) With decarboxylation	Na^+ in some bacteria
(bb) Secondary active	
(bba) In symport with H^+	Lactose in bacteria
(bbb) In symport with Na^+	Glycine in tumours
3. Via group translocation	
(a) Phosphorylating	Sugars in bacteria
(b) Hydrolyzing and ligating	Amino acids on γ-glutamyltranspeptidase
C. Transport of macromolecules	
1. Without vesiculation	Some toxins in target cells
2. With vesiculation	
(a) By fluid-phase endocytosis	Nutrients in amoebae
(b) Via coated pit internalization	Transferrin in liver cells
(c) Via exocytosis	Neurotransmitters at nerve end plates
3. By an intermediate mechanism	Influenza virus in target cell

The terms *uniport*, *symport* and *antiport* used in the table are operational terms denoting transport of a single solute (uniport), of a solute accompanied by another, usually driving, ion (symport), and of a solute in one direction in exchange for a solute in the opposite direction (antiport).

3.3 NONSPECIFIC PERMEATION

3.3.1 General aspects

This is the type of transport that takes place across most, if not all, natural as well as artificial membranes, during which *no specific interaction* between the permeating molecule or ion and the membrane components takes place. It is obvious that interactions of physico-chemical nature are a part of the translocation process but they are never so selective as to distinguish between optical isomers, positional isomers or enantiomers. The only general limitation may be the size of the molecule but this again is highly dependent on the particular mode of permeation, of which there may be several different kinds.

The nonspecific permeation mechanisms have often been termed simple diffusion but this may be a misleading designation as some of the mechanisms are complicated indeed and most of them involve diffusion that may be so described kinetically but not mechanistically. In fact, some of them, such as the single-file diffusion, cannot be described by linear equations relating the rate of unidirectional flow to concentration of permeating solute.

Generally, however, the process is characterized by first-order kinetics where the rate of transport of an uncharged solute is proportional to the difference in concentrations at the two membrane sides.

To appreciate the significance of the rate constants used in the subsequent text an excursion into the physical chemistry of diffusion may be of help.

3.C PRINCIPLES OF TRANSPORT IN THE CONTINUOUS PHASE

The simplest molecular mechanism of diffusion of electrically neutral particles of a solute in a liquid phase is based on the quasicrystalline structure of the solvent. The solute molecules, which are situated at the nodes of the rapidly changing crystalline structure of the liquid or in interstitial positions of this structure, oscillate around equilibrium positions. After several hundreds of vibrations they attain an activation energy sufficient for a jump to a neighbouring position in the quasicrystalline lattice, provided there is a free position for them. Thus, the jump frequently depends on the activation energy necessary for releasing the particle from its equilibrium position and on the activation energy for forming a vacancy in the neighbouring position in the solvent. The frequency of these jumps is between 10^{10}–$10^{11}\ \mathrm{s}^{-1}$ at room temperature. If the process of diffusion occurs in a space of dimensions sufficiently larger than the length of the molecules ($\simeq 1$ nm) it is described by Fick's First Law for material flux $\mathbf{J}_{\mathrm{diff}}$ ($\mathrm{mol\,m^{-2}\,s^{-1}}$)

$$\mathbf{J}_{\mathrm{diff,i}} = -\operatorname{grad}(D_i c_i) \tag{3.C1}$$

where D_i is the generally concentration-dependent diffusion coefficient ($\mathrm{m^2\,s^{-1}}$) and c_i the concentration of the *ith* component of the system. If D_i does not depend on concentration the usual form of Fick's First Law is obtained, viz.

$$\mathbf{J}_{\mathrm{diff,i}} = -D_i \operatorname{grad} c_i \tag{3.C2}$$

or, if concentration only changes along the x-coordinate,

$$\mathbf{J}_{\mathrm{diff,i}} = -D_i\, \mathrm{d}c_i/\mathrm{d}x. \tag{3.C3}$$

For a more general treatment of transport processes on the basis of linear irreversible thermodynamics (see Section 3.E) it is convenient to introduce the chemical potential gradient of the solute as the driving force of diffusion. Equation (3.C2) is then transformed to

$$\mathbf{J}_{\mathrm{diff},i} = -u_i c_i \operatorname{grad} \mu_i \tag{3.C4}$$

where u_i is the mobility of the particle ($\mathrm{m\,s^{-1}\,N^{-1}}$, i.e. velocity at unit force) and μ_i the chemical potential of the ith component

$$\mu_i = \mu_i^{\circ} + RT \ln a_i \tag{3.C5}$$

where R is the gas constant, T the absolute temperature, a_i the activity of the ith component, for a dilute solution numerically equal to its concentration, and μ_i° the standard chemical potential, independent of c_i. In Einstein's very simple treatment the Stokes–Einstein equation is obtained

$$u_i \simeq N_A / 6\pi\eta r_i \tag{3.C6}$$

where N_A is the Avogadro constant, η the viscosity of the medium and r_i the radius of the (supposedly spherical) particle.

In a system with membranes simple diffusion of an electrically neutral solute may be expected if it proceeds:

(i) from the bathing solution to the cell surface, provided that convection does not interfere with it,
(ii) from the intercellular space to the membrane surface (if the membranes are not in direct contact) or, possibly,
(iii) from the protoplast to the membranes of cell organelles or to the plasma membrane.

In the case of diffusion across a natural or model lipid membrane the process as a whole cannot be exactly described by Fick's First Law valid for a homogeneous system as the solute particle has to cross the region of polar heads of phospholipids and then to diffuse through the 'hydrocarbon' region. Doubtless this part of the process as a whole requires a more detailed treatment, for example, by Eyring's approach (see p. 143).

On applying the condition of material balance and the theorem of divergence to Eq. (3.C2) Fick's Second Law is obtained, written both in vectorial form and in Cartesian coordinates,

$$\frac{\partial c_i}{\partial t} = D_i \operatorname{div} \operatorname{grad} c_i = D_i\left(\frac{\partial^2 c_i}{\partial x^2} + \frac{\partial^2 c_i}{\partial y^2} + \frac{\partial^2 c_i}{\partial z^2}\right) \tag{3.C7}$$

or, for the case of linear diffusion,

$$\frac{\partial c_i}{\partial t} = D_i \frac{\partial^2 c_i}{\partial x^2}. \tag{3.C8}$$

For diffusion translocation along a plane, e.g. a membrane surface, we have

$$\frac{\partial \Gamma_i}{\partial t} = D_i^{\mathrm{p}}\left(\frac{\partial^2 \Gamma_i}{\partial x^2} + \frac{\partial^2 \Gamma_i}{\partial y^2}\right) \tag{3.C9}$$

where the Γ's are surface concentrations of the ith component.

For an appropriate description of a given system and a given process, initial and boundary conditions are required in addition to the partial differential Eqs (3.C7) and (3.C8). For example, let us have a planar interface (e.g. a membrane surface) from which the liquid phase extends to regions unaffected by diffusion. When the diffusion process starts the liquid contains no solute while its concentration at the interface has a definite constant value during the diffusion process. Then, Eq. (3.C8) is supplemented by the initial condition

$$t = 0, \qquad x > 0 \; : \; c_i = 0 \tag{3.C10}$$

and by two boundary conditions,

$$t > 0, \qquad x = 0 \; : \; c_i = c^0 \tag{3.C11}$$

$$t > 0, \qquad x \to \infty \; : \; c_i = 0. \tag{3.C12}$$

The concentration distribution as a function of time and distance from the interface is obtained by solving the system of Eqs (3.C8) to (3.C11) in the form

$$c_i = c^0[1 - \mathrm{erf}(x/2D_i^{1/2}t^{1/2})]. \tag{3.C13}$$

The error function erf is defined as

$$\mathrm{erf}(z) = 2/\pi^{1/2} \int_0^z \mathrm{e}^{-y^2}\,\mathrm{d}y = 1 - 2/\pi^{1/2} \int_z^\infty \mathrm{e}^{-y^2}\,\mathrm{d}y. \tag{3.C14}$$

It is often important to know the material flux across the interface, i.e., the concentration gradient $(\partial c_i/\partial x)_{x=0}$ This is readily obtained by differentiation of Eq. (3.C13) as

$$(\partial c_i/\partial x)_{x=0} = -c^0/(\pi D_i t)^{1/2}. \tag{3.C15}$$

The diffusion flux then is

$$\mathbf{J}_{\mathrm{diff,i}} = -D_i(\partial c_i/\partial x)_{x=0} = c^0(D_i/\pi t)^{1/2}. \tag{3.C16}$$

The concentration gradient is the slope of the tangent drawn at a given point of the curve describing the dependence of concentration on distance, in the present case, for $x = 0$. The corresponding tangent intersects the x-coordinate at a distance of

$$\delta_D = (\pi D_i t)^{1/2} \tag{3.C17}$$

from the origin. This distance is termed the effective *diffusion layer thickness* (cf. Fig. 3.1). It is a measure of the region affected by diffusion.

The mean displacement of particles in the direction of one, two or all three coordinate axes can be derived on the basis of the following consideration. Due to the complete randomness of the diffusion process there is as much movement forward as back and the arithmetical mean of displacements for any greater number of particles must approach zero. However, the square root of the arithmetic mean of squared displacements yields the required result. An elegant solution of the problem was given in a concise way by Einstein and subsequently elaborated by Fürth (cf. 'Further reading').

For the description of the diffusion process Eq. (3.C8) will again be used. In contrast to the situation described by the initial and boundary conditions (Eqs (3.C10–12)), the case of an instantaneous planar source situated in the origin will be considered, i.e. the material subject to diffusion is concentrated in a tiny planar layer at $x = 0$. Thus the initial condition is

$$t = 0, \qquad x = 0 \; : \; c_i \to \infty \tag{3.C18}$$

$$t = 0, \qquad x \neq 0 \; : \; c_i = 0. \tag{3.C19}$$

The boundary condition given in an integral form describes the fact that the number of particles in the system is limited

$$t > 0 \; : \; \int_{-\infty}^{+\infty} c_i\,\mathrm{d}x = n \tag{3.C20}$$

where n is the number of moles per unit area. The solution of this system of equations is

$$c_i = n/(4\pi Dt)^{1/2} \exp(-x^2/4Dt). \tag{3.C21}$$

Division of this expression by the total number of particles yields the probability of an individual particle being displaced by a distance x during time t. Multiplication of the probabilities by squares of the corresponding displacements and integrating between minus infinity and plus infinity provides a definition of the arithmetic mean of the squares of displacements (the mean squared displacement) $\overline{x^2}$:

$$\overline{x^2} = \frac{1}{r}\int_{-\infty}^{+\infty} c(x,t)\,\mathrm{d}x = \frac{1}{(4\pi Dt)^{1/2}}\int_{-\infty}^{+\infty} \exp(-x^2/4Dt)x^2\,\mathrm{d}x = 2Dt. \tag{3.C22}$$

The mean displacement in the direction of the x-axis is thus

$$\sqrt{\overline{x^2}} = \sqrt{2Dt} \tag{3.C23}$$

and if the movement of particles were limited to one dimension only it would be equal to the total displacement. When, however, the diffusion occurs in three-dimensional space the mean total displacement $\sqrt{\overline{r^2_{\mathrm{xyz}}}}$ will be

$$\sqrt{\overline{r^2_{\mathrm{xyz}}}} = \sqrt{\overline{x^2} + \overline{y^2} + \overline{z^2}} \tag{3.C24}$$

and since in isotropic space $\overline{x^2} = \overline{y^2} = \overline{z^2}$

$$\sqrt{\overline{r^2_{\mathrm{xyz}}}} = \sqrt{3\overline{x^2}} = \sqrt{6Dt}. \tag{3.C25}$$

The typical value of the diffusion coefficient for small molecules in an aqueous solution is about $10^{-9}\,\mathrm{m^2\,s^{-1}}$. Hence, the mean total displacement $\sqrt{6Dt}$ for such a molecule is 2.45 μm in 1 ms, 77 μm in 1 s, 600 μm in 1 min and 4.6 mm in 1 h.

The mean total displacement in two-dimensional diffusion (such as lateral diffusion in a membrane) may be calculated analogously:

$$\overline{r^2_{\mathrm{xy}}} = \overline{x^2} + \overline{y^2} = 4Dt. \tag{3.C26}$$

In all cases is the displacement proportional to the square root of the time rather than to time itself and, as a result of this, the diffusing particles overcome short distances very rapidly and longer ones rather slowly. Over the short distances of cellular dimensions, the amount of substance transferred by diffusion is quite substantial. Thus, for a millimolar concentration of a substance given in the above example, moving across a planar membrane surface ($A = 100\,\mu\mathrm{m}^2$) the amount of substance transferred through this surface between $t = 1$ ms and $t = 2$ ms is

$$M = AD_{\mathrm{i}}^{1/2}\int_{t_1}^{t_2} c^{\circ}/(\pi t)^{1/2}\,\mathrm{d}t = 2Ac^{\circ}D_{\mathrm{i}}^{1/2}\pi^{-1/2}(t_2^{1/2} - t_1^{1/2}) = 4.6\cdot 10^{-16}\,\mathrm{mol}$$

which compares with the total amount of substance of 1 mM concentration in a cell with a volume of 500 $\mu\mathrm{m}^3$. In the case of rapid membrane processes (for example, with excitable cells, photosensitive cells to which short illumination pulses are applied, etc.) diffusion alone can supply a surprisingly large amount of material.

Over short distance such as between two neighbouring cells the model of diffusion in semi-infinite space described above is inadequate. In these cases steady-state diffusion in a limited space is rapidly established. If the concentration at one interface is c_1^0 and at the other, situated at distance d, c_2^0, then Eq. (3.C8) for the steady-state condition $\partial c/\partial t = 0$ gives (cf. Eq. (3.2))

$$J_{\mathrm{diff}} = D(c_2^0 - c_1^0)/l. \tag{3.C27}$$

Again, for the small distance between the interfaces the diffusion flux attains rather high values.

So far a resting medium has been considered but this assumption has its grave limitations. Biological systems are, generally speaking, always in a state of motion. In transport process terminology the motion (streaming) of a medium evoking transport of matter or energy is called *convection*. A medium may be put in motion by an external mechanical force, for example, in a large system, such as a rotating microbial fermentor or the very small system of cilia at a cell surface. This case is termed *forced convection*. On the other hand, *natural convection* is a frequent phenomenon connected with formation of even slight density gradients due to other transport phenomena (diffusion, heat conduction). In this case the hydrodynamic flow is driven by the force of gravity. Its cause is frequently in the imperfect temperature maintenance of the liquid phase which is indeed hard to bring about.

Transport of matter by convection to an interface which is impenetrable for medium flow cannot work alone, without cooperation of diffusion. Close to the interface the streamlines of the medium are parallel to the interface but immediately at the interface the velocity of the medium is the same as the velocity of the adjacent other phase (if this is at rest the medium velocity is also zero). The decrease of the velocity of motion of the medium in the direction perpendicular to the interface is due to counterbalancing of the momentum of the moving medium and of the viscous forces. Under these conditions the final step of transport is diffusion (termed *convective diffusion*).

Very frequently the presence of a 'stagnant layer' of the medium at the membrane surface is assumed where diffusion alone should occur. This idea comes from Nernst but was later corrected by investigators of fluid dynamics. As already pointed out a layer of decreasing medium velocity is formed at the interface (the *hydrodyamic* or *Prandtl layer*). The motion of the medium can remove the concentration gradients at the 'outer edge' of the hydrodynamic layer while closer to the interface a *diffusion* or *Nernst layer* is formed which is characterised by a drop or increase of concentration (Fig. 3.1).

The material flux due to convection is described by the equation

$$\mathbf{J}_{\text{conv}} = c_i \mathbf{v} \tag{3.C28}$$

where $\mathbf{v}$ is the velocity of the medium (m s^{-1}). By combining Eqs (3.C3) and (3.C28) and by

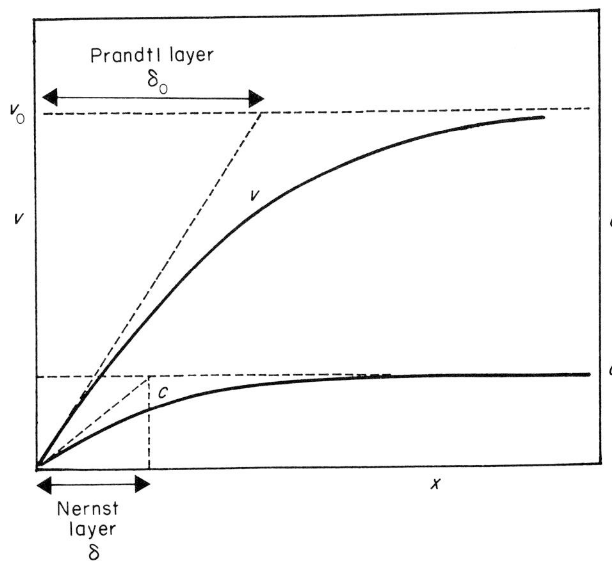

Fig. 3.1. Schematic representation of flow rate v, concentration of solute c_s at distance x from an immobile phase boundary

applying the theorem of divergence to the result a partial differential equation of convective diffusion is obtained, which, if complete incompressibility of the medium is assumed (div $\mathbf{v} = 0$), yields

$$\partial c_i/\partial t = D_i \operatorname{div} \operatorname{grad} c_i - \mathbf{v} \operatorname{grad} c_i$$
$$= D_i\left(\frac{\partial^2 c_i}{\partial x^2} + \frac{\partial^2 c_i}{\partial y^2} + \frac{\partial^2 c_i}{\partial z^2}\right) - v_x \frac{\partial c_i}{\partial x} - v_y \frac{\partial c_i}{\partial y} - v_z \frac{\partial c_i}{\partial z} \quad (3.C29)$$

where v_x, v_y and v_z are the velocity components in the direction of the coordinates of the system. In general, the solution of this equation for a specified case requires the knowledge of the velocity $\mathbf{v}$ as a function of coordinates. This should be obtained from the fundamental relationship of hydrodynamics, the Navier–Stokes equation, but the solutions of such partial differential equations are a very difficult task. Numerical approximate approaches are usually indispensable. The assumption of a steady state prevails in this analysis.

As a comparatively simple example the convective diffusion to a planar plate of width l in the direction of flow in a streaming liquid will be considered. In this case the thickness of the Prandtl layer is

$$\delta_0 \simeq \sqrt{\eta l/\rho v_0} \quad (3.C30)$$

where η is the viscosity of the medium, ρ its density and v_0 the velocity of the medium at a large distance from the interface; the thickness of the Nernst layer, to be identified with the so-called unstirred layer

$$\delta_N \simeq (\eta l^3 D_i^2/\rho v_0^3)^{1/6} \quad (3.C31)$$

i.e. about one-tenth of the Prandtl layer thickness in the case of an aqueous solution. Thus, the convective diffusion flux is given by

$$J_{cd} = D_i(c_i^0 - c_i^*)/\delta \quad (3.C32)$$

where c_i^* is the concentration close to the interface.

The effect of this stagnant layer on permeability values measured experimentally will be discussed on p. 149 and on p. 210.

As already stated, in the transient case of diffusion the diffusion-layer thickness is very small at the initial stage. Under these conditions diffusion proceeds in the region where, due to the proximity of the interface, the motion of the medium does not affect it and the simple laws of diffusion can be applied. An analogous situation arises in the case of transport across thin intercellular compartments or in not exceedingly thin pores, say, of a cell wall (the case of channels with cross-sections of molecular dimensions is discussed in Section 3.4). Here, like in other small volumes, very often the viscous forces prevail over the external forces to such a degree that simple diffusion is possible.

In the case of biological important liquid phases which are more or less homogeneous the *transport of electric charge* is exclusively connected with the transport of matter (electrolytic transport) as practically no homogeneous electronically conducting phase is known in biological systems (the tiny magnetite magnets present in membranes of some bacteria being an exception, see p. 321). Electron-transfer phenomena in membrane enzyme systems will be treated in Section 4.3.

In the presence of an electric field in a phase containing dissolved ions the molecular meachanism of transport of the ions is the same as described for electrically neutral solutes but the jumps in the direction of the field are slightly preferred to the jumps in other directions. The influence of the field in comparison with the thermal motion is small but steady. For example, an external field of 1 V m^{-1} causes in an aqueous solution an effective translocation of ions at a rate of the order of 50 nm s^{-1} while the instantaneous velocities of the thermal motion of ions are of the order of 100 m s^{-1}.

The transport of protons in continuous phase is connected with their properties as a quantum particle. In an aqueous solution they are transferred by tunnelling between neighbouring water molecules. The activation of this kind of transport depends on the rotation of the donor oxonium (H_3O^+) and water particles. Nevertheless, due to tunnelling, proton transport is considerably faster than it would be if the whole H_3O^+ particle were to move.

The electric fields appearing in homogeneous parts of biological systems are usually rather weak. On the other hand, in the diffuse electric layer, which is a part of the electric double layer at the membrane stretching into the interior of the adjacent solution, very strong electric fields are formed. The treatment of this case has already been discussed in Section 2.7.

The fundamental equation describing electrodiffusion, i.e. simultaneous diffusion and ion transport in an electric field (migration) is formulated in the same way as Eq. (3.C4) if the electrochemical potential $\tilde{\mu}_i$ is substituted for the chemical potential μ_i. In order to avoid confusion which deplorably appears in some biological texts let us first present an appropriate definition of this quantity. Guggenheim's electrochemical potential $\tilde{\mu}_i(\alpha)$ is the reversible work connected with releasing 1 mol of the ith charged component from the interior of phase α and transferring it to an infinitely diluted gaseous phase. If the total Gibbs energy (both chemical and electric) of phase α is denoted $\tilde{G}$ and the amount of the ith component is n_i, then

$$\tilde{\mu}_i(\alpha) = (\partial \tilde{G}/\partial n_i)_{p,T,n_{j \neq i}} \tag{3.C33}$$

where p denotes pressure, T absolute temperature and $n_{j \neq i}$ the amount of a component different from the ith one.

Usually, the electrochemical potential is separated into its chemical component (the chemical potential) and the electric component (the inner electric potential ϕ of phase α multiplied by the product of the charge number of component z_i and the Faraday constant F),

$$\tilde{\mu}_i(\alpha) = \mu_i(\alpha) + z_i F \phi(\alpha). \tag{3.C34}$$

In the case of transport, i.e. of a system out of equilibrium, the electric potential ϕ as well as the concentration c_i are functions of the coordinates of the system. Then on substituting $\tilde{\mu}_i(\alpha)$ from Eq. (3.C34) for μ_i in Eq. (3.C4) and in view of Eq. (3.C5) the Nernst–Planck equation is obtained,

$$\begin{aligned} J_{ed,i} &= -u_i c_i \,\mathrm{grad}\, \tilde{\mu}_i = -RTu_i \,\mathrm{grad}\, c_i - z_i F u_i \,\mathrm{grad}\, \phi \\ &= -D_i \,\mathrm{grad}\, c_i - (z_i/|z_i|) U_i \,\mathrm{grad}\, \phi. \end{aligned} \tag{3.C35}$$

Thus, the diffusion coefficient D_i is given by

$$D_i = RTu_i \tag{3.C36}$$

and the electrolytic mobility U_i is

$$U_i = |z_i| F u_i. \tag{3.C37}$$

These quantities are connected by the Nernst–Einstein equation

$$D_i/U_i = RT/|z_i|F. \tag{3.C38}$$

In general, the transport of charged particles may occur in the presence of electric current flowing through the system as well as in its absence. The partial current density $\mathbf{j}_i$ corresponding to the transport of an individual ionic species, is equivalent to its electrodiffusion flux

$$\mathbf{j}_i = z_i F \mathbf{J}_{ed,i}. \tag{3.C39}$$

The total current density then is

$$\mathbf{j} = \sum \mathbf{j}_i = \sum z_i F \mathbf{J}_{ed,i} \tag{3.C40}$$

which is a general formulation of Faraday's Law. In the absence of concentration gradients, charge transport proceeds by electric conduction alone,

$$\mathbf{j} = -\sum z_i F c_i u_i \operatorname{grad} \phi = -\sum \lambda_i c_i \operatorname{grad} \phi = -\kappa \operatorname{grad} \phi \tag{3.C41}$$

where λ's are individual ionic conductivities and κ is the conductivity of the electrolyte. Equation (3.C40) is also valid in the absence of electric current, giving

$$\mathbf{j} = 0 = \sum z_i F \mathbf{J}_{ed,i} = -\sum z_i F D_i \operatorname{grad} c_i - \kappa \operatorname{grad} \phi_{diff}. \tag{3.C42}$$

Thus, for the diffusion potential, ϕ_{diff}, dependent only on the x-coordinate,

$$\frac{\partial \phi_{diff}}{\partial x} = -\frac{\sum z_i F D_i \mathrm{d}c_i/\mathrm{d}x}{\kappa}. \tag{3.C43}$$

The theory of the diffusion potential played an important role particularly in the pioneering stage of membranology, when the membrane potential was considered as a sort of diffusion potential. While at present this approach is still used a molecular kinetic treatment is mostly preferred.

3.3.2 Diffusion of nonelectrolytes across membranes

Although the equations that describe the rate of nonspecific transport are akin to Fick's First Law of diffusion, on the one hand, and to Eyring's equation, on the other, depending on the molecular mechanism, there are a number of features that distinguish diffusion across a membrane from that in the continuous phase.

Let us examine first the permeability constant P (related to the diffusion coefficient) of a transported substance. As shown in Fig. 3.2 diffusion of solute S

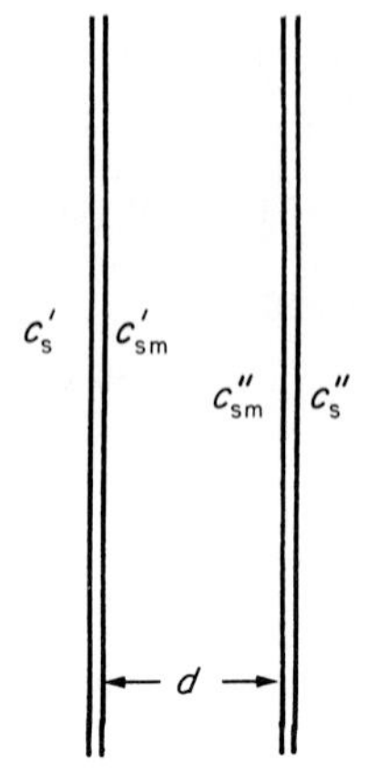

Fig. 3.2. Concentrations of a solute at the solution-membrane interface

across a membrane occurs from concentration c'_{sm} to concentration c''_{sm}. Fick's First Law for the steady state in a membrane (cf. Eq. (3.C27)) can be written as

$$J_s = -D\frac{\Delta c_{sm}}{x} = -D\frac{c''_{sm} - c'_{sm}}{d} = D\frac{c'_{sm} - c''_{sm}}{d} \tag{3.2}$$

where d is the membrane thickness and D the diffusion coefficient. From the equilibrium assumption, the concentration just inside the membrane surface is related to the external concentration by the partition coefficient

$$c'_{sm} = \beta c'_s, \qquad c''_{sm} = \beta c''_s \tag{3.3}$$

so that

$$J = \frac{\beta D}{d}(c'_s - c''_s) = P(c'_s - c''_s). \tag{3.4a}$$

If the initial rate of transport is measured such that

$$c''_s = 0, \qquad J_{so} = Pc'_s, \tag{3.4b}$$

hence a single linear relationship between rate and concentration is obtained.

This interpretation of the permeability constant (with $DM_r^{1/3}$ = constant, from the Stokes–Einstein equation for spherical molecules) was found adequate for several small, medium-hydrophilic molecules within a narrow range of molar refraction values where a linear relationship was found between $\log PM_r^{1/3}$ and $\log \beta$.

This Stokesian diffusion, however, has not been really satisfactory in explaining the rates of diffusion of molecules, unless they were relatively hydrophobic. In most cases the dependence of permeability constant on molecular volume is much steeper and one speaks of non-Stokesian diffusion which is characterized by the fact that the solvent molecules here cannot flow freely around the diffusing molecule.

As in a membrane the diffusion process is apparently governed both laterally and transversely by the size and number of vacancies attainable by the diffusing molecule and as the sizes of these holes are obviously distributed with a bias toward small holes the overall diffusion coefficient is given by

$$D = k\exp(-V/\bar{V}) \tag{3.5}$$

where V is the volume of the diffusing molecule and $\bar{V}$ is the mean volume of the vacancy.

From this the diffusion coefficient can vary with different proportionality constants for different membranes but can change quite steeply with molecular volume. In human erythrocytes a molecule with a molar volume of $2\cdot 10^{-5}\,\mathrm{m^3\,mol^{-1}}$ will have a D of $2\cdot 10^{-10}\,\mathrm{m^2\,s^{-1}}$ while one with a molar volume of $4\cdot 10^{-5}\,\mathrm{m^3\,mol^{-1}}$ will have a D of $10^{-11}\,\mathrm{m^2\,s^{-1}}$.

Before proceeding to an analysis of the microscopic parameters of a transmembrane diffusion process the partition of molecules between the membrane and the surrounding solution should be described in some detail.

3.D PARTITION EQUILIBRIUM

Two liquid phases in contact, no matter how different in solubility properties they are, are never completely immisicible so that water, a strongly polar solvent, slightly dissolves even in a typical nonpolar solvent, such as liquid hydrocarbons. When a nonelectrolyte *i* dissolves in such two phases and the distribution (partition) equilibrium is reached the chemical potential of this substance must be equal in both phases,

$$\mu_{i,w} = \mu_{i,o} \tag{3.D1}$$

giving, for a dilute solution, Nernst's distribution law,

$$\frac{c_{i,o}}{c_{i,w}} = \exp[(\mu^{\circ}_{i,w} - \mu^{\circ}_{i,o})/RT] = \beta_i^{o,w} \tag{3.D2}$$

where *w* and *o* denote the aqueous and the organic phase, respectively, c_i's are the concentrations of the species *i*, μ's are their standard chemical potentials and $\beta_i^{o,w}$ is the partition coefficient of *i*.

As the standard chemical potentials of a substance in different media differ exclusively because of its interaction with each of the media the difference $\mu^{\circ}_{i,w} - \mu^{\circ}_{i,o}$ is equal to the difference of standard (molar) Gibbs energies of solvation of the substance. This difference is termed the standard (molar) Gibbs transfer energy of *i*

$$\Delta G^{\circ,w\to o}_{tr,i} = \mu^{\circ}_{i,o} - \mu^{\circ}_{i,w}. \tag{3.D3}$$

Similarly, when an electrolyte BA is distributed between water and another high-permittivity solvent immiscible with water the distribution equilibrium is described by

$$\begin{aligned}\beta^{o,w}_{BA} &= \exp[(\mu^{\circ}_{BA,w} - \mu^{\circ}_{BA,o})/RT] \\ &= \left(\frac{a_{B^+,o}\, a_{A^-,o}}{a_{B^+,w}\, a_{A^-,w}}\right)^{1/2} = \frac{a_{\pm,o}}{a_{\pm,w}}\end{aligned} \tag{3.D4}$$

where the *a*'s are the activities of ions B^+ and A^- in water (*w*) and in the other solvent (*o*), and the $a_{\pm}$'s are the mean activities of the electrolyte BA. For a dilute solution

$$\beta^{o,w}_{BA} = \left(\frac{c_{B^+,o}\, c_{A^-,o}}{c_{B^-,w}\, c_{A^-,w}}\right)^{1/2} = \frac{c_{BA,o}}{c_{BA,w}}. \tag{3.D5}$$

Analogously, like in the preceding case,

$$\Delta G^{\circ,w\to o}_{tr,BA} = \mu^{\circ}_{BA,o} - \mu^{\circ}_{BA,w}. \tag{3.D6}$$

To find the change of energy connected with the transfer of an individual ionic species from one medium to the other requires a definite extrathermodynamic procedure for separating the quantity $\Delta G^{\circ,w\to o}_{tr,BA}$ into the components describing the solvation change of these ions following the transfer. For membranology, the most useful is Parker's TATB assumption stating that the standard Gibbs transfer energy of tetraphenylarsonium cation and tetraphenylborate anion is equal between any pair of solvents, in the present case,

$$\Delta G^{\circ,w\to o}_{tr,TPAS^+} = \Delta G^{\circ,w\to o}_{tr,TPB^-} = \Delta G^{\circ,w\to o}_{tr,TPASPB}. \tag{3.D7}$$

The value of the standard Gibbs transfer energy of the salt, tetraphenylarsonium tetraphenylborate can easily be determined, for example, by a partition equilibrium study or by electrolysis at the interface at two immiscible electrolyte solutions (ITIES). Once the standard Gibbs energies for TPA^+ and TPB^- are known these quantities are easily determined for any ion by estimating the partition coefficient of the salt of that particular ion with TPA^+ or TPB^-. In this way scales of standard Gibbs transfer energies are built up (Table 3.5). As the standard Gibbs transfer energy is supposed to be equal to the difference of solvation Gibbs energies in the media concerned the zero of such a scale corresponds to

Table 3.5. Standard Gibbs energies of ion transfer from water to nitrobenzene

Ion	$-\Delta G_{tr,i}^{\circ,w\to o}$ (kJ mol^{-1})
H^+	−32.5
Li^+	−38.4
Na^+	−34.5
K^+	−24.3
Rb^+	−19.9
Cs^+	−15.5
Mg^{2+}	−71.4
Ca^{2+}	−68.3
Sr^{2+}	−67.2
Ba^{2+}	−63.3
NH_4^+	−27.4
Choline$^+$	−11.3
Acetylcholine$^+$	−4.8
Me_4N^+	−3.4
Et_4N^+	5.8
Pr_4N^+	15.5
Bu_4N^+	24.2
Ph_4As^+	35.9
Cl^-	−30.5
Br^-	−28.5
NO_3^-	−24.4
I^-	−18.8
CNS^-	−15.5
ClO_4^-	−8.7
2,6-Dinitrophenolate$^-$	−8.0
2,4-Dinitrophenolate$^-$	−7.3
Lauryl sulphate$^-$	−4.1
Picrate$^-$	4.6
Tetraphenylborate$^-$	35.9
Dipicrylaminate$^-$	39.3
Dicarbollylcobaltate$^-$	50.2

the ion whose solvation energy is the same in both media. Then, if one of the media is water, one may call the ions with $\Delta G_{tr,i}^{\circ,w\to o} > 0$ *hydrophilic* and those with $\Delta G_{tr,i}^{\circ,w\to o} < 0$, *hydrophobic*. It should be stressed that this designation is related to the given solvent and can change from solvent to solvent. In common usage, however, this sorting of ions for a definite solvent, say, nitrobenzene, is preserved for general designation of ions. Thus, e.g., the lithium ion is called hydrophilic, the tetrabutyl-ammonium hydrophobic and picrate semi-hydrophobic.

The inaccuracy of this way of speaking is obvious when ion transfer to very low-permittivity solvents is considered. As these solvents solvate the ionic solutes very weakly for all ions $\Delta G_{tr,i}^{\circ,w\to o} > 0$, i.e. all are hydrophilic according to the above definition. Some of the salts of rather hydrophobic ions in the common way of speaking are dissolved in such solvents but they form *ion pairs*, the concentration of free ions remaining very low. It is this low equilibrium concentration of free ions that determines the standard Gibbs transfer energy of the ions.

Association of ions to higher structures, such as ion pairs, triple ions, etc. occurs in aqueous solutions at very high concentrations of little interest to biology. In organic solvents of medium or low permittivity it is a common phenomenon.

For the sake of simplicity consider a uni-univalent electrolyte BA dissolved in a medium where ion pairing occurs. The constant of ion-pair formation is defined for a dilute solution as

$$K_{ip} = \frac{c_{BA}}{c_{B^+} \cdot c_{A^-}}. \tag{3.D8}$$

When analyzing the electrostatics of the interactions among ions Bjerrum found that ion-pair formation depends, at a given salt concentration, on the probability for oppositely charged ions to reach a definite mutual distance. This probability function depends on permittivity of the solvent as well as on ion radii. In a simplified form the constant of ion-pair formation is given as

$$\ln K_{ip} = \ln(4\pi N_A d^3) + \frac{e^2}{4\pi k_B T d} \cdot \frac{1}{\varepsilon} \tag{3.D9}$$

where d is the distance of ion centres in closest approach. Thus, in a solvent with relative permittivity $\varepsilon_r \approx 2$ a salt formed from large ions, such as the valinomycin complex of potassium cation and tetraphenylborate anion ($d \approx 0.96$ nm) present in a concentration of 10^{-6} mol dm^{-3} is associated to 99.4 per cent.

The potential difference between two phases in contact is obtained from the equality of electrochemical potentials of the potential-determining ion i in the phases,

$$\tilde{\mu}_{i,w} = \tilde{\mu}_{i,o}. \tag{3.D10}$$

Thus, the potential difference is given by the Nernst–Donnan potential equation

$$\Delta_o^w \phi = (\mu_{i,o}^\circ - \mu_{i,w}^\circ)/z_i F + (RT/z_i F) \ln(a_{i,o}/a_{i,w}). \tag{3.D11}$$

In view of Eq. (3.D6) we have

$$\begin{aligned} \Delta_o^w \phi &= \Delta G_{tr,i}^{\circ, w \to o}/z_i F + (RT/z_i F) \ln(a_{i,o}/a_{i,w}) \\ &= \Delta_o^w \phi_i^\circ + (RT/z_i F) \ln(a_{i,o}/a_{i,w}) \end{aligned} \tag{3.D12}$$

where $\Delta_o^w \phi_i^\circ$ is the standard electric potential difference of transfer of the ion i from water to the organic phase (an analogy of the standard electrode potential). It is standardized in the same way as $\Delta G_{tr,i}^{\circ, w \to o}$.

An electrolyte in distribution equilibrium between two solvents in contact changes, in general, the electric properties of the interface. When a simple case of a univalent electrolyte and a dilute solution ($a_i \approx c_i$) is considered the following relationship is obtained

$$\begin{aligned} \Delta_o^w \phi_{distr} &= \Delta_o^w \phi_+ + (RT/F) \ln(c_{+,o}/c_{+,w}) \\ &= \Delta_o^w \phi_- - (RT/F) \ln(c_{-,o}/c_{-,w}) \end{aligned} \tag{3.D13}$$

subscripts + and − denoting quantities referring to the cation and to the anion, respectively. Thus the distribution potential $\Delta_o^w \phi_{distr}$ is given as

$$\Delta_o^w \phi_{distr} = \tfrac{1}{2}(\Delta_o^w \phi_+^\circ + \Delta_o^w \phi_-^\circ). \tag{3.D14}$$

Obviously, for ions differing only slightly in their hydrophobicity ($\Delta_o^w \phi_+^\circ \simeq -\Delta_o^w \phi_-^\circ$), the distribution potential approaches zero.

Particle-solvent interaction

In processes where ions are transferred into and across membranes solvation of ions plays a paramount role. Solvation originates from the decrease of the Gibbs energy of a particle caused by its transfer from a gaseous phase into a medium with a relative permittivity $\varepsilon_r > 1$ and by interaction of the particle with solvent dipoles, resulting in a change of structure of the medium. The simplest approach to ion solvation is the Born equation for solvation Gibbs energy where the medium is viewed as a structureless dielectric with a relative permittivity ε_r,

$$\Delta G^{\circ}_{s,i} = -\frac{z_i e^2 N_A}{8\pi\varepsilon_0 r_i}(1 - 1/\varepsilon_r) \tag{3.D15}$$

where r_i is the radius of the ion (or, more exactly, the radius of the spherical cavity around the ion). In spite its simplicity, this equation characterizes some important features of solvation. Small ions (e.g. Li^+) are solvated more than large ions of the same type (e.g. Cs^+). Solvation energy increases with the charge number of the ion. In low-permittivity solvents solvation is weaker and its dependence on size and charge of ions is smaller.

Due to ion-solvent interaction the neighbourhood of the ion is changed. The solvent dipoles are oriented towards the centre of the ion, forming thus a primary solvation shell. This structure may be considered as a sort of coordination compound, either of outer-sphere or, in the case of some transition-metal ions, of inner-sphere nature. The influence of the ion on more distant solvent molecules results in two different effects. The situation when the solvent structure is maintained or even more ordered is termed the structure-forming effect. In contrast the destructive influence of ions upon the solvent structure is called the structure-breaking effect.

The increase or decrease of the order of the solvent results in corresponding changes of the entropy of the system. When attached to a molecule dissolved in water nonpolar groups, such as long alkyl chains, increase its quasicrystalline structure. Thus, any removal of these ordering groups increases disorder and results in an entropy increase of the system. This group of phenomena has the generic name 'hydrophobic effect'. Thus, the folding of proteins causes burying of the originally exposed hydrophobic parts of the macromolecule. The hydrophobic effect appearing with amphiphilic substances containing *one* long alkyl chain results in the formation of micelles and accumulation of these particles at the water/air interface. In the presence of an indifferent electrolyte both these processes are assisted by the salting-out effect. The electric field squeezes out a low-permittivity dielectric (the alkyl chain) from its vicinity and favours its substitution by a high-permittivity dielectric (water).

Formation of bilayer lipid membranes (BLM) is a significant property of amphiphilic molecules with *two* long-chain alkyl groups. Only in this case will the energy balance favour the bilayer formation as the hydrophobic effect of these amphiphilic molecules and the short-range attraction force of the long alkyl group (C_{18} groups are best suited) must be sufficiently larger than the repulsion of the closely packed dipoles in BLM.

Rate theory of membrane transport

The membranes is conceived here as a series of barriers, each of them being characterized by a Gibbs function (free energy) and by rate constants, as shown in Fig. 3.3. The rate of transport across the first barrier in the forward direction (v_f), which may be also called unidirectional influx $\vec{J}$ is given by

$$v_f = \vec{J} = k_1 c_1 \lambda_1 \tag{3.6}$$

where λ is the lattice distance, the length of the region from which particles may jump across the barrier and c is the local concentration. Analogously, the rate of transport in the back direction (unidirectional efflux) is described by

$$v_b = \overleftarrow{J} = k'_2 c_2 \lambda_2. \tag{3.7}$$

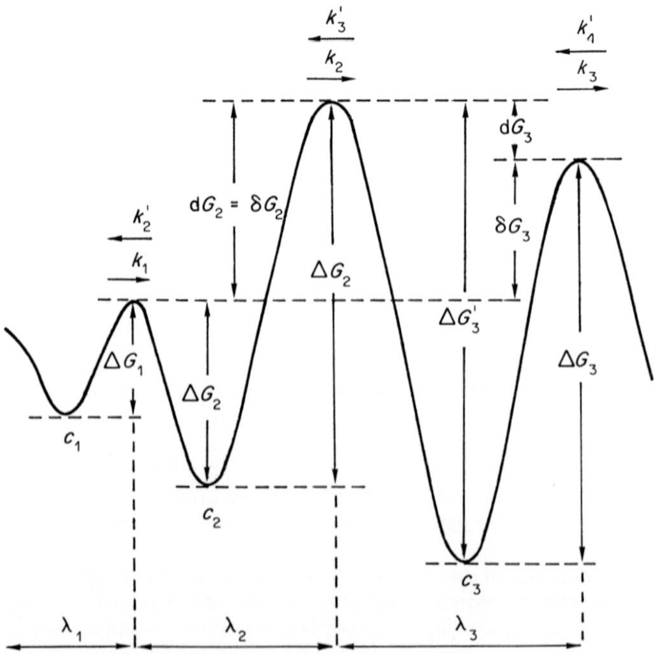

Fig. 3.3. Membrane as a series of barriers. λ's are lattice distance, c's are local concentrations. Note that $dG_i = \Delta G_i - \Delta G_i'$ and $\delta G_i = \sum_{i=2}^{i}(\Delta G_i - \Delta G_i') = \sum_{i=2}^{i} dG_i$

The net flux J is the difference between the influx and efflux

$$J = v_f - v_b = k_1 c_1 \lambda_1 - k_2' c_2 \lambda_2. \tag{3.8}$$

In general, the net flux across the ith barrier is

$$J = k_i c_i \lambda_i - k_{i+1}' c_{i+1} \lambda_{i+1}. \tag{3.9}$$

In the simplest case (resembling simple diffusion in a solution) all rate constants and lattice distances are equal

$$k_i = k_{i+1}' = k, \qquad \lambda_i = \lambda_{i+1} = \lambda$$

so that

$$J = k\lambda(c_i - c_{i+1}) = -k\lambda^2 \frac{c_{i+1} - c_i}{\lambda} = -k\lambda^2 \frac{dc}{dx}. \tag{3.10}$$

Comparison with Fick's First Law

$$J = -D\frac{dc}{dx} \tag{3.11}$$

yields

$$D = k\lambda^2. \tag{3.12}$$

Even when the individual barriers in the series are different, in a steady state the net flux J across every barrier is the same:

$$\begin{aligned} J &= k_1c_1\lambda_1 - k_2'c_2\lambda_2 \\ J &= \quad k_2c_2\lambda_2 - k_3'c_3\lambda_3 \\ J &= \quad\quad k_3c_3\lambda_3 - k_4'c_4\lambda_4 \\ &\vdots \\ J &= \quad\quad\quad k_{n-1}c_{n-1}\lambda_{n-1} - k_n'c_n\lambda_n. \end{aligned} \tag{3.13}$$

Multiplying both sides of the second equation by k_2'/k_2, of the third by $k_2'k_3'/k_2k_3$, etc. (so that the last equation is multiplied by $k_2'k_3'k_4' \ldots k_{n-1}'/k_2k_3k_4 \ldots k_{n-1}$) we obtain on adding all equations

$$J = \frac{k_1c_1\lambda_1 - \dfrac{k_2'k_3'k_4' \ldots k_{n-1}'}{k_2k_3k_4 \ldots k_{n-1}} k_n'c_n\lambda_n}{1 + \dfrac{k_2'}{k_2} + \dfrac{k^{2\prime}k_3'}{k_2k_3} + \dfrac{k_2'k_3'k_4'}{k_2k_3k_4} + \cdots + \dfrac{k_2'k_3'k_4' \ldots k_{n+1}'}{k_2k_3k_4 \ldots k_{n+1}}}. \tag{3.14}$$

When

$$k_1 = k_2 = k_2' = \cdots k_n = k, \qquad \lambda_1 = \lambda_2 = \cdots \lambda_n = \lambda$$

we obtain again Fick's law in the form

$$J = \frac{k\lambda(c_1 - c_n)}{n} = k\lambda^2 \frac{c_1 - c_n}{n\lambda} = -k\lambda^2 \frac{\Delta c}{\Delta x}. \tag{3.15}$$

Now, according to the theory of rate processes

$$k_i = \kappa \frac{k_B T}{h} \exp\left(-\frac{\Delta G_i}{RT}\right). \tag{3.16}$$

Here κ is the transmission coefficient expressing the probability that the transition (activated) state will actually proceed toward the product side. It lies for most reactions between 0.9 and 1.0 and can be assumed to be equal to unity for most but the finest calculations, k_B is Boltzmann's constant (the gas constant per molecule), h is Planck's constant and T the absolute temperature. ΔG_i is the free energy change corresponding to the transition of the molecule from the stable equilibrium position at the bottom of the barrier to its top.

What makes the theory of absolute reaction rates applicable to the diffusion process is that there, too, the particles in the transition state (corresponding to the 'activated complex') may be considered to be in equilibrium with those at the preceding minimum of free energy; their concentration at the top of the barrier, $c_i^{\neq}$ being related to the concentration on its left c_i by an equilibrium constant

$$c_i^{\neq}/c_i = K^{\neq}. \tag{3.17}$$

Further, one of the vibrations of the particle in the transition state is considered as a translational degree of freedom (again in analogy with a chemical reaction, where the translation occurs along the reaction coordinate). The frequency v of the vibration at temperature T is given by $v = k_B T/h$, so that, e.g.,

$$v_f = \vec{J} = k_i c_i = v c_i^{\neq} k_B T c_i^{\neq}/h \tag{3.18}$$

and thus

$$k_i = k_B T c_i^{\neq}/h c_i = k_B T K^{\neq}/h \tag{3.19}$$

Since $K^{\neq}$ is an equilibrium constant, it is related to the free energy of the transport to the transition state by

$$RT \ln K^{\neq} = -\Delta G_i. \tag{3.20}$$

From this

$$K^{\neq} = \exp(-\Delta G_i/RT) \tag{3.21}$$

so that

$$k_i = \frac{k_B T}{h} \exp\left(-\frac{\Delta G_i}{RT}\right) \tag{3.22}$$

which is, apart from the transmission coefficient κ, our formula for the rate constant

Let us now, referring again to Fig. 3.3, express various combinations of rate constants by differences of free energy, defined in the figure:

$$\frac{k_2'}{k_2} = \exp\left(\frac{\Delta G_2 - \Delta G_2'}{RT}\right) = \exp\left(\frac{\delta G_2}{RT}\right),$$

$$\frac{k_2' k_3'}{k_2 k_3} = \exp\left[\frac{(\Delta G_2 - \Delta G_2') + (\Delta G_3 - \Delta G_3')}{RT}\right] = \exp\left(\frac{\delta G_3}{RT}\right),$$

and since

$$\frac{k_n'}{k_1} = \exp\left(\frac{\Delta G_1 - \Delta G_n'}{RT}\right) \tag{3.23}$$

we can write

$$\frac{k_2' k_3' \ldots k_{n-1}' k_n'}{k_1 k_2 k_3 \ldots k_{n-1}} = \exp\left(\frac{\Delta G_1 + \delta G_{n-1} - \Delta G_n}{RT}\right) = \exp\left(\frac{\Delta G_n}{RT}\right)$$

where ΔG_n is the difference in free energy between positions 1 and n. Introducing these expressions into the equation for net flux J (Eq. 3.14) we obtain

$$J = \frac{\lambda_1 k_1 [c_1 - (\lambda_n/\lambda_1)\{\exp(\Delta G_n/RT)\} c_n]}{1 + \exp\left(\frac{\delta G_2}{RT}\right) + \exp\left(\frac{\delta G_3}{RT}\right) + \cdots + \exp\left(\frac{\delta G_{n-1}}{RT}\right)}. \tag{3.24}$$

If all barriers are of the same height

$$\delta G_2 = \delta G_3 = \cdots = \delta G_{n-1} = 0$$

the denominator becomes n and the equation for net flux reduces to

$$J = \frac{\lambda_1^2 k_1}{\lambda_1 n}\left[c_1 - \frac{\lambda_n}{\lambda_1}\left\{\exp\left(\frac{\Delta G_n}{RT}\right)\right\} c_n\right] = \frac{D}{d}\left[c_1 - \frac{\lambda_n}{\lambda_1}\left\{\exp\left(\frac{\Delta G_n}{RT}\right)\right\} c_n\right], \tag{3.25}$$

where D is the diffusion coefficient and d the membrane thickness. If, moreover, $\lambda_1 = \lambda_n$ and $\Delta G_n = 0$ the last condition is satisfied if a membrane separates two ideal solutions in the same solvent (in the rate theory free energy does not contain a concentration term). The steady-state flux across the membrane then is

$$J = (D/d)(c_1 - c_n) = P(c_1 - c_n) \tag{3.26}$$

where P is the permeability constant. Clearly, the classical equation describing permeation of a nonelectrolyte across a membrane by simple diffusion is actually an equation of the rate theory adapted to particularly simple conditions. In this treatment phenomena at the interfaces between the membrane and the two solutions are neglected. A more realistic description is as follows (see Fig. 3.4):

The equation for net flux is extended by rate constants $k_1 = k_{sm}$ (solution-membrane), $k_2' = k_{ms}$ (membrane-solution), $k_2 = k_m \ldots$, etc. From this then

$$J = \frac{k_{sm} c_1 \lambda - \dfrac{k_{ms} k_m \ldots k_m}{k_m k_m \ldots k_{ms}} k_{sm} c_n \lambda}{1 + \dfrac{k_{ms}}{k_m} + \dfrac{k_{ms}}{k_m}\dfrac{k_m}{k_m} + \cdots + \dfrac{k_{ms}}{k_m}\dfrac{k_m}{k_{ms}}}$$

$$= \frac{k_{sm}\lambda(c_1 - c_n)}{2 + m\dfrac{k_{ms}}{k_m}} = \frac{\lambda k_{sm} k_m}{2k_m + m k_{ms}}(c_1 - c_n) = P(c_1 - c_n) \tag{3.27}$$

where

$$P = \lambda k_{sm} k_m / (2k_m + m k_{ms}) \tag{3.28}$$

and

$$1/P = 2/\lambda k_{sm} + m k_{ms}/\lambda k_{sm} k_m. \tag{3.29}$$

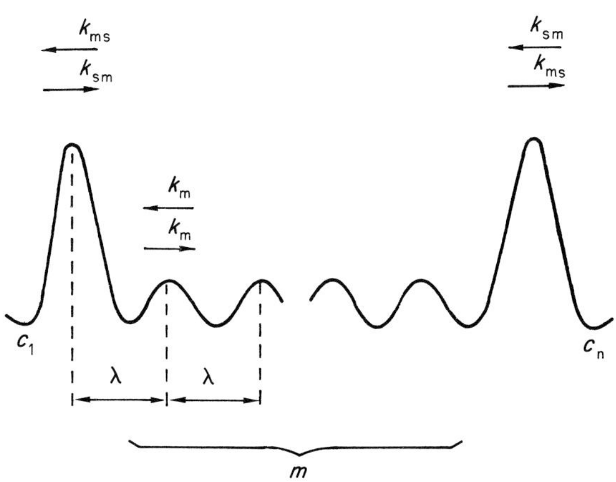

Fig. 3.4. Membrane with barriers at the interfaces between solution and membranes

The partition coefficient β is then defined as

$$\beta = k_{\mathrm{sm}}/k_{\mathrm{ms}} \tag{3.30}$$

whereas the diffusion coefficient in the membrane proper D_{m} will be

$$D_{\mathrm{m}} = k_{\mathrm{m}}\lambda^2 \tag{3.31}$$

and that for the interface

$$D_{\mathrm{sm}} = k_{\mathrm{sm}}\lambda^2. \tag{3.32}$$

The thickness of the membrane is $d = m\lambda$ so that

$$1/P = 2\lambda/D_{\mathrm{sm}} + d/D_{\mathrm{m}}\beta \tag{3.33}$$

and it is immediately seen that there are two limiting cases. When the transport across the membrane proper is rate-limiting

$$P = D_{\mathrm{m}}\beta/d. \tag{3.34}$$

If, on the other hand, the resistance of the membrane proper is negligible as compared with those at the two interfaces

$$P = D_{\mathrm{sm}}/2\lambda. \tag{3.35}$$

Molecular dynamics

Kinetic models of diffusion across membranes based on the rate theory are realistic and make it possible to interpret transport characteristics found experimentally in terms of the number of potential barriers and their heights. However, there is little doubt that a physical description of interactions of diffusing particles with the real structure of the transport pathway or channel will provide deeper insight into the permeation process, its selectivity and rate. Such a description is in principle possible by using methods of molecular dynamics, provided the structure of the channel is known in sufficient detail.

The principal aim of molecular dynamics applied to a transport process is to find coordinates and momenta of all particles in the studied system over a given time interval so that both the position and velocity of each is known at any specified time, and then to calculate from them the measurable macroscopic properties, such as, e.g., the diffusion coefficient.

If at some time t the coordinates $q_1, \ldots, q_{3\mathrm{n}}$ and the momenta $p_1, \ldots, p_{3\mathrm{n}}$ are known for a system of n particles, coordinates q_1, q_2, q_3 give the position of the first particle, q_4, q_5, q_6 of the second, etc. The positions and momenta at time $t + \mathrm{d}t$ can be calculated as

$$q_1 + \dot{q}\,\mathrm{d}t, \ldots, q_{3\mathrm{n}} + \dot{q}_{3\mathrm{n}}\,\mathrm{d}t \quad \text{and} \quad p_1 + \dot{p}_1\,\mathrm{d}t, \ldots, p_{3\mathrm{n}} + \dot{p}_{3\mathrm{n}}\,\mathrm{d}t$$

where the dots designate derivatives with respect to time. The numerical values of these derivatives can be obtained from Hamilton's canonical equations of motion

$$\dot{q}_{\mathrm{i}} = \frac{\partial \mathscr{H}}{\partial p_{\mathrm{i}}} \quad \text{and} \quad p_{\mathrm{i}} = -\frac{\partial \mathscr{H}}{\partial q_{\mathrm{i}}} \tag{3.36}$$

where $\mathscr{H}$ is the Hamiltonian of the system, corresponding to its total energy (the sum of the its kinetic and potential energies). As soon as the Hamiltonian can be expressed as a function of coordinates and momenta, the values of the derivatives are available and the trajectory of the system (i.e., coordinates and momenta at any time) can be calculated by a step-by-step numerical integration. A random combination of coordinates and momenta is usually chosen as the initial condition. The whole simulation requires an advanced computer, such as was used to study sodium ion diffusion through the gramicidin A channel. From the calculated ion trajectories it was seen that ions oscillate back and forth many times in their potential energy minima ('trapping sites') with the oscillation frequency of 10^{12} to $10^{13}\,s^{-1}$ before they jump over a barrier (with a frequency of some $10^7\,s^{-1}$). The hopping distance being about 0.4 nm, the diffusion coefficient (the square of the distance multiplies by unidirectional barrier-crossing frequency) was calculated to be $8.4 \cdot 10^{-13}\,m^2\,s^{-1}$, i.e. about three orders of magnitude smaller than that for free diffusion in water.

Impressive as the molecular dynamics approach is, its value still rests mainly in methodology—it provides a method that can be applied as soon as the actual energy profiles corresponding to transmembrane pathways are known with sufficient precision.

This can only be done by a computation taking into account intra- and intermolecular interactions as a sum of various energy components, viz. electrostatic, polarization, repulsion, dispersion and torsion. Such a highly demanding enterprise is obviously possible only with membrane-crossing paths that are precisely defined structurally, such as the case of gramicidin A channels (see p. 218). Description of diffusion of particles through the highly mobile lipid core of membranes must rest on a more or less phenomenological approach which abstracts from the individual energy barriers as concrete entities.

Effects of the so-called unstirred layers

When evaluating the permeability properties of membranes with respect to rapidly permeating substances it is often necessary to take into account the effects of transport from the intercellular space or from the inside of the cell towards the membrane (see Section 3.C). If the intracellular or intercellular medium is moving its relative motion with respect to the membrane decreases in the Prandtl layer and stops at the membrane/medium interface. Transport by diffusion alone is practically restricted to the Nernst layer which is much thinner than the Prandtl layer and very often is called the unstirred layer (in fact, strictly speaking, the motion of the medium only disappears at the interface). The effective thickness of these unstirred layers (thickness of layers with linear concentration profiles that would produce the same effects) may be as much as 500 μm and cannot be reduced below some 20 μm even by vigorous mechanical stirring (Fig. 3.5, see also p. 136). Typical diffusion coefficients of small solute particles in water being about $10^{-9}\,m^2\,s^{-1}$ one can express the 'permeability coefficient' of an unstirred layer 100 μm thick as $10^{-9}\,m^2\,s^{-1}/10^{-4}\,m = 10^{-5}\,m^{s-1}$. Whenever the measured permeability coefficients are as high as this it is clear that the transport processes is rate-limited by diffusion across the unstirred layers and the true permeability coefficient of the membrane proper is still higher. When the measured permeability coefficients are of the order of $10^{-6}\,m\,s^{-1}$ a correction is necessary since the concentration difference across the membrane proper is less than that between the bulks of the solutions at the two sides Δc, by the amount represented by concentration differences across the two unstirred layers $(\Delta c)_1$ and $(\Delta c)_2$. Thus, when the measured flux is J,

$$P_{\text{measured}} = J/\Delta c \tag{3.37}$$

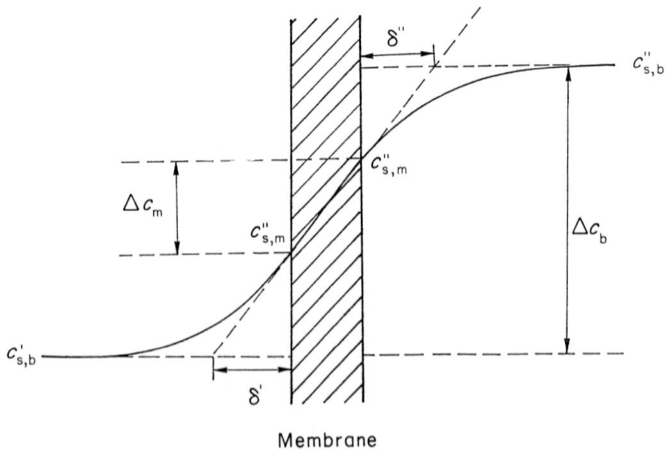

Fig. 3.5. Concentration profile of a solute permeating across a membrane. δ', δ'' – unstirred (stagnant) layer thickness

and

$$P_{\text{true}} = J/[\Delta c - (\Delta c)_1 - (\Delta c)_2]. \tag{3.38}$$

Each of the two concentration drops is related to the thickness of the unstirred layer δ and to the appropriate diffusion coefficient

$$(\Delta c)_1 = J/(D_1/\delta_1) \quad (\Delta c)_2 = J/(D_2/\delta_2) \tag{3.39}$$

so that

$$P_{\text{true}} = 1/(1/P_{\text{measured}} - \delta_1/D_1 - \delta_2/D_2). \tag{3.40}$$

3.3.3 Electrodiffusion and membrane potentials

If the solute diffusing passively across a membrane carries an electrical charge, the entire treatment of fluxes gains a new dimension. The fluxes are substantially influenced by the electrical field in a membrane which, in their turn, they may help to generate as will be shown in the discussion of the membrane potential.

In the following the membrane potential will be first rigorously defined and then the passive transport of ions across membranes will be discussed from the point of view of the rate theory which gives a vivid picture of the permeation process. This will be compared with an approach based on the theory of electrodiffusion from which the widely used Goldman formula for the membrane potential will follow. Finally, a simple treatment of the problem from the point of view of the thermodynamics of irreversible processes will be presented and shown to be very well suited for thin biological membranes in which electrogenic pumps often operate.

If one mole of *charged* particles of species i is to be transferred from a given reference state, further specified by the absence of an electric field, into the bulk of an electrically charged phase, then work must be expended to overcome both the chemical binding forces and the electric forces. This work is equivalent to the electrochemical potential $\Delta\tilde{\mu}_i$ (see Section 3.A)

$$\tilde{\mu}_{i,\alpha} = \tilde{\mu}_{i,\alpha} + z_i F \phi_\alpha. \tag{3.41}$$

In this equation, the second term describes purely electric work, connected with an infinitely slow transfer of charge from infinity in a vacuum into the bulk of the phase concerned, i.e. to a point with the electric potential ϕ_α. Here, only electric charge is transferred, not a 'material' species with which it may be connected. The ratio of this work to the transferred charge is equal to the inner potential ϕ of the given phase.

The difference of inner potentials of two phases (see p. 97) is the *Galvani potential difference*. This quantity is not accessible to measurement by purely thermodynamic procedures, with the exception of the potential difference between chemically identical phases (two pieces of the same metal or two electrolyte solutions with the same composition, for example). However, it is the quantity most often used in membranology. Consider, for example, a membrane which is permeable only to cations (a cation-exchanging membrane). If it separates two solutions of the same cation at different concentrations c_1 and c_2, the equilibrium is characterized by the equality of electrochemical potentials (see p. 121) of this cation

$$\begin{gathered}\tilde{\mu}_{+,1} = \tilde{\mu}_{+,2}\\ \mu_+ + RT\ln a_{+,1} + z_+F\phi_1 = \mu_+ + RT\ln a_{+,2} + z_+F\phi_2.\end{gathered} \tag{3.42}$$

The membrane potential, $\Delta\phi_m$, is then given by the equation

$$\Delta\phi_m = \phi_2 - \phi_1 = \frac{RT}{z_+F}\ln\frac{a_{+,1}}{a_{+,2}} = \frac{RT}{z_+F}\ln\frac{\gamma_{+,1}c_{+,1}}{\gamma_{+,2}c_{+,2}}. \tag{3.43}$$

No such equation can be written for the anion as the membrane is not permeable to anions (they do not 'communicate' across the membrane). The determination of the membrane potential is achieved using a suitable extrathermodynamical approach. For this measurement usually a potential probe is used, which is a suitable reference electrode with a negligible liquid-junction potential between its internal solution and the solution under investigation. The electrolyte of the reference electrode (calomel or Ag/AgCl) is usually a saturated KCl solution which fulfils the requirement of the negligible liquid junction potential, because the mobilities of K^+ and Cl^- are almost equal. If the diffusion of potassium ions into the solution under investigation is to be avoided (for example, in nerve cells) another electrolyte of this property must be used, such as sodium formate.

The membrane potential as shown above can be determined by an ionic species which can communicate, i.e., be exchanged across the membrane. A more complicated case occurs when several ionic species, cations as well as anions, can cross the membrane with different but comparable rates. This case will be dealt with later (p. 154).

The rate theory was found useful in describing simple diffusion of nonelectrolytes across a membrane. To apply it to the transport of ions one must introduce, in addition to the concentration gradient, also the electrical field (the negative of the electrical potential gradient) in the membrane. Assuming a single barrier, the electrical potential to the left and to the right of the barrier in Fig. 3.6 is ϕ_i and ϕ_{i+1}, respectively. If $\phi_i = \phi_{i+1}$, $J = k_i^0 c_i \lambda_i - k_{i'+1}^0 c_{i+1} \lambda_{i+1}$ as before. If, however, $\phi_i \neq \phi_{i+1}$

$$J = k_i^0 c_i \lambda_i \exp\left[\frac{-\alpha z F(\phi_{i+1} - \phi_i)}{RT}\right] - k_{i+1}^{0\prime} c_{i+1} \lambda_{i+1} \exp\left[\frac{(1-\alpha) z F(\phi_{i+1} - \phi_i)}{RT}\right] \tag{3.44}$$

where α (coefficient of charge transfer) will be $\frac{1}{2}$ if the barrier is symmetrical. If, moreover $k^0 = k_{i'+1}^0 = k_{i+1}^0 = \cdots k^0$ and $\lambda_i = \lambda_{i+1} = \cdots \lambda$, the membrane potential $\Delta\phi_m$ will be evenly distributed across the individual barriers, so that $\phi_{i+1} - \phi_i = \Delta\phi_m / n$ and

$$J = k^0 c_i \lambda \exp\left[-\frac{zF\,\Delta\phi_m}{2nRT}\right] - k^0 c_{i+1} \lambda \exp\left[\frac{zF\,\Delta\phi_m}{2nRT}\right]. \tag{3.45}$$

Setting $\exp[zF\,\Delta\phi_m/nRT] = \xi$,
the individual rate constants may be expressed as

$$k_1 = k^0 \exp\left[-\frac{zF\,\Delta\phi_m}{2nRT}\right] = k^0 \xi^{-1/2} = k_2 = k_3 = \cdots = k_{n-1}$$

$$k_2' = k^0 \exp\left[\frac{zF\,\Delta\phi_m}{2nRT}\right] = k^0 \xi^{1/2} = k_3' = k_4' = \cdots = k_n'$$

so that

$$k_2'/k_2 = \frac{\xi^{1/2}}{\xi^{-1/2}} = \xi, \qquad \frac{k_2' k_3'}{k_2 k_3} = \xi^2, \quad \text{etc.}$$

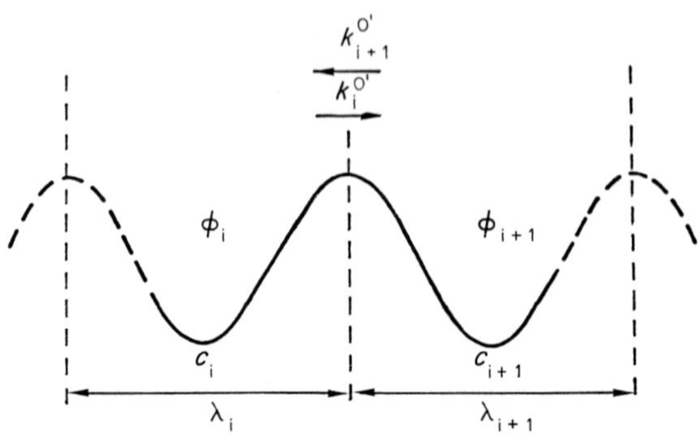

Fig. 3.6. Barrier with a jump of electric potential

Introducing these expressions into the equation for net flux J (Eq. (3.14)) one obtains

$$J=\frac{k^0c_1\lambda\xi^{-1/2}-k^0c_n\lambda\xi^{n-1}\xi^{1/2}}{1+\xi+\xi^2+\cdots+\xi^{n-1}}$$

$$=\frac{k^0\lambda\xi^{-1/2}(c_1-c_n\xi^n)}{1+\xi+\xi^2+\cdots+\xi^{n-1}}. \tag{3.46}$$

Since

$$1+\xi+\xi^2+\cdots+\xi^{n-1}=\frac{1-\xi^n}{1-\xi}$$

$$J=\frac{k^0\lambda\xi^{-1/2}(1-\xi)}{1-\xi^n}(c_1-\xi^n c_n)$$

$$=k^0\lambda\frac{\xi^{-1/2}-\xi^{1/2}}{1-\xi^n}(c_1-\xi^n c_n)\frac{\xi^{-n/2}}{\xi^{-n/2}}$$

$$=k^0\lambda\frac{\xi^{-1/2}-\xi^{1/2}}{\xi^{-n/2}-\xi^{n/2}}(\xi^{-n/2}c_1-\xi^{n/2}c_n). \tag{3.47}$$

In the rather unusual case that the membrane potential is very small ($\Delta\phi_m \ll 2RT/zF = 50\,\mathrm{mV}$, for $z=1$) the approximation $\exp(\pm x)=1\pm x$ may be used to simplify this expression further and

$$\frac{\xi^{-1/2}-\xi^{1/2}}{\xi^{-n/2}-\xi^{n/2}}=\frac{\exp-(zF\,\Delta\phi_m/2nRT)-\exp(zF\,\Delta\phi_m/2nRT)}{\exp-(zF\,\Delta\phi_m/2RT)-\exp(zF\,\Delta\phi_m/2RT)}$$

$$\approx\frac{1-\dfrac{zF\,\Delta\phi_m}{2nRT}-1-\dfrac{zF\,\Delta\phi_m}{2nRT}}{1-\dfrac{zF\,\Delta\phi_m}{2RT}-1-\dfrac{zF\,\Delta\phi_m}{2RT}}=\frac{\dfrac{zF\,\Delta\phi_m}{nRT}}{\dfrac{zF\,\Delta\phi_m}{RT}}=\frac{1}{n}. \tag{3.48}$$

Setting $P=k^0\lambda/n$ it holds that

$$J=P(c_1\exp-(zF\,\Delta\phi_m/2RT)-c_n\exp(zF\,\Delta\phi_m/2RT))$$

$$\approx P\left(c_1-c_1\frac{zF\,\Delta\phi_m}{2RT}-c_n-c_n\frac{zF\,\Delta\phi_m}{2RT}\right)$$

$$=P\left[(c_1-c_n)-\tfrac{1}{2}(c_1+c_n)\frac{zF\,\Delta\phi_m}{RT}\right] \tag{3.49}$$

where the first term on the right-hand side describes simple diffusion across membrane, the second term migration in a homogeneous electrical field under the condition that the concentration profile of the electrodiffusing ion is linear. In the next section we shall see that diffusion and migration should be in general superimposed in a more refined way.

Goldman's solution of electrodiffusion

Derivation of Goldman's equations for ion fluxes and the membrane potential is not free of contradictions; on the one hand, one must assume that the membrane is so thin that the electrical potential profile in it may be considered as linear (constant electrical field); on the other hand, integration of the differential equation across the thickness of the membrane should be physically meaningful. This already circumscribes the validity of these equations.

The equation describing the flow of ion j at any point in the membrane is the Nernst–Planck or electrodiffusion Eq. (3.C35)

$$J_j = -RT\mu_j \frac{dc_j}{dx} - z_j F \mu_j c_j \frac{d\phi}{dx} \tag{3.50}$$

in which J_j is the flow of j (e.g. in mol m^{-2} s^{-1}) and μ_j is its mobility (in m^{-2} s^{-1} J^{-1} mol). The other symbols have their usual meaning. Its first term describes diffusion (RTu_j is Einstein's interpretation of the diffusion coefficient D_j) and the second term describes migration of charged particles in an electrical field. The first term thus represents Fick's law, the second Ohm's law; laws in differential forms may be simply superimposed.

The Nernst–Planck equation is a special case of Teorell's general transport equation according to which

$$\text{Flux} = \text{mobility} \cdot \text{concentration} \cdot \text{total driving force.}$$

The natural choice of the driving force for an ion is the fall (the negative of the gradient) of its electrochemical potential—if the electrochemical potential is the same everywhere, equilibrium has been reached. Thus

$$J_j = -u_j c_j \frac{d\tilde{\mu}_j}{dx} \tag{3.51}$$

and since the electrochemical potential can be expressed as

$$\tilde{\mu}_j = \mu_j^\circ + RT \ln a_j + z_j F \phi \tag{3.52}$$

$$J_j = -u_j c_j \left(RT \frac{d \ln a_j}{dx} + z_j F \frac{d\phi}{dx} \right). \tag{3.53}$$

From elementary differential calculus

$$d \ln a_j = \frac{da_j}{a_j} = \frac{d(\gamma_j c_j)}{\gamma_j c_j} = \frac{\gamma_j \, dc_j}{\gamma_j c_j} = \frac{dc_j}{c_j} \tag{3.54}$$

so that the Nernst–Planck equation again results (Eq. (3.50))

$$J_j = -RTu_j \frac{dc_j}{dx} - z_j F u_j c_j \frac{d\phi}{dx}.$$

Let us note that the activity coefficient γ_j was not entirely neglected; it was enough to assume its independence of x.

Finally, the Nernst–Planck equation can be viewed as a special simplified form of Schlögl's equation

$$J_j = c_j J_V - RTu_j \frac{dc_j}{dx} - z_j F u_j c_j \frac{d\phi}{dx}$$

$$-RTu_j \frac{d \ln \gamma_j}{dx} - u_j M_j c_j \left(\frac{\bar{V}_j}{M_j} - \frac{\bar{V}_w}{M_w} \right) \tag{3.55}$$

in which the first term shows how the ion is entrained by volume flow J_V, the next two terms correspond to diffusion and migration in an electrical field, the fourth describes ion migration in a gradient of its activity coefficient and the last term (in which M_j and M_w are the molar masses of the ion and of the solvent, respectively, $\bar{V}_j$ and $\bar{V}_w$ their partial molal volumes and p hydrostatic pressure) describes pressure diffusion, in which the more voluminous particles tend to accumulate in regions of lower hydrostatic pressure. Schlögl's equation was derived by methods of the thermodynamic of irreversible processes and represents a general description of the transport of ions which interact neither mutually, nor with a chemical reaction. If volume flow, the gradient of the activity coefficient and the pressure gradient are neglected the Nernst–Planck equation is again obtained.

To proceed from local derivatives to measurable differences over finite distances such as the membrane thickness the differential equations must be integrated. It is not an easy task with the Nernst–Planck equation (let alone Schlögl's equation); even when a stationary state prevails and the flux J_j is thus a constant, there are two unknown functions of x the concentration c_j and the electrical potential ϕ. Of the various simplifying assumptions (such as local electroneutrality postulated by Planck or linear concentration profiles in a mixed boundary of Henderson) that of Goldman seems to be best suited for biological membranes. The electrical potential ϕ is assumed to be a linear function of the coordinate x so that

$$d\phi/dx = (\phi'' - \phi')/d \tag{3.56}$$

where d is the membrane thickness. Introducing this condition into the Nernst–Planck equation the variables may be separated and the integration can be carried out from $x = 0$ where $c_j = c_j'$ to $x = d$ where $c_j = c_j''$

$$\int_0^d dx = -RTu_j \int_{c_j'}^{c_j''} \frac{dc_j}{J_j + z_j F u_j [(\phi'' - \phi)/d] c_j}$$

i.e.

$$d = \frac{RTd}{z_j F(\phi'' - \phi')} \ln \frac{J_j + z_j F u_j [(\phi'' - \phi')/d] c_j''}{J_j + z_j F u_j [(\phi'' - \phi')/d] c_j'}$$

and solving the equation for J_j

$$J_j = z_j F u_j \frac{\phi'' - \phi'}{d} \frac{c_j'' - c_j' e^{-z_j F(\phi'' - \phi')/RT}}{e^{-z_j F(\phi'' - \phi')/RT} - 1} \tag{3.57}$$

which is the well-known constant-field equation for ion flux. Such equations can be written for the fluxes of principal ions influencing by their high concentrations and high permeabilities the membrane potential, viz. K^+, Na^+ and Cl^-, and to use the condition that when the potential is steady and there is no external circuit there is no electrical current across the membrane, i.e.

$$J_{K^+} + J_{Na^+} = J_{Cl^-} \tag{3.58}$$

The electrical potential difference across the membrane proper can then be expressed as

$$\phi'' - \phi' = \frac{RT}{F} \ln \frac{u_{K^+}c'_{K^+} + u_{Na^+}c'_{Na^+} + u_{Cl^-}c''_{Cl^-}}{u_{K^+}c''_{K^+} + u_{Na^+}c''_{Na^+} + u_{Cl^-}c'_{Cl^-}} \tag{3.59}$$

The overall potential difference between the bulks of the solutions at the two sides of the membrane involves two surface potentials ψ' and ψ''. An equilibrium distribution of ions at the two membrane surfaces will be assumed so that the concentrations just inside the membrane (c'_{K^+}, c''_{K^+}, etc.) be related to the concentrations in solution ($c'_{K^+,s}, c''_{K^+,s}$, etc.) by relations

$$\begin{aligned} c'_{K^+} &= k_{K^+}c'_{K^+,s}e^{-F\psi'/RT} \\ c''_{K^+} &= k_{K^+}c''_{K^+,s}e^{F\psi''/RT} \\ &\vdots \\ c'_{Cl^-} &= k_{Cl^-}c'_{Cl^-,s}e^{F\psi'/RT} \\ c''_{Cl^-} &= k_{Cl^-}c''_{Cl^-,s}e^{-F\psi''/RT}. \end{aligned} \tag{3.60}$$

Then using the identity

$$\psi' + \psi'' = \frac{RT}{F} \ln (e^{F\psi'/RT} \cdot e^{F\psi''/RT}) \tag{3.70}$$

one can derive the overall potential difference $\Delta\phi_m = \psi' + \phi'' - \phi' + \psi''$ as

$$\Delta\phi_m = \frac{RT}{F} \ln \frac{P_{K^+}c'_{K^+,s} + P_{Na^+}c'_{Na^+,s} + P_{Cl^-}c''_{Cl^-,s}e^{F(\psi'-\psi'')/RT}}{P_{K^+}c''_{K^+,s} + P_{Na^+}c''_{Na^+,s} + P_{Cl^-}c'_{Cl^-,s}e^{F(\psi'-\psi'')/RT}} \tag{3.71}$$

where $P_{K^+} = RTu_{K^+}k_{K^+}/d$, etc.

This equation is known as Goldman's equation for membrane potential; the exponentials accompanying anion concentrations are usually omitted, this being equivalent to the assumption that $\psi' = \psi''$.

Piek's extension of the Goldman equation

It is not generally known that Goldman's equation for membrane potential may involve ions of different valencies (Piek's equation). Let us illustrate Piek's procedure by a simple case in which the membrane potential is governed by two ionic species only, K^+ and Mg^{2+}. The constant-field equations for the currents carried by two ions across a membrane will be

$$j_{K^+} = z_{K^+}FJ_{K^+} = z_{K^+}^2F^2u_{K^+}\frac{\phi''-\phi'}{d}\frac{c''_{K^+}-c'_{K^+}e^{-z_{K^+}F(\phi''-\phi')/RT}}{e^{-z_{K^+}F(\phi''-\phi')/RT}-1} \tag{3.72}$$

and

$$j_{Mg^{2+}} = z_{Mg^{2+}}FJ_{Mg^{2+}} = z_{Mg^{2+}}^2F^2u_{Mg^{2+}}\frac{\phi''-\phi'}{d}\frac{c''_{Mg^{2+}}-c'_{Mg^{2+}}e^{-z_{Mg^{2+}}F(\phi''-\phi')/RT}}{e^{-z_{Mg^{2+}}F(\phi''-\phi')/RT}-1}. \tag{3.73}$$

When there is no net current across the membrane ($j_{K^+} + j_{Mg^{2+}} = 0$) and setting $z_{K^+} = 1$, $z_{Mg^{2+}} = 2$ one obtains

$$\frac{c''_{K^+}-c'_{K^+}e^{-F(\phi''-\phi')/RT}}{e^{-F(\phi''-\phi')/RT}-1} = \frac{4\gamma c''_{Mg^{2+}}-4\gamma c'_{Mg^{2+}}e^{-2F(\phi''-\phi')/RT}}{e^{-2F(\phi''-\phi')/RT}-1} \tag{3.74}$$

where $\gamma = u_{Mg^{2+}}/u_{K^+}$. The denominator of the right-hand side can be written as

$$e^{-2F(\phi''-\phi')/RT}-1 = (e^{-F(\phi''-\phi')/RT}-1)(e^{-F(\phi''-\phi')/RT}+1) \tag{3.75}$$

so that

$$(c''_{K^+}-4\gamma c''_{Mg^{2+}})e^{2F(\phi''-\phi')/RT} + (c''_{K^+}-c'_{K^+})e^{F(\phi''-\phi')/RT} - (c'_{K^+}+4\gamma c'_{Mg^{2+}}) = 0 \tag{3.76}$$

which can be solved as a quadratic equation for $e^{F(\phi''-\phi')/RT}$:

$$\phi''-\phi' = \frac{RT}{F}\ln\frac{-(c''_{K^+}-c'_{K^+})+[(c''_{K^+}-c'_{K^+})^2+4(c''_{K^+}+4\gamma c''_{Mg^{2+}})(c'_{K^+}+4\gamma c'_{Mg^{2+}})]^{1/2}}{2(c''_{K^+}+4\gamma c''_{Mg^{2+}}).} \tag{3.77}$$

It may be seen that for $\gamma \ll 1$

$$\phi''-\phi' \simeq \frac{RT}{F}\ln\frac{c'_{K^+}}{c''_{K^+}} \tag{3.78}$$

whereas for $\gamma \gg 1$

$$\begin{aligned}\phi''-\phi' &\simeq \frac{RT}{F}\ln\frac{(64c''_{Mg^{2+}}c'_{Mg^{2+}})^{1/2}}{8c'_{Mg^{2+}}}\\ &= \frac{RT}{F}\ln\frac{(64c''_{Mg^{2+}}c'_{Mg^{2+}})^{1/2}}{(64c''^2_{Mg^{2+}})^{1/2}} = \frac{RT}{2F}\ln\frac{c'_{Mg^{2+}}}{c''_{Mg^{2+}}}.\end{aligned} \tag{3.79}$$

In a more general case in which K^+, Cl^-, Na^+, Mg^{2+} and Ca^{2+} are present

$$\phi''-\phi' = \frac{RT}{F}\ln\frac{-b+(b^2-4ac)^{1/2}}{2a} \tag{3.80}$$

where

$$\begin{aligned} a &= c''_{K^+} + \alpha c'_{Cl^-} + \beta c''_{Na^+} + 4\gamma c''_{Mg^{2+}} + 4\delta c''_{Ca^{2+}} \\ b &= c''_{K^+} - c'_{K^+} - \alpha(c''_{Cl^-} - c'_{Cl^-}) + \beta(c''_{Na^+} - c_{Na^+}) \\ c &= c'_{K^+} - \alpha c''_{Cl^-} - \beta c'_{Na^+} - 4\gamma c'_{Mg^{2+}} - 4\delta c'_{Ca^+} \end{aligned} \tag{3.81}$$

and α, β, γ and δ are permeability coefficients for chloride, sodium, magnesium and calcium ions, respectively, divided by the permeability for potassium ions. If surface potentials are taken into account the formulas are prohibitively complicated.

In special instances the Goldman formalism may take into account even the presence of electrogenic pumps; however, treatment of such cases by thermodynamics of irreversible processes is preferable for its simplicity (p. 159).

The Hodgkin–Horowicz equation and its extensions

An alternative to the Nernst–Planck equation describing the passive flux of an ion (Eq. 3.51)

$$J_j = -c_j u_j \frac{d\tilde{\mu}_j}{dx}$$

uses finite differences of the electrochemical potentials across the membrane

$$J_j = L_j(\tilde{\mu}''_j - \tilde{\mu}'_j). \tag{3.82}$$

As discussed elsewhere in this book (p. 172), equations in finite differences across thin biological membranes should be considered as basic laws equivalent to differential equations but not operationally derivable from them. The electrical current carried across the membrane by ion j may be then expressed as follows:

$$\begin{aligned} j_j = z_j F J_j &= -z_j F L_j(RT \ln c''_j + z_j F\phi'' - RT \ln c'_j - z_j F\phi') \\ &= z_j^2 F^2 L_j\left[\frac{RT}{z_j F}\ln\frac{c'_j}{c''_j} - (\phi'' - \phi')\right] = G_j(E_j - \Delta\phi_m) \end{aligned} \tag{3.83}$$

where G_j is the conductivity of the membrane for the jth ion, E_j is its Nernst–Donnan potential and E the membrane potential. Let us carry the rest of the derivation on a specific example; it is not difficult to generalize the final formulas.

Let passive fluxes of potassium, sodium and chloride ions take place across a membrane; active fluxes are either absent or electrically silent as a result of coupling. The steady-state open-circuit membrane potential can be then expressed from the condition that the total electrical current across the membrane is zero

$$j_{K^+} + j_{Na^+} + j_{Cl^-} = 0 \tag{3.84}$$

or

$$\Delta\phi_m = \frac{G_{K^+}E_{K^+} + G_{Na^+}E_{Na^+} + G_{Cl^-}E_{Cl^-}}{G_{K^+} + G_{Na^+} + G_{Cl^-}} \tag{3.85}$$

which can be written as

$$\Delta\phi_m = T_{K^+}E_{K^+} + T_{Na^+}E_{Na^+} + T_{Cl^-}E_{Cl^-} \tag{3.86}$$

where T's are transference numbers. The equation is known as the Hodgkin–Horowicz equation and appears to describe most cell membrane potentials better than the corresponding Goldman equation as it predicts correctly the linear dependence of membrane potential on the logarithm of concentration of a given ion, as found experimentally. The success of this equation indicates that biological membranes should be considered as networks of parallel selective channels rather than thin continuous layers and that the approach of relating fluxes linearly to finite differences of electrochemical potentials across these membranes should be preferred to an integration of the differential electrodiffusion equation.

Addition of an electrogenic pump

Let us now imagine that an electrogenic sodium pump transporting actively sodium ions operates in the membrane. According to the principles of the thermodynamics of irreversible processes (see Section 3.E) the flux of sodium ions will be expressed as

$$J_{Na^+} = -L_{Na^+}(\tilde{\mu}''_{Na^+} - \tilde{\mu}'_{Na^+}) + L_{Na^+r}A_r \tag{3.87}$$

where A_r is the affinity of the reaction driving the electrogenic sodium pump. The corresponding current will be

$$j_{Na^+} = L_{Na^+}F^2(E_{Na^+} - \Delta\phi_m) + L_{Na^+r}F^2A_r/F = G_{Na^+}(E_{Na^+} - \Delta\phi_m) + G^{act}_{Na^+}E^{act}{}_{Na^+} \tag{3.88}$$

where $G^{act}_{Na^+}$ stands for $L_{Na^+r}F^2$ and $E^{act}_{Na^+}$ for A_r/F ($z_{Na^+} = 1$). The steady-state open-circuit membrane potential then will be

$$\begin{aligned}\Delta\phi_m &= \frac{G_{Na^+}E_{Na^+} + G_{K^+}E_{K^+} + G_{Cl^-}E_{Cl^-} + G^{act}_{Na^+}E^{act}_{Na^+}}{G_{Na^+} + G_{K^+} + G_{Cl^-}}\\ &= T_{Na^+}E_{Na^+} + T_{K^+}E_{K^+} + T_{Cl^-}E_{Cl^-} + \Delta\end{aligned} \tag{3.89}$$

where $T_{Na^+} + T_{K^+} + T_{Cl^-} = 1$.

$$\Delta = \frac{G^{act}_{Na^+}E^{act}_{Na^+}}{G_{Na^+} + G_{K^+} + G_{Cl^-}} = \frac{L_{Na^+r}FA_r}{G_{Na^+} + G_{K^+} + G_{Cl^-}} \tag{3.90}$$

represents the contribution of the electrogenic sodium pump to the membrane potential. It is of interest that Ripoll's equation for membrane potential across a thick asymmetrical membrane, derived by rigorous methods of irreversible thermodynamics has an analogous form.

If now the sodium pump is only partly electrogenic, transporting e.g. α sodium ions per one potassium ion in the opposite direction ($\alpha > 1$) the active transport must be split into an electrogenic and electrically silent part such that

$$L_{Na^+r}A_r = (1 - 1/\alpha)L_{Na^+r}A_r + (1/\alpha)L_{Na^+r}A_r. \tag{3.91}$$

The current carried by sodium ions then will be

$$j_{Na^+} = G_{Na^+}(E_{Na^+} - \Delta\phi_m) + (1 - 1/\alpha)L_{Na^+r}F^2 A_r/F. \tag{3.92}$$

Defining $G^{act}_{Na^+} = (1 - 1/\alpha)L_{Na^+r}F^2$, the formula for the membrane potential remains valid. For $\alpha \to \infty$ (a purely electrogenic pump) the two definitions of $G^{act}_{Na^+}$ coincide.

Let us now consider a special case; a balanced state in which $J_{Na^+} = 0$ (active transport of sodium ions and their leak just balance). Then

$$-L_{Na^+}(\tilde{\mu}''_{Na^+} - \tilde{\mu}'_{Na^+}) = -L_{Na^+r}A_r \tag{3.93}$$

so that

$$L_{Na^+}F^2(E_{Na^+} - \Delta\phi_m) = -L_{Na^+r}F^2 A_r/F = -L_{Na^+r}F^2 E^{act}_{Na^+}. \tag{3.94}$$

The sodium current across the membrane can then be written as

$$\begin{aligned} j_{Na^+} &= L_{Na^+}F^2(E_{Na^+} - \Delta\phi_m) + (1 - 1/\alpha)L_{Na^+r}F^2 E^{act}_{Na^+} \\ &= L_{Na^+}F^2(E_{Na^+} - \Delta\phi_m) - (1 - 1/\alpha)L_{Na^+}F^2(E_{Na^+} - \Delta\phi_m) \\ &= (1/\alpha)L_{Na^+}F^2(E_{Na^+} - \Delta\phi_m) = (1/\alpha)G_{Na^+}(E_{Cl^-} - \Delta\phi_m). \end{aligned} \tag{3.95}$$

Thus

$$\begin{aligned} j_{Na^+} + j_{K^+} + j_{Cl^-} = {} & (1/\alpha)G_{Na^+}(E_{Na^+} - \Delta\phi_m) \\ & + G_{K^+}(E_{K^+} - \Delta\phi_m) + G_{Cl^-}(E_{Cl^-} - \Delta\phi_m) \end{aligned} \tag{3.96}$$

and

$$\Delta\phi_m = \frac{G_{Na^+}E_{Na^+} + \alpha G_{K^+}E_{K^+} + \alpha G_{Cl^-}E_{Cl^-}}{G_{Na^+} + \alpha G_{K^+} + \alpha G_{Cl^-}}. \tag{3.97}$$

For α very large (a purely electrogenic pump)

$$\Delta\phi_m = \frac{G_{K^+}E_{K^+} + G_{Cl^-}E_{Cl^-}}{G_{K^+} + G_{Cl^-}}. \tag{3.98}$$

Formulas applicable to diverse situations can be constructed analogously.

Ussing's flux ratio

The type of transmembrane flow is conveniently characterized by the flux ratio first introduced by Ussing. In a simple electrodiffusion of an ion it is defined as

$$J'_i/J''_i = c'_i/c''_i e^{-zF\Delta\phi_m/RT} \tag{3.99}$$

It will be seen that for a nonelectrolyte ($z = 0$) the exponent will be zero and the fluxes will be directly proportional to actual concentrations.

If Ussing's flux ratio is obeyed it indicates that the unidirectional flows are independent of each other. In nonspecific permeation this is certainly the case except, perhaps, in single-file diffusion. It will be seen later that carrier-type saturated fluxes deviate from Ussing's ratio under general conditions.

It should be clear from the preceding discussion that in a 'simple diffusion' system, net flow ceases ($J_j = 0$) when

$$c_j''/c_j' = e^{-zF\Delta\phi_m/RT} \tag{3.100}$$

which, for nonelectrolytes and/or for an uncharged membrane, reduces to

$$c_s''/c_s' = 1. \tag{3.101}$$

3.3.4 Mechanisms of simple diffusion

Table 3.4 enumerates three mechanisms of nonspecific permeation, these being of two principally different types. The first is passage through the nonpolar parts of the membrane, the second is passage through aqueous channels or pores.

Passage through nonpolar membrane parts

The obvious mechanism here is partitioning between the aqueous solution and the membrane as described in Section 3.D followed by genuine diffusion within the membrane.

Any relatively small molecule or ion, as long as it is soluble in water as well in the membrane lipid can use this pathway; e.g. lower alcohols, esters, organic acids, aromatic and heterocyclic compounds (most artificial drugs), and also gases. Once in the membrane they move within and across it by diffusion which, however, is non-Stokesian in nature as the microenvironment in the membrane is not homogeneous and the underlying mechanism of movement is apparently by jumping to vacancies that occur in the membrane structure. It may be recalled that lateral diffusion of lipids along the membrane plane followed an analogous pattern.

Clearly, the formation of kinks in the fatty acyl chains of membrane lipids (see p. 62) would substantially assist such movement and would extend its possibilities even to rather hydrophilic molecules or ions (glycols, urea, Cl^-, NO_3^-) as one can visualize the kink pockets to be filled with microdroplets of external solution. Still, the significance of the kinks for the transport of polar molecules is not as great as might be suspected—in the erythrocyte no more than 20 per cent of water passes across the plasma membrane by this mechanism.

Larger nonpolar molecules (higher alcohols, fatty acids, and the like) apparently move through the lipid regions of the membrane but there, owing to their size, the passage across the membrane will be certainly accompanied by much lateral movement as they will undergo lateral diffusion just like genuine membrane lipid components. This type of nonspecific permeation still awaits its theoretical analysis which might show that this transport has some elements of the flip-flop movement of lipids.

None of these mechanisms permits the transmembrane movement of highly hydrophilic and charged molecules, such as simple sugars or amino acids.

Transmembrane pores with effective diameters below 0.5 nm appear to be present in or at membrane-spanning proteins, the pores being either part of a specific mechanism (as will be considered in the next chapter) or simply water-filled channels permitting the passage of small hydrophilic molecules, foremost among them water itself.

Considering the size of these water-filled pores it is most likely that molecules move in them one by one, in a single file. This puts some constraints on their movement and thus alters the kinetic description of their rates of movement. While in the foregoing section the Eyring rate theory described adequately the microscopic movement of individual independent particles, such as apparently exists during passage through the lipid domains of the membrane or through nonconvective but broad enough aqueous pores (perhaps the situation described as reversible electric breakdown in Section 2.5.5), the kinetics of flow through long narrow channels has some elements of saturation, as will be discussed in detail in Section 3.5.

To illustrate the point let us consider the following situation.

1. The channel through which single-file diffusion takes place is a sequence of L potential wells separated by identical potential barriers.
2. Each well can be either empty or occupied by a single solute particle.
3. The particles do not interact beyond their mutual exclusion from occupying the same well or site.
4. The translocation of a particle from its site to an adjacent free site is defined by the probability of such translocation in a given time interval (v).
5. The solution to the left will only contain substance A at concentration c_A, the solution to the right only substance B at concentration c_B.

The occupancy of intramembrane sites can be defined by a paired correlation function $F(X_1X_2)$ where $X = 1$ if the site is occupied and $X = 0$ if the site is vacant. If $\vec{v}$ is the rate at which the solute particle is translocated from site n to site $n+1$, and $\overleftarrow{v}$ the rate of translocation from $n+1$ to n the flux through the system will be given by

$$J = \vec{v}\,F(10) - \overleftarrow{v}\,F(01). \tag{3.102}$$

The probability of occupancy of site n and of site $n+1$ is given by $F(10) + F(11) = \theta_1$ and $F(01) + F(11) = \theta_2$, respectively. If A and B are both considered separately then $\theta_A = F(\mathrm{A0}) + F(\mathrm{AA}) + F(\mathrm{AB})$ and $\theta_B = F(\mathrm{B0}) + F(\mathrm{BB})$, $F(\mathrm{BA})$ being zero because A and B can never switch places in the channel.

The unidirectional flow of A from left to right will be given by

$$J_A = v[F(\mathrm{A0}) - F(\mathrm{0A})] = -D_A[\mathrm{d}\theta_A/\mathrm{d}x + F(\mathrm{AB})/\delta] \tag{3.103}$$

where $D_A = v\delta$ with δ the distance between sites. By analogy, proceeding from right to left,

$$J_B = D_B[\mathrm{d}\theta_B/\mathrm{d}x - F(\mathrm{AB})/\delta]. \tag{3.104}$$

Now the edge sites of the channel are assumed to be in equilibrium with the adjacent solutions so that at the left end the probability of occupancy will be $\theta_A = c_A/(c_A + \beta)$ while at the right end it will be $\theta_B = c_B/(c_B + \beta)$ where β is the partition coefficient of A or B between the channel site and the solution.

The cross function $F(\mathrm{AB})$ has been identified in the first approximation with $\theta_A\theta_B$ so that

$$J_A = -D_A(\mathrm{d}\theta_A/\mathrm{d}x + \theta_A\theta_B/\delta). \tag{3.105}$$

For low occupancies, i.e. low solute concentration, the rate of flow will be as in classical diffusion, for higher ones a saturation term will decrease the flow in a nonlinear way.

Taking into account the number of sites in the channel, for low occupancies J_A is approximated by $D_A c_A/\beta L$, hence a linear dependence of unidirectional flow on concentration, like in classical diffusion. For high occupancies ($\theta_A \to 1$)

$$J_A = \left[\frac{1}{e^L - 1} - \frac{e^L(L-2)+2}{(e^L-1)^2}(1-\theta_A)\right] D_A \tag{3.106}$$

so that a hyperbolic relationship may be expected between J_A and c_A. The rate of unidirectional flow at limiting occupancy is equal to

$$J_A = D_A/(e^L - 1) \tag{3.107}$$

i.e., a constant independent of concentration (analogous to maximum rate of transport) but steeply inversely proportional to the number of sites along the channel.

To conclude the section on aqueous pores, the statistical pores assumed to exist continuously in membranes (see Section 2.5.5) should be mentioned. It is likely that these pores are nonconductive except when a short and/or small electrical pulse is applied. However, they may contribute to volume flow of water as well as to diffusional flow of small uncharged solutes.

3.3.5 Membrane currents

Various formalisms can be used to describe the transport of ions in dependence on the membrane electric potential and to express it as steady-state current flowing through the membrane, the current being considered as a sum of individual steady-state ionic currents. It may be seen that, if the concentrations of ions at the two sides of the membrane are identical, i.e. the system is symmetrical, and the mobilities of ions do not depend on potential, the Goldman formalism predicts an ohmic behaviour, i.e. a linear dependence of current on voltage. In the Hodgkin–Horowicz approach extended in the sense of the thermodynamics of irreversible processes the ohmic behaviour is predicted throughout as long as the individual conductances remain constant. Finally, the Eyring rate theory can obviously explain various nonlinearities of current-voltage characteristics by operating with the number and heights of individual barriers, but it seems that only the molecular dynamics approach may remove certain arbitrariness from such procedures. Be this as it may, it appears that in many cases additional factors must be invoked, influencing the shape of current-voltage characteristics.

It is a common experience that already the current-voltage characteristics of artificial lipid bilayer membranes are nonlinear. Such membranes behave as ohmic resistances in the region of some ± 20 or 30 mV and then their current-voltage characteristic is superlinear (Fig. 3.7), the current increasing more rapidly than the voltage. Various hypotheses explaining the superlinearity were tested. One of them was ion injection into the membrane at higher voltages but it was found that the effect gives a significant contribution only at low ionic strengths (< 1 mM) of the aqueous medium. Another possibility was the effect

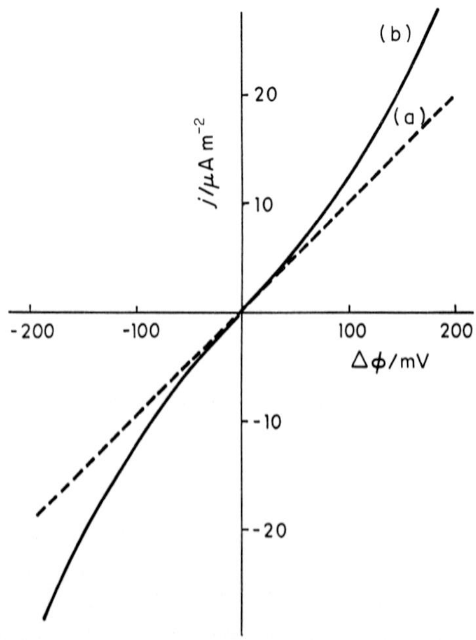

Fig. 3.7. Current-voltage relationship as predicted by the Goldman equation, showing an ohmic behaviour (a), and the superlinear type observed experimentally (b)

of image charges in the membrane but even there satisfactory agreement was found only at low ionic strengths and low voltages. The dissociation field effect (the second Wien effect) was introduced as another possible explanation of the superlinearity; in electric fields of a high intensity the dissociation of ion pairs is substantially increased. It was found that the effect is appreciable only when the rate-limiting step for the ion transport is the passage of the ion across the interface. Still more successful was the description involving simply a barrier modified by the applied electric field. These two last explanations of the superlinearity of the current-voltage characteristics (the Wien effect and/or the modification of the barrier by the electric field) then again require unrealistically high distribution coefficients of permeating ions between the lipid and the aqueous phases. A recent hypothesis is based on the transport of ions through transient pores as postulated by the theory of reversible electric breakdown of bilayer lipid membranes (see Section 2.5.5). The nonlinearity is thus explained by the shifting of pore distribution towards wider radii with increasing applied voltage.

In contrast, hydrophobic molecules and ions pass quite readily through the cell or model membrane—to name a few: tetraphenylborate, dipicrylaminate, triiodide and tetraalkylammonium ions with long alkyl chains. The transfer of hydrophilic ions through a lipid membrane can either proceed by a diffusion-type process or by a carrier- or channel-mediated process.

In the case of diffusion, the current passing through the membrane is governed by its surface concentrations of ions c_s' and c_s'' (affected by the surface potential) and by the translocation rate (a function of the energy barriers to be overcome) and is given by equations of the type of Eq. (3.45) where flows in both directions are considered; thus

$$j = Fk[c_s' \exp(F\,\Delta\phi/2RT) - c_s''(-F\,\Delta\phi/2RT)]. \qquad (3.108)$$

In the case of a mobile carrier for ions (such as valinomycin for K^+) an analogous equation will hold with c_s' and c_s'' being determined by the association-dissociation equilibrium between carrier, ligand and the carrier-ligand complex (cf. Section 3.5.3).

In the two cases discussed so far a continuous distribution of the species participating in the ion transport connected with electric current flow could be assumed. The situation is different when the membrane contains built-in ion-conducting channels and the individual molecular events must be accounted for. Such processes belong to the most important in membrane transport but, because of their complexity, only few could be elucidated in detail. The best example is the channel-forming peptide, gramicidin A (Fig. 3.8).

The rate of ion transport through the channel is rather high, amounting, for the example of Cs^+ in a gramicidin channel, to $6 \cdot 10^7$ ions per second at a potential of 0.1 V imposed on an artificial membrane containing such channels. The mechanism of ion transport through the channel is discussed on p. 218 but the main interest has so far been focused on the mechanism and rate of channel formation and dissociation. The formation (and, at the same time, the opening) of the gramicidin channel may be due to the dimerization

$$2G \underset{k_c}{\overset{k_o}{\rightleftharpoons}} G_2. \qquad (3.I)$$

In this way, a channel of a length of 2.5–3.0 nm can be formed which is capable of bridging the hydrocarbon region of the membrane. Three approaches to the

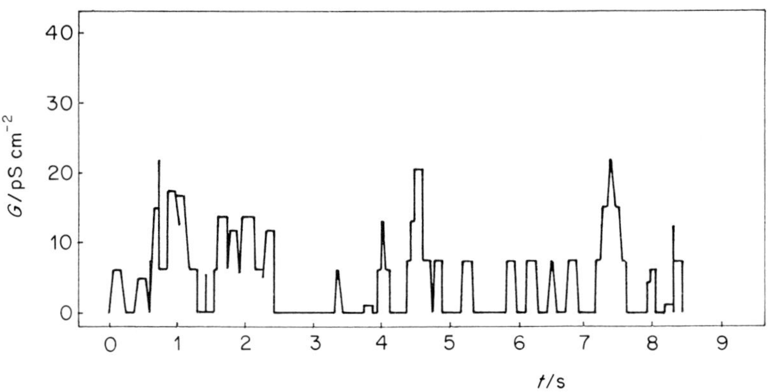

Fig. 3.8. Conductivity of gramicidin A channels observed in an artificial membrane

investigation of channel forming and opening have so far been used, based on the relaxation of a multichannel system, on relaxation of a single channel using voltage jump and temperature-jump methods and finally, on treating this process as a noise phenomenon.

In a multichannel system the current $j(t)$ fluctuates around a mean value $\bar{j}$. The amplitude of these fluctuations is described by the variance

$$\sigma^2 = \langle [j(t) - \bar{j}]^2 \rangle \tag{3.109}$$

where the brackets $\langle\ \rangle$ denote the average value. Both $\bar{j}$ and σ^2 are given by the equations

$$\bar{j} = iNk_c/(k_o + k_c) \tag{3.110}$$

$$\sigma^2 = i^2 Nk_o k_c/(k_o + k_c) \tag{3.111}$$

where i is the single-channel current and N the number of channels in the membrane. From a Fourier analysis of current fluctuation the spectral density function $S(f)$ is obtained which, for the present model, is given by the equation

$$S(f) = \frac{4\sigma^2}{k_o + k_c + 4\pi^2 f^2} \tag{3.112}$$

where f is the frequency of channel opening. From the frequency distribution characterized by $S(f)$ the constants k_o and k_c are obtained. The constants determined by noise analysis are identical with those obtained from relaxation experiments.

The behaviour of many cell membranes under natural conditions is more complicated, doubtless due to presence of ionic gradients and to saturable electrogenic primary and secondary active transports.

Both primary and secondary ion transport systems, i.e., respectively, those where the current-carrying ion is pumped across a membrane in conjunction with a coupled chemical reaction and those where it serves as a symported species along with a nonelectrolyte (or another ion), can be described formally as is done in Section 3.5, the various formulas corresponding to different modes of coupling with energy. The potential-dependent steps are those in which charge is translocated across a membrane and gives rise to a current. In all but the very esoteric models these steps are the ones where the 'active' complex of a carrier with the ion moves and the whole treatment can thus be simplified by singling out these two steps while all the other voltage-independent steps are grouped in two kinetic constants. This gives rise to a very simple model, such as

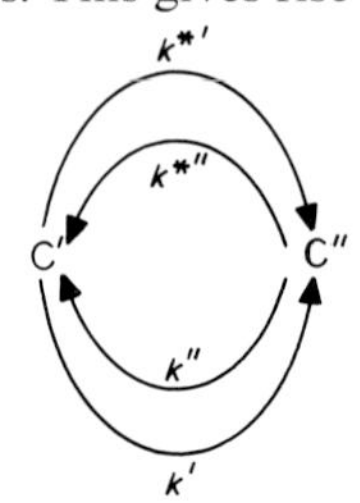

(3.II)

where the charge-carrying steps are asterisked, the others shown as simple k's; c' and c'' are concentrations of the 'active' membrane-crossing complex of carrier with ion (in primary) or of carrier with solute and ion (in secondary) transport.

If Eyring's symmetrical barrier concept is again used for dividing the membrane potential the $k^{*\prime}$s can be interpreted as

$$\begin{aligned} k^{*\prime} &= k'_{\mathrm{o}}\,\mathrm{e}^{-zF\Delta\phi/2RT} \\ k^{*\prime\prime} &= k''_{\mathrm{o}}\,\mathrm{e}^{zF\Delta\phi/2RT}. \end{aligned} \qquad (3.113)$$

The equation expressing the dependence of current on voltage for an electrogenic ion-transport system is given by

$$j = zFc_{\mathrm{E_t}}(k'k^{*\prime\prime} - k''k^{*\prime})/(k^{*\prime\prime} + k^{*\prime} + k' + k'') \qquad (3.114)$$

where $c_{\mathrm{E_t}}$ is the total carrier concentration (corrected for the so-called concealed reserve factors).

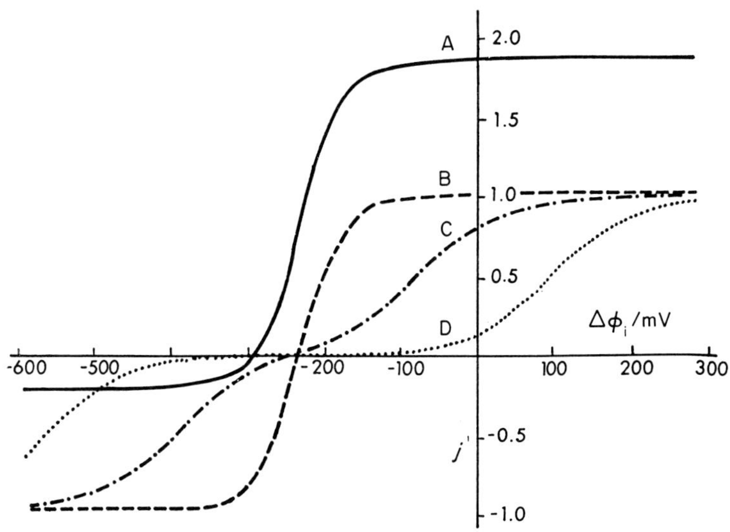

Fig. 3.9. Current-voltage curves calculated on the basis of the two-state model described in the text where j' is the reduced current equal to $2jr'r''/zFNk'k''$ where N is the total amount of carrier and $r' - 1$ and $r'' - 1$ are the carrier forms that are not involved in the current-carrying step. The following values were used to plot the individual curves

k'	k'	$k^{*\prime}$	$k^{*\prime\prime}$
0.1	1000	0.003	0.03
0.1	1000	0.03	0.03
0.1	1000	1000	100
0.001	10	30	30

The reversal potentials of the individual cases, i.e. potentials when $\exp(zF\Delta\phi/RT) = k'k^{*\prime\prime}/k''k^{*\prime}$, are seen to be -296 mV for case A and -236 mV for B, C, and D

Remarkably, a variety of j–V curves can be obtained by simply varying the numerical values of k's and $k^{*\prime}$s, so that virtually all existing types of relationship can be displayed (Fig. 3.9).

An analysis of the behaviour of such systems under nonsteady-state conditions including transients of electric parameters has been made and is highly revealing (see 'Further reading').

3.3.6 Specifics of water transport

Transmembrane movement of water deserves special attention as it can proceed both by a diffusion type mechanism and by a bulk or mass flow. The interactions between water and solute flows are best described by an approach based on nonequilibrium but near-equilibrium thermodynamics, known also as steady-state thermodynamics or thermodynamics of irreversible processes.

3.E THERMODYNAMICS OF IRREVERSIBLE PROCESSES

The fundamental relations of this thermodynamics are linear phenomenological equations. Some simple linear phenomenological equations are in fact common knowledge: Ohm's law, according to which the current density in a conductor is proportional to the gradient of the electric potential; Fick's law, according to which a diffusional flow is proportional to the concentration gradient. On the other hand, both electric current and diffusion flow can be driven by other forces, e.g., by a gradient of temperature. Such phenomena are known as thermoelectricity and thermodiffusion, respectively.

According to the thermodynamics of irreversible processes (known also as steady-state or nonequilibrium thermodynamics) each flow J_i is linearly dependent on all forces X_k present in a system

$$J_i = \sum_{k=1}^{n} L_{ik} X_k \qquad (i = 1, 2, \ldots, n) \tag{3.E1}$$

For the case of two flows and two forces then

$$\begin{aligned} J_1 &= L_{11} X_1 + L_{12} X_2 \\ J_2 &= L_{21} X_1 + L_{22} X_2. \end{aligned} \tag{3.E2}$$

It must be tested in each specific case whether the linear relations are a satisfactory description of reality. If it is so and the flows and forces are correctly chosen, the matrix of the phenomenological coefficients L is symmetrical (it remains the same when columns and rows are interchanged). Thus

$$L_{ik} = L_{ki} \qquad (i \neq k) \tag{3.E3}$$

and, in the simple case of two flows and two forces,

$$L_{12} = L_{21}. \tag{3.E4}$$

This important law is named after Lars Onsager and it is a useful relationship, since from the dependence of flow J_i on force X_k one can predict the dependence of flow J_k on force X_i which may not be directly measurable.

The flows and forces are chosen correctly if the sum of the products of individual flows with their conjugate forces gives the production of entropy in the system, the production of entropy being a measure of the spontaneity of irreversible processes

$$\sum_{i=1}^{n} J_i X_i = \frac{d_i S}{dt}. \tag{3.E5}$$

For isothermal systems it is preferable to use the so-called dissipation function Φ, which is the production of entropy multiplied by absolute temperature

$$\sum_{i=1}^{n} J_i X_i = T\frac{d_i S}{dt} = \Phi. \tag{3.E6}$$

The entropy production (and thus also the dissipation function) must be positive. In an open system the total entropy change may be either positive or negative, since the flow of entropy across the boundaries of the system may result in a decrease of its entropy, but the change of entropy resulting from irreversible (i.e. spontaneous) processes in the system is always positive.

In the present simple system with two flows and two forces the following equations may be easily verified

$$\begin{aligned}\Phi &= J_1X_1 + J_2X_2 = L_{11}X_1^2 + 2L_{12}X_1X_2 + L_{22}X_2^2\\ &= L_{11}\left(X_1 + \frac{L_{12}X_2}{L_{11}}\right)^2 + \left(L_{22} - \frac{L_{12}^2}{L_{11}}\right)X_2^2\\ &= L_{22}\left(X_2 + \frac{L_{12}X_1}{L_{22}}\right)^2 + \left(L_{11} - \frac{L_{12}^2}{L_{22}}\right)X_1^2 > 0.\end{aligned} \tag{3.E7}$$

Since the expressions must be positive for all values of X, both positive and negative, the coefficients at each of the squares have to be positive. Hence, the straight coefficients are to be positive ($L_{11} > 0$, $L_{22} < 0$), whereas the cross coefficients relating the flows to nonconjugate forces may be either positive or negative, as long as they satisfy the condition $L_{12}^2 < L_{11}L_{22}$. In some systems $L_{12}^2 = L_{11}L_{22}$ since in these systems equilibrium can be reached even when the forces are different from zero. In such systems $\Phi \geqslant 0$.

Osmotic equilibrium

Equilibrium distribution of water or of a solvent in general can be obtained with an ideal semipermeable membrane, permeable to the solvent and impermeable to the solute. Transport of solute across such a membrane is completely prevented by a large kinetic barrier so that a static equilibrium of the solute is established. On the other hand, water is transported across the membrane from the solution where there is less water and the solute is more concentrated. There are two possibilities: either the water flow proceeds until the concentration differences vanish, or, in the more interesting case, a hydrostatic pressure is built up, preventing further water transport. Thermodynamic equilibrium is thus attained at which the chemical potentials of water at the two sides of the membrane are equal. Then

$$\mu_0' + \bar{V}_w p' + RT\ln X_w' = \mu_0'' + \bar{V}_w p'' + RT\ln X_w'' \tag{3.115}$$

where μ_0's are the standard chemical potentials, $\bar{V}_w$ is the partial molal volume of water (about 0.018 $dm^3\,mol^{-1}$) and x_w its mole fraction. Using mole fractions instead of activities implies that the solutions are ideal. In an ideal solution the interactions between the solute and solvent particles are equal to those between the particles of the solvent themselves, so that the activities of the two components depend only on their relative abundances. The mole fraction of water in a two-component system can be expressed as

$$x_w = n_w/(n_s + n_w) \tag{3.116}$$

where n_s and n_w are numbers of moles of solute and water, respectively, in an arbitrarily chosen volume of the solution; for a volume of 1 dm^3

$$x_w = c_w/(c_s + c_w) \tag{3.117}$$

where c's are molarities.

If, for simplicity, pure solvent (water) is assumed to be outside, so that $x'_w = 1$ and $RT \ln x'_w = 0$, and the system is isothermal, so that $\mu'_0 = \mu''_0$ the equilibrium difference of the hydrostatic pressure preventing the solvent from entering into the inside solution can be obtained as the osmotic pressure π

$$\pi = p'' - p' = -\frac{RT}{\bar{V}_w} \ln x''_w. \tag{3.118}$$

Using the approximations valid for dilute solutions, that

$$\ln(1 + c_s/c_w) \simeq c_s/c_w \quad (\text{since } c_s/c_w \ll 1) \quad \text{and} \quad c_w \bar{V}_w \simeq 1$$

the volume of a dilute solution is practically equal to the volume of its solvent, and writing c'_w for c''_w, etc.,

$$\pi = -\frac{RT}{\bar{V}_w} \ln X_w = -\frac{RT}{\bar{V}_w} \ln \frac{c_w}{c_s + c_w} = \frac{RT}{\bar{V}_w} \ln \frac{c_s + c_w}{c_w}$$

$$= \frac{RT}{\bar{V}_w} \ln\left(1 + \frac{c_s}{c_w}\right) \approx \frac{RT}{\bar{V}_w} \frac{c_s}{c_w} \approx RTc_s. \tag{3.119}$$

In a general case osmotic equilibrium at an ideal semipermeable membrane is given by

$$p'' - p' = \pi'' - \pi' = RT(c''_s - c'_s)$$

or

$$\Delta p = \Delta \pi = RT \Delta c_s. \tag{3.120}$$

The equations are valid for dilute solutions of nonelectrolytes. In the case of salts, their dissociation and osmotic coefficients, determined best cryoscopically in an osmometer, must be taken into account.

The approximation obtained above is identical with the original van't Hoff formula for osmotic pressure, derived from experimental measurements.

In a certain sense the excellence of this approximation has been unfortunate, since van't Hoff's formula served as the basis for various ill-formulated bombardment theories of osmosis. Van't Hoff's formula is an analogy to the ideal gas equation, since the molarity of the solute (mol dm^{-3}) is the reciprocal of the volume occupied by one mole ($dm^3\ mol^{-1}$). From this, it might seem that solute particles behave in all respects as ideal gas molecules and that their impacts on the walls of a vessel, on a semipermeable membrane or on the free surface of the solution develop the same pressure as such molecules would. However, the analogy between molecules of the ideal gas (which do not interact) and those of a solute (which, in dilute solutions also do not interact with one another) cannot be stretched that far. The solute molecules in dilute solutions do not interact with one another but they interact with molecules of the solvent, in ideal solutions exactly as much as the solvent molecules interact with one another, and for this reason they cannot develop the same pressure as molecules of an ideal gas.

Actually, the pressure exerted by a solution may be expressed as

$$p = T\left(\frac{\partial p}{\partial T}\right)_V - \left(\frac{\partial U}{\partial V}\right)_T = T\frac{\alpha}{\beta} - \left(\frac{\partial U}{\partial V}\right)_T \tag{3.121}$$

where U is the internal energy, T the absolute temperature, V the volume, α the coefficient of thermal expansion,

$$\alpha = \frac{1}{V}\left(\frac{\partial V}{\partial T}\right)_p \tag{3.122}$$

and β the coefficient of compressibility,

$$\beta = -\frac{1}{V}\left(\frac{\partial V}{\partial p}\right)_T. \tag{3.123}$$

The term $T(\partial p/\partial T)_V$ is called *thermal pressure* and arises from the thermal energy of the molecules. On the other hand, the term $(\partial U/\partial V)_T$, *internal pressure*, is related solely to the variation of intermolecular potential energy with volume. The two pressures are related to the entire system of solute and solvent particles in solution; there is no partial pressure of the solute and no partial pressure of the solvent. Partitioning of this kind is valid only for mixtures of ideal gases with no intermolecular forces. For this reason one cannot accept the much debated solvent tension theory of osmosis of Hammel and Scholander, according to which the solute in a solution exerts pressure RTc_s and the solvent is under tension (negative pressure) $p' = p - RTc_s$, where p is the ambient pressure.

Flow of water and of solute

Most membranes are not ideally semipermeable; they are permeable, although not to the same degree, to both water and solutes. During its passage across such membranes, the flow of solvent may interact with the flow of solute and this is the reason why, on the one hand, the solute is entrained by the solvent flow and, on the other, the osmotic pressure developed by a permeant solute may be considerably less than the theoretical value. The interactions between the flow of solvent J_w and the flow of solute J_s can be best described by the methods of nonequilibrium thermodynamics as follows:

$$\begin{aligned} J_{\mathrm{w}} &= L_{\mathrm{ww}}\,\Delta\mu_{\mathrm{w}} + L_{\mathrm{ws}}\,\Delta\mu_{\mathrm{s}} \\ J_{\mathrm{s}} &= L_{\mathrm{sw}}\,\Delta\mu_{\mathrm{w}} + L_{\mathrm{ss}}\,\Delta\mu_{\mathrm{s}} \end{aligned} \tag{3.124}$$

Each of the two flows is linearly dependent not only on its own conjugate driving force, the difference of the chemical potential of the transported substance across the membrane, but also on the force conjugate to the other flow. The equations are written directly in terms of finite differences of chemical potentials across thin biological membranes, rather than in terms of differential coefficients, which would cause difficulties in integrating them across such thin nonhomogeneous layers. The final justification for such a procedure are the experimentally found linear relations. However, the two flows should be first expressed in terms of directly measurable quantities, viz. differences of solute concentrations Δc_{s}, and of hydrostatic pressure Δp. To this end we shall assume that our solutions are ideal, i.e. that the chemical potential differences depend only on differences of mole fractions and of hydrostatic pressure

$$\begin{aligned} \Delta\mu_{\mathrm{w}} &= \bar{V}_{\mathrm{w}}\,\Delta p + RT\,\Delta\ln x_{\mathrm{w}} \\ \Delta\mu_{\mathrm{s}} &= \bar{V}_{\mathrm{s}}\,\Delta p + RT\,\Delta\ln x_{\mathrm{s}}. \end{aligned} \tag{3.125}$$

since

$$x_{\mathrm{w}} = c_{\mathrm{w}}/(c_{\mathrm{w}} + c_{\mathrm{s}}) \qquad x_{\mathrm{s}} = c_{\mathrm{s}}/(c_{\mathrm{w}} + c_{\mathrm{s}})$$

it follows that

$$\begin{aligned} \Delta\ln x_{\mathrm{w}} &= \Delta\ln\frac{c_{\mathrm{w}}}{c_{\mathrm{w}} + c_{\mathrm{s}}} = -\Delta\ln\frac{c_{\mathrm{w}} + c_{\mathrm{s}}}{c_{\mathrm{w}}} \\ &= -\Delta\ln\left(1 + \frac{c_{\mathrm{s}}}{c_{\mathrm{w}}}\right) \approx -\Delta\frac{c_{\mathrm{s}}}{c_{\mathrm{w}}} = -\frac{\Delta c_{\mathrm{s}}}{\bar{c}_{\mathrm{w}}} \end{aligned} \tag{3.126}$$

where $\tilde{c}_{\mathrm{w}}$ is the mean concentration of water in the two solutions. Similarly

$$\begin{aligned} \Delta\ln x_{\mathrm{s}} &= \Delta\ln\frac{c_{\mathrm{s}}}{c_{\mathrm{w}} + c_{\mathrm{s}}} \approx \Delta\ln\frac{c_{\mathrm{s}}}{c_{\mathrm{w}}} \\ &= \Delta(\ln c_{\mathrm{s}} - \ln c_{\mathrm{w}}) \approx \Delta\ln c_{\mathrm{s}} = \frac{\Delta c_{\mathrm{s}}}{\bar{c}_{\mathrm{s}}} \end{aligned} \tag{3.127}$$

where $\bar{c}_{\mathrm{s}}$ is the mean concentration of the solute in the two solutions. Thus,

$$\begin{aligned} \Delta\mu_{\mathrm{w}} &= \bar{V}_{\mathrm{w}}\,\Delta p - \frac{RT\,\Delta c_{\mathrm{s}}}{\bar{c}_{\mathrm{w}}} \\ \Delta\mu_{\mathrm{s}} &= \bar{V}_{\mathrm{s}}\,\Delta p + \frac{RT\,\Delta c_{\mathrm{s}}}{\bar{c}_{\mathrm{s}}}. \end{aligned} \tag{3.128}$$

It should be noted that any combination of flows and forces is legitimate as long as the sum of products of flows with their conjugate forces gives the correct dissipation function. For instance, using the above defined flows of solvent and solute and their conjugate forces

$$\frac{\mathrm{d_i}S}{\mathrm{d}t}T = J_\mathrm{w}\Delta\mu_\mathrm{w} + J_\mathrm{s}\Delta\mu_\mathrm{s}$$

$$= J_\mathrm{w}\left(\bar{V}_\mathrm{w}\Delta p - \frac{RT\Delta c_\mathrm{s}}{\bar{c}_\mathrm{w}}\right) + J_\mathrm{s}\left(\bar{V}_\mathrm{s}\Delta p + \frac{RT\Delta c_\mathrm{s}}{\bar{c}_\mathrm{s}}\right)$$

$$= (J_\mathrm{w}\bar{V}_\mathrm{w} + J_\mathrm{s}\bar{V}_\mathrm{s})\Delta p + \left(\frac{J_\mathrm{s}}{\bar{c}_\mathrm{s}} - \frac{J_\mathrm{w}}{\bar{c}_\mathrm{w}}\right)RT\Delta c_\mathrm{s}$$

$$= J_\mathrm{V}\Delta p + J_\mathrm{D}RT\Delta c_\mathrm{s}. \tag{3.129}$$

The new flows thus defined are the volume flow J_V

$$J_\mathrm{V} = J_\mathrm{w}\bar{V}_\mathrm{w} + J_\mathrm{s}\bar{V}_\mathrm{s} \tag{3.130}$$

which is conjugate to the force Δp (hydrostatic pressure difference) and the exchange flow J_D

$$J_\mathrm{D} = J_\mathrm{s}/\bar{c}_\mathrm{s} - J_\mathrm{w}/\bar{c}_\mathrm{w} \tag{3.131}$$

conjugate to the force $RT\,\Delta c_\mathrm{s}$ (osmotic pressure difference). The two new flows have dimensions of velocity; the volume flow expressed, e.g., in $\mathrm{cm^3\,cm^{-2}\,s^{-1}}$ corresponds at the same time to the velocity of flow in $\mathrm{cm\,s^{-1}}$ and the exchange flow gives the velocity of solute relative to water in the membrane. The phenomenological equations written in terms of new flows and forces are

$$J_\mathrm{V} = L_\mathrm{p}\,\Delta p + L_\mathrm{pD}RT\,\Delta c_\mathrm{s} \tag{3.132a}$$

$$J_\mathrm{D} = L_\mathrm{Dp}\,\Delta p + L_\mathrm{D}RT\,\Delta c_\mathrm{s} \tag{3.132b}$$

where $L_\mathrm{pD} = L_\mathrm{Dp}$ (Onsager's law) whenever the equations are linear *sensu stricto*, i.e. when all coefficients L are constants. It is seen that the coefficient L_pD is a measure of the volume flow brought about by a difference of osmotic pressure and the coefficient L_Dp is a measure of ultrafiltration brought about by a difference of hydrostatic pressure.

At one extreme, in very coarse membranes (e.g. a sintered-glass disc) there is very little osmosis and ultrafiltration and the cross coefficients L_pD and L_Dp are almost zero. (There is always some osmosis and ultrafiltration, unless the diffusion coefficients of solute and solvent are identical.)

Another limiting case is an ideally semipermeable membrane. The flow of solute J_s, is zero across such membrane, so that

$$J_\mathrm{V} = J_\mathrm{w}\bar{V}_\mathrm{w} \tag{3.133}$$

(the volume flow across an ideally semipermeable membrane is equal to the volume flow of the solvent alone) and

$$J_\mathrm{D} = -J_\mathrm{w}/\bar{c}_\mathrm{w} = J_\mathrm{w}\bar{V}_\mathrm{w} \tag{3.134}$$

(velocity of solute is zero and in a dilute solution $J/\bar{c}_\mathrm{w} \simeq \bar{V}_\mathrm{w} \simeq 0.018\,\mathrm{dm^3\,mol^{-1}}$).

Thus for an ideally semipermeable membrane

$$J_V + J_D = 0 \tag{3.135}$$

or, introducing the phenomenological equations (3.132a) and (3.132b) together with Onsager's law $L_{pD} = L_{Dp}$

$$(L_p + L_{pD})\,\Delta p + (L_D + L_{pD})RT\,\Delta c_s = 0. \tag{3.136}$$

The equation has to be satisfied for all possible hydrostatic and osmotic pressure differences and this can be achieved only if

$$L_p = -L_{pD} = L_D \tag{3.137}$$

so that an ideally semipermeable membrane is completely characterized by a single phenomenological coefficient, the hydraulic conductivity L_p. Equation (3.132a) may thus be written for an ideally semipermeable membrane as

$$J_V = L_p(\Delta p - RT\,\Delta c_s). \tag{3.138}$$

In an ideally semipermeable membrane the volume flow depends on the difference of osmotic pressure by the same coefficient L_p as on the hydrostatic (sometimes called hydraulic) pressure. This fact implies that volume flow across an ideally semipermeable membrane proceeds always by the same mechanism, irrespective of its hydraulic or osmotic origin. In porous membranes the hydraulic flow involves a mass or bulk flow (convection; p. 135), during which solvent molecules encounter less friction than when they diffuse separately. In order to produce mass flow across a porous membrane the osmotic flow itself has to generate a drop of hydrostatic pressure at the interface between the solution and the pure solvent in membrane pores. Jumps of water molecules from the higher concentration in the pores are more frequent than in the opposite direction and hence leave behind more vacancies. As a result of this the density of water and hence also its hydrostatic pressure is lower in this part of the pore than in the surroundings. This pressure drop counteracts the concentration fall of water at the pore opening and, at the same time, it brings about a bulk flow from the other end of the pore, i.e., osmosis. Calculations using the admittedly still crude kinetic theory of liquids provided the right order of magnitude of the pressure.

Membranes permeable not only to water but also to the solute require for their characterization two more coefficients, L_D and L_{pD} or, preferably ω and σ, derived from them as follows. From Eq. (3.131)

$$J_s = \bar{c}_s(J_D + J_w/\bar{c}_w) \tag{3.139}$$

but $J_w/\bar{c}_w \simeq J_w\bar{V}_w \simeq J_V$, since with dilute solutions $\bar{V}_w \simeq 1/\bar{c}_w$ and the volume flow is almost entirely due to the volume flow of the solvent. Thus

$$J_s = \bar{c}_s(J_D + J_V). \tag{3.140}$$

If solute flow J_s is measured with tracers in the absence of volume flow

$$J_s = \bar{c}_s J_D \tag{3.141}$$

and from Eq. (3.132a) it follows for $J_V = 0$ that

$$\Delta p = -\frac{L_{pD}}{L_p} RT\,\Delta c_s. \tag{3.142}$$

Replacing L_{Dp} with L_{pD} and using the definition of Δp in Eq. (3.142), J_D defined by Eq. (3.132b) can be introduced into 3.141 to yield

$$J_s = \frac{L_p L_D - L_{pD}^2}{L_p} \bar{c}_s RT\,\Delta c_s \tag{3.143}$$

and defining the coefficient ω as

$$\omega = (L_p L_D - L_{pD}^2)\bar{c}_s / L_p \tag{3.144}$$

one can write

$$J_s = RT\omega\,\Delta c_s \tag{3.145}$$

where $RT\omega$ is the permeability coefficient of a nonelectrolyte. The reflection coefficient σ is defined as

$$\sigma = -L_{pD}/L_p \tag{3.146}$$

so that Eq. (3.132a) can now be written as

$$J_V = L_p(\Delta p - \sigma RT \Delta c_s) \tag{3.147}$$

and Eq. (3.142), valid for zero volume flow, can be written

$$\Delta p = \sigma RT \Delta c_s. \tag{3.148}$$

Thus the volume flow is seen to stop when the observed osmotic pressure Δp is equal to the theoretical osmotic pressure $RT\,\Delta c_s$ multiplied by the reflection coefficient σ. With an impermeant solute the membrane behaves as an ideally semipermeable membrane for which $L_{pD} = -L_p$, so that $\sigma = 1$. As already discussed, osmotic pressure is then fully equivalent to hydrostatic pressure. Equation (3.147) may thus be generalized for the presence of impermeant with the concentration difference Δc_i as

$$J_V = L_p(\Delta p - RT\,\Delta c_i - \sigma RT\,\Delta c_s) \tag{3.149}$$

and the reflection coefficient σ can be measured in the absence of a hydrostatic pressure difference; the volume flow is zero if

$$\sigma RT\,\Delta c_s = -RT\,\Delta c_i \tag{3.150}$$

so that

$$\sigma = -\Delta c_i / \Delta c_s. \tag{3.151}$$

The principal reason why the reflection coefficients of solutes are less than one is frictional interaction between the flow of solute molecules and the opposite osmotic flow of solvent. Low values of reflection coefficients are used as evidence for water-filled pores permeable to solute.

For permeant nonelectrolytes, the reflection coefficient lies between zero and unity, thus, for example, for urea (with a radius of 0.27 nm), it is 0.62 in erythrocytes, 0.70 in squid axon, 0.82 in sarcolemma and 1.00 in the plasmalemma of *Chara corralina*; for glycerol (with a radius of 0.28 nm), it is 0.88 in erythrocytes, 0.96 in squid axon and 0.86 in sarcolemma.

However, it should be remembered that the reflection coefficient of a permeating solute is always less than one, even when the solute pathway and the solvent pathway are distinct. Let us treat a solute-impermeable pathway for the solvent; the reflection coefficient for this ideally semipermeable pathway is one and the volume flow across it is given by Eq. (3.138)

$$J_w \bar{V}_w = L_p(\Delta p - RT\,\Delta c_s).$$

On the other hand, the volume flow of solute across the solute-permeable and solvent-impermeable pathway is, according to Eq. (3.145)

$$J_s \bar{V}_s = \bar{V}_s RT\omega\,\Delta c_s.$$

The reflection coefficient is equal to $\Delta p/RT\,\Delta_s$ when the total volume flow $J_V = J_w \bar{V}_w + J_s \bar{V}_s$ is zero, i.e.

$$\sigma = \Delta p/RT\,\Delta c_s = 1 - \bar{V}_s\omega/L_p \tag{3.152}$$

for a membrane with distinct solute and solvent pathways. If there is a common pathway for solute and solvent

$$\sigma < 1 - \bar{V}_s\omega/L_p. \tag{3.153}$$

Another possibility how to demonstrate a common pathway for solute and solvent is to show that solute particles are entrained by the volume flow. The appropriate equation may be derived as follows.

Introducing Eqs (3.146) into (3.144), we get

$$\omega = (L_D - L_p\sigma^2)\bar{c}_s. \tag{3.154}$$

Then, combining Eqs (3.140), (3.147) and (3.132b) for $L_{Dp} = L_{pD}$ with Eqs (3.146) and (3.154) we obtain

$$J_s = \bar{c}_s L_p(1-\sigma)\,\Delta p + [\omega - \bar{c}_s L_p(1-\sigma)\sigma]RT\,\Delta c_s. \tag{3.155}$$

Finally, combining Eqs (3.155) and (3.147)

$$J_s = RT\omega\,\Delta c_s + \bar{c}_s(1-\sigma)J_V. \tag{3.156}$$

A few words of caution concerning the effect of unstirred layers are required here. It would seem, on the basis of the above equation, that whenever e.g. $\Delta c_s = 0$ at $J_V \neq 0$, then $J_s \neq 0$ means a frictional interaction of solute particles in water-filled pores. However, even when there is no concentration difference between the bulks of the solutions at the two sides of the membrane, there can and indeed must be a concentration difference across the membrane, between two adjacent unstirred layers, as a result of ultrafiltration and sweeping effect of the solvent flow on the solute concentration profiles. Net transport of solute is thus brought about even when the solute and solvent pathways across the membrane are distinct.

Electro-osmosis

The phenomenon involving the flow of water across a membrane separating two salt solutions under the influence of passing current is called electro-osmosis. In fact, electro-osmosis and related processes are more important in technical and laboratory practice than in living systems but they provide a good example of the usefulness of Onsager's law. Electro-osmosis (electro-osmotic volume flow) takes place in the system shown in Fig. 3.10. In the middle of the chamber a membrane separates two salt solutions; pores in the membrane are permeable both to water and to ions. In this case, the pores are more permeable to cations than to anions since negative charges (dissociated acid groups) line the pore walls and, by electrostatic repulsion, limit the entry of anions into the pores. The system is completed by two electrodes and a source of direct current. The electrical current is carried across the membrane predominantly by cations and thus cations entrain more water than the smaller amount of anions transported in the opposite direction.

Volume flow J_V can be measured by a shift of menisci in calibrated capillaries joined to the chamber. It has the same direction as the electric current, driven by potential difference U registered by the ammeter A. If positive fixed charges lined the pores, volume flow would proceed against the direction of the electric current.

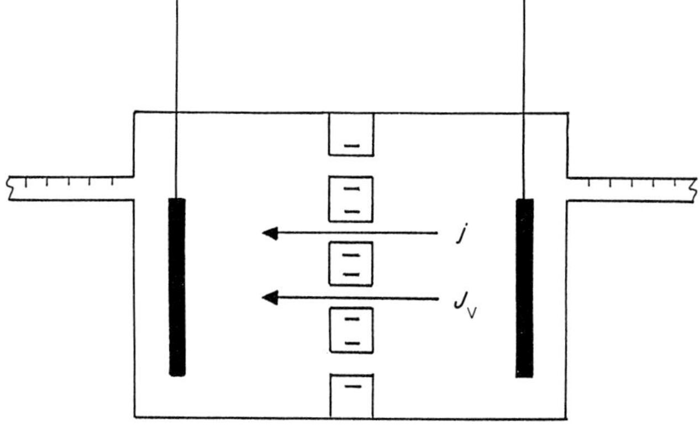

Fig. 3.10. Chamber for electro-osmotic volume flow

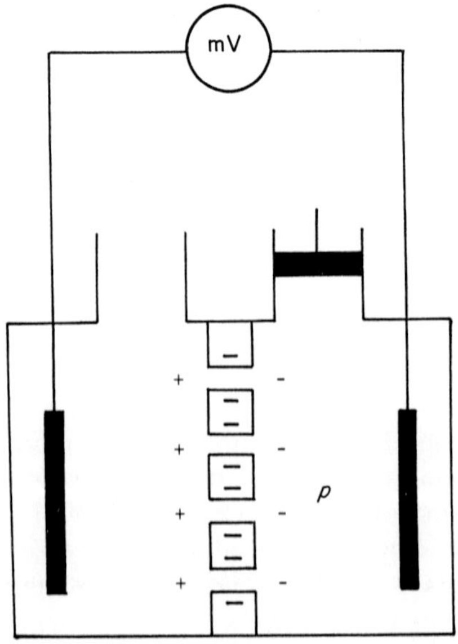

Fig. 3.11. Chamber for measuring streaming potentials

In a system with the same membrane and identical salt solutions streaming potentials can be measured instead of electro-osmotic flow (Fig. 3.11). In this case pressure is applied to the solution at one side of the membrane, e.g. by a piston. The liquid pushed through membrane pores entrains predominantly cations and thus an excess of positive charge prevails at the opposite side of the membrane. The resulting difference of electric potential can be measured by a pair of electrodes and a millivoltmeter and is called streaming potential.

Phenomenological equations describing this system express the fact that volume flow J_V is driven not only by its conjugate force, pressure p, but also by the difference of electric potential across the membrane U. Analogously, electric current j is driven both by U and pressure p. Thus

$$J_V = L_p p + L_{pU} U \tag{3.157}$$

$$j = L_{Up} p + L_U U \tag{3.158}$$

Coefficient L_p is hydraulic conductivity of the membrane and L_U its electric conductivity. Coefficients L_{pU} and L_{Up} are equal (Onsager's law) since flows and forces are properly chosen (the sum of products of flows with their conjugate forces gives the dissipation of energy in the system) and the equations are linear, as can be demonstrated experimentally. These cross-coefficients describe how much the volume flow and the electric current interact in a given system.

In *electro-osmosis* the pressure difference will be zero, the terms containing p will vanish and

$$(J_V/j)_{p=0} = L_{pU}/L_U. \tag{3.159}$$

When measuring a *streaming potential* the electric current j is zero (millivoltmeter with a high input resistance is used). From the second equation we then get for $j=0$

$$(U/p)_{j=0} = -L_{Up}/L_U. \tag{3.160}$$

Since Onsager's law is valid the right-hand sides of the last two equations are the same (except for the minus sign). This means that for a given membrane system the ratio of the electro-osmotic volume flow to the corresponding electric current is the same as the ratio of the streaming potential to the corresponding pressure.

The result

$$(J_V/j)_{p=0} = -(U/p)_{j=0} \tag{3.161}$$

was known already at the end of the last century under the name of Saxen's relationship, but at that time it was derived in a more complicated way, on the basis of assumptions about the mechanism of electrokinetic phenomena. Steady-state thermodynamics reaches such results more easily. Thus, for a current entrained by volume flow and measured by a pair of electrodes with an ammeter we can find the relation

$$(j/J_V)_{U=0} = L_{Up}/L_p. \tag{3.162}$$

On the other hand, for the electro-osmotic pressure preventing further volume flow entrained by electric current into a compartment of fixed volume, in which pressure is built up, one can write

$$(p/U)_{J_V=0} = -L_{pU}/L_p \tag{3.163}$$

and, as a result of Onsager's law, we have

$$(j/J_V)_{U=0} = -(p/U)_{J_V=0}. \tag{3.164}$$

FURTHER READING

Abidor, I. G., Arakelyan, V. B., Chernomordik, L. V., Chizmadzhev, Yu. A., Pastushenko, V. F., and Tarasevich, M. R. Electric breakdown of bilayer lipid membranes. I. The main experimental facts and their qualitative discussion, *Bioelectrochem. Bioenergetics*, **6**, 37–52 (1979).

Aityan, S. Kh. Single-file diffusion of uncharged particles, *Biologicheskie Membrany*, **1**, 524–30 (1984).

Aityan, S. Kh., and Chizmadzhev, Yu. A. Simulation of molecular dynamics of water movement in ion channels, *Gen. Physiol. Biophys.*, **5**, 213–29 (1986).

Barry, P. H., and Diamond, J. M. Effects of unstirred layers on membrane phenomena, *Physiol. Rev.*, **64**, 763–872 (1984).

Brynda, E., Nešpůrková, L., and Nešpůrek, S. Superlinearity of current-voltage characteristics of oxidized cholesterol bilayer membranes, *Czech. J. Phys.*, **B35**, 873–82 (1985).

Dainty, J. Osmotic flow. In: *The State and Movement of Water in Living Organisms. Symp. Soc. Exptl. Biol.*, **19**, 75–85 (1964).

De Levie, R. Mathematical modeling of transport of lipid-soluble ions and ioncarier complexes through lipid bilayer membranes, *Adv. Chem. Phys.*, **37**, 99–137 (1978).

Einstein, A. Über die von der molekularkinetischen Theorie der Wärme geforderte Bewegung von in ruhenden Flüssigkeiten suspendierender Teilchen, *Ann. Physik*, **17**, 549–60 (1905).

Fischer, W., Brickmann, J., and Läuger, P. Molecular dynamics study of ion transport in transmembrane protein channels, *Biophys. Chem.*, **13**, 105–16 (1981).

Fürth, R. (ed.). *Investigations on the Theory of Brownian Movement*, pp. 86–119, Dover Publications, New York (1956).

Goldman, D. E. Potential, impedance and rectification in membranes, *J. Gen. Physiol.*, **27**, 37–60 (1943).

Hammel, H. T., and Scholander, P. F. Osmosis and Tensile Solvent. Springer-Verlag, Berlin (1976).

Hansen, U.-P., Gradmann, D., Sanders, D., and Slayman, C. L. Interpretation of current-voltage relationships for 'active' ion transport systems. I. Steady-state reaction kinetic analysis of class-I mechanisms, *J. Membr. Biol.*, **63**, 165–90 (1981).

Hansen, U.-P., Tittor, J., and Gradmann, D. Interpretation of current-voltage relationships for 'active' ion transport systems. II. Nonsteady-state reaction kinetic analysis of class-I mechanisms with one slow time constant, *J. Membr. Biol.*, **75**, 141–69 (1983).

Haydon, D. A., and Hladky, S. B. Ion transport across thin lipid membranes. A critical discussion of mechanisms in selected systems, *Quart. Rev. Biophys.*, **5**, 187–282 (1972).

Heckmann, K. Single-file diffusion. In: *Passive Permeability of Cell Membranes* (F. Kreuzer, and J. F. G. Slayers, eds), pp. 127–53, Plenum Press, New York (1972).

House, C. R. *Water Transport in Cells and Tissues*, Edward Arnold, London (1974).

Jaffe, L. F. The interpretation of voltage-concentration relations, *J. Theor. Biol.*, **48**, 11–18 (1974).

Johnson, F. H., Eyring, H., Polissar, M. J. *Kinetic Basis of Molecular Biology*, pp. 515–603, John Wiley, New York (1954).

Kedem, O., and Katchalsky, A. Thermodynamic analysis of the permeability of biological membranes to non-electrolytes, *Biochim. Biophys. Acta*, **27**, 229–46 (1958).

Läuger, P., Benz, R., Stark, G., Bamberg, E., Jordan, P. C., Fahr, A., and Brock, W. Relaxation studies in lipid bilayer membranes, *Quart. Rev. of Biophys.*, **14**, 513–98 (1981).

Lieb, W. R., and Stein, W. D. Non-Stokesian nature of transverse diffusion within human red cell membranes, *J. Membr. Biol.*, **92**, 111–19 (1986).

Marcus, Y. *Ion Solvation*, John Wiley, Chichester (1985).

Mauro, A. Forum on osmosis. III. Comments on Hammel and Scholander's solvent tension theory and its application to the phenomenon of osmotic flow, *Am. J. Physiol.*, **237**, R110–R113 (1979).

Morf, W. E. *The Principles of Ion-Selective Electrodes and Membrane Transport*, Elsevier, Amsterdam (1981).

Piek, T. Ionic properties. In: *Insect Muscle* (P. N. R. Usherwood, ed.), pp. 282–7, Academic Press, New York (1975).

Polymeropoulos, E. E., and Brickmann, J. Molecular dynamics of ion transport through transmembrane model channels, *Annu. Rev. Biophys. Biophys. Chem.*, **14**, 315–30 (1985).

Schafer, J. A., and Andreoli, T. E. Water transport in biological and artificial membranes, *Arch. Int. Med.*, **129**, 279–92 (1972).

Schlögl, R. *Stofftransport durch Membranen*, Dr. Dietrich-Steinkopf-Verlag, Darmstadt (1964).

Shapiro, E., and Ohki, S. The interaction energy between hydrocarbon chains, *J. Colloid Interface Sci.*, **47**, 38–49 (1974).

Stein, W. D. Permeability for lipophilic molecules. In: *Membrane Transport* (S. L. Bonting, and J. J. H. H. M. de Pont, eds), pp. 1–28, Elsevier/North Holland, Amsterdam (1981).

Tanford, C. *The Hydrophobic Effect*, John Wiley, New York (1980).

Teorell, T. Transport processes and electrical phenomena in ionic membranes, *Progr. Biophys. Biophys. Chem.*, **3**, 305–69 (1953).
Thellier, M., Ripoll, C., Demarty, M., and Lasalles, J.-P. Thermodynamic model of ion transport. Flux/force relationships and biological implications. In: *Biophysics of Membrane Transport* (J. Kuczera, C. Grygorczyk, and S. Przestalski, eds), Vol. 2, pp. 31–57, Publ. House Agricult. University, Wrocław (1981).
Thomas, R. C. Electrogenic sodium pump in nerve and muscle cells, *Physiol. Rev.*, **52**, 563–94 (1972).

3.4 MOLECULAR BASIS OF AQUEOUS CHANNELS

Several aqueous channels have been defined in molecular terms, some of them being utterly nonspecific, some showing a certain degree of selectivity. Several examples of the less specific ones will be described in some detail.

3.4.1 Connexin channels

The ones that are truly nonselective pores and where no necessity for single-file diffusion exists are the gap junction connexons. Connexons are made of six molecules of connexin (M_r 27 000) arranged hexagonally at each apposed membrane, thus forming a channel from cell to cell, composed of 12 connexin molecules. The channel is 1.2–1.5 nm wide and permits the passage of globular molecules of up to M_r 1500. The connexin molecules under normal conditions are tilted by about 120 degrees from the surface but the tilt can be abolished by exposure to distilled water. In this conformation the central channel disappears and one may speculate that this is the conformation in which the connexons are effectively closed, e.g. when the Ca^{2+} concentration rises above 0.05 mM (Fig. 3.12).

3.4.2 Porins

The outer membrane of gram-negative bacteria, such as *Escherichia coli*, and in keeping with the endosymbiont theory, of mitochondria and chloroplasts,

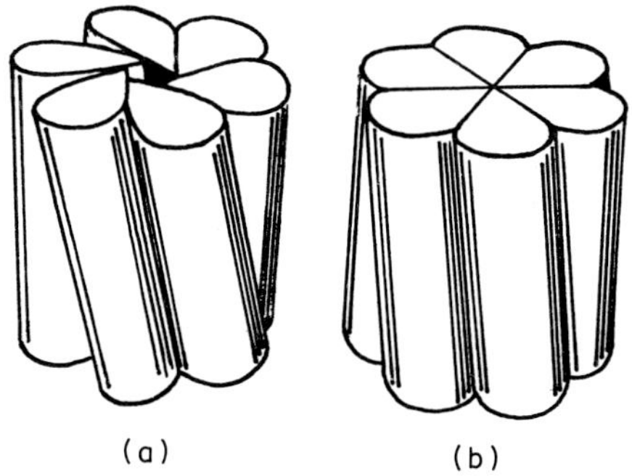

Fig. 3.12. Connexon of a gap junction, open (a) and closed (b)

Table 3.6. Peptidoglycan-related porins of gram-negative bacteria

Species	Porin designation	Effective channel diameter (nm)	Molecular weight	Permeability ratio cations/anions	Note
E. coli	OmpF	1.1–1.2	37 000	3.6–3.9	—
	OmpC	1.0	36 000	26	Preferentially expressed in high-osmolarity media
	PhoE	1.1	36 800	0.3	Induced by phosphate limitation
	K	1.0	40 000	7.6	Found in encapsulated strains
	NmpC	1.0	38 000	0.3	Found in revertants of porin-deficient mutants
S. typhimurium	OmpF	1.2	39 300	14	
	OmpC	1.3	39 800	41	
	OmpD	1.3	38 000	23	

[a] All these porins act as receptors for phages and/or bacteriocins of the colicin type

contain several membrane-spanning complexes about 6 nm long, composed of proteins of M_r 30 000–80 000, the so-called porins.

(*a*) *Prokaryotic porins*

Bacteria contain porins in numbers of 10^5 per cell which are always arranged in trimers and constitute a transmembrane channel which, in some cases, may have three openings to the outside of the cell which merge into a single channel opening into the intermembrane (or periplasmic) space. However, in *Rhodopseudomonas sphaeroides* and *Anabaena variabilis* porin trimers appear to contain three independent channels all the way through the membrane.

Bacterial porins are acidic proteins with up to 14 net negative charges which are probably instrumental in binding together the monomers in the complex. Some of the porins are present in all strains of a given species, some are formed inducibly under special growth or nutrition conditions. The first group of porins, all of which are immunologically related and all of which are linked *in situ* to the peptidoglycan layer, are listed in Table 3.7.

In addition to these porins several pore-forming proteins are known from other prokaryotes, thus (in parentheses the designation, the channel diameter,

Table 3.7. Some more specific pore proteins of *Escherichia coli* outer membrane

Designation	Molecular weight	Possible function in transport of
T6 phage receptor	26 000	Nucleosides and deoxynucleosides
λ-phage receptor ('maltoporin')	47 000	Maltose and maltotriose (*in vitro* acts as a general pore)
ButB	60 000	Cyanocobalamin
Cir	74 000	Complexed Fe^{3+}
FhuA	78 000	Fe^{3+}-ferrichrome
FeC	80 500	Fe^{3+}-citrate
FepA	81 000	Fe^{3+}-enterochelin

the permeability ratio for cations/anions): *Pseudomonas aeruginosa* (F; 2.0 nm; 2.7), *Yersinia pestis* (1.1 nm; ?), *Neisseria gonorrhoeae* (protein I; 1.1 nm; 0.28), *Anabaena variabilis* (1.6 nm; 1.7), *Rhodopseudomonas sphaeroides* (1,2 nm; ?).

In keeping with the channel diameter, the exclusion limit differs from case to case, being 800 for the normal *E. coli* pores but 5000 for the F porin of *P. aeruginosa.* They apparently represent nonselective paths through the outer membrane for all solutes of molecular weights up to the exclusion limit. In fact, the low temperature dependence (activation energies of the order of 10 kJ mol^{-1}) indicate that the passage occurs by simple diffusion through the aqueous phase filling the pores. The ion selectivity displayed by some porins is not comparable with that in true specific channels, being simply due to the presence of positively or negatively charged groups exposed in the channel. The greatest resemblance to a specific channel is displayed by the P porin of *Pseudomonas aeruginosa* where the critical part of the channel is formed by the ε-amino groups of three lysine residues and hence displays a remarkably great affinity for anions that show an intrinsically three-fold symmetry, such as phosphate (a K_D of 0.3 mM compared with roughly 40 mM for Cl^-).

However, a certain selectivity is observed with pore proteins that are not antigenically related to the Omp group of *Escherichia coli.* These may serve as specific channel-type transport systems but no kinetic data are available on their function (Table 3.7). These again, with the possible exception of FeC, can serve as receptors for phages and colicins.

(*b*) *Organelle porins*

One type of porin has been isolated from the outer membrane of mitochondria of yeasts, fungi, protozoans, plants and animals. It is a protein of M_r 35 500 (in mammalian mitochondria) and 30 000 (in fungal mitochondria) and represents a major protein constituent of the outer membrane (20 per cent of total protein). The pores formed by this porin have been estimated to have diameters between 1.7 and 1.9 nm, corresponding to molecular weight exclusion limit of about 2 500.

Like bacterial porins, the mitochondrial pore protein shows no specificity and only a slight preference for anions over cations (except the porin from *Paramecium*) and, like in bacteria, the porin is presumably organized in trimers although each of the monomers can act as an independent diffusion channel.

A notable property of the mitochondrial porin is that it can exist in several states of 'open' or 'closed', the closing being voltage-dependent.

The outer chloroplast membrane appears to contain a porin-like channel which, however, has a large diameter (3 nm) and, correspondingly, permits the passage of compounds up to 10 000 molecular weight.

3.4.3 Extraneous pore formers

(a) The membrane-attack complex of complement

Complement is a group of serum proteins active in a complicated set of reactions leading, among other things, to phagocytosis and membrane damage of foreign cells. Of the 18 proteins involved the ones designated C5b, C6, C7, C8, and C9, can form a complex containing two sets of the five proteins with an overall M_r of 1 700 000 (the so-called membrane-attack complex) with extremely tight mutual binding (dissociation constants for the individual steps of the reaction sequence $C5b.C6 \rightleftharpoons C5b.C6.C7 \rightleftharpoons C5b.C6.C7.C8 \rightleftharpoons C5b.C6.C7.C8.C9$ are all about 10^{-12} M) and a surprisingly hydrophobic surface so that the complex readily inserts itself into the membrane and forms there a huge pore about 10 nm in diameter which, however, does not permit the passage of large molecules from the cell. The insertion of the membrane-attack complex results in osmotic lysis of the cell.

(b) Polyene antibiotics and saponins

A number of cyclic antibiotics with several conjugated double bonds are known to form transmembrane pores, among them nystatin, filipin and amphotericin B. Likewise, saponins, such as digitonin or tomatin, disrupt the membranes in a nonselective fashion.

Although the reactions of these substances with cell membranes may differ in detail all of them appear to associate with membrane sterols in a manner resembling that of phospholipids. They are pronouncedly amphipathic and can generate oligomeric complexes with sterols which give rise to water-filled pores 0.7–1.1 nm in diameter, slightly preferring anions.

Because of their affinity for sterols, the two groups of substances are ineffective with inner mitochondrial, chloroplast and bacterial membranes.

The 'pore' formation by both polyene and saponin compounds is accompanied by microscopically observable generation of membrane microdomains such that the membrane appears to contain impressions or protrusions about 60–80 nm across, hence very much larger than a single aggregate with sterols could be.

(*c*) *Bacteriocins*

Bacteriocins are proteins produced by various bacteria that are harmful to other bacteria, often of the same species but different strain. Many of them are produced by *Escherichia coli* and are named colicins. They act (1) by inhibiting protein synthesis, (2) by blocking DNA synthesis, and (3) by depolarizing the membrane and causing leakage of some intracellular components. It is the third class that is interesting in the present context as they apparently form pores in the outer membrane of *E. coli*. It was mentioned in the section on porins that most of them act as receptors for colicins and that also other, more specific, pores can act as such receptors. Several of the colicins, the El, K, Ia, Ib, and A, bind to receptors (e.g., for vitamin B_{12} or for complexed Fe^{3+}) and form a membrane channel not only in the outer but also in the inner membrane of *E. coli*. The El colicin contains 522 amino acids and is thus of the same size as its receptor protein with M_r of 60 000. Apparently, upon binding to the receptor, it disrupts the outer membrane and is then drawn, by virtue of its hydrophobic sections, into the inner membrane where it is anchored and spans it to form a narrow aqueous channel, about 0.62 nm in diameter.

Other colicins form channels of similar magnitude with a slight preference for cations over anions.

SYNOPSIS

Aqueous channels or pores exist naturally in most membranes but can also be formed by adding extraneous pore formers to natural or artificial membranes.

Among the natural pores that have been described in some detail are the connexons of intercellular gap junctions and the porins of outer bacterial, mitochondrial and chloroplast membranes.

Channels can be formed in membranes, among other agents, by the membrane-attack complex of complement, by polyene antibiotics of the nystatin type, and by saponins. Some of the bacteriocin peptides are also pore formers.

FURTHER READING

Benz, R. Porin from bacterial and mitochondrial outer membranes, *CRC Critical Rev. Biochem.*, **19**, 145–90 (1985).

Gomperts, D. *The Plasma Membrane: Models for Structure and Function*, Academic Press, London (1977).

Revel, J. P., Yancey, S. B., and Nicholson, B. J. The gap junction proteins, *Trends Biochem. Sci.*, **11**, 375–7 (1986).

3.5 PROTEIN-MEDIATED TRANSPORT

3.5.1 Introduction

In contrast to the nonspecific, at most partly selective, transport discussed in the previous sections, whereby small hydrophilic molecules and ions, as well as large or small lipophilic molecules and ions can cross biomembranes, most substances of physiological interest to the cell, both charged and uncharged, have their special vehicles built into the membranes to carry them across. Invariably, these vehicles are integral proteins, sometimes aided by small peripheral proteins, in eukaryotic cells apparently bearing sugar residues on the noncytoplasmic face of the membrane.

As the solute particle traverses the membrane it must at one stage interact with a protein binding site and, while associated with it, gains access to the other side of the membrane.

This very general statement can take on more precisely defined forms, depending on the molecular nature of the translocation process as will be discussed below but, in all its versions, it includes an analogy to an enzyme reaction where the substrate interacts with the active site of the enzyme and, after chemical modification, is released as a product. In protein-mediated transport, the 'product' of the reaction is identical with the substrate but it appears in a separate compartment so that it is distinguishable from the substrate. To describe the situation in Cleland's terminology of enzyme kinetics, the usual protein-mediated transport reactions are of the isouni-uni type. If two different substrates are involved the reaction is isobi-bi.

This analogy of transport reactions with enzyme-catalyzed chemical conversion has as its corollary the possibility of describing the kinetic quantities of protein-mediated transport rates by expressions formally equivalent to those of enzyme kinetics, typically in terms of Michaelis or *half-saturation constants* K_T and of *maximum rates of transport* ($V \equiv J_{max}$) which are usually different for the one and the other direction of transmembrane flow, hence $K'_T, J'_{max}, K''_T, J''_{max}$.

Even the more subtle kinetic features of protein-mediated transport, such as countertransport, competitive acceleration or equilibrium unidirectional flows can be demonstrated with enzyme reactions, although there they are not commonly studied and, if so, carry different designations.

At the basis of quantitative descriptions of specific transport mechanisms are the rates of transport of solute J_s, usually expressed in moles (or their fractions) per time (min, s, or fractions of a second) per amount of membrane or transport protein (as area, weight or protein content), the rates becoming maximum as the solute concentration is raised beyond any limit (in practice, 10 times the value of K_T will do). A more accurately expressed J_{max} is the turnover number J_f defined as the maximum number of solute particles passing through a transport unit in a second—the correct expression for this quantity is *molecular activity*.

A useful expression is the *unidirectional flow* $\vec{J}_s$ or $\overleftarrow{J}_s$ under various conditions of saturation which is conveniently estimated with the aid of radioactive tracers. Occasionally, a *permeability constant* P is used even in specific transport like it was in nonspecific permeation, indicating here the rate of transport at limitingly low concentrations divided by solute concentration.

The specific transport can proceed in accordance with the electrochemical potential gradient of the solute—then no input of energy into the reaction beyond the obvious thermal energy is necessary and the solute will eventually reach equilibrium which, for an uncharged molecule, will be $c''_s/c'_s = 1$ while for an ion it will be $c''_s/c'_s = \exp(-nF\,\Delta\phi/RT)$, exactly as in the nonspecific permeation systems.

However, in most cases of specific transport, the movement of solute proceeds uphill and the final distribution between cell and medium, designated as *maximum accumulation ratio*, $\omega = (c''_s/c'_s)_\infty$ will be very different from unity even for nonelectrolytes and will thus represent another specifying feature for a transport under given conditions.

Until quite recently, a sharp distinction had always been drawn between specific transport of the carrier type and that of the channel type. In both cases a finite number of specific solute-binding sites is available in the membrane but with the carrier these binding sites are accessible *either* from the starting (*cis*) side *or* from the target (*trans*) side but not simultaneously from both sides. With a channel, on the other hand, the sites are open to both membrane faces simultaneously.

It will be shown below that an approach based on the rate theory of transport brings together the two extremes in a unified view encompassing both channel-like and carrier-like transport. Before proceeding to this, it may be useful to derive equations for a typical specific channel transport and for a typical carrier transport, bearing in mind the distinctions outlined above.

3.5.2 Rate of transport through a specific channel

If a pore-like channel is envisaged, carrying a single binding site for transported substrate located centrally in the membrane, substrate molecules can reach it from both sides with equal ease (Fig. 3.13). A kinetic diagram of this situation, with E the enzyme-like binding site and S the substrate, can be drawn as follows

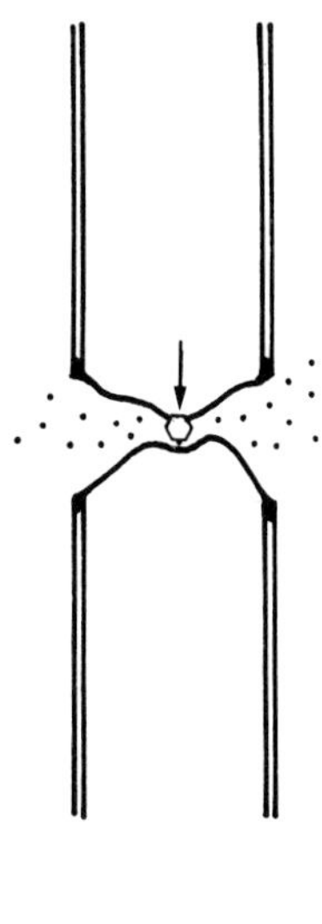

Fig. 3.13. Specific channel with a single binding site accessible from both sides simultaneously, and a single major energy barrier

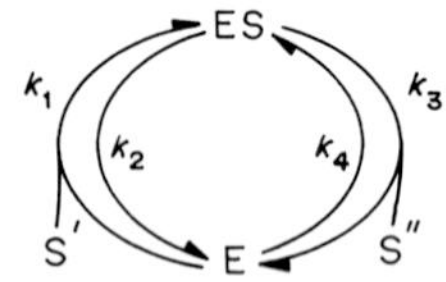

(3.III)

Assuming E and ES to be in steady state such that $\mathrm{d}c_{\mathrm{E}}/\mathrm{d}t = 0$ and $\mathrm{d}c_{\mathrm{ES}}/\mathrm{d}t = 0$ while $c_{\mathrm{E_t}} = c_{\mathrm{E}} + c_{\mathrm{ES}}$, one can derive expressions for unidirectional fluxes of S, $\vec{J}_{\mathrm{s}}$ from left to right, $\overleftarrow{J}_{\mathrm{s}}$ from right to left

$$\vec{J}_{\mathrm{s}} = c_{\mathrm{E_t}} \frac{k_1 k_4 c'_{\mathrm{s}}}{k_2 + k_4 + k_1 c'_{\mathrm{s}} + k_3 c''_{\mathrm{s}}} \tag{3.165a}$$

and

$$\overleftarrow{J}_{\mathrm{s}} = c_{\mathrm{E_t}} \frac{k_2 k_3 c''_{\mathrm{s}}}{k_2 + k_4 + k_1 c'_{\mathrm{s}} + k_3 c''_{\mathrm{s}}} \tag{3.165b}$$

so that the net flux will be

$$J_{\mathrm{s}} = c_{\mathrm{E_t}} \frac{k_1 k_4 c'_{\mathrm{s}} - k_2 k_4 c''_{\mathrm{s}}}{k_2 + k_4 + k_1 c'_{\mathrm{s}} + k_3 c''_{\mathrm{s}}}. \tag{3.165c}$$

For a situation where no energy is introduced into the system, the expression is constrained by $k_1 k_4 = k_2 k_3$, so that in equilibrium when $J_{\mathrm{s}} = 0$, $c''_{\mathrm{s}}/c'_{\mathrm{s}} = 1$.

Expressed in enzyme-kinetic terms

$$J_{\mathrm{s}} = J'_{\max} \frac{c'_{\mathrm{s}} - c''_{\mathrm{s}}/K_{\mathrm{eq}}}{K'_{\mathrm{T}}(+ c''_{\mathrm{s}}/K''_{\mathrm{T}}) + c'_{\mathrm{s}}} \tag{3.165d}$$

where $J'_{\max} = k_4 c_{\mathrm{E_t}}$, $K'_{\mathrm{T}} = (k_2 + k_4)/k_1$, $K''_{\mathrm{T}} = (k_2 + k_4)/k_3$, $K_{\mathrm{eq}} = k_1 k_4 / k_2 k_3 = 1$.

The initial rates of flow then obey Michaelis-type expressions (for $c''_{\mathrm{s}} = 0$).

If an ion moves through such a channel the rate constant governing its movement from left to right will be given by $k_1 \xi^{1/2}$, that from right to left by $k_3 \xi^{-1/2}$, as follows from Eyring's interpretation of rates of charged particles in an electric field, with $\xi = \exp(-zF\Delta\phi/RT)$. Then

$$J_{\mathrm{s}} = c_{\mathrm{E_t}} \frac{k_1 k_4 \xi^{1/2} c'_{\mathrm{s}} - k_2 k_3 \xi^{-1/2} c''_{\mathrm{s}}}{k_2 + k_4 + k_1 \xi^{1/2} c'_{\mathrm{s}} + k_3 \xi^{-1/2} c''_{\mathrm{s}}} \tag{3.166}$$

so that, for $J_{\mathrm{s}} = 0$,

$$c''_{\mathrm{s}}/c'_{\mathrm{s}} = k_1 k_4 \xi / k_2 k_3 \tag{3.167}$$

and, with no energy pumped in ($k_1 k_4 = k_2 k_3$)

$$c''_{\mathrm{s}}/c'_{\mathrm{s}} = \xi. \tag{3.168}$$

Now, although the passage through a specific channel involves the interaction with a binding site, the individual unidirectional fluxes are independent as follows from the flux ratio $\vec{J}_{\mathrm{s}}/\overleftarrow{J}_{\mathrm{s}} = c'_{\mathrm{s}}/c''_{\mathrm{s}}$. It should be noted that even a channel with a large number of potential barriers that must be crossed by the particle before it reaches its specific site (Fig. 3.14), shows the properties of the simple channel,

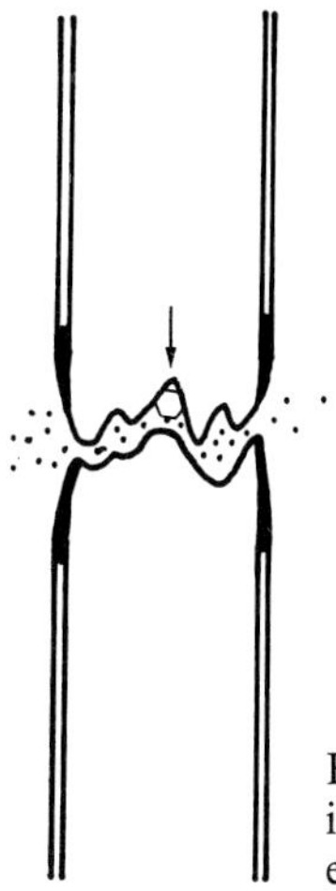

Fig. 3.14. Composite specific channel with several energy barriers

the rate of flow of S being expressed by an expression of the type

$$J_s = J_{max} \frac{c_s' - c_s'' \xi^{-1}}{A + Bc_s' + Cc_s'' \xi^{-1}} \tag{3.169}$$

A characteristic of channel-type transports is that the denominator of the rate equation does not contain a term combining c_s' and c_s'', a diagnostic feature of carrier transports. This, among other things, prevents the occurrence of countertransport, trans-acceleration and similar phenomena in a channel mechanism.

As mentioned in Table 3.4, some channels are either voltage- or ligand-gated. This means simply that a supraliminal change of potential or increase of ligand concentration or both open the access to the channel so that ions can pass through with a certain selectivity. While the macroscopic maximum rate of transport will depend on the time the channel is open the other properties remain unchanged. A more thorough discussion of such channels will be found in Sections 5.1 and 5.2.

3.5.3 Rate of transport by a specific carrier

Contrary to earlier views, a carrier (often termed a mobile carrier) is not to be understood as a vehicle moving translationally along and across the membrane and able to bind and carry the solute from one aqueous medium to the other. It is rather a binding site within an integral membrane protein which has the property of being alternately accessible from one or the other membrane side. As it flips from a left-facing position to the right-facing one it manages to move the bound solute 'across' the membrane (Fig. 3.15).

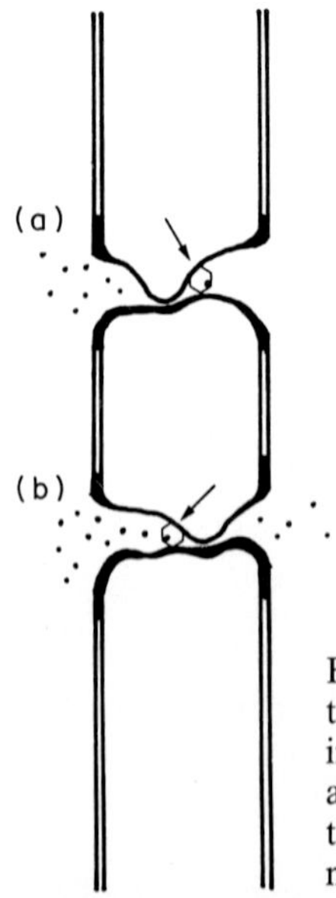

Fig. 3.15. A carrier-type transport in which the binding site (arrow) is exposed alternately to the one or to the other side of the membrane

A kinetic diagram of this situation is

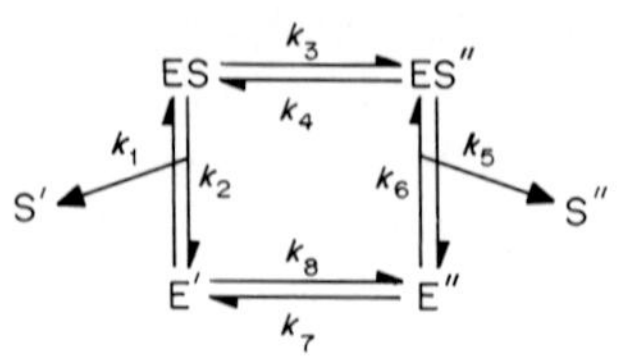

(3.IV)

Assuming again that $dc'_E/dt = dc'_{ES}/dt = dc''_E/dt = dc''_{ES}/dt = 0$ and applying the conservation equation $c_{E_t} = c'_E + c'_{ES} + c''_E + c''_{ES}$, we obtain for the simplest case that

$$J_s = c_{E_t} \frac{k_1 k_3 k_5 k_7 c'_s - k_2 k_4 k_6 k_8 c''_s}{(k_7 + k_8)(k_2 k_4 + k_2 k_5 + k_3 k_5) + k_1[k_7(k_4 + k_5) + k_3(k_5 + k_7)]c'_s + k_6[k_4(k_2 + k_8) + k_8(k_2 + k_3)]c''_s + k_1 k_6 (k_3 + k_4) c'_s c''_s}$$

$$= c_{E_t} \frac{A c'_s - B c''_s}{C + D c'_s + E c''_s + F c'_s c''_s} \tag{3.170a}$$

or, in an enzyme-kinetic form,

$$J_s = J'_{max} \frac{c'_s - c''_s / K_{eq}}{K'_T (1 + c''_s / K''_T) + c'_s (1 + c''_s / K_i)} \tag{3.170b}$$

where

$K'_T = C/D, \quad K''_T = C/E, \quad J'_{max} = A/D, \quad K_i = D/F, \quad K_{eq} = A/B$ (here = 1).

Now the individual unidirectional fluxes will be defined by

$$\vec{J}_s = c_{E_t} \frac{k_1 k_3 k_5 c_s'[k_7 + k_2 k_4 k_6 c_s''/(k_2 k_4 + k_2 k_5 + k_3 k_5)]}{\text{denominator}} \tag{3.171a}$$

and

$$\overleftarrow{J}_s = c_{E_t} \frac{k_2 k_4 k_6 c_s''[k_8 + k_1 k_3 k_5 c_s'/(k_2 k_4 + k_2 k_5 + k_3 k_5)]}{\text{denominator}} \tag{3.171b}$$

(the denominator is the same as in the net-flow equation above). It may be seen that their simple subtraction does not yield the net flow formula which indicates they are not independent (this would be so only for a complete symmetry of the dissociation constants and identity of the translocation constants). Consequently, the flux ratio

$$\frac{\vec{J}_s}{\overleftarrow{J}_s} = \frac{k_1 k_3 k_5 c_s'[k_7(k_2 k_4 + k_2 k_5 + k_3 k_5) + k_2 k_4 k_6 c_s'']}{k_2 k_4 k_6 c_s''[k_8(k_2 k_4 + k_2 k_5 + k_3 k_5) + k_1 k_3 k_5 c_s']} = \frac{c_s'}{c_s''} \cdot \frac{K_{eq} a + b c_s''}{a + b c_s'}$$

where

$$a = k_2 k_4 + k_2 k_5 + k_3 k_5 \quad \text{and} \quad b = k_1 k_3 k_5 / k_8,$$

and hence $(\vec{J}_s/\overleftarrow{J}_s):(c_s'/c_s'') \neq 1$. Without any input of energy $K_{eq} = 1$ and, in equilibrium when $c_s' = c_s''$, $\vec{J}_s/\overleftarrow{J}_s = 1$.

At any time point of transport at very high concentrations $\vec{J}_s/\overleftarrow{J}_s = 1$, while at very low concentrations $\vec{J}_s/\overleftarrow{J}_s = c_s'/c_s''$, like in simple diffusion. At any intermediate concentration generally $\vec{J}_s/\overleftarrow{J}_s < c_s'/c_s''$.

Still, the initial rate of carrier transport obeys again Michaelis kinetics and is indistinguishable from transport through a specific channel.

If an ion is transported by such a system k_3 and k_4 will be altered to $k_3\xi^{1/2}$ and $k_4\xi^{-1/2}$, respectively, and the rate equation acquires the form

$$J_s = J_{max} \frac{c_s' - \xi^{-1} c_s''/K_{eq}}{K_T'(1 + c_s'' K_T'') + c_s'(1 + c_s''/K_i)}. \tag{3.172}$$

If the carrier is otherwise fully symmetrical, the denominator is not affected by ξ. However, if the individual mirror constants differ and are constrained only by the condition that $k_1 k_3 k_5 k_7 = k_2 k_4 k_6 k_8$, all the kinetic parameters in the denominator are subject to electric potential influence.

The flux ratio will be given by

$$\frac{\vec{J}_s}{\overleftarrow{J}_s} = \frac{c_s'}{c_s''} \cdot \frac{K_{eq}\xi a + b c_s''}{a + b c_s'} \tag{3.173}$$

where

$$a = k_2 k_4 \xi^{-1/2} + k_2 k_5 + k_3 k_5 \xi^{1/2} \quad \text{and} \quad b = k_1 k_3 k_5 \xi^{1/2}/k_8.$$

The limitations are similar to the case of uncharged particles. At very high values of c_s it will be equal to unity, independently of the membrane potential. At low values of c_s it will be just like for simple diffusion of ions, viz. $(c_s'/c_s'')\xi$.

It should be observed here that the four-state carrier model is not the minimum model required to account for systems with no input of energy (mediated or facilitated diffusion). Kinetically equivalent is a three-state model where

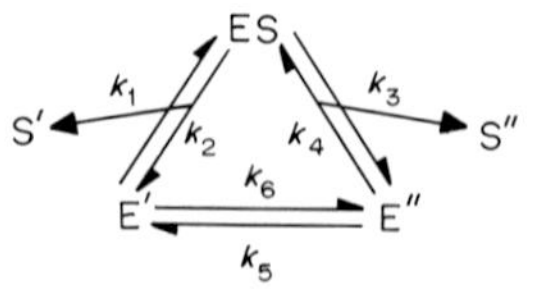

(3.V)

This model yields the rate equation

$$J_s = c_{E_t} \frac{k_1 k_3 k_5 c_s' - k_2 k_4 k_6 c_s''}{(k_2 + k_3)(k_5 + k_6) + k_1 c_s'(k_3 + k_5) + k_4 c_s''(k_2 + k_6) + k_1 k_4 c_s' c_s''}. \quad (3.174)$$

However, for active transport systems (to be treated below) it does not offer as many possibilities of manipulating the system as the above four-state model does.

Another simplification often introduced into carrier rate equations is the assumption that the surface reactions (governed by k_1, k_2, k_5, k_6 in the basic model) proceed much faster than the translocations—the rapid-equilibrium approach. This is probably justified in most cases but the rapid equilibrium is not to be taken for granted.

An example based on reaction kinetics may help the reader appreciate the appropriateness of the rapid-equilibrium assumption.

Consider the situation of solute A of about the same size as the surrounding water molecules which, upon encounter with its receptor on a carrier molecule, does not bounce back but is bound (= it reacts). This is a permissible assumption because of the cage effect of water and it is even more justified in the case of a carrier which is assumed to be always present in the proper orientation for reaction. The rate constant has then the character of a diffusion-controlled reaction and is given by $k = 4N_A\pi(r_A + r_B)(D_A + D_B)$ (see p. 125). The reaction rate is then $v_r = kc_Ac_B = Z_{AB}/N_A$, where Z_{AB} is the number of collisions in m^3 per second.

Consider the constitutive monosaccharide carrier of the yeast *Saccharomyces cerevisiae* and D-glucose at a concentration corresponding to 95 per cent saturation, about 50 mmol dm^{-3}. In a suspension containing 3 mg dry wt. ml^{-1} there are approximately $1.5 \cdot 10^8$ cells in one millilitre. Each of the cells was found to contain on its surface 10^6 carrier binding sites for D-glucose and let 50 per cent of them be oriented outward. There are then $7.5 \cdot 10^{13}$ such binding sites per ml, corresponding to an apparent concentration of 112 nmol dm^{-3} (this is a crude approximation as the binding sites are not distributed homogeneously in the suspension but rather clustered on each individual cells but the error introduced by this is not greater than one order of magnitude).

Combining the above equations and taking $r_A = 0.3$ nm, $r_B = 1$ nm, $D_A = 6.7 \cdot 10^{-10}\,m^2\,s^{-1}$, $D_B = 0$, we obtain for Z_{AB} $2.2 \cdot 10^{31}$ encounters per m^3 per second. From this $v_r = 3.67 \cdot 10^5\,M\,s^{-1}$. The measured maximum rate of D-glucose uptake J_{max} is 2.8 nmol s^{-1} (mg dry wt.)$^{-1}$, at our suspension density then 8.4 μmol s^{-1} dm^{-3}. Then $J_{max}/v_r = 2.39 \cdot 10^{-11}$, i.e. one collision in $4 \cdot 10^{10}$

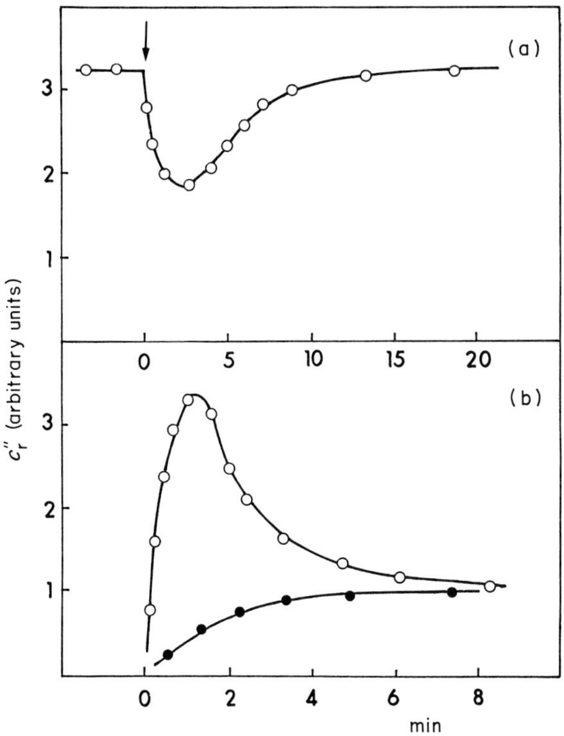

Fig. 3.16. Countertransport in a carrier system. (a) At arrow, a high concentration of nonlabelled S was added to cells preloaded with R; (b) cells were preincubated in buffer (•) or in nonlabelled (○) and at time zero labelled R was added. For details see the text

will result in translocation. Even if a large factor for tortuosity is introduced not because of the cell wall which does not reduce the rate of transport but to account for the sterically hampered accessibility to the binding site, there will be many more collisions with the carrier than translocations across the membrane. Thus it appears that this glucose transport system is probably in rapid equilibrium with the external solution.

Going back to the rapid-equilibrium simplification we can introduce $k_2/k_1 = K'$ and $k_5/k_6 = K''$, as well as $k_3 = k_4 = k_{es}$; $k_7 = k_8 = k_e$, and obtain

$$J_s = c_{E_t} \frac{k_e k_{es}(K'' c_s' - K' c_s'')}{(k_e K' K'' + k_{es} c_s' c_s'') + (k_e + k_{es})(K'' c_s' + K' c_s'')}. \tag{3.175}$$

If, moreover, all the translocation constants are identical and $K' = K'' = K$ as the case appears to be with the monosaccharide transport in rabbit erythrocytes the simplest transport equation emerges as

$$J_s = c_{E_t}\left(\frac{c_s'}{c_s' + K} - \frac{c_s''}{c_s'' + K}\right). \tag{3.176}$$

Perhaps it is by using this simple equation that the phenomenon of *countertransport* can be elucidated. According to its proper definition, countertransport is the flow of solute against its concentration difference which takes place at the expense of energy stored in the opposite gradient of a related solute using the same carrier.

There are two ways of demonstrating countertransport, one according to Wilbrandt, the other according to Miller (Fig. 3.16).

In the first, a labelled solute R is permitted to enter cells to reach equilibrium whereupon a nonlabelled solute S at a higher concentration is added. This solute enters the cells rapidly and elicits a transient efflux of the labelled solute from cells. Using the above simple equation and knowing that R and S are chemically identical and hence act as perfect mutually competitive inhibitors, we can write for the flow of R in the presence of S

$$J_{\mathrm{r(s)}} = c_{\mathrm{E_t}}\left(\frac{c'_{\mathrm{r}}}{c'_{\mathrm{r}} + c'_{\mathrm{s}} + K} - \frac{c''_{\mathrm{r}}}{c''_{\mathrm{r}} + c''_{\mathrm{s}} + K}\right). \tag{3.177}$$

At the moment of adding S to the system $c'_{\mathrm{r}} = c''_{\mathrm{r}}$ and $c''_{\mathrm{s}} = 0$ so that

$$J_{\mathrm{r(s)}} = c_{\mathrm{E_t}}\left(\frac{c_{\mathrm{r}}}{c_{\mathrm{r}} + c_{\mathrm{s}} + K} - \frac{c_{\mathrm{r}}}{c_{\mathrm{r}} + K}\right) \tag{3.178}$$

and hence the movement is negative, i.e., by convention, out of cells.

In the other arrangement, cells are preincubated with unlabelled solute S and transferred to S-free medium containing labelled solute R. While in a non-preincubated cell, a short instant after adding R the rate will be

$$J_{\mathrm{r(1)}} = c_{\mathrm{E_t}}\left(\frac{c'_{\mathrm{r}}}{c'_{\mathrm{r}} + K} - \frac{c''_{\mathrm{r}}}{c''_{\mathrm{r}} + K}\right) \tag{3.179}$$

in the preincubated one it will be

$$J_{\mathrm{r(2)}} = c_{\mathrm{E_t}}\left(\frac{c'_{\mathrm{r}}}{c'_{\mathrm{r}} + K} - \frac{c''_{\mathrm{r}}}{c''_{\mathrm{r}} + c''_{\mathrm{s}} + K}\right) \tag{3.180}$$

and, clearly, $J_{\mathrm{r(1)}} < J_{\mathrm{r(2)}}$.

3.5.4 Kinetics of a two-site carrier

Multisite carriers are known to exist and it is a general observation that the relationship between the binding sites is a kinetic cooperativity, in most cases positive. Such systems have been treated by enzyme kineticists in connection with the allosteric enzymes and, in principle, the same kinetic formalism can be applied to a cooperative transport system. Its peculiarities are best grasped from an equation for the transport of two competing solutes (best again, a labelled and a nonlabelled form of the same compound). Using the simplest carrier equation for a single substrate one can develop it into the following for two binding sites

$$J_{s(r)} = c_{E_t}\frac{x''y' - x'y''}{x''w' + x'w''} \tag{3.181}$$

where

$$x = c_{s_o}(k_{es} + k_{ess}c_{s_{oo}} + k_{esr}c_{r_{ooo}}) + c_{r_o}(k_{er} + k_{err}c_{r_{oo}} + k_{ers}c_{s_{ooo}}) + k_e$$
$$y = c_{s_o}(k_{es} + k_{esr}c_{r_{ooo}} + 2k_{ess}c_{s_{oo}}) + k_{esr}c_{s_{ooo}}c_{r_o}$$
$$w = 1 + c_{s_o}(1 + c_{s_{oo}} + c_{r_{ooo}}) + c_{r_o}(1 + c_{r_{oo}} + c_{s_{ooo}}).$$

Here $c_{s_o} = c_s/K_{CS}, c_{s_{oo}} = c_s/K_{CSS}, c_{s_{ooo}} = c_s/K_{CSR}$, and, analogously, for c_r. It should be observed that for the transport of a single solute the x, y and w are defined similarly but omitting all terms containing c_r. Obviously, for S = R chemically, $c_{s_{oo}} = c_{s_{ooo}}$ $c_{r_{oo}} = c_{r_{ooo}}$.

The equation predicts several interesting phenomena.

1. The dependence of initial rate on concentration will be sigmoid (cf. Fig. 5.7).

2. If cells are equilibrated with different concentrations of S and then R is added, the initial uptake rate of R plotted against the logarithm of c_s will yield a curve with a maximum (rather than having a zero slope at low values of c_s), and have a slope of -1 at higher concentrations.

3. In an experiment with countertransport of the first type using again two forms of the same substrate, the movement of preequilibrated R at the moment of adding S will be given by

$$J_r = c'_s c''_r(1 - c''_{r_o}c'_{s_{oo}} - c''_{r_{oo}}c''_{r_{ooo}})a/D \tag{3.182}$$

where a is a complex positive constant and D the denominator containing only positive terms. Hence, for relatively high concentrations of S and R, $J_r < 0$ which defines countertransport while for very low concentrations of S and R, $J_r > 0$ which leads to the so-called cotransport. In this context cotransport is the transport of a solute exceeding its normal control rate caused by a kinetic advantage of having another molecule of substrate bound—this can be either in terms of faster translocation of the doubly occupied complex or in terms of a much lower dissociation constant for the second substrate molecule than for the first one. This definition having nothing to do with the energy input, it is suggested that driven flows where a solute moves together with H^+ or Na^+ (see Section 3.5.6) be called symports rather than cotransports to indicate the thermodynamic distinction between the two.

3.5.5. Kinetic identity of carriers with multiconformational channels

Before proceeding to the treatment of active transport it should be useful to compare the channels with the carriers in their kinetic properties. Inspection of the simple drawings in Figs 3.14 and 3.15 shows that the mechanistic difference between a channel and a carrier may be rather subtle and that the rigorous distinction based on the accessibility of the binding site from the two membrane sides may rather represent two limiting cases of a general model.

If, in a channel with at least two energy barriers enclosing the binding site, conformational change is possible such that in one state a high barrier will be to the left, in another to the right (Fig. 3.17) the situation begins to resemble a carrier. It can be modelled as a four-state system of two protein states E_q and E_r, either empty or binding the solute S.

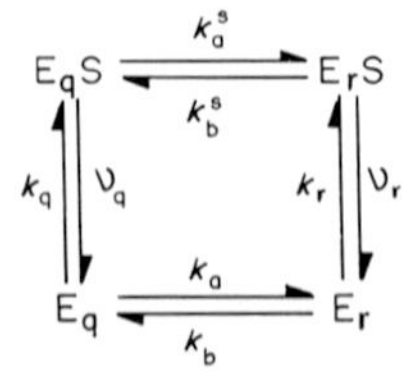

(3.VI)

While the transition constants k_a, k_b, k_a^s and k_b^s as well as the breakdown constants k_q and k_r are monomolecular constants, the jump rates v_q and v_r are concentration-dependent, such that

$$v_q = v_q' + v_q'' = c_s'\rho_q' + c_s''\rho_q'' \quad \text{and} \quad v_r = v_r' + v_r'' = c_s'\rho_r'' + c_s''\rho_r'',$$

ρ standing for the probability of a given jump. It should also be observed that both the v's and the k's are different, depending on direction of movement, so that also $k_q = k_q' + k_q''$ and $k_r = k_r' + k_r''$.

A steady-state analysis, observing the detailed balance condition that

$$\frac{v_r' k_r'}{v_r'' k_r''} = \frac{v_q' k_q'}{v_q'' k_q''} = \frac{v_r' k_q' k_a k_b^s}{v_q' k_r'' k_b k_a^s} = \frac{c_s'}{c_s''}\xi$$

and the conservation equation of the four substates, $c_{E_t} = c_{E_q} + c_{E_r} + c_{E_qS} + c_{E_rS}$, yields

$$J_s = \frac{c_{E_t}(1 - c_s''\xi^{-1}/c_s')[v_r'k_r'(v_q k_a^s + k_a^s k_a + k_q k_a) + v_q' k_q'(v_r k_a^s + k_b^s k_b + k_r k_b) + v_r' k_q' k_a k_b^s + v_q' k_r' k_b k_a^s]}{(k_a + k_b)(k_r k_q + k_r k_a^s + k_q k_b^s) + (k_a^s + k_b^s)(v_r v_q + v_r k_a + v_q k_b) + v_r k_q(k_a + k_b^s) + v_q k_r(k_a^s + k_b)} \qquad (3.183)$$

This unwieldy expression may be converted into Eq. (3.170a) by assuming the alternating high barrier to be insurmountable

$$(k_r' = 0, v_r' = 0, k_q'' = 0, v_q' = 0)$$

and by setting

$$v_r' = k_2, \quad k_r'' = k_1, \quad v_q' = k_5, \quad k_q' = k_6, \quad k_b = k_3, \quad k_a = k_4, \quad k_b^s = k_7, \quad k_a^s = k_8.$$

In this manner, the carrier mechanism may be viewed as a limiting case of a channel with alternating barriers.

All the rate equations discussed in this section are nonlinear in their dependence of flow on concentration of solute—in fact, this constitutes the principal difference between nonspecific and specific transports. However, for low concentrations, when $\bar{c}_s' < K_T$, all the equations simplify substantially and

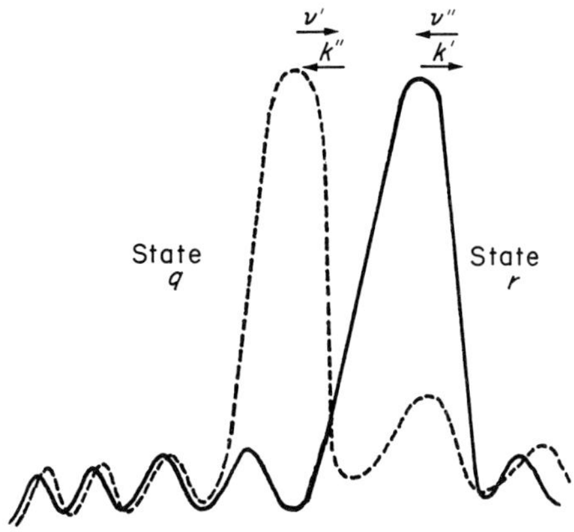

Fig. 3.17. A multiconformational channel energy profile as it alternates between state q and state r

resemble those of simple diffusion. Thus, for instance, Eq. (3.170a) reduces to

$$J_s = c_{E_t}(Ac_s' - Bc_s'')/C \tag{3.184}$$

and, for initial rate,

$$J_{s(0)} = c_{E_t}Ac_s'/C \tag{3.185}$$

which is identical with Eq. (3.4b).

3.5.6 Active transport

All the saturable transport mechanisms discussed so far could operate without any input of energy into the system which had the consequence that the equilibrium concentration of a solute in cell water was the same as in the external solution, unless ions were transported whereupon the accumulation ratio was a function of the membrane potential.

Now most transports of physiologically important ions proceed at the expense of a coupled exergonic reaction, either chemical or photochemical (Table 3.8). This type of transport is called *primary active*. However, a great many anions (both organic and inorganic) and most nonelectrolytes of nutritional value are

Table 3.8. Sources of energy for primary active transports

Source of energy	Organism or membrane	Transported solute
Oxidation of substrate	Mitochondria, bacteria	H^+, Na^+
Capture of photon	Chloroplasts, bacteria	H^+, Cl^-
Hydrolysis of ATP	Nearly all organisms	H^+, Na^+, K^+, Ca^{2+}, HCO_3^-, $HAsO_4^{2-}$
Decarboxylation of oxo acids	Alkalophilic bacteria	Na^+

Table 3.9. Examples of secondary active transport

Organism of membrane	Co-transported solute
H^+-driven	
Escherichia coli	Galactose
	Alanine
	Glycine
	Glucose-6-phosphate
	Lactate
Staphylococcus aureus	Amino acids
Saccharomyces cerevisiae	Purines, pyrimidines
	Amino acids
	Disaccharides
	Sulphate
	Phosphate
Rhodotorula glutinis	Monosaccharides
Chlorella vulgaris	Monosaccharides
Mitochondria	Citrate
Vesicles in nerve endings	Acetylcholine
Chromaffin granules	Catecholamines
Na^+-driven	
Escherichia coli	Glutamic acid
	Melibiose
Halobacterium halobium	Amino acids
Salmonella typhimurium	Pantothenate
	Thiomethylgalactoside
Liver cells	Thiamin
Intestinal epithelium	Ascorbate
	Monosaccharides
	Amino acids
Kidney tubules	Amino acids
	Monosaccharides
Tumour cells	Amino acids
Avian erythrocytes	Amino acids

transported at the expense of a difference of the electrochemical potential of either H^+ or Na^+ by the so-called *secondary active transport* (Table 3.9).

Thermodynamic approach

Nonequilibrium thermodynamics is a useful tool particularly for distinguishing between a primary and a secondary active transport and for determining the degree of coupling between the source reaction and the driven flux. In this context it is useful to transform Eqs (3E.2) and (3E.3) where the L coefficients have the character of conductances, into

$$\begin{aligned} X_1 &= R_{11}J_1 + R_{12}J_2 \\ X_2 &= R_{21}J_1 + R_{22}J_2 \end{aligned} \tag{3.186}$$

in which the coefficients R have the character of resistances. The coefficients of the inverse matrix can be calculated from the L coefficients as

$$R_{11} = L_{22}/|L|, \qquad R_{12} = -L_{12}/|L|, \qquad R_{21} = -L_{21}/|L|, \qquad R_{22} = L_{11}/|L|$$

where

$$|L| = L_{11}L_{22} - L_{12}L_{21}.$$

It is obvious that Onsager's law is valid also for the new coefficients R.

These phenomenological equations can be manipulated in various ways, the flows being expressed as functions of other flows. Flows of solutes $J_j\,(j = 1, \cdots n)$ and the flow of a metabolic reaction J_r at the membrane are described by

$$J_j = -\frac{1}{R_{jj}}\left(\Delta\mu_j + R_{jr}J_r + \sum_{k=1}^{n} R_{jk}J_k\right) \tag{3.187}$$

$$J_r = -\frac{1}{R_{rr}}\left(-A_r + \sum_{j=1}^{n} R_{jr}J_j\right) \tag{3.188}$$

where $\Delta\tilde{\mu}_j$ is the difference of the electrochemical potential across the membrane and A_r is the affinity of the chemical reaction (negative change of Gibbs free energy in the reaction). The first equation shows that the flow of an ion or molecule need not depend on its conjugate force alone but also on the flow of a metabolic reaction J_r and on flows of other substances $J_{k \neq j}$ $(k = 1, \cdots n)$. Analogously, the flow of the metabolic reaction J_r depends not only on its affinity but also on flows with which the reaction is coupled. The presence of coupling manifests itself in a nonzero coefficient R. When $R_{jr} \neq 0$, the substance j is subject to primary active transport. If $R_{jr} = 0$ but $R_{jk} \neq 0$ and $R_{kr} \neq 0$ we are dealing with secondary active transport of j. This sort of transport is designated secondary because it uses as a driving force the gradient of k generated by its primary transport.

Coupling of transport with a chemical reaction requires special comment as it involves processes of different tensorial order. The phenomenological equations we have just discussed are, in fact, global counterparts of local phenomenological equations, such as

$$\mathrm{d}\tilde{\mu}_j/\mathrm{d}x = r_{jj}J_j + r_{jr}J_r, \qquad A_r = r_{jr}J_j + r_{rr}J_r. \tag{3.189}$$

It is assumed here that only one substance is transported, denoted by j, and that the transport is potentially coupled to a single reaction with affinity A_r, that the coefficients r are constant and that Onsager's law $r_{jr} = r_{rj}$ is valid. Now, $\mathrm{d}\tilde{\mu}_j/\mathrm{d}x$ and J_j are vectors, whereas A_r and J_r are scalars. Since vector · vector and scalar · scalar products are scalar while vector · scalar is a vector, r_{jj} and r_{rr} must be scalars whereas r_{jr} is a vector. The vectorial character of r_{jr} may result only from anisotropy, i.e., directional bias of matter at the point of coupling. If the local conditions are isotropic there is no directional bias, the coefficient r_{jr} cannot be a vector and must therefore be zero. No coupling between a vectorial process and a scalar process is possible at a point where isotropy prevails, in accordance with the so-called Curie principle.

The above conclusions may be directly applied to thin cell membranes. Equations, such as

$$\Delta\tilde{\mu}_j = R_{jj}J_j + R_{jr}J_r, \qquad A_r = R_{jr}J_j + R_{rr}J_r \tag{3.190}$$

written with finite differences for cell membranes should be considered as basic laws equivalent to their local counterparts, but not operationally derivable from them, since the interiors of membranes of thickness of the order of molecular dimensions cannot properly be treated as a microscopic phase in the thermodynamic sense. And indeed, in cell membranes performing active transport the transporting enzymes are organized anisotropically.

The situation is rather different in thick model enzyme membranes, for which the above global equations are obtained by integration of local phenomenological equations. Such membranes are usually locally isotropic but display a spatially inhomogeneous distribution of transport and reaction parameters across the membrane. A global analogue of the Curie principle derived for these membranes shows that if such a membrane is symmetrical, whether locally isotropic or not, the coefficient R_{jr} is zero and no coupling between the vectorial transport and scalar chemical reaction is possible.

The *degree of coupling* between the flow of an actively transported substance and the flow of a chemical reaction is defined as

$$q = -R_{jr}/\sqrt{R_{jj}R_{rr}} = L_{jr}/\sqrt{L_{jj}L_{rr}}. \tag{3.191}$$

If there is no coupling between the two flows R_{jr} and hence also q is equal to zero and the transport is not primary active. If coupling is complete $q = \pm 1$; in such stoichiometric coupling the maximum $\Delta\tilde{\mu}_j$ may be sustained without an energy dissipation and the fixed ratio of fluxes is independent of either driving force. This can be shown by substituting $L_{jr} = \sqrt{L_{jj}L_{rr}}$ into the flux Eqs (3E.2) and (3E.3), thus

$$J_j = L_{jj}X_j + \sqrt{L_{jj}L_{rr}}\cdot X_r$$
$$J_r = L_{rr}X_r + \sqrt{L_{jj}L_{rr}}\cdot X_j.$$

Multiplication of J_j with $\sqrt{L_{rr}/L_{jj}}$ yields

$$J_j\sqrt{L_{rr}/L_{jj}} = \sqrt{L_{jj}L_{rr}}\cdot X_j + L_{rr}X_r = J_r. \tag{3.192}$$

This complete coupling is not realistic in natural membranes; even if there is no slipping in the transport mechanism itself, parallel leaks decrease the degree of coupling. A linear relation between the flow of a transported substance and the flow of a chemical reaction does not mean stoichiometric coupling!

The degree of coupling can be estimated in a number of ways.

1. From the mutual dependence of fluxes:

$$q^2 = (\partial J_j/\partial J_r)_{X_j}\cdot(\partial J_r/\partial J_j)_{X_r}. \tag{3.193}$$

2. From the driving force at which each flux is zero:

$$q^2 = (X_r)_{J_j=0}/(X_r)_{J_r=0}. \tag{3.194}$$

3. From the flux of one species when either flux or force of the other is zero:

$$q^2 = 1 - (J_r)_{J_j=0}/(J_r)_{X_j=0}. \tag{3.195}$$

Whenever a substance is transported against a substantial fall of μ_j or $\tilde{\mu}_j$ the *efficiency of energy utilization* η (energy output divided by energy input) may be expressed as

$$\eta = J_j \, \Delta\tilde{\mu}_j / J_r A_r. \tag{3.196}$$

The efficiency of coupling is zero at the situation called *static head* (when the force, here $\Delta\tilde{\mu}_j$, is largest and the net flux zero), assumes finite values at intermediate positions but is again zero at the situation called *level flow* (when the driving force is zero, here $\Delta\tilde{\mu}_j = 0$) (Fig. 3.18).

To obtain an estimate of the coupling at static head, one can measure the *efficacy of force*

$$\varepsilon_{fo} = \Delta\tilde{\mu}_j / J_r A_r \tag{3.197}$$

or the *efficacy of flow*

$$\varepsilon_{fl} = J_j / J_r A_r. \tag{3.198}$$

Measurements of fluxes at level flow are often done in such a way that the naturally acting force is compensated by an externally applied force, e.g. the potential difference caused by the flow of Na^+ across an epithelial layer can be offset by an external electric current—the short-circuit current. In this situation the potential difference is zero and the flow of Na^+ then measured is level flow.

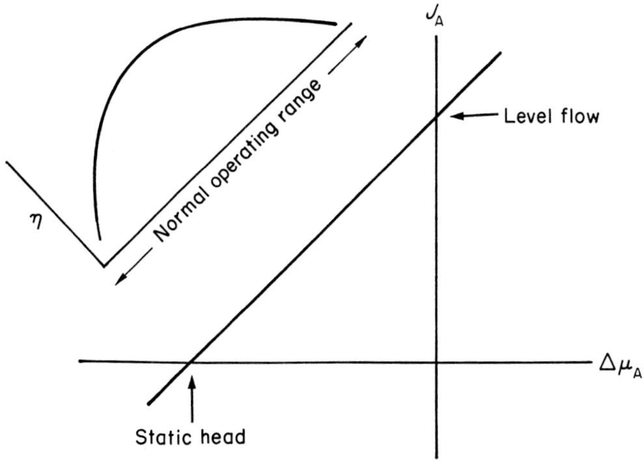

Fig. 3.18. Relationship between a force $\Delta\mu_A$ and a conjugate flow J_A. The oblique insert shows the efficiency η at different values of J_A

Isolated epithelial layers provide a fine object on which the use of nonequilibrium thermodynamics can be demonstrated. The net flow of Na^+ (difference between two unidirectional fluxes determined, e.g., by using ^{22}Na and ^{24}Na) or J_{Na} is driven (not directly but in terms of measurable quantities) by the oxidative reactions taking place in mitochondria so that the consumption of oxygen can be identified with J_r so that

$$J_{Na} = L_{Na} X_{Na} + L_{Nar} A_r \tag{3.199}$$

$$J_r = L_{Nar} X_{Na} + L_r A_r \tag{3.200}$$

where $X_{Na} = -\Delta\tilde{\mu}_{Na}$, the negative of the electrochemical potential difference. The relation between sodium flow and each of the two components of the electrochemical potential (and hence also the total force X_{Na}) was found to be linear. A linear relation was also found to hold between J_r and X_{Na}, demonstrating, at the same time, that A_r was a constant. The partial differential coefficient $(\partial J_r/\partial X_{Na})_{A_r} = L_{Nar}$, which is simply the slope of the straight line experimentally found, gives the value of the cross-coefficient L_{Nar}. As the following step, A_r can be calculated from the equation for J_{Na} for the experimentally easily available condition $X_{Na} = 0$ (see above at level flow)

$$A_r = (J_{Na}/L_{Nar})_{X_{Na}=0} \tag{3.201}$$

The affinity A_r can be thus determined under various conditions, such as inhibition or regulation by various agents. Thus, 2-deoxyglucose decreases the affinity A_r, since it inhibits glucose metabolism. On the other hand, ouabain (G-strophantin) inhibits the transport mechanism itself and the affinity remains unchanged.

Linear relations of irreversible thermodynamics were found to hold in various biological objects, even under situations quite far from equilibrium, such as in mitochondria. It is believed that some regulatory mechanism, a product of evolution, ensures the validity of linear laws far from equilibrium since the efficiency of energy transformation in the linear mode is superior to that in a nonlinear mode.

Kinetic approach

From the kinetic point of view active transports, accomplished always by a carrier-type mechanism, can be analysed as was done for saturable transport without input of energy (mediated diffusion) but with additional factors included.

Primary active transport

In primary active transport, in the simple scheme on p. 190 an energy-releasing reaction can couple onto any of the four partial reactions whereupon the appropriate constant may acquire a coefficient corresponding to 'concentration' of the energy source. In addition to entering into all the kinetic parameters of that model ($K'_T, K''_T, \vec{J}_{max}, \overleftarrow{J}_{max}$) it affects the maximum accumulation ratio $\bar{c}''_s/\bar{c}'_s$, an important characteristic of all active transports. This ratio is easily computed from the condition when $J_s = 0$ and, in this particular case, would be equal to $k_1k_3k_5k_7$[source of energy]$/k_2k_4k_6k_8$.

More complicated schemes can be devised for primary active transport, some of them reflecting the molecular properties of a given system; e.g., an ATPase that can exist in six states, where ATP is bound during the process, might be depicted as

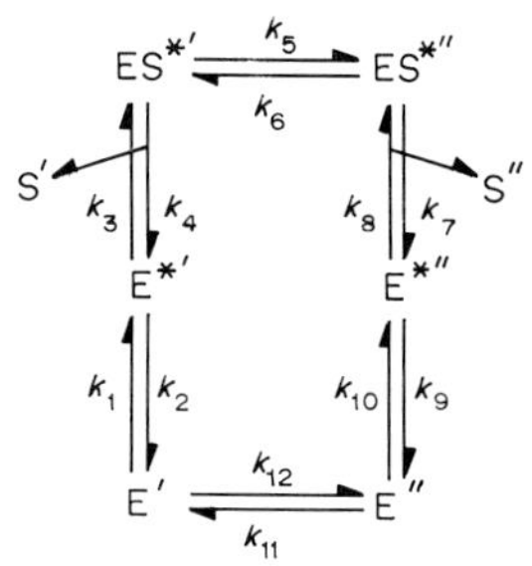

(3.VII)

Here, energy is most likely to couple onto the scheme between E′ and E*′ and the accumulation ratio of S (e.g., an H^+ ion) would be given in a somewhat simplified version as

$$\frac{\bar{c}''_s}{\bar{c}'_s} = \frac{k_1k_3k_5k_7k_9k_{11}}{k_2k_4k_6k_8k_{10}k_{12}} \cdot \frac{c_{ATP}/c_{ADP}c_P}{\bar{c}_{ATP}/\bar{c}_{ADP}\bar{c}_P} \tag{3.202}$$

the $\bar{c}$'s referring to equilibrium concentrations.

It is a general feature of primary active transport systems that they exist in two or more conformational states with different affinities for substrate and it is generally inferred from this observation that a change in affinity must take place as the carrier reorientates in the membrane. However, this is not a necessary condition. Inspection of the expression for the accumulation ratio shows that any of the constants there may be associated with the input of energy. Indeed, it can easily be the reorientation itself (rate constant k_5 or k_{11}), with the consequence that the given carrier form is more frequently exposed on the one than on the other side so that its effective concentration there is increased (rather than its affinity decreased), allowing for a higher concentration of the free ligand. Expressed as a Haldane-type ratio

$$\frac{c''_s}{c'_s} = K_{eq} = \frac{\vec{J}_{max}K''_T}{\overleftarrow{J}_{max}K'_T} \tag{3.203}$$

and here it can be any of the kinetic parameters that is 'coupled' with energy and thus shifts the K_{eq} to higher values.

Secondary active transport

In secondary active transport it is generally assumed that the carrier can transiently bind both the transported solute (often a nonelectrolyte) and the driving ion (either H^+ or Na^+). The energy used for the transport is available from a gradient of the driving ion oriented oppositely to that formed by the solute during transport, and is defined as

$$\Delta\tilde{\mu}_{\text{ion}} = \Delta\phi\, F + RT \ln \frac{c''_{\text{ion}}}{c'_{\text{ion}}}. \tag{3.204}$$

If the use of this electrochemical gradient is tightly coupled to the generation of a solute gradient, such that

$$-RT \ln \frac{c''_{\text{s}}}{c'_{\text{s}}} = n\left(\Delta\phi\, F + RT \ln \frac{c''_{\text{ion}}}{c'_{\text{ion}}}\right) \tag{3.205}$$

where n is the number of ions transported per cycle, one can derive for the accumulation ratio

$$\frac{\bar{c}''_{\text{s}}}{\bar{c}'_{\text{s}}} = \left(\frac{\bar{c}'_{\text{ion}}}{\bar{c}''_{\text{ion}}}\right)^{\text{n}} \exp(-nF\,\Delta\phi/RT). \tag{3.206}$$

The same, but somewhat refined, result can be reached by a kinetic analysis of carrier mechanisms, such as those depicted by the schematic models below where E is the carrier, H the driving ion and S the transported solute.

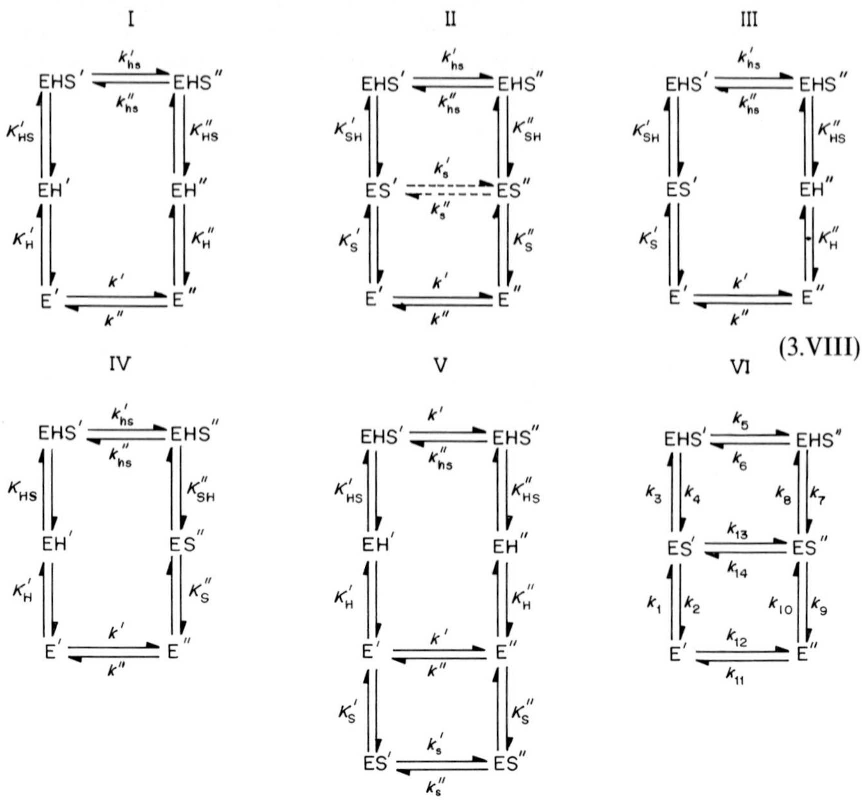

(3.VIII)

In all these models EH was taken to be immobile as otherwise it would wastefully short-circuit the electrochemical potential gradient of the ion. The models have been analysed both by steady-state approaches (according to Model VI) and by the rapid-equilibrium approach where the translocation constants were taken to be smaller than the dissociation constants at the surface. The fundamental kinetic parameters K_T and J_{max} may depend on the concentrations of the driving ion, both at the starting (*cis*) and the target (*trans*) side, as well as on the membrane potential. As an example, the values for Model V, in rapid equilibrium, are given:

$$J_{max} = c_{E_t} \frac{k'' K''_H(k_{hs}\xi^{1/2} K'_s c'_{ion} + k'_s K'_H K'_{HS})}{\text{denominator}} \tag{3.207}$$

$$K_T = \frac{(K''_H + c''_{ion})k' K'_H + (K'_H + c'_{ion})k'' K''_H}{\text{denominator}} \tag{3.208}$$

where

$$\text{denominator} = (k'_s K'_H K'_{HS} + k'_{hs} K'_S \xi^{1/2})(K''_H + c''_{ion}) + k'' K''_H(K'_H K'_{HS} + K'_S c'_{ion}).$$

The models shown probably cover the entire range existing in nature as manipulation of the various rate constants permits one to derive situations where either c'_{ion} or ξ affect either J_{max} or K_T or both in a positive or a negative way.

The accumulation ratios derived by these procedures are listed in an abbreviated form in Table 3.10. These ratios represent the static head situation which, however, does not necessarily reflect the thermodynamic maximum of the system, viz. in the cases where ES is mobile and represents an internal slip of the system. Of all the models tested only V in a steady-state approach shows an effect of the external concentration of S on the accumulation ratio (some mixed rapid-equilibrium and steady-state models also do that). Here the ratio, which can only be arrived at numerically, tends toward a constant value at high concentrations of S, and toward unity if the system is symmetrical except for the $EHS' \rightleftharpoons EHS''$ step.

The trend toward an accumulation ratio equal to one is usually attributed to an external parallel leak (by simple or mediated diffusion) of the substrate through the membrane. Clearly, this need not be so—an internal leak can account for this phenomenon.

Another observation is often made with secondary active transports of nonelectrolytes where the intracellular 'equilibrium' concentration tends toward constancy at high values of c'_s. This cannot be explained by any of the models in the present form. One must assume here that the available concentration of the driving ion is reduced by its being bound to the carrier. The cubic equation arising from considering that

$$c'_{ion,free} = c'_{ion,total} - c'_{EH} - c'_{EHS}$$

Table 3.10. Accumulation ratios $\bar{c}''_s/\bar{c}'_s$ predicted by various models of secondary transport with H^+ for a neutral carrier

Model	Rapid equilibrium*		Steady state	
	ES mobile	ES immobile	ES mobile	ES immobile†
I	—	$\frac{k'' K''_H c'_H}{k' K'_H c''_H}\xi$	—	
II	$\frac{k'_s K'_{HS} + k_{hs}\xi^{1/2} c'_H}{k''_s K''_{HS} + k_{hs}\xi^{-1/2} c''_H}$	$\frac{c'_H}{c''_H}\xi$	$\frac{k_1 k_9 k_{11}[k_4 k_{13}(k_6 + k_7) + k_5 k_7 \xi^{1/2}(k_{13} + k_3 c'_H)]}{k_2 k_{10} k_{12}[k_7 k_{14}(k_4 + k_5) + k_4 k_6 \xi^{-1/2}(k_{14} + k_8 c''_H)]}$	
III	—	$\frac{k'' K''_H c'_H}{k' K'_H c''_H}\xi$	—	$\frac{\Pi_c c'_H}{\Pi_{cc} c''_H}\xi$
IV	—	$\frac{k'' K''_H c'_H}{k' K'_H c''_H}\xi$	—	
V	$\frac{(K'_H K'_{HS} k'_s + K'_S k_{sh} \xi^{1/2} c'_H)(K''_H K''_{HS} K''_S k'')}{(K''_H K''_{HS} k''_s + K''_S k_{sh} \xi^{-1/2} c''_H)(K'_H K'_{HS} K'_S k'_s)}$	$\frac{k'' K''_H c'_H}{k' K'_H c''_H}\xi$	$\frac{(2K_1 c'_s + K_2 + K_3 c'_H)}{(2K_4 c''_s + K_5 + K_6 c''_s)}$	

* Assuming that $\Pi k'_{ij} = \Pi k''_{ij}$
† Π_c product of all rate constants going clockwise; Π_{cc} product of all rate constants going counterclockwise

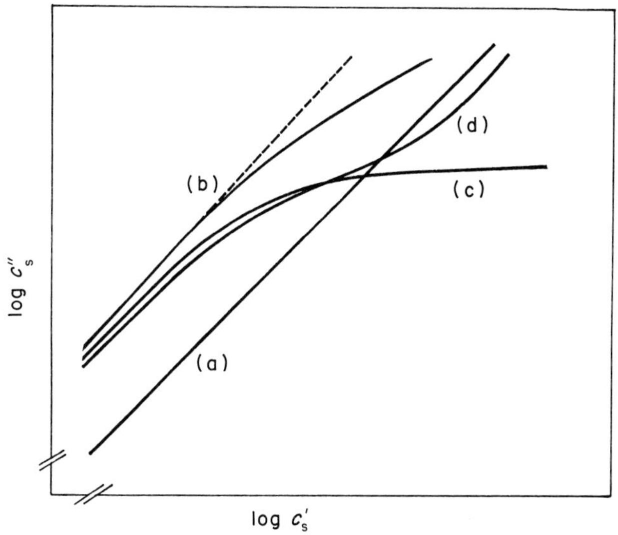

Fig. 3.19. Steady-state accumulation of an actively transported nonmetabolized solute (c'—outside, c''—inside concentration). (a) No energy input; (b) active transport with a (mediated) diffusional leak; (c) active transport with energy limitation at higher solute concentrations if energy coupling takes place prior to substrate binding; (d) the same but energy coupling takes place after substrate binding. The dashed line shows the theoretical (constant) steady-state accumulations ratio such as is found at low solute concentrations

and from substituting the $c'_{\text{ion,free}}$ in the steady-state carrier cycle, permit one to derive the dependences shown in Fig. 3.19.

Antiport systems

In these mechanisms the carrier presumably cannot move when unoccupied and, moreover, it changes its affinity for substrate as it reorientates in the membrane. The simplest kinetic diagram of an antiport with substrates A and B is

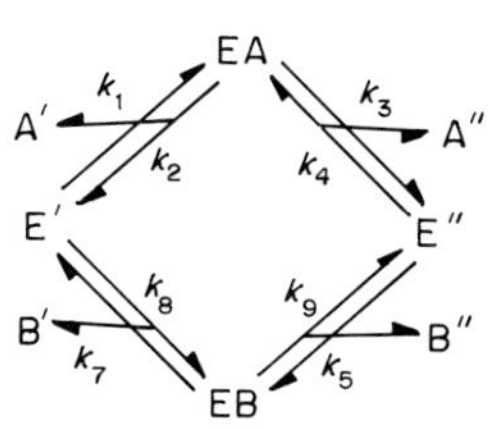

(3.IX)

The forms EA and EB can be decomposed into EA′⇌EA″ and EB′⇌EB″ but this adds no new quality to the model. The rate equation, proceeding from a steady-state assumption and the appropriate conservation equation, is

$$J_A = c_{E_t} \frac{k_1k_3k_5k_7c'_Ac''_B - k_2k_4k_6k_8c'_Bc''_A}{\begin{aligned}&(k_6+k_7)(k_1k_3c'_A + k_2k_4c''_A + k_1k_4c'_Ac''_A) +\\ &(k_2+k_3)(k_6k_8c'_B + k_5k_7c''_B + k_5k_8c'_Bc''_B) +\\ &k_4k_8(k_2+k_6)c''_Ac'_B + k_1k_5(k_3+k_7)c'_Ac''_B.\end{aligned}} \tag{3.209}$$

If no energy is put into the system (then $k_1k_3k_5k_7 = k_2k_4k_6k_8$), obviously

$$\bar{c}'_A/\bar{c}''_A = \bar{c}''_B/\bar{c}'_B. \tag{3.210}$$

However, here again energy can couple to any of the constants so that the full ratio will be

$$\bar{c}'_A/\bar{c}''_A = (k_1k_3k_5k_7\bar{c}''_B/k_2k_4k_6k_8\bar{c}'_B)\cdot(c'_H/c''_H) \tag{3.211}$$

where c_H stands for a source of energy in general. The same ratio will hold for the case if both A and B carry the same charge; if, however, A is a cation the ratio must be multiplied by $\xi = \exp(-z\Delta\phi F/RT)$.

3.5.7 Properties of channel and carrier systems

Permeability

Although this may vary substantially, channels are considered to be rather more permeable than carriers, with average catalytic numbers of 10^6–10^9 particles per second for the former against 10^2–10^4 particles per second for the latter. The catalytic numbers of carriers are surpassed by most soluble and membrane-bound enzymes and the low figure is usually taken to indicate that it is a conformational change of the protein that is rate-limiting, such changes being known to be relatively slow even with enzymes.

Dependence on temperature

Here again, the dependence of carrier systems is steeper than with channels, apparently because the conformational change of the carrier has a higher activation energy than the presumably static binding site in a channel. Activation energies of up to $100\,\mathrm{kJ\,mol^{-1}}$ are found with carriers. However, this is no rigorous rule. Even simple diffusion in which a solute particle must be stripped of its hydration shell may display rather high activation energies.

An interesting property of Arrhenius plots of specific membrane transport is a break at (presumably) the transition temperature of the membrane which could be expected if the transport protein were freely dissolved in the membrane but should not be there if it were a fixed structure. Such situation is approximated by the case of valinomycin as a true mobile carrier (unlike the natural ones) and gramicidin as a true channel. Valinomycin practically ceases to operate below the transition temperature while gramicidin is not appreciably affected (cf. Fig. 2.22). The natural membrane carriers are certainly relatively inflexible structures as a whole but, apparently, the ease with which the binding site changes its conformation during transport is affected by the microviscosity of the surrounding lipids.

Dependence on pH

Like perhaps all enzymes, the transport proteins show a more or less expressed dependence of their activity on pH, in most cases an optimum being observed at a value intermediate between the two pH extremes which is indicative of there being at least two protonatable groups in or near the active site of the transport proteins. In other words, the transport systems operate only at a certain degree of protonation. The dependence on pH can be found both for J_{max} and for K_T and, from the breaks on their logarithmic plots, the dissociable groups can be assessed (Fig. 3.20).

In contrast with soluble enzymes, the pH dependence of membrane carrier

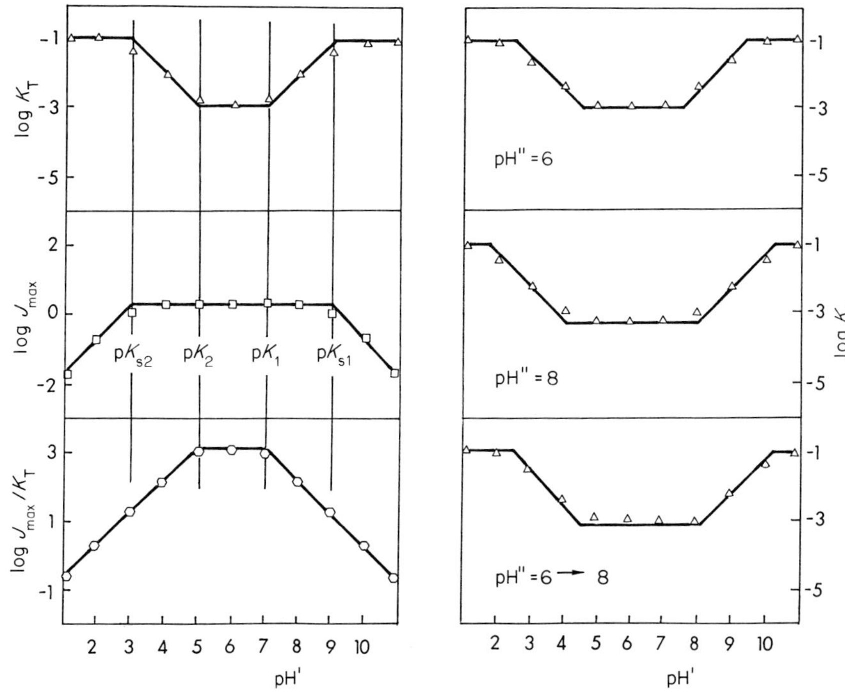

Fig. 3.20 pH dependence of saturable transport parameters J_{max} and K_T plotted in three different ways. K_1 is the dissociation constant of the carrier-H$^+$ complex, K_2 that of the (abortive) carrier-H$^+$–H$^+$ complex, K_{s1} that of the carrier-substrate-H$^+$ complex, K_{s2} that of the (abortive) carrier-substrate-H$^+$–H$^+$ complex. The left-hand panel shows the idealized dependence for an enzyme-like protein exposed to a uniform pH. The right-hand panel shows the situation for a carrier at intracellular pH of 6 and 8, and one where the intracellular pH is assumed to change from 6 to 8 at extracellular pH equal to 7

activity is not simple as the pH at both sides of the membrane should be taken into account which may smear out the dependence somewhat.

An important comment should be made in this context. The pH effect discussed here is a purely kinetic effect. It is assumed that protonation of the active site takes place at a different subsite than that for H^+ in a symport mechanism. This H^+ binding does not prevent the mobility of the carrier, unlike in the case of the 'thermodynamic' binding during symport (cf. p. 205).

Effect of surface potential

While the role of the transmembrane potential was taken into account in the foregoing section where it was shown to influence the distribution of ions in equilibrium, the surface potential derived in Section 2.7 can exert a substantial kinetic effect on all transports involving ions. This is because it alters the concentration of ions at the membrane surface, attracting cations and repelling anions. Equation (2.34) can be rewritten as

$$c_{i,s} = c_{i,b} \exp(z_i F \psi_s / RT) = c_{i,b} y^{z_i} \tag{3.212}$$

to make it easier to manipulate. Table 3.11 shows the modified parameters for different cases of ion-involving transport.

Effect of unstirred layers

Like with simple diffusion mechanisms, saturable transport systems are affected by the presence of stagnant, unstirred layers of solution adjacent to the cell surface.

A substance that can be translocated by a saturable process across the membrane displays a flux through the unstirred layer defined by the same

Table 3.11. Effect of decreasing the surface potential on membrane transport parameters

Transport of (order of binding)	K_m	J_{max}
Anion alone	↓	—
Cation alone	↑	—
Neutral solute with cation		
(cation first)	↑	—
(solute first)	↑	↓
Cation with anion		
(anion first)	↑	—
(cation first)	↑	↑
Anion with cation		
(anion first)	↓	↓
(cation first)	↓	—

expression as one moving by diffusion even through the membrane, viz.

$$J_{\text{ul}} = (D/\delta)(c_{\text{s,b}} - c_{\text{s,m}}) \tag{3.213}$$

where $c_{\text{s,b}}$ is its concentration in the bulk phase, $c_{\text{s,m}}$ its concentration at the membrane, D is the diffusion coefficient in the unstirred layer and δ its thickness. As in a steady state this flux must be equal to its transport rate across the membrane which, for initial conditions, is equal to $J_{\text{max}} c_{\text{s,m}}/(c_{\text{s,m}} + K_{\text{T}})$ we can solve for $c_{\text{s,m}}$ from the two equations and introducing the root of the square equation into Eq. (3.213) obtain

$$J_{\text{s}} = P[0.5(K_{\text{T}} + c_{\text{s,b}} + J_{\text{max}}/P) \pm \sqrt{0.25(K_{\text{T}} - c_{\text{s,b}} + J_{\text{max}}/P) + c_{\text{s,b}} K_{\text{T}}}. \tag{3.214}$$

If $c_{\text{s,m}} \ll K_{\text{T}}$ then $J_{\text{s}} = c_{\text{s,b}}/(1/P + K_{\text{T}}/J_{\text{max}})$. If $c_{\text{s,m}} \gg K_{\text{T}}$ then $J_{\text{s}} = J_{\text{max}}$.

While J_{max} is not affected by the unstirred layers, the observed half-saturation constant is changed thus:

$$K'_{\text{T}} = K_{\text{T}} + 0.5 J_{\text{max}}/P \tag{3.215}$$

so that the measured value proceeding from bulk concentration is higher than the true constant by $0.5 J_{\text{max}}/P$. Here everywhere $P = D/\delta$.

3.5.8 Transport-associated chemical reaction

The special cases where the transported solute is chemically modified during the transport step belong into this category. This situation is different in molecular terms from that in primary active transport where a chemical reaction can drive a transport mechanism because the solute at the left here is obligatorily different from the solute at the right (group translocation).

However, both kinetically and thermodynamically, it is equivalent to a primary active transport, following the kinetic diagram

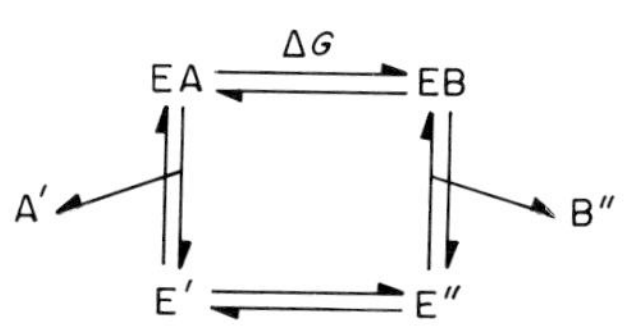

(3.X)

where ΔG fully governs the equilibrium ratio of $\bar{c}_{\text{B}}/\bar{c}_{\text{A}}$.

SYNOPSIS

Specificity in transmembrane transport is achieved by the involvement of a protein in the process, the protein possessing a binding site for the substance to be transported. In some cases, the protein is part of a channel and can be approached from both membrane faces with equal ease. In other cases, the access is alternately permitted from one or the other membrane face—this is the mechanism described as a carrier. Microscopically, the two situations may merely be limiting cases of a channel-like transport agency with a conformation-dependent gating unit.

Specific transport can proceed in an energy-independent mode (facilitated or mediated diffusion) or by an energy-linked process (active transport). The coupling with a free energy source may involve a chemical or photochemical reaction—then it is primary active transport, or comprise a regulated dissipation of an electrochemical potential gradient of H^+ or Na^+—then it is secondary active transport.

In addition to these transports in which the 'substrate' taken up at the starting (*cis*) side is released unchanged at the target (*trans*) side, several transport processes are vectorially coupled with a chemical reaction so that the substrate emerges as a chemically different product at the other membrane side.

FURTHER READING

Barts, P. W. J. A., and Borst-Pauwels, G. W. F. H. Effects of membrane potential and surface potential on the kinetics of solute transport, *Biochim. Biophys. Acta*, **813**, 51–60 (1985).

Blaustein, M. P., and Lieberman, M. (eds). *Electrogenic Transport. Fundamental Principles and Biological Implications*, Raven Press, New York (1984).

De Felice, L. *Introduction to Membrane Noise*, Plenum Press, New York (1981).

Essig, A. Energetics of active transport processes, *Biophys. J.*, **15**, 651–65 (1975).

Essig, A., and Caplan, S. R. Energetics of active transport processes, *Biophys. J.*, **8**, 1434–57 (1968).

Hladky, S. B. The carrier mechanism, *Curr. Topics Membr. Transport*, **12**, 53–164, Academic Press, New York (1979).

Kedem, O. Criteria of active transport. In: *Membrane Transport and Metabolism* (A. Kleinzeller, and A. Kotyk, eds), pp. 87–93, Academic Press, London (1961).

Kotyk, A., and Stružinský, R. Effect of high substrate concentrations on active transport parameters, *Biochim. Biophys. Acta*, **470**, 484–91 (1977).

Läuger, P. Kinetic properties of ion carriers and channels, *J.* Membr. Biol., **57**, 163–78 (1980).

Läuger, P. Thermodynamic and kinetic properties of electrogenic pumps, *Biochim. Biophys. Acta*, **779**, 307–41 (1984).

Läuger, P. Mechanism of biological ion transport—Carriers, channels and pumps in artificial lipid membranes, *Angew. Chemie* (Internat. Ed. Engl.), **24**, 905–23 (1985).

Mierle, G. A method for estimating the diffusion resistance of the unstirred layer of microorganisms, *Biochim. Biophys. Acta*, **812**, 824–34 (1985).

Pietrobon, D., Zoratti, M., Azzone, G. F., Stucki, J. W., and Walz, D. Non-equilibrium thermodynamic assessment of redox-driven H^+ pumps in mitochondria, *Eur. J. Biochem.*, **127**, 483–94 (1982).

Sanders, D., Hansen, U.-P., Gradmann, D., and Slayman, C. L. Generalized kinetic analysis of ion-driven cotransport systems. A unified interpretation of selective ionic effect on Michaelis parameters, *J. Membr. Biol.*, **77**, 123–52 (1984).

Silverman, M., and Goresky, C. A. A unified kinetic hypothesis of carrier-mediated transport, *Biophys. J.*, **5**, 487–509 (1965).

Stark, G. Carrier-mediated transport across thin lipid membranes. In: *Membrane Transport in Biology* (G. Giebisch, D. C. Tosteson, H. H. Ussing, eds), Vol. 1, pp. 447–73, Springer-Verlag, Berlin (1978).

Stark, G. Hydrophobic ions, carriers and pore formers in lipid bilayers as studied by fast kinetic methods. In: *Biomembranes, Dynamics and Biology* (R. M. Burton, and F. C. Guerra, eds), pp. 193–224, Plenum Press, New York (1984).

Stein, W. D. Concepts of mediated transport. In: *Membrane Transport* (S. L. Bonting, and J. J. H. H. M. de Pont, eds), pp. 123–58, Elsevier/North Holland, Amsterdam (1981).

Tanford, C. Mechanism of free energy coupling in active transport, *Annu. Rev. Biochem.*, **52**, 379–409 (1983).

3.6 SPECIFIC CHANNEL AND CARRIER PROTEINS

Perhaps a hundred different channel and carrier proteins are now known to exist in cell membranes and only the ones that are heuristically significant will be described in this section. However, proteins catalyzing an energy transduction, i.e. a transformation from one form of energy to another, such as the primary active transport proteins, as well as channels or carriers that are functional in the transfer of information, will be described in Sections 4 and 5, respectively.

3.6.1 Band 3 protein

This working designation is meant for a major erythrocyte membrane protein which, in classical polyacrylamide gel electrophoresis, places third in the final band distribution. Alternative names for this protein are *anion transport protein* or *capnophorin.*

The amount of band 3 protein is spectacular: Its $1.2 \cdot 10^6$ molecules per erythrocyte may account for 30 per cent of total membrane protein. Moreover, it does not seem to be unique for the erythrocyte since a great number of other animal cells (in blood, liver, lung epithelium) contain membrane proteins cross-reacting with antiband 3 antibodies.

The protein has a molecular weight of 97 000 and its amino acid sequence is known to contain 929 residues. It has two distinct structural domains. The first, roughly comprising residues 1–420, is hydrophilic, with two hydrophobic regions at residues 100–150 and 250–290 which, however, are not believed to insert into the membrane. The second, from 420 to the C terminus, contains 14 regions that could be candidates for spanning the membrane. However, as shown in Fig. 2.19, some of the regions are submerged in the membrane without appearing on the hydrophilic surface.

While the membrane-inserted domain is obviously the channel part of the protein, the hydrophilic segment is presumably involved in interactions with cytosolic proteins, specifically with ankyrin which, in its turn, forms a bridge to the underlying spectrin network. Presumably, several enzymes of glucose metabolism, as well as hemoglobin, can bind to this band 3 domain.

The band 3 protein is likely to exist in the form of dimers which might offer more possibilities for constructing the channel with polar residues facing into the channel and the hydrophobic ones into the lipid core.

The principal, even if not particularly specific, function of band 3 is the transport of anions. Under normal conditions the system operates in an exchange mode so that no net current can be observed and it can be described as electrically silent. In an equilibrium exchange experiment the flux of Cl^- ions represents $5 \cdot 10^{-8}$ mol cm^{-2} s^{-1} which, considering the presence of 10^6 protein molecules per cell and the erythrocyte surface area of 100 μm^2, yields a turnover of about 30 000 Cl^- ions per band 3 protein per second. The system is saturable, the apparent dissociation constants for anion binding being (in mM): F^- 88, Cl^- 67, Br^- 32, I^- 10, HCO_3^- 10, SO_4^{2-} 30–40, HPO_4^- 68. Moreover, at high anion concentrations, an excess-substrate inhibition is observed, the K_i of the 'excess' or modifier site being (again in mM): F^- 340, Cl^- 330, Br^- 160, I^- 60, HCO_3^- 600, SO_4^{2-} 350–600, HPO_4^{2-} 200.

A general finding has been that divalent anions are cotransported with a proton.

The transport of anions involves here conformational changes of the protein molecule with different representation of the two (or more) conformers at the two membrane faces and thus corresponds better to the definition of a carrier rather than channel where, typically, the binding sites are exposed simultaneously to both membrane faces. A gate or double-gate mechanism is obviously the model of choice here.

At the molecular level, there is good evidence that the binding of anions is effected by several lysine residues of the band 3 protein and that carboxyl groups are involved in divalent anion transport. Lysine residues in positions 561, 608, 649 and 657 have been implicated either directly or indirectly in the process and are assumed to form part of the channel which may be lined by segments 5, 6 and 7 which span the membrane.

3.6.2 The ADP/ATP carrier of mitochondria

The most abundant protein of the mitochondrial inner membrane (14 per cent of mitochondrial membrane proteins), the ADP/ATP carrier is an innovative feature of mitochondria as compared with the bacteria they have evolved from. In bacteria there is no need to transport ATP anywhere, in mitochondria it is carried out to the cytosol in exchange for ADP brought into mitochondria. This exchange *per se* does not require energy but it is assisted by the mitochondrial membrane potential as ATP carries one more negative charge than does ADP and, therefore, has a tendency to follow the electric potential oriented with its positive side toward the cytosol.

It is a dimeric protein of 2 × 297 amino acids (total M_r 64 000), very much homologous to the uncoupling protein of brown fat mitochondria. Each monomer spans the membrane six times and its principal feature is that it exists in two conformations. As it faces the matrix the *m* conformation predominates and it binds ATP which is produced on the matrix side. This conformation is distinguished by being able to bind the inhibitor bongkrekic acid (a product of a false coconut pulp fermentation due to *Pseudomonas cocovenenans*). In the *c* conformation which predominates in facing the intermembrane space it binds ADP (which is abundantly present in the cytosol) and is able to interact with the inhibitor atractyloside or even better carboxyatractyloside (extremely toxic products of the Mediterranean plant *Atractylis gummifera*). All the K_D values of the inhibitors are of the order of 5–20 nM and the inhibition is noncompetitive.

The K_D for ADP in the *m* form is 0.1 – 0.5 μM, in the *c* form 4–7 μM; that for ATP is 0.6 μM in the *m* form and 12 μM in the *c* form. The p*I* of the protein is at pH 10. Its turnover number is 10–30 s^{-1}.

In spite of its existing in the dimeric form it has a single binding site for the nucleotide, a phenomenon ascribed to half-site reactivity, such as is frequently found with oligomeric allosteric enzymes. In fact, there is now evidence that the carrier may be tetrameric, with two functioning sites per tetramer, one being exposed to the outside, the other to the inside of the mitochondrion. The two binding sites are not identical and their flipping from one exposure to the other corresponds to the carrier being able to bind either bongkrekic acid or carboxyatractyloside. According to this model, ADP may be bound at the *c* side at the same time as ATP is bound at the *m* side and only then are they exchanged through a double-gate mechanism.

3.6.3 The lactose transport system of bacteria

Lactose and other galactosides are transported in *Escherichia coli* by a carrier protein now known to be a single polypeptide of M_r equal to 46 500 whose amino acid sequence is known. Hydrophobicity data would predict 14 membrane crossings of the polypeptide—this would be probably a record number among membrane proteins. The four N-terminal and the two C-terminal helices are extremely hydrophobic and are almost definitely embedded in the lipid membrane while helices 7–12 are more hydrophilic and it is there that the binding site for galactoside is located.

The system apparently functions as an intrinsically symmetrical carrier for galactosides, artificial substrates, such as galactosyl-1-β-thio-D-galactoside, showing a much higher affinity for it than lactose itself. The K_D for lactose is the same at both membrane sides but the imposition of $\Delta\tilde{\mu}_{H^+}$ brings about active accumulation in cells. The affinity for H^+ is extremely high at both membrane sides, the K_D being $5 \cdot 10^{-11}$ M.

A system for melibiose transport has also been characterized in *Escherichia coli*. The carrier has a M_r of 52 000, its molecule spans the membrane perhaps ten times and, besides binding melibiose, appears to have a binding site for H^+ and another for Na^+ (and Li^+) as both $\Delta\tilde{\mu}_{H^+}$ and $\Delta\tilde{\mu}_{Na^+}$ contribute, in an additive way, to the rate of melibiose transport.

Like all bacterial transport systems studied so far the carriers are inserted during biogenesis into the membrane without any N-terminal signal sequence or presequence, such as is present in secretory and in mitochondria- or chloroplast-targetted proteins.

3.6.4 Binding protein systems in bacteria

The periplasmic space of bacteria contains perhaps a dozen binding proteins which are known to be components of transport systems that are not clearly driven by a H^+ gradient. The histidine system is the best known of all. Others include branched amino acids, oligopeptides, maltose, β-methylgalactoside, ribose, arabinose, phosphate and sulphate. These transport systems are composed of several proteins, one or two of them being binding proteins that are easily removed from cells by an osmotic shock, others being membrane-bound.

The binding proteins range in molecular weights from 25 000 to 56 000, are heat-stable and display a high affinity for substrate. All contain two functionally and genetically separable domains.

An important property of the binding proteins is that they undergo a conformational change on binding their substrate. When the substrate is bound the protein can interact with the membrane-anchored components of the system. They are bean-shaped with a cleft at one end which closes somewhat upon substrate binding.

The membrane components are present in much smaller amounts but are known in sufficient detail to show distinct homology of certain regions of the amino acid sequence. The membrane proteins that interact with the binding protein (hisP, rbsA, malK, pstB, oppD9) all contain two sites, at near positions 50 and 190 of hisP that are highly homologous with parts of the β subunits of the mitochondrial H-ATPase. Moreover, it is at the conserved homology site that ATP is bound to these membrane proteins although no clear role has yet been assigned to it.

The other membrane-bound proteins have possibly a regulatory role but then again may act as the ‘channels’ for the binding protein ligand to get into the cell (Fig. 3.21).

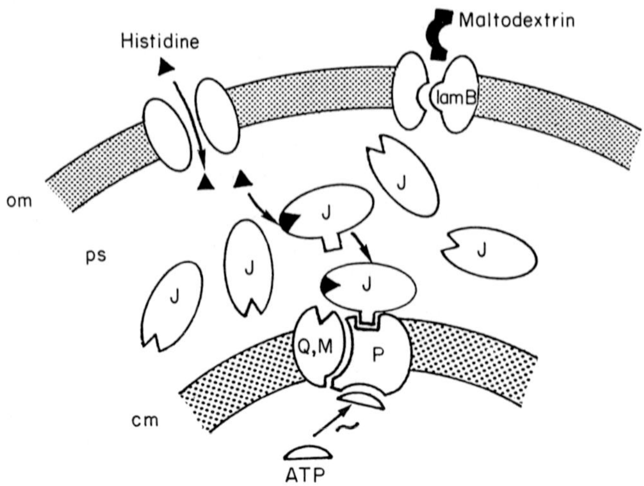

Fig. 3.21. A model of histidine transport in *Salmonella typhimurium* through the outer membrane (om), the periplasmic space (ps) where histidine encounters the binding protein J, and through the cytoplasmic membrane where an interaction with Q, M, and P proteins is required for efficient transport

3.6.5 Sodium-glucose symport of the intestinal brush border

Of the many sodium-driven systems of animal cells (see Table 3.9) the glucose transport system of the small intestine has been probably most thoroughly studied.

It is composed of a single polypeptide (M_r 75 000) which binds glucose and Na^+ at distinct sites. The 'gate' of the carrier molecule bears several negative charges of amino acid carboxyls and application of a membrane potential apparently causes the gate to move so as to preferably position the glucose and Na^+ binding sites toward the external (luminal) face of the membrane. Binding of these ligands results in a reorientation of the binding sites toward the cell interior.

It appears that the mechanism is one of (iso)random sequential Bi-Bi type, with a K_T of 0.2 mM at low concentrations of Na^+ and of 0.08 mM at high c'_{Na^+}. The turnover of the carrier is 3–6 s^{-1}. It has a specific binding site for phlorizin with a K_D of 2 μM.

3.6.6 Extraneous carrier-like molecules

Of the several hundred peptides, both cyclic and linear, depsipeptides, macrotetrolides, nigericin-like penta- or hexacyclic molecules, and synthetic crowns, the best known and most frequently employed is the depsipeptide *valinomycin.*

It was isolated from the prokaryotic *Streptomyces fulvissimus* and its structure is (–Lac–L-Val–HIV–D-Val–)$_3$, where Lac stands for lactic acid and HIV for D-hydroxyisovaleric acid. In solution, valinomycin exists in three different conformations (Fig. 3.22). The A form has all the NH groups linked by a hydrogen bond to the nearest carbonyl oxygen and is the predominant form in nonpolar solvents (in membranes), the B form shows the binding only for the D-valine groups and the C form, which is the principal form in water, forms hydrogen bonds only with the solvent.

The B form readily reacts with univalent cations to form a stable A_2 complex (Table 3.12). The highest permeability is found with Rb^+ and K^+ but the stability constants may be quite high (Ag^+) even if transport catalysis is extremely low. This illustrates the important point that a high affinity for a carrier must be accompanied by a reasonable rate of movement in the membrane to have a given carrier function properly.

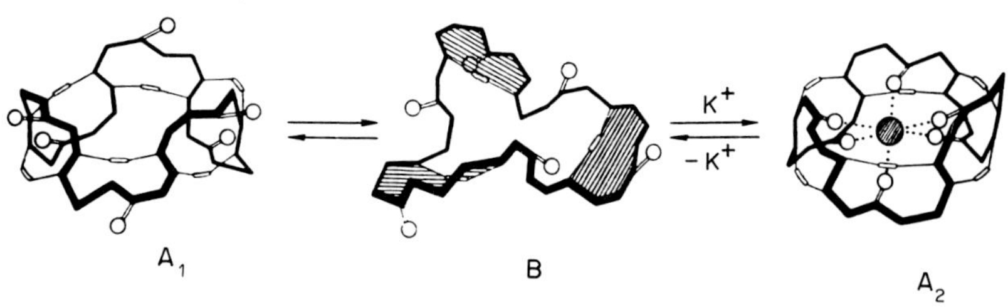

Fig. 3.22. Three different forms of valinomycin. For details see the text

Table 3.12. Constant of stability (in M^{-1}) of valinomycin with univalent cations in methanol and its effect on ionic permeability of an artificial membrane

Ion	$K_{stability}$	P_{ion}/P_{K^+}
Li^+	5.0	$4.2 \cdot 10^{-6}$
Na^+	4.5	$3.6 \cdot 10^{-5}$
K^+	$7.9 \cdot 10^4$	1.00
Rb^+	$2.0 \cdot 10^5$	1.80
Cs^+	$2.5 \cdot 10^4$	0.76
Ag^+	$7.9 \cdot 10^3$	$5.6 \cdot 10^{-6}$
Tl^+	$5.0 \cdot 10^3$	0.11
NH_4^+	50	0.04

The valinomycin-ion complex apparently moves freely in the plane of the membrane and transmits its ligand to another valinomycin when it aligns itself properly in the direction normal to the membrane. Some ions may be carried sandwiched between two molecules of valinomycin. The frequency of transmembrane movement of empty valinomycin was estimated to be $10^4\,s^{-1}$ while with K^+ it is $10^5\,s^{-1}$. The movement of K^+ bound to valinomycin, although relatively fast by carrier standards, is still much slower than in aqueous solution—in the membrane about 3 nm μs^{-1}, in water about 250 nm μs^{-1}. This supports the statement made on p. 208 that carrier transport is much slower than channel transport.

3.6.7 Extraneous specific channel formers

Several oligopeptides have been shown to act as channel formers in both artificial and natural membranes. The prototype of these is *gramicidin A* (isolated from *Bacillus brevis*), two molecules of which join head to head to form a transmembrane $\beta^{6.3}$-left-handed helical dimer with a central channel about 0.4 nm in diameter.

The amino acid sequence of gramicidin A is HCO–L-Val–Gly–L-Ala–D-Leu–L-Ala–D-Val–L-Val–D-Val–L-Trp–D-Leu–L-Trp–D-Leu–L-Trp–$NHCH_2OH$. Replacement of Trp-11 with Phe gives rise to gramicidin B, with Tyr to gramicidin C while multiple replacements lead to gramicidin D.

The gramicidin A channel is probably the most thoroughly studied transport passage at a submolecular level. Taking into account all the components of interactive energy (electrostatic, repulsion, polarization, dispersion, and torsion) a transmembrane energy profile could be established (Fig. 3.23). There are two characteristic minima, one at a distance of 0.5 nm, another about 1 nm from the centre. The pattern is retained and the minima are even more pronounced if the individual amino acid chains are included in the computation. The tryptophan residues play a special role in bonding water molecules to their free carbonyls. Even the peptide carbonyls of the L-amino acids play a substantial role.

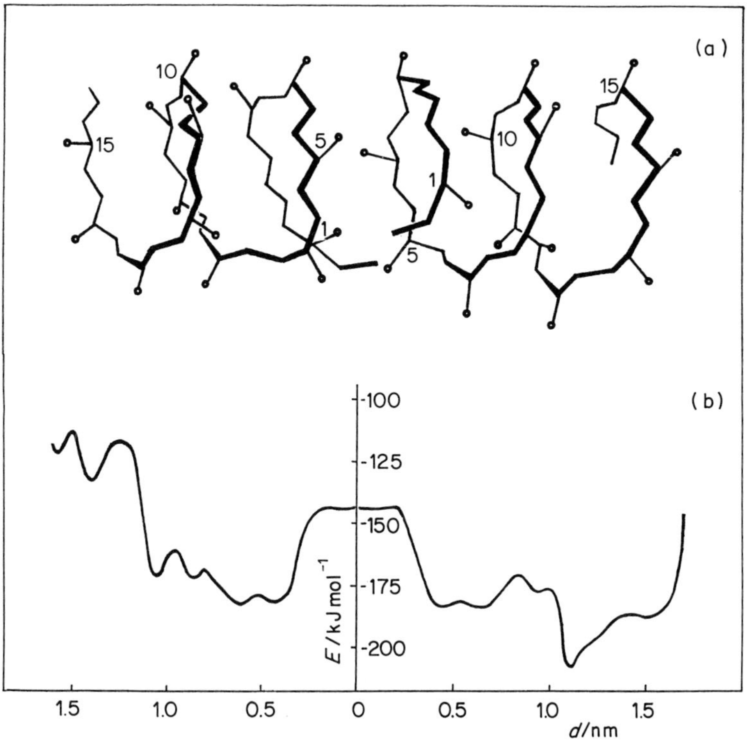

Fig. 3.23. Gramicidin A channel in an artificial bilayer membrane. (a) Two molecules of gramicidin are arranged head to head across the membrane. (b) Energy profile of the dimer computed by considering the contribution of every atom of the channel, including water and a Na^+ ion. In the left-hand part the most stable conformation of the ethanolamine tail was taken to be fixed, in the right-hand part it was assumed to be fully mobile. It will be seen that the absolute minimum or interaction energy is shifted from the region about 0.5 nm from the central barrier to about 1.1 nm

The energy profiles obtained in the presence of different alkaline ions showed the central energy barrier to be highest for Na^+, lower for K^+ and lower still for Cs^+. Likewise, the enthalpies associated with the barriers at the channel mouth were highest for Na^+, followed by K^+ and Cs^+.

These values are in fine agreement with macroscopically determined membrane conductances in the presence of gramicidin (cf. p. 165) at a membrane potential of 0.1 V. In relative values $g_{Na^+} = 1.0$, $g_{K^+} = 1.8$, $g_{Cs^+} = 2.9$.

Passage through the gramicidin channel is a typical single-file diffusion where the ion interacts with only one molecule of water ahead and one molecule of water behind. The total number of water molecules in the channel proper is probably 20–25 depending on the precise definition of the channel boundaries, and the rate at which K^+ ions traverse the channel may be as high as $10^8\ s^{-1}$.

Some of the ion-active peptides are cyclic, such as gramicidin S, tyrocidins, surfactin, polymyxins and alamethicin. All these are related to antamanide, whose affinity for Ca^{2+} is remarkable (complexation constant in methyl nitrile $10^5\,M^{-1}$) and for Na^+ is rather high (10^3–$10^4\,M^{-1}$) but it is unable to act as a channel in the membrane.

Alamethicin channels are probably more complicated than the gramicidin ones. Their conductance depends on the applied voltage, their selectivity is influenced by the surface potential and changes in conductivity determined from current fluctuations in the membrane do not occur in integral numbers but rather at ratios of 1.0:1.8:2.1:2.3:2.8. As a whole, however, the total conductance due to alamethicin is greater than that due to gramicidin A.

SYNOPSIS

Every cell contains transport proteins in its membranes, the greatest number being probably present in the bacterial membrane where it may range between 50 and 80. Differentiated cells of multicellular organisms possess lower numbers of different transport agencies.

Some of the transport proteins are involved in energy transduction, others in signal transmission. The pure transport proteins include a variety of types, such as single-peptide systems (apparently most of the Na^+ and H^+ symport carriers), oligomeric clusters (e.g., the ADP/ATP carrier of mitochondria, the band 3 anion exchange protein of erythrocytes) or arrays of different proteins (the binding-protein systems of bacteria). Extraneous channel formers (gramicidin A, alamethicin) and carrier-like ionophores (valinomycin) are much used for short-circuiting ionic gradients in both natural and artificial membranes.

FURTHER READING

Etchebest, C., and Pullman, A. The gramicidin A channel: Energetics and structural characteristics of the progression of a sodium ion in the presence of water, *J. Biomol. Structure Dynamics*, **3**, 805–825 (1986).

Klingenberg, M. Principles of carrier catalysis elucidated by comparing two similar membrane translocators from mitochondria, the ADP/ATP carrier and the uncoupling protein, *Ann. N.Y. Acad. Sci.*, **456**, 279–88 (1985).

Kotyk, A. (ed.). *Molecular Basis of Membrane Transport*, UNESCO, Prague (1986).

Mao, B., Pear, M. R., and McCammon, J. A. Hinge-bending in L-arabinose-binding protein, *J. Biol. Chem.*, **257**, 1131–33 (1982).

Passow, H. Molecular aspects of band 3 protein-mediated anion transport across the red blood cell membrane, *Rev. Physiol. Biochem. Pharmacol.*, **103**, 61–223 (1986).

Semenza, G., Kessler, M., and Schmidt, U. The small-intestinal sodium-glucose cotransporter(s), *Ann. N.Y. Acad. Sci.*, **456**, 83–90 (1985).

Urry, D. W. The gramicidin transmembrane channel: A proposed $\pi_{(L,D)}$ helix, *Proc. Nat. Acad. Sci. U.S.A.*, **68**, 672–6 (1971).

Wright, J. K., Dornmair, K., Mitaku, S., Möröy, T., Neuhaus, J. M., Seckler, P., Vogel, H., Weigel, U., Jähnig, F., and Overath, P. Lactose: H^+ carrier of Escherichia coli: Kinetic mechanism, purification and structure, *Ann. N.Y. Acad. Sci.*, **456**, 326–41 (1985).

3.7 TRANSPORT OF MACROMOLECULES

3.7.1 Introduction

Large molecules and even microscopically visible particles or aggregates can cross the cell membranes in two principally different ways.

(a) By binding to a receptor and subsequent internalization either alone or together with its receptor, without vesiculation.

(b) By endo- or exocytosis of various types, either involving binding to a receptor or not.

These mechanisms are distinct from all nonspecific permeation modes as well as from the specific channel or carrier systems in that the rates of transport cannot be treated here in terms of concentrations (often a single large molecule is translocated into a cell) and in that they involve a transient disturbance of the lipid membrane structure.

3.7.2 Mechanisms without vesiculation

Deatiled evidence lacking, some proteins are thought to bind to membrane receptors and undergo a shift from the outside to the inside of a cell by either of two hypothetical mechanisms. In the first the receptor—macromolecule complex is internalized by the action of the cytoskeletal components, mainly actin microfilaments. In the second, the bound ligand slides through an opening in the membrane arising through local tension in the receptor region.

The cholera toxin which catalyzes the ADP-ribosylation of the G protein of the adenylate cyclase assembly and causes its continuous activation is one of the proteins that might cross the membrane in this fashion. Others are colicins E2 and E3. Humic acids are probably taken up by soil fungi, such as *Aspergillus niger*, without vesiculation. Indeed, organelle proteins that are synthesized in the cytosol must do so (see Section 2.9).

However, evidence keeps accumulating that most extraneous large molecules are taken up by a vesiculation process as discussed below.

A mechanistically interesting case is the penetration of proteins through the nuclear pore (see Fig. 1.14) which, with a diameter of about 15 nm, should let through most proteins by simple aqueous-phase diffusion. Still, there exists a selective recognition of the sequence Pro-Lys-Lys-Lys-Arg-Lys-Val prior to the penetration. The process thus resembles passage through a specific channel.

Somewhat surprisingly, all tRNAs synthesized in the nucleus are transferred in a saturable way with competition between different tRNA species. Hence, again a specific channel mechanism is applicable to this transport through the nuclear pore.

A special case of direct macromolecular transport is that of DNA penetrating into bacterial cells during the process of transformation (of competent cells), transduction or transfection, as well as during artificial introduction of DNA on a suitable vector during gene manipulation. In all these cases the recipient

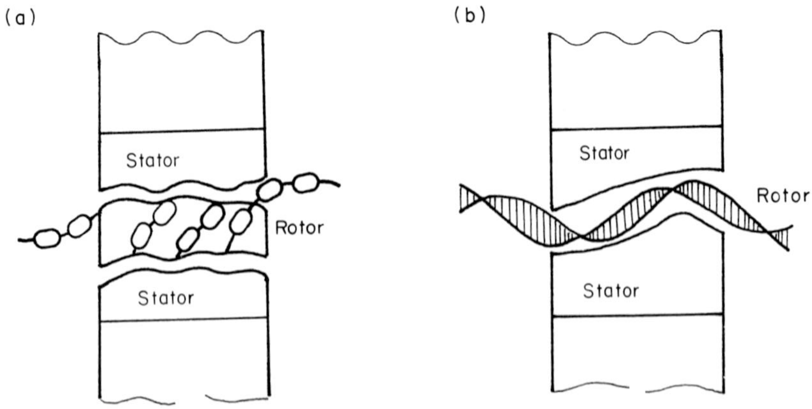

Fig. 3.24. Hypothetical views of nucleic acid transport across a biomembrane. (a) The molecule rides on a rotating protein within the membrane; (b) the molecule itself rotates its way through the membrane along a proton potential gradient

cells must possess an apparatus for bringing the DNA in. The competence of bacterial cells appears relatively suddenly during culture growth but it can be induced, e.g. by $CaCl_2$, in a defined temperature regime. The appearance of competence is connected with the synthesis of an endogenous protein on another cell (M_r 10 000) which then binds to a surface receptor and brings about synthesis of RNA and proteins, some of them (in the form of an agglutinin) then emerging on the surface. DNA is attached to this agglutinin, one of the strands is partly degraded and the other is pulled into the cell. This mechanism appears to involve a rotation of the single DNA strand attached to a 'pilot' protein in a membrane 'stator' formed by one of more proteins associated with the generation of competence in the cell. The energy for the rotational pull is provided by the protonmotive force (Fig. 3.24) which can be understood from the analogy between the DNA as a polyanionic (polyphosphate) molecule and inorganic phosphate which is transported in symport with H^+ or Na^+ ions in fact as an undissociated molecule.

The mechanism is apparently similar in cases where double-stranded DNA crosses the membrane, such as during transformation of *Neisseria gonorrhoeae* or during secretion of such DNA from bacterial cells. There the DNA may rotate as such (without the pilot proteins), each rotation pulling a base pair across. If the energy for DNA secretion is also provided by the protonmotive force one would have to assume the existence of an additional system that would somehow couple the antiport of OH^- ions to the extrusion of DNA.

3.7.3 Mechanisms with vesiculation

Most large molecules and particles cross biomembranes in the form of drops of solution or suspension that involve centripetal invagination, fusion and separation in the inward movement, and contact, fusion and release of content in the outward movement.

The inward-oriented process or endocytosis is of two types, the constitutive endocytosis, or fluid-phase pinocytosis, and the 'inducible', receptor-mediated or adsorptive endocytosis.

Fluid-phase endocytosis

Constitutive endocytosis is a process almost ubiquitous in animal cells and apparently common in plant and fungal cells. Parts of the outer membrane are continually ingested and, in the form of endocytotic vesicles, fuse with primary lysosomes, are processed (perhaps in the Golgi apparatus) and eventually reinserted into the outer membrane, representing thus a major portion of membrane traffic, apparently essential for maintaining the membranes ready for their various functions (cf. Fig. 1.18). Apart from the membrane recycling, constitutive endocytosis serves other purposes in different cells.

1. In macrophages and polymorphonuclear leukocytes (components of the immune system) this is a process whereby foreign material is removed from the body fluids, whether by ingestion of droplets of solution (pinocytosis) or of solid particles (phagocytosis).

2. In amoebae and some epithelial cells, especially in the intestinal lining, the process facilitates ingestion of nutrients. The amount of plasma membrane that undergoes the endocytotic process can be quite substantial (Table 3.13). The endocytotic vesicles can range from 50 to 500 nm in diameter and can be further enlarged by fusion with the primary lysosomes.

Table 3.13. Pinocytotic flow displayed by various cells

Cell type	Cell volume (μm^3)	Cell surface (μm^2)	Plasma membrane equivalents internalized per hour
Mouse peripheral macrophage	395	825	1.9
Guinea-pig alveolar macrophage	1750	2250	5.3
Mouse fibroblast	1765	2100	0.5
Acanthamoeba castellanii	2540	2390	9
Dictyostelium discoideum	4200	6300	1.5

Endocytosis via coated pits and vesicles

While much of the constitutive endocytosis is effected by pinching off smooth membrane vesicles, some of it, in many cells a substantial part, is based on vesicles that are 50–250 nm in diameter, are coated, and arise from membranes that prior to vesicle formation, appear as coated pits (Fig. 3.25). The coat contains several proteins, the major one being *clathrin*, a fibrous protein (M_r 180 000) which forms basket-like coats on the vesicles. The coats are built from three-pronged oligomers of three molecules of clathrin and three molecules of a smaller (M_r 35 000) polypeptide, the so-called triskelions, in a structure composed of 12 pentagons and 8 or more hexagons. The structure is unusually rigid, resembling the geodesic dome organization of the recently discovered form of elementary carbon, provisionally called buckminsterfullerene.

The invagination of coated pits is suspected to be associated with the underlying network of microfilaments and possibly microtubules (some effects of cytochalasin B and colchicine, respectively). However, in artificial membranes, pinocytosis can be elicited by binding some compounds or particles. Still, in the living cell the process requires metabolic energy, from either oxidative or fermentative metabolism. There is some evidence that the pinching off of the vesicles is preceded by a short-lived intermediate stage in which the vesicle to be separated is connected by a long membranous neck with the plasma membrane, possibly a hexagonal (H_I) tube that may be thermodynamically quite stable.

The process of coated pit endocytosis can be saturable—due to interaction of the solute or particle with a finite number of coated sites and due to the limiting ability of the cell to replace the membrane lost by vesicles pinched off. The maximum rate of pinocytosis expressed as volume of integral solution is surprisingly similar in many cells (leukocytes, tumour cells, macrophages), being about $1.5 \cdot 10^{-17}$ litre per cell per second.

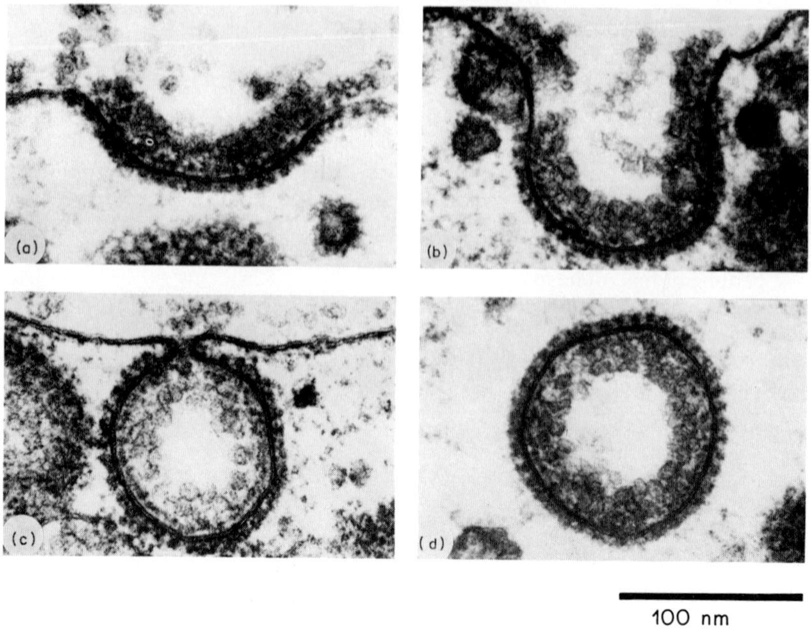

Fig. 3.25. Sequence of events as a coated pit on the cell surface (a) invaginates (b), forms a round vesicle intracellularly (c) which is finally released (d)

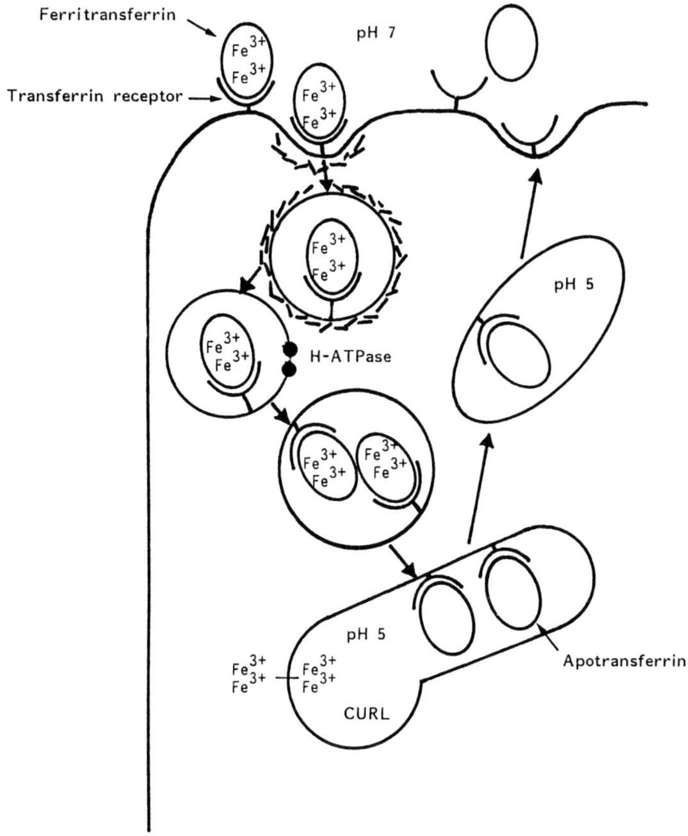

Fig. 3.26. Function of the transferrin receptor in bringing Fe^{3+} into cells. The CURL (*c*ompartment of *u*ncoupling *r*eceptor from *l*igand) is often a part of the lysosome scenario shown in Fig. 1.18

Likewise, the size of pinosomes, the primary endocytotic vesicles, is rather uniform in different membranes (about 0.3 μm in diameter) which apparently has to do with the chemical composition and physicochemical properties of membranes in general.

Receptor-mediated endocytosis

In addition to constitutive endocytosis many cells are capable of receptor-mediated endocytosis which takes place exclusively via the above-described coated pits but only after adsorption to specific sites that are spread over most of the membrane and migrate to the coated pits upon binding their specific ligand. The case of transferrin may serve as an example.

Transferrin mediates the transport of ferric ions by blood from the intestine where ion is reabsorbed to the liver where it is stored and to various cells in the body. Transferrin is a glycoprotein (M_r 77 000) that binds two Fe^{3+} ions and has a high affinity for a two-site receptor (cf. Fig. 2.19) that is found on the surface of all growing cells (K_D 6 nM at pH 5). Upon binding it is internalized together with the receptor, sheds its Fe^{3+} ions in a lysosome and returns to the plasma membrane while still bound to its receptor. There it is released (Fig. 3.26).

The induced endocytosis is a much more efficient process than the spontaneous one. A case in point is the uptake of cholesterol by animal cells, the cholesterol travelling in blood in the form of low-density lipoprotein (LDL) particles 22 nm in diameter that contain well over 1000 cholesterol molecules and are surrounded by a lipid bilayer with a single kind of protein (two copies per particle, M_r 250 000).

Another case of uptake acceleration by induction was observed in the large amoeba *Chaos chaos* where as many as 10^8 molecules of protein can be taken up by the cell per second, roughly 10^6 times as many as before induction. However, the usual enhancement is by a factor of 5–50. Clearly, more than with constitutive endocytosis, the receptor-mediated one requires membrane recycling to replenish the material lost by pinching off the vesicles (Table 3.14).

Exocytosis

The outward-oriented pinocytotic process is called exocytosis and forms an integral part of the function of all exocrine and endocrine gland systems, as well as of presynaptic nerve terminals. Even other cells, including plant secretory cells and eukaryotic microorganisms, are capable of this process if large molecules are to be secreted (Fig. 3.27).

It was described in the section on membrane biogenesis how membrane proteins are synthesized in the endoplasmic reticulum membranes and, anchored in a vesicle membrane, are transferred step by step toward the target, either the plasma membrane or some intracellular membrane. The fate of the secretory protein is similar except that as its translation is terminated it is released into the ER lumen. Then it moves through that section of the Golgi complex which specializes in proteins to be secreted and, in the form of secretory vesicles, fuses with the external, plasma, membrane to open its contents into the external medium. In a pancreatic acinar cell the turnover of the vesicles at the Golgi complex is truly enormous. It is estimated that 20 000 vesicles are formed and re-fuse in a second in such a cell, there being the underlying vectorial movement of those that contain the substance to be secreted toward the periphery. In such an actively secreting cell the extent of membrane added to the plasma membrane may temporarily exceed the resting state extent by a factor of 30 (Fig. 3.28).

The preponderance of secretory membrane elements in an exocrine cell is documented by data on the surface area of various membranes in such a cell.

Table 3.14. Rates of receptor-mediated endocytosis in some cells

Cell type	Receptor for	Dissociation constant of ligand with receptor (nM)	Number of receptors per cell	Lifetime on cell surface (min)	Cycle duration (min)
Human hepatoma	Transferrin	5	50 000	4	16
	Asialoglycoprotein	7	200 000	2.3	16
Freshly isolated rat hepatocyte	Asialoglycoprotein	5.6	450 000	6.3	3.7
Rat alveolar macrophage	Mannose residue of glycoprotein	10	100 000	3	9
Human skin fibroblast	LDL	1	10 000	6	12
Rabbit alveolar macrophage	α-Macroglobulin	1	120 000	6	18
Cultured fibroblast	α-Macroglobulin (low-affinity type)	40–70	700 000	2	20

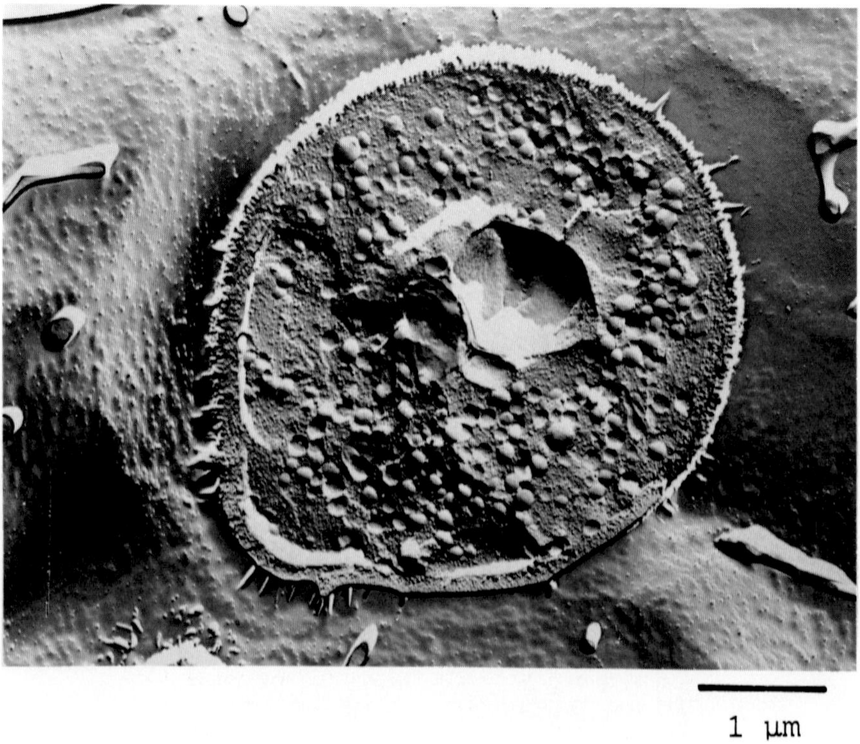

Fig. 3.27. A freeze-fracture view of a *Saccharomyces cerevisiae sec 1* mutant which has a defect in the exocytosis step for secretion of cell proteins. The secretory vesicles, for technical reasons not seen in the wild type, accumulate here and form stable structures intracellularly. (Courtesy of Professor O. Nečas, J. E. Purkyně University, Brno)

In a pancreatic acinar cell that may be 10–15 μm across there are 8000 μm^2 of rough endoplasmic reticulum, 1300 μm^2 of Golgi apparatus, 150 μm^2 of condensing vacuoles and 900 μm^2 of secretory vesicles (zymogen granules).

The whole process of synthesis and movement toward the plasma membrane may take tens of minutes but the actual fusion with the membrane is extremely rapid, taking approximately 1 ms. It requires metabolic energy at several stages and particularly the exocytosis itself is an energy-dependent process. While in some cells it proceeds more or less continuously, others must be triggered to accomplish it. Generally, Ca^{2+} ions are thought to be instrumental in the process as exocytosis sets in only at Ca^{2+} concentrations above 0.1 mM. The fusion of the secretory vesicles with the plasma membrane is preceded by the formation of an annulus in the membrane, about 300 nm in diameter, made of specific proteins. It is within this annulus that secretory vesicles fuse with the membrane.

In contrast with the secretory protein scenario, small molecules, such as

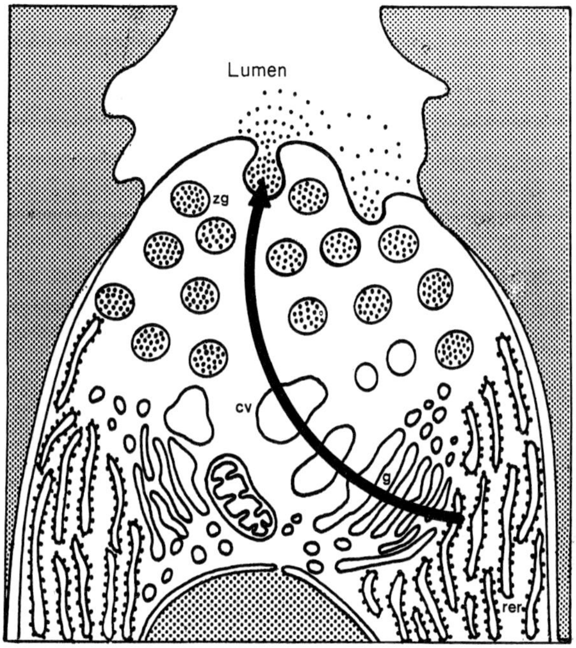

Fig. 3.28. Exocytotic secretion of proteins in an acinar cell of the pancreas, proceeding from synthesis at the rough endoplasmic reticulum (rer), through the Golgi complex (g), condensing vacuoles (cv) and zymogen granules (zg)

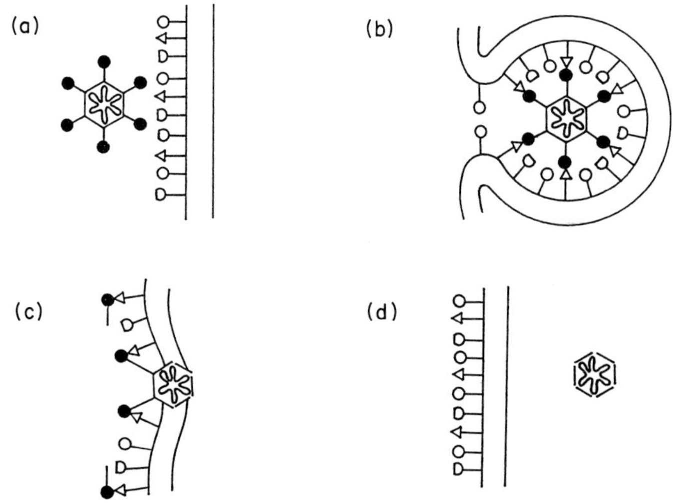

Fig. 3.29. Penetration of a naked virus through a cell plasma membrane

hormones or neurotransmitters, are transported into secretory vesicles already formed and undergo in the vesicle only the terminal part of the centrifugal movement.

Secretory vesicles with small molecules play a decisive role in the nerve impulse transmission across the synaptic cleft. In the nerve cell the substances enclosed in the vesicle travel from the soma toward the end plate and there, upon arrival of an action potential which opens the voltage-gated channels for Ca^{2+}, and subsequent entry of Ca^{2+} into the cell, the vesicles are induced to fuse with the presynaptic membrane. The whole process represents a 10^4-fold increase of the exocytotic rate over the resting state.

Although superficially exocytosis appears like the reverse of endocytosis, triggering of these two processes even at the membrane level is quite different. In endocytosis the noncytoplasmic monolayers of the membrane fuse while in exocytosis the cytoplasmic monolayers merge.

3.7.4 Intermediate mechanisms

Viruses enter cells in different ways that involve membrane rearrangement but not vesicle formation.

The membraneless picorna and adenoviruses apparently merge with the membrane via interaction of surface receptor groups with the membrane and appear in the cell stripped of these receptor groups (Fig. 3.29).

Membrane-coated viruses, such as influenza, paramyxo, pox and oncorna viruses, fuse with the host cell membrane, whereupon the virion is released into the cell interior. The fusion is preceded by a more or less specific binding of the virus (especially the influenza and paramyxo viruses) to neuraminic acid in the cell membrane (Fig. 3.30).

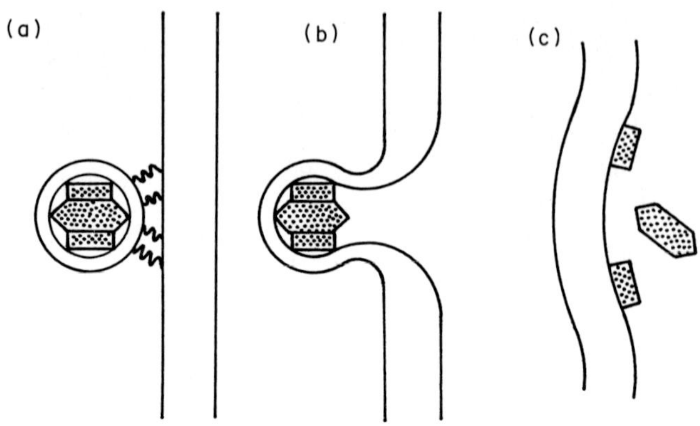

Fig. 3.30. Penetration of an influenza virus through a cell plasma membrane

3.7.5 Cell fusion

All the endocytotic as well as exocytotic processes have in common the phenomenon of membrane fusion. In endocytosis the external monolayer of the plasma membrane fuses first while in exocytosis the cytoplasmic face comes first into contact.

It should be noted, however, that membrane fusion is also possible at the

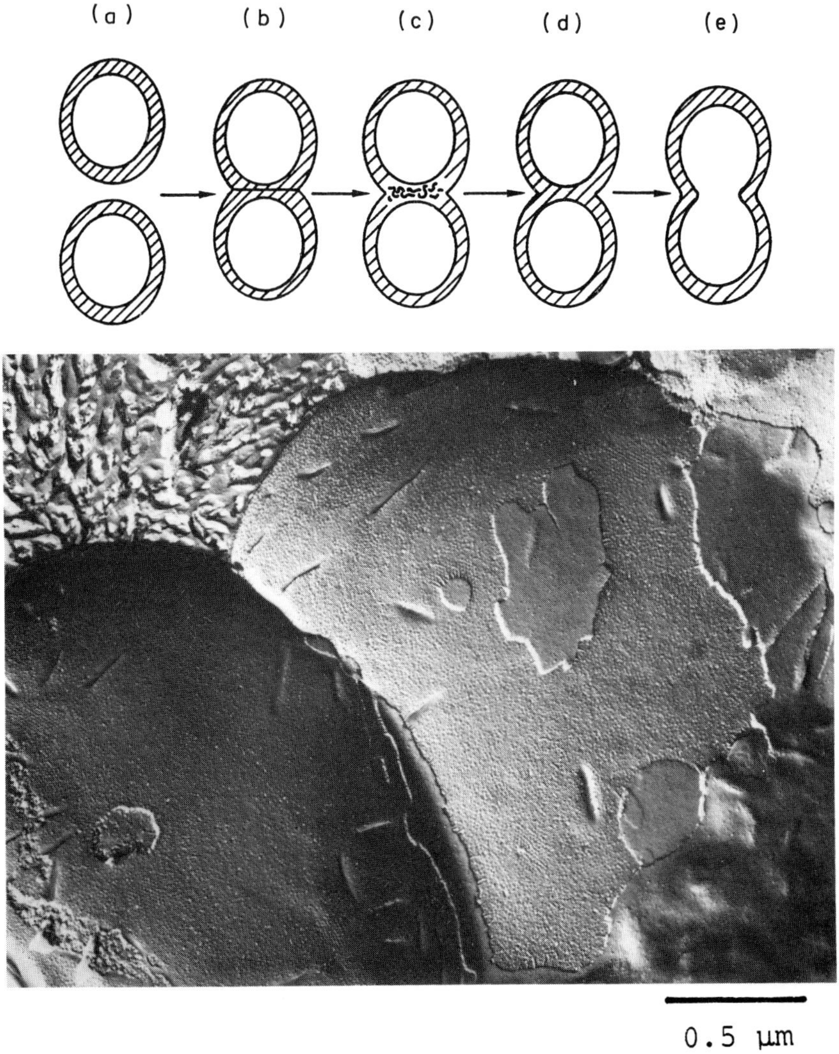

Fig. 3.31. Train of events taking place during cell fusion. The critical stage when the cell membranes come into close contact (b) and (c) is depicted in the freeze-fracture preparation of two *Saccharomyces cerevisiae* protoplasts undergoing fusion in 40 per cent polyethylene glycol. (Courtesy of Professor A. Svoboda, J. E. Purkyně University, Brno)

cellular level and often is used as such for the production of vegetative hybrid cells. Here two different plasma membranes fuse, a process assisted by the presence of polyethylene glycol, Li^+ ions and electric discharge across two or more aligned cells (see Section 2.5.5).

The probable sequence of events during fusion includes (a) bilayer apposition or adherence, (b) rapid reorganization of lipid molecules in the area of contact, with possible transient hexagonal structure, (c) extension of a bilayer between the membranes to be separated, and (d) breaking of the bilayer (Fig. 3.31).

The crucial moment appears to be the adherence without any intervening water or solute molecules. One should consider three forces acting on the membranes during the approach of cells, (1) van der Waals attraction, (2) electrostatic repulsion, (3) hydration repulsion. It is the water molecules relatively firmly bound to the lipids—up to 15 H_2O molecules per lecithin at the closest approach in a lamellar structure—that prevent fusion under normal conditions. If they are efficiently removed, such as may be the case with polyethylene glycol or even with Li^+ ions (they are heavily hydrated) the chance of direct apposition is increased.

Synopsis

Large molecules, particularly those of proteins and nucleic acids, can cross biomembranes in different ways. 1. By receptor-mediated transport involving internalization of the complex with ligand and its intracellular dissociation. 2. By some type of pinocytosis, either fluid-phase, which is nonspecific and serves especially the internalization of nutrients, or coated-pit-mediated which may be either constitutive and not very specific, or triggered, induced, involving a specific receptor. In some cells the membrane thus internalized per hour can exceed the total plasma membrane area by an order of magnitude. Exocytosis is a process taking place in the opposite direction to endocytosis and is found in all secretory cells as well as in presynaptic membranes. Membrane fusion underlies both endocytosis and exocytosis. Whole cells, stripped of their polysaccharide envelopes, can fuse. This is a process in which water molecules must be displaced from the plane of adherence of two membranes whereupon fusion proceeds spontaneously.

FURTHER READING

Dautry-Varsat, A., and Lodish, H. F. How receptors bring proteins and particles into cells. *Sci. Amer.*, **250**, 52–8 (1984).

Dingwall, C. The accumulation of proteins in the nucleus. *Trends Biochem. Sci.*, **10**, 64–6 (1985).

Doxsey, S. J., Sambrook, J., Helenius, A., and While, J. An efficient method for introducing macromolecules into living cells, *J. Cell Biol.*, **101**, 19–27 (1985).

Harrison, S. C., and Kirchhausen, T. Clathrin, cages and coated vesicles, *Cell*, **33**, 650–2 (1983).

Lucy, J. A. Molecular aspects of membrane fusion. In: *Membrane Processes* (Gh. Benga. H. Baum, and F. A. Kummerow, eds), pp. 27–48, Springer-Verlag, New York (1984).

Schneider, W. J., Beisiegel, U., Goldstein, J. L., and Brown, M. S. Purification of the low-density lipoprotein receptor, an acidic glycoprotein of 164 000 molecular weight, *J. Biol. Chem.*, **257**, 2644–73 (1982).

Schneider, C., Owen, M. J., Banville, D., and William, J. G. Primary structure of human transferrin receptor deduced from the mRNA sequence, *Nature*, **311**, 675–8 (1984).

Wilschut, J., and Hoekstra, D. Membrane fusion: From liposomes to biological membranes, *Trends Biochem. Sci.*, **9**, 479–83 (1984).

Zasloff, M. tRNA transport from the nucleus in a eukaryotic cell: Carrier-mediated translocation process, *Proc. Nat. Acad. Sci. U.S.A.*, **80**, 6436–40 (1983).

4 *Membranes as Convertors of Energy*

4.1 FLOWSHEET OF BIOENERGETICS

All the energy flowing through living systems or stored in them has reached the Earth in the form of radiation energy from the Sun (the amount of heat absorbed by thermophilic organisms in geysers or that taken up by bathers in a thermal pool is certainly negligible), the rate at which it keeps arriving being 10^{22} J per day, which corresponds to a 'solar constant' of $1.35\,\mathrm{kW\,m^{-2}}$.

Not this whole amount of energy reaches the surface (Table 4.1). Of the 43 per cent that reach the Earth about one-half is reradiated and the other half is used to warm the surface and, secondarily, is converted into other forms of energy in the atmosphere and the hydrosphere. Only 0.05 per cent of the energy is trapped by plants and directly converted into other forms of energy. This is all that is required for maintaining the existing steady state in the biosphere, with its annual production of some $1.7 \cdot 10^{14}$ kg organic dry matter.

Photosynthesis as a whole is a process with a relatively low efficiency overall—perhaps 20 per cent of the absorbed (not incident) light energy is converted into combustible matter. The energy trapped by photosynthetic organisms is then channelled according to diverse patterns. It may be instructive to consider the flowsheet of energy in the biosphere as outlined, certainly with a substantial margin of error, in Table 4.1.

Table 4.1. Distribution of incident solar radiation on Earth

	%	%
Diffuse scattering		
in the atmosphere	25	
Portion thereof that reaches		
Earth surface		16
Absorption in clouds	8	
Reflexion from clouds	33	
Absorption in the		
atmosphere	7	
Part that reaches Earth surface	27	27
	100	43

This being a book on biomembranes, we shall omit from our considerations all bioenergetic processes that do not involve membranes, the major ones being oxidations of substrates in solution (at the receiving end) and mechanical work (at the giving-away end). They will only occasionally be mentioned for the sake of comparison.

Before entering into the varied realm of energy conversions it might be useful to mention briefly some of the basic ideas involved in such conversions.

The structural wealth which is characteristic for biological membrane systems is paralleled by the variety of the types of energy transformation where the biological systems are substantially ahead of those encountered in conventional chemistry.

Chemical reactions, in a narrow sense of the word, come about through forming and breaking chemical bonds, i.e. by changing orbital interactions in the reacting molecules. Consequently, these changes in electron configurations are accompanied by chemical bond energy changes, but other energy variations also play a role. These include changes in vibration and rotation energy of the

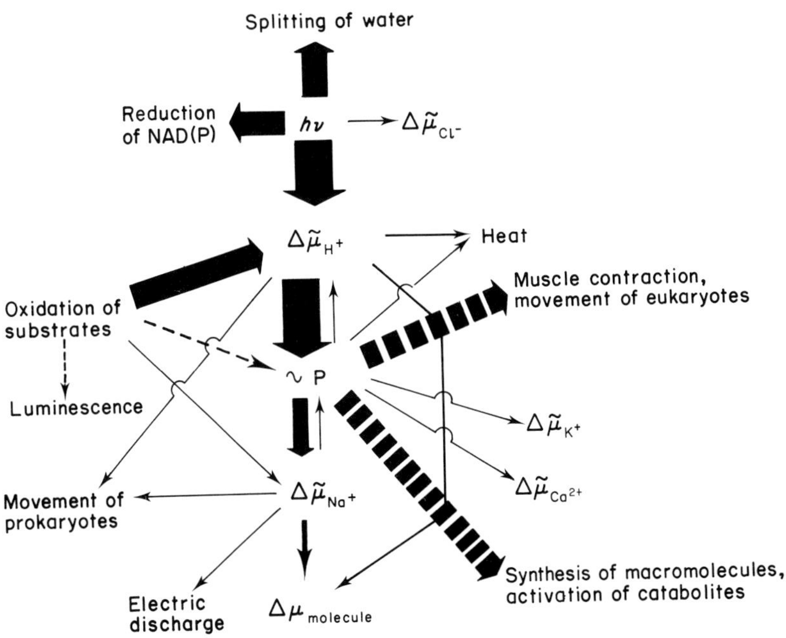

Fig. 4.1. Schematic flow of energy in biological systems. The primary source of energy, the solar radiation, is utilized mainly for the production of $\Delta\tilde{\mu}_{H^+}$, for the generation of reducing power, and for oxidation of water to oxygen. The central positions of $\Delta\tilde{\mu}_{H^+}$ (and, analogously, in a minor way) of $\Delta\tilde{\mu}_{Na^+}$, as well as of high-energy phosphate bonds, mainly in the form of ATP, are well seen in the schematic. Processes that are primarily or exclusively of nonmembrane character are shown in broken arrows. The thickness of arrows reflects the quantitative importance of the given process

bonds, of various groups of which the molecule is composed, and of the molecules themselves, together with their translational energy, as well as energy changes connected with electrostatic interactions between ionic and dipolar groups in the molecules. Furthermore, the energies of interaction of the molecules with their surroundings must be taken into account. They include various changes of energies connected with interactions in crystalline or liquid-crystalline structures, with dipolar, various short-distance (particularly van der Waals) and association-bonding interactions in liquids and, moreover, with solvation interactions in solutions. The sum of these energies should result in a change of the thermal energy content of the reaction system. However, when the reaction, as usual, proceeds at constant temperature and pressure, the process is acompanied by a heat exchange between the reacting system and its surroundings and by some, often negligible, volume work. This amount of heat is termed the reaction enthalpy ΔH_r (positive when the heat is transferred to the reacting system and *vice versa* for the negative value). If the reaction proceeds to unit extent, i.e. if the amount of reacting substances (in moles) corresponds to the reaction equation as it is written, it is called molar reaction enthalpy and, particularly if the reactants as well as the products appear in their standard states, the term standard reaction enthalpy is used. It should be noted that in chemical thermodynamics this quantity corresponds to the process which starts with isolated reactants in their standard states, proceeds by transferring them to the reaction mixture and ends with isolation of the products, again in their standard states.

As biochemical reactions usually proceed in solution (or at phase boundaries of solutions) and at least some of the reactants show acid-base properties it is advantageous to consider as reference (better than standard) states of participating species their solutions at specified concentrations and a given pH value (usually pH 7). The temperature of 37 °C is frequently taken as reference instead of the standard temperature of 25 °C. The corresponding energetic terms are then called conditional or formal instead of standard.

To calculate reaction enthalpies it is useful to employ the formation enthalpies of various substances. These constants are usually obtained from thermal data but when these are not available bond enthalpies may be calculated as averaged values of known formation enthalpies. However, it is often misleading to expect that these values indicate the degree of stability of a given compound. Thus, the α- and β-phosphoryl bonds in adenosine triphosphate are often called 'energy-rich phosphate bonds'. In fact, according to its bond enthalpy, the P–O bond is not particularly weak but the necessity to form the compound ATP from a very stable molecule $H_2PO_4^-$ is the cause of the high reaction enthalpy of formation of ATP from ADP and $H_2PO_4^-$.

The change of reaction Gibbs energy may be defined in similar terms. This quantity is connected, in contrast to reaction enthalpy, with the reversible course of the process. Again, the choice of the reference state is very important. Consider as an example the hydrolysis of ATP

$$ATP^{4-} + H_2O \rightleftharpoons ADP^{3-} + HPO_4^- .$$

The standard Gibbs energy change of this reaction is $-28\,\text{kJ mol}^{-1}$ at 37 °C which is a value independent of pH. However, at pH 7, ATP^{4-}, $ATPH^{3-}$, ADP^{3-}, $ADPH^{2-}$, HPO_4^{2-} and $H_2PO_4^-$ are simultaneously present in an aqueous solution. Under these conditions the conditional Gibbs energy change corresponding to the transformation of 1 mole of ATP to 1 mole each of ADP and inorganic phosphate, all present in their coexistent ionization forms, is $-33.0\,\text{kJ mol}^{-1}$ at 37 °C. Considerable differences between the standard Gibbs energy changes and the conditional Gibbs energy changes at pH 7 show how important a detailed specification of reaction conditions can be. (In a concrete biological system where the actual Gibbs energies are computed for the splitting or synthesizing ATP, the concentrations of all reactants and products will of course affect the value far beyond the changes of standard Gibbs energy considered here.)

For biological energy transformations the processes of charge separation are particularly important. There are two characteristic cases of this type.

1. Electron-hole pair formation during harvesting of light energy in the photosynthetic systems where, eventually, the energy is stored in the form of energy-rich molecules.

2. Translocation of ions and/or of electrons across membranes where the energy is accumulated in an increase of the electrochemical potential of the species transferred.

In the first example described in detail in Section 4.2.2 it is within the thylakoid membrane that electrons and holes are formed by excitation of pigments, especially chlorophyll. Once separated, the two charged species are channelled to the electron-transfer chain and to the water-oxidizing system.

The conversion of light energy to other energy types is, at its initial stage, a strongly nonisothermal process. Originally, light is emitted from the surface of the Sun ($T \simeq 6800$ K) but as it reaches the Earth surface this temperature is lowered to approximately 1100 K. This is still a high temperature compared with that of the photosynthetic system ($T \simeq 300$ K). Thus, at this stage, Carnot's thermal cycle yields an efficiency of about 73 per cent. In the subsequent step, which is light absorption in the antenna pigments and translocation of electrons and holes to their destination in specific enzyme systems, a considerable amount of energy is trapped by and dissipated in carotenoids which protect chlorophyll molecules from photochemical destruction and by fluorescence and nonradiative transitions. These losses of useful energy are not important in view of the abundance of light energy as a total.

The second case is the translocation of electrons along and of protons across the mitochondrial membrane to the intermembrane space. This process of charge separation resulting in accumulation of osmotic and electric energy gained in overall oxidation of carbon 'fuels' by oxygen proceeds in small oxidation-reduction steps in the electron transfer chain. This stepwise process is the basis of an almost reversible course of the sequence of reactions. Sometimes this process is likened to the oxidation of carbon compounds in a fuel cell but this parallel is not well taken as no such stepwise process takes place either in the anodic reaction of fuel oxidation or in the cathodic reduction of oxygen.

Similar but simpler are the charge separations caused by ATPase-type transport enzymes in electrogenic transport across biological membranes.

Dissipation of Gibbs energy of a membrane system often occurs through passive transport (cf. Section 3.3.2). It should be noted, however, that no thermal effect accompanying this energy dissipation is to be expected when diffusion occurs in dilute solutions of nonelectrolytes. In an ideal solution the enthalpy of the system does not depend on dilution and the decrease of the Gibbs energy of the system is only balanced by the increase of the $T\Delta S$ term. However, if an ion moves passively across a membrane a thermal effect of the type of Joule's heat ensues. Here, for the rate of heat production one may write

$$\mathrm{d}Q/\mathrm{d}t = \sum_{\mathrm{i}} |j_{\mathrm{i}}||\Delta\phi_{\mathrm{m}}| \tag{4.1}$$

where j_{i}'s are partial current densities corresponding to the fluxes of the components i through the membrane with a membrane potential of $\Delta\phi_{\mathrm{m}}$. In this way, for example, heat is produced in brown-fat mitochondria.

SYNOPSIS

Practically all the energy stored and turning over in the biosphere has arrived on Earth in the form of solar radiation. Of the total energy arriving from the Sun only some 0.05 per cent is utilized by green plants and converted into other forms of energy, by a process of photosynthesis. The major forms of stored energy that is readily convertible are electrochemical potential gradients of H^+ and Na^+ and various high-energy phosphate compounds.

The energy of this type can then be used for active membrane transport, for locomotion and, in special cases, for heat production.

FURTHER READING

Gerischer, H., and Klotz, J .J. (eds). *Light-induced Charge Separation in Biology and Chemistry*, Verlag Chemie, Weinheim (1979).

Klotz, I. M. *Energy Changes in Biochemical Reactions*, Academic Press, New York (1976).

Wiedenroth, E.-M. *Das grüne Kraftwerk*, Urania-Verlag, Jena (1981).

4.2 RADIATION ENERGY CONVERSION INTO ELECTROCHEMICAL POTENTIAL GRADIENTS

The title of this chapter is equivalent to 'photochemical part of photosynthesis' and, accordingly, will deal with those events in photosynthesis that convert the energy of electromagnetic radiation into differences of electrochemical potentials of ions which, in their turn, can then be used for further energy transformations, e.g., such as are required during synthesis of ATP.

The conversion of radiation energy into an osmotic and an electric gradient can proceed either directly—by what has been termed photosynthesis without chlorophyll, or via intermediate steps involving movement of electrons within a membrane—by what is known as photosynthesis with chlorophyll. Both types of photosynthetic systems (and it will be seen later that all systems where light is absorbed by membranes as an information signal) contain coloured compounds with conjugated double bonds, the carotenoid pigments. Only the more advanced photosynthesis where CO_2 is subsequently fixed, requires chlorophyll. Incidentally both carotenoids and chlorophylls are synthesized in cells from common precursor molecules of isopentenyl diphosphate (see Section 2.8.1).

4.2.1 Photosynthesis without chlorophyll

This type of light absorption and processing occurs in the extremely halophilic bacteria of the genus *Halobacterium*, containing several species living in salt-beds and concentrated brine, capable of maintaining an intracellular concentration of 1 M Na^+ against 4 M NaCl outside, along with nearly 3 M K^+ intracellularly against 50 mM KCl outside. They synthesize their ATP at the expense of a proton electrochemical potential gradient ($\Delta\tilde{\mu}_{H^+}$) generated by either a photochemical process or by a set of oxidative reactions. It appears that the $\Delta\tilde{\mu}_{H^+}$ is converted into a large concentration difference of K^+ which serves to generate a diffusion potential which can then drive ATP synthesis.

The envelope of Halobacteria is a simple membrane containing some di- and tetra-ether phospholipid plus pseudomurein as a vestige of cell wall. The peculiar feature of the cell membrane is the occurrence of crystalline patches, the so-called purple membrane, covering a sizable portion of the whole cell surface. These patches are composed of trimers of a polypetide called *bacteriorhodopsin* (bR), organized in a hexagonal lattice of 6.3 nm unit cell distance. These patches contain some 25 per cent lipid, corresponding to about ten lipid molecules per bacteriorhodopsin. Each bR monomer traverses the membrane seven times (cf. Fig. 2.19), the α-helices being organized in a cylindrical unit, and it can function as such, although organization in trimers gives it some cooperative advantage. The molecule contains 248 amino acid residues and is synthesized by translation from mRNA with a short (13 amino acids) presequence very different from the usual signal sequences of membrane or secretory proteins.

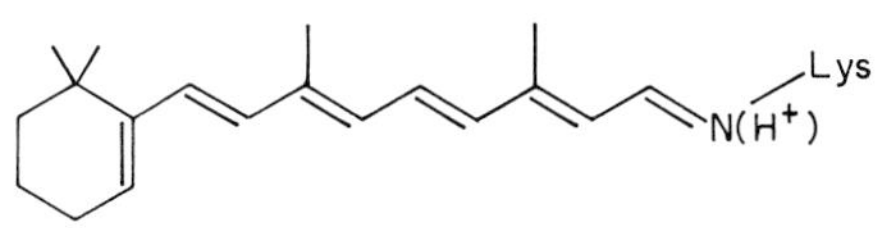

The distinctive feature of bR is that it carries a retinal attached as a Schiff base to Lys 216 at an angle of 23 degrees with respect to the membrane plane. The retinal absorbs light at 568 nm and this triggers a photocycle, characterized by at least five further forms as shown in Fig. 4.2. The function of the sequence of intramolecular reactions is obviously to transport a proton from the intracellular compartment to the outside medium, the quantum yield of the process being as high as 0.6. The photocycle is irreversible in principle but its operation can be short-circuited by the back-pressure of the proton gradient

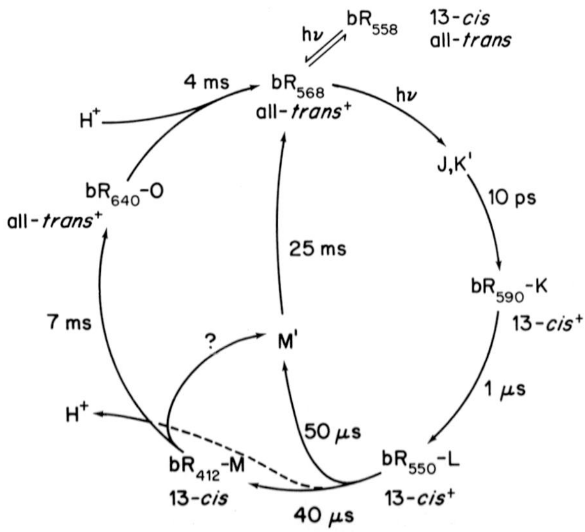

Fig. 4.2. The reaction photocycle of bacteriorhodopsin (bR). The approximate relaxation times of the intermediates are shown

formed whereby the photocycle returns to bR_{568} without releasing or attaching H^+ ions.

The actual movement of the H^+ ions remains unclear but it appears to follow this scenario.

1. Protonated all-*trans* bR_{568} absorbs a photon which, by affecting the resonance of the conjugated double bonds, causes isomerization to protonated 13-*cis* bR_{590}.

2. After a bR_{550} intermediate, a major structural alteration of the environment, involving Tyr-26 and Tyr-64, as well as Trp-182 and Trp-189, cause the proton to be released from the N atom and transferred to a neighbouring dissociable group on an amino acid while the 13-*cis* bR_{412} forms.

3. As bR_{412} proceeds to the bR_{640} form which is again all-*trans*, the proton is released from the membrane, possibly after a hopping sequence along dissociable groups of amino acids.

4. The bR_{640} is now ready to be reprotonated, taking the proton from a neighbouring dissociable group located at the internal membrane surface.

Although still speculative, there is evidence from measuring electric currents evoked by a laser flash that the translational distances a proton moves during the photocycle in the outward direction are as follows:

$bR_{568} \rightarrow K$	-0.13 nm	
$K \rightarrow L$	-0.02 nm	
$L \rightarrow M$	$+0.5$ nm	
$M \rightarrow O$	$+3.1$ nm	
$O \rightarrow bR_{568}$	$+1.5$ nm	(this is another proton!)

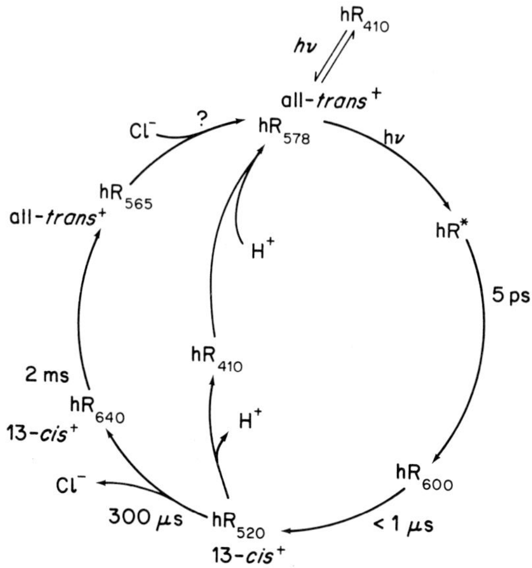

Fig. 4.3. The reaction photocycle of halorhodopsin (hR). The approximate relaxation times of the intermediates are shown

The total distance of nearly 5 nm tallies with the thickness of the purple membrane. A complication for the above scenario would arise if two rather than one proton were transported per cycle as some measurements suggest. A novel theory assumes that the shift from all-*trans* to 13-*cis* mechanically disturbs the G α-helix of the apoprotein and, through a piezoelectric effect, drives two H^+ to the outside.

The pumping capacity of the bR molecule is about 20 protons per second.

Not in the purple membrane but abundant in some mutants of *Halobacterium* species, another photo-activated protein was found and called *halorhodopsin* (hR). Like bR, it is likely to occur in an oligomeric form and, judging from its hydropathy plot, may cross the membrane several times. It contains about 250 amino acids with an approximately 50 per cent sequence homology with bR. Its gene contains information for some 20 amino acids at the N terminus that do not appear in final hR.

It also contains a lysine-bound retinal which, upon absorbing light (λ_{max} = 578 nm), undergoes a photocycle with several stages (Fig. 4.3). The principal function of hR is to pump chloride anions into *Halobacterium* cells, this time with no deprotonation accompanying the all-*trans* to 13-*cis* transition. (However, in a slow reaction involving the hR_{410} intermediate, a proton is released and recaptured.) The quantum yield of the chloride transport is 0.34.

4.2.2 Chlorophyll-associated photosynthesis

Beginning about 3.3 billion years ago, some bacteria acquired the ability to utilize electron donors other than the fermentable organic molecules presumably available to the first anaerobic bacteria. The newly utilizable donors were hydrogen sulphide and water but their reducing power was extremely low for assimilation purposes and a way had to be found to increase the reducing power.

Green sulphur bacteria appear to have been the first to achieve this by the stratagem of using light to activate a tetrapyrrole metal-liganded molecule in such a way as to make it a strong reductant which then was capable to generate coenzymes, such as NADPH, to catalyze anabolic pathways in the cell. About a billion years later, these green sulphur bacteria were joined by green filamentous bacteria which thrive in an oxygen-rich atmosphere by being able to oxidize organic substrates.

Another branch of the primitive fermentors that solved the problem of generating reducing power were the purple sulphur bacteria which, in addition, contained a carbon-fixation (Calvin) cycle in which CO_2 is incorporated into ribulose 1,5-bisphosphate by the enzyme ribulose-bisphosphate carboxylase (4.1.1.39) which has proved its worth throughout the two billion years of its existence by being the most abundant enzyme in the biosphere.

The purple sulphur bacteria were accompanied by another related evolutionary branch, the oxygen-utilizing purple nonsulphur bacteria, and by another, more avantgarde, branch, that acquired the ability to use water instead of H_2S as the source of reducing power, giving rise to the so-called oxygenic photosynthesis that changed the face of the Earth by providing virtually all the gaseous oxygen on this planet, thereby making it possible to introduce oxidizing, rather than dehydrogenating, reactions into biochemistry.

It is useful to classify the present-day photosynthetic organisms in order to understand the differences between the ways light is trapped and its energy utilized (Table 4.2). It should be noted that the photosynthetic reaction centres of algae and higher plants are very similar to those of cyanobacteria (blue–green algae) but contain chlorophyll *a* and *b*, and some algae, chlorophyll *c* and *d*. The fundamental structures of different chlorophylls are shown in Fig. 4.4.

General organization of photosynthetic systems

All the reaction components of a photosynthetic system, ranging from green sulphur bacteria to higher plants, are incorporated in membranes, either the plasma membrane with its associated chlorosomes, or thylakoids of one type or another (Fig. 4.5). The photosynthetic enzymes and accessory molecules are organized in a characteristic and functional way to achieve their tasks of (1) generating a proton electrochemical potential gradient and (2) reducing NADP to NADPH, or NAD to NADH, the core of the array being one or two photosynthetic reaction centres. They belong to individual photosystems which, in organisms with oxygenic photosynthesis, are abbreviated PS I and PS II.

Irrespective of their morphology, thylakoids always contain components required for light capturing and those that mediate electron transfer to the terminal electron acceptor. The complete sequence, such as is found in oxygenic photosynthesis (cyanobacteria, algae and higher plants) is as follows:

Table 4.2. Properties of photosynthetic systems of bacteria

Taxonomic group	Chlorobiaceae	Chloroflexaceae	Chromatiaceae	Rhodospirillaceae	Cyanobacteria	Prochlorophyta
English equivalent	Green sulphur bacteria	Green nonsulphur bacteria	Purple sulphur bacteria	Purple nonsulphur bacteria	Blue–green algae	—
Typical genera	*Chlorobium* *Pelodictyon* *Chlorochromatium* *Chloroherpeton* *Prosthecochloris*	*Chloroflexus*	*Chromatium* *Thiospirillum* *Thiocystis* *Thiocapsa* *Ectothiorhodospora*	*Rhodopseudomonas* *Rhodospirillum* *Rhodocyclus* *Rhodomicrobium*	*Synechococcus* (= *Anacystis*) *Gloeocapsa* *Pleurocapsa* *Spirulina* *Oscillatoria* *Anabaena* *Nostoc* *Fischerella*	*Prochloron*
Locale of photosynthetic pigments	Chlorosomes	Chlorosomes	Mostly vesicular thylakoids (chromatophores)	Thylakoids of different forms	Thylakoids with phycobilisomes	Thylakoids without phycobilisomes
Type of photosynthesis	Anoxygenic	Anoxygenic	Anoxygenic	Anoxygenic	Oxygenic	Oxygenic
Type of catabolism	Anaerobic	Facultative	Anaerobic	Facultative	Facultative	Facultative
Fixation of CO_2 through Calvin cycle	No	No	Yes	Yes	Yes	Yes
Type of movement	Immobile or (*Chloroherpeton*) gliding	Gliding	Mainly flagellar	Flagellar (except *Rhodocyclus*)	Immobile or gliding	Immobile?
Type of chlorophyll	BChl *a* and *c* or *d* or *e*	BChl *a* and *c*	BChl *a* or *b*	BChl *a* or *b*	Chl *a*	Chl *a* and *b*

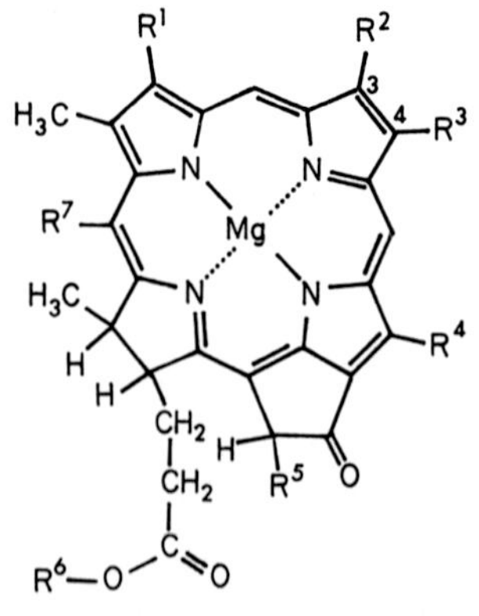

Fig. 4.4. Structures of the principal types of chlorophyll and bacteriochlorophyll. The R^1–R^7 substituents are as follows

		R^1	R^2	R^3	R^4	R^5	R^6	R^7	Long-wave length absorption peak when membrane-bound
Chlorophyll	*a*	$-CH{=}CH_2$	$-CH_3$	$-CH_2CH_3$	$-CH_3$	$-CO-O-CH_3$	phytyl	$-H$	678
	b	$-CH{=}CH_2$	$-CHO$	$-CH_2CH_3$	$-CH_3$	$-CO-O-CH_3$	phytyl	$-H$	630
	c	$-CH{=}CH_2$	$-CH_3$	$-CH{=}CH_2$	$-CH_3$	$-CO-O-CH_3$	phytyl	$-H$	
	d	$-CHO$	$-CH_3$	$-CH_2CH_3$	$-CH_3$	$-CO-O-CH_3$	$-H$	$-H$	
Bacterio-chlorophyll*	*a*	$-CO-CH_3$	$-CH_3$	$-CH_2CH_3$	$-CH_3$	$-CO-O-CH_3$	phytyl,	$-H$	805
							geranylgeraniol		860
	b	$-CO-CH_3$	$-CH_3$	$=CH-CH_3$	$-CH_3$	$-CO-O-CH_3$	phytyl	$-H$	842
	c	$-CHOHCH_3$	$-CH_3$	$-CH_2CH_3$	$-CH_3$	$-H$	phytadienyl		1025
				$-C_3H_7$	$-C_2H_5$		farnesyl	$-CH_3$	745—760
	d	$-CHOHCH_3$	$-CH_3$	$-CH_2CH_3$	$-CH_3$	$-H$	stearyl		
				$-C_2H_{11}$	$-C_2H_5$	$-H$	farnesyl	$-H$	735
	e	$-CHOHCH_3$	$-CHO$	$-CH_2CH_3$	$-C_2H_5$	$-H$	farnesyl	$-CH_3$	720
	g	$-CH{=}CH_2$	$-CH_3$	$=CH-CH_3$	$-CH_3$	$-CO-O-CH_3$	geranyl-geraniol	$-H$	788

*The double bond between C atoms 3 and 4 is reduced to a single bond in all bacteriochlorophylls

(Chla = chlorophyll *a*, Ph = pheophytin, PQ = plastoquinone, PC = plastocyanin, Fd = ferredoxin, FNR = ferredoxin-NADP$^+$ reductase, 1.18.1.2):

$$\begin{array}{l} \text{P 680} \rightarrow \text{Chla} \rightarrow \text{Ph} \rightarrow \text{quinone-Fe} \rightarrow \text{PQ} \rightarrow [\text{Fe–S}]_2 \rightarrow \text{cyt } f \rightarrow \text{PC} \rightarrow \\ \text{P 700} \rightarrow \text{Chla} \rightarrow [\text{Fe–S}]_x \rightarrow \text{cyt } b_6 \text{—}\!\uparrow \\ \qquad\qquad\quad \hookrightarrow \text{Fd} \rightarrow \text{FNR} \rightarrow \text{NADP}^+. \end{array}$$

In green bacteria the electron-transfer sequence is (BChl = bacteriochlorophyll, BPh = bacteriopheophytin):

$$\begin{array}{l} \text{P 840} \rightarrow \text{BChla} \rightarrow \text{BPh} \rightarrow [\text{Fe–S}]_x \dashrightarrow \text{Fd} \dashrightarrow \text{NAD}^+ \\ \uparrow \qquad\qquad\qquad\qquad\qquad \downarrow \\ \text{└──────── cyt } c \leftarrow \text{cyt } b. \end{array}$$

In purple bacteria then (UQ = ubiquinone):

$$\begin{array}{l} \text{P 870} \rightarrow \text{BChla} \rightarrow \text{BPh} \rightarrow \text{quinone-Fe} \rightarrow \text{UQ} \rightarrow \text{cyt } b \dashrightarrow \text{NAD}^+ \\ \uparrow \qquad\qquad\qquad\qquad\qquad\qquad\qquad\qquad \downarrow \\ \text{└──────────────────── cyt } c. \end{array}$$

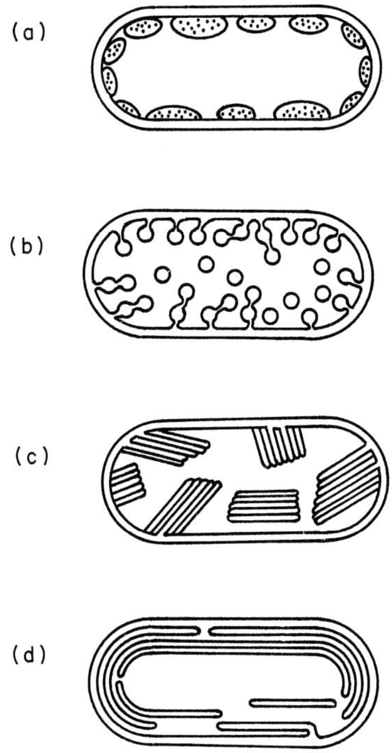

Fig. 4.5. Different structures of membranes carrying photosynthetic pigments. (a) Chlorosomes in *Chlorobium limicola*, (b) vesicular thylakoids and chromatophores in *Rhodospirillum rubrum*, (c) tubular thylakoids in *Rhodopseudomonas viridis*, (d) lamellar thylakoids in *Rhodomicrobium vannielii*

Expressed on a scale of redox potentials the various components are positioned as shown in Fig. 4.23. The components of the electron-transport chain are apparently very well organized in the membrane so that minimum losses of energy are incurred in transit (Figs 4.6 and 4.7).

Organization of the reaction centres

The parts of the photosynthetic systems where the energy of the captured light induces the primary photochemical reactions are called reaction centres. They are organized in a unique way, containing a variety of proteins, pigments and ancillary factors, both of organic and inorganic nature.

Schematic representations of the photosystems of the different groups of organisms are shown in Figs 4.8–4.12. The location of the photosynthetic

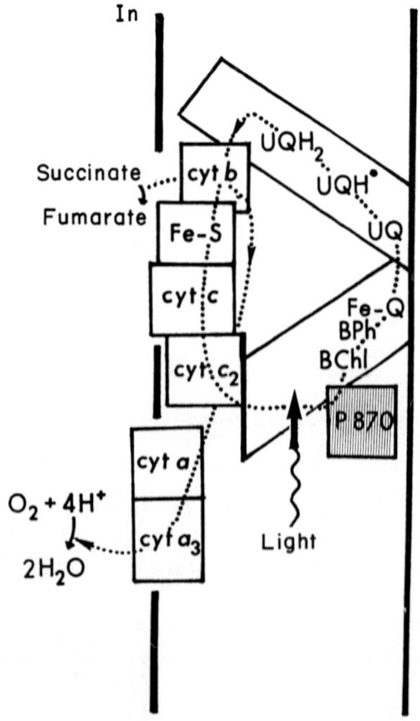

Fig. 4.6. Functional organization of electron transport in a chromatophore of *Rhodopseudomonas sphaeroides*. In addition to (or replacing) the photochemical excitation and electron lifting to a higher reductive power level, oxidation of various substrates generates electrons that travel via cytochrome *b* to cytochrome *a* and cytochrome a_3. BPh—bacteriopheophytin, UQ, $UQH^{\cdot}$, UQH_2—oxidized, half-reduced and fully reduced forms of ubiquinone

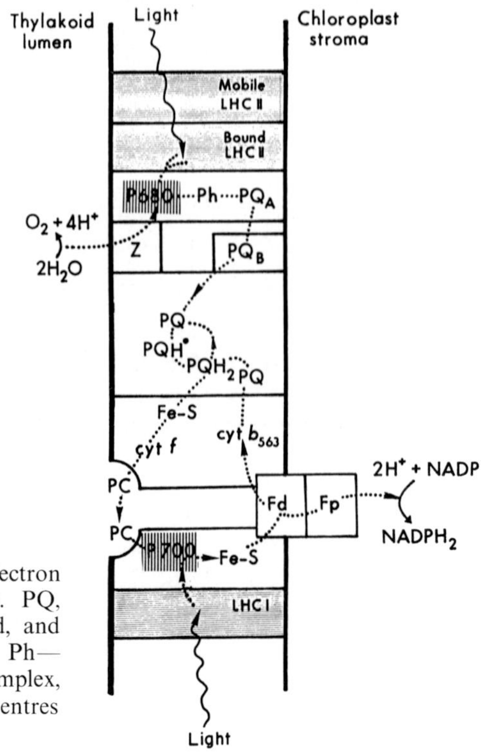

Fig. 4.7. Functional organization of electron transport in a thylakoid membrane. PQ, $PQH^{\cdot}$, PQH_2—oxidized, half-reduced, and fully reduced forms of plastoquinone, Ph—pheophytin, LHC—light-harvesting complex, P680, P700—photosynthetic reaction centres

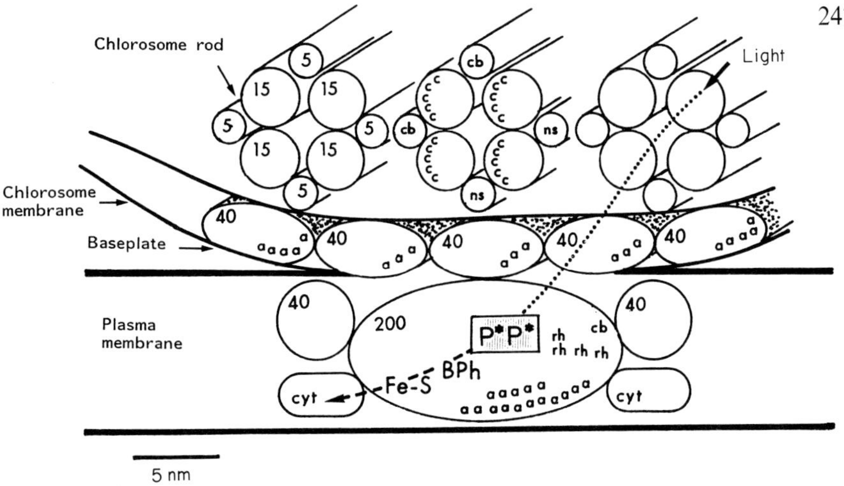

Fig. 4.8. Probable disposition of photosystem components in the green sulphur bacterium *Prosthecochloris aestuarii*. Numbers indicate the molar masses (in 10^3) of polypeptides or protein complexes. a—Bacteriochlorophyll *a* ($\lambda_{max} = 805$ nm, plasma membrane; $\lambda_{max} = 809$ nm, baseplate), c—bacteriochlorophyll *c*, BPh—bacteriopheophytin ($\lambda_{max} = 674$ nm), P*—primary electron donor pigment ($\lambda_{max} = 865$ nm), cb—chlorobactene (a carotenoid), ns—neurosporene (a carotenoid), rh—rhodopin (a xanthophyll), cyt—cytochrome c_{550}. The paths of excitation transfer and electron movement are shown by heavy dotted and dashed lines, respectively. The number of individual symbols reflects their approximate stoichiometry with respect to the reaction centre P*. The individual chlorosome rods are composed of several polypeptides of antenna nature (12 in *Chloroflexus aurantiacus*) each carrying seven BChl *c* molecules. The chlorosome membrane appears to be composed of a single layer of sugar-containing lipids

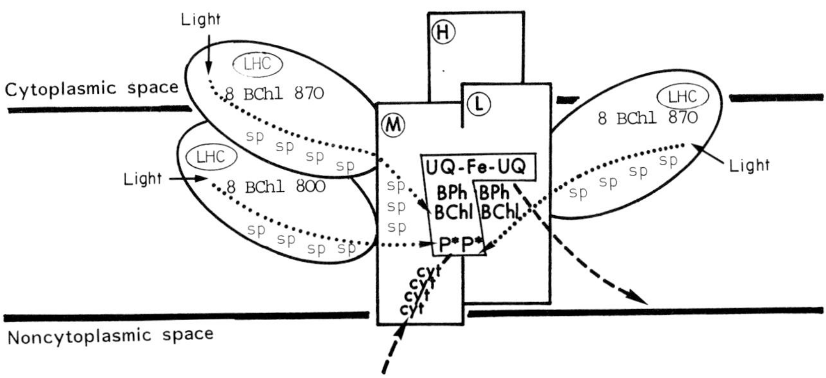

Fig. 4.9. Probable disposition of the photosystem components in the purple nonsulphur bacterium *Rhodopseudomonas sphaeroides*. P*—primary electron donor bacteriochlorophyll *b*-860, sp—spheroidene (a xanthophyll), LHC–light-harvesting complexes, cyt–cytochrome c_2; H, L, and M are the core proteins with M_r 28 000, 30 000 and 34 000, respectively. The light-harvesting bacteriochlorophyll proteins are generally heterodimers of α- and β-polypeptides, each containing 60–80 amino acids. Bacteriochlorophyll as such is apparently bound to them as a pair at juxtaposed histidine residues (His-29 and His-37). Around the reaction centre the BChl heterodimers are clustered in groups of 6 $\alpha\beta$ or 12 $\alpha\beta$ chains and it is here that the 'special chlorophyll pair' is bound

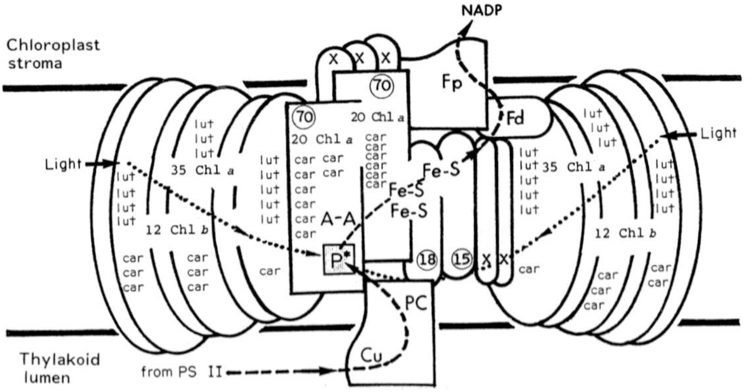

Fig. 4.10. Probable disposition of photosystem I components in oxygenic organisms. Numbers in circles indicate relative molar masses of polypeptides. X—membrane-spanning polypeptide of unknown function, Fd—ferredoxin, Fp—flavin-containing ferredoxin:NADP reductase, PC—plastocyanin, Fe-S—iron–sulphur centres, A—special chlorophyll acceptors of electrons, car—β-carotene, lut—lutein (a xanthophyll), P*—primary electron donor chlorophyll *a*-700. All the 'stoichiometries', expressed either as multiples of chlorophyll molecules or as the number of abbreviations of pigments, are averages from different estimations

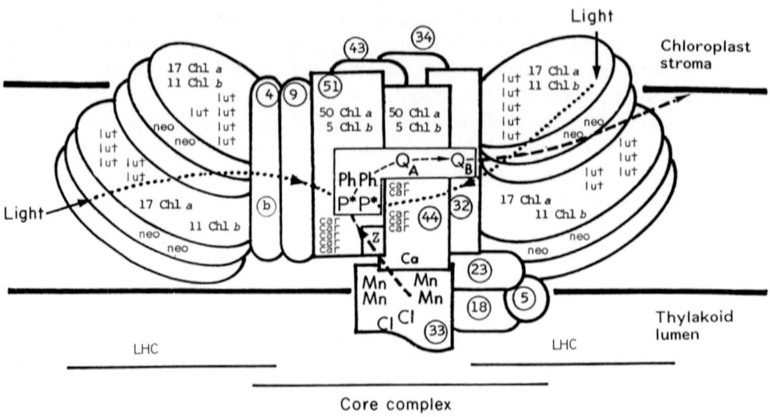

Fig. 4.11. Probable disposition of the reaction centre components of PS II of higher plants (e.g., spinach chloroplasts). b—heterodimer of cytochrome b_{559}, P*—primary electron donor Chl *a*-680, car—β-carotene, lut—lutein, neo—neoxanthin (a xanthophyll), LHC—light-harvesting complexes, a total of four triple-dimer pigment protein units—there exist 16 different polypeptides in the LHC. All the 'stoichiometries' preceding symbols or in terms of numbers of abbreviations are rough averages from different estimations. It now appears that perhaps a half of the LHC are mobile in the membrane while the other half are bound to the core complex. Several of the subunits have now been analyzed and their localization in the membrane is known. Thus the (51) peptide spans the membrane seven times and has a binding site for Q_A on the IVth helix; the (44) peptide is very similar with seven spans and a binding site for Q_A on the IVth helix; the (32) peptide (again seven spans) binds Q_B on its IVth helix, the (34) peptide (seven spans) apparently binds the primary electron donor plastoquinone cation radical (Z) on its Vth helix. There is much similarity between (51) and (44), on the one hand, and (32) and (34), on the other, a distinctive feature of all being symmetrical histidine residues in adjacent α-helices at the membrane surface which can apparently accommodate a heme molecule between them

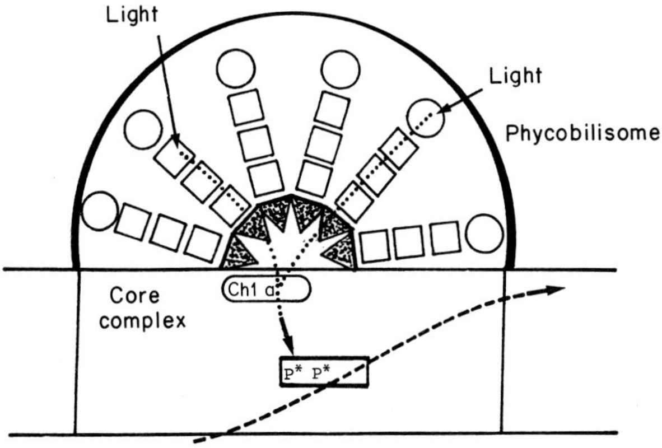

Fig. 4.12. Probable disposition of the photosystem II components in cyanobacteria and red algae. Stippled triangles—allophycocyanin; squares—phycocyanin; circles—phycoerythrocyanin. The phycobilisome contains 300–800 phycobiliprotein molecules. The phycobiliproteins consist of two polypeptide chains with 160–170 amino acids each, with cysteine S-attached prosthetic group of tetrapyrrole linear molecules of the above pigments which are akin to the pheophytins. The phycobiliproteins occur as monomers $\alpha\beta$, trimers $(\alpha\beta)_3$ or hexamers $(\alpha\beta)_6$

reaction centres in the thylakoid membrane is not isotropic even in the two dimensions of the membrane. The chlorosome (in green bacteria) and phycobilisome (in cyanobacteria and some algae) geometry and association with the reaction centers is quite clear. In purple bacteria the distribution of the centers may be more or less homogeneous. In stacked thylakoids, however, a considerable heterogeneity exists in the arrangement of photosystems and of the ATP synthase (Fig. 4.13).

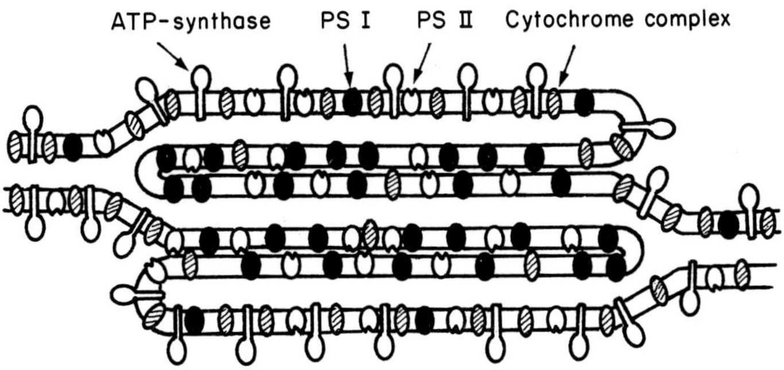

Fig. 4.13. Illustration of the lateral heterogeneity of chloroplast thylakoid membranes. The appressed regions contain both photosystems PS I and PS II and the cytochrome b_6–f complex but no ATP-synthase. The unstacked regions contain virtually no PS II

A new type of photosystem was now discovered in *Heliobacterium chlorum*. It contains BChl *g* ($\lambda_{max} = 798$ nm) and no extrinsic light-harvesting proteins. It is considered to be a primitive form of PS I.

Transfer of excitation energy in photosynthetic membranes

Let us now proceed to a discussion of how a photon is captured by the systems and how its energy is propagated through the photosynthetic system.

To begin with, it should be realized that photosynthetic membrane components can absorb light of different wavelengths (Fig. 4.14) and they manage to cover the entire visible range plus a stretch in the near infrared. For this reason they contain the light-harvesting complexes usually arranged in clusters around the reaction centre core with a preserved concentric organization, starting with pigments absorbing at shorter wavelengths (i.e., higher-energy photons) proceeding to the longer wavelengths all the way to the key molecule in the reaction centre which undergoes special excitation that gives then rise to charge separation (Fig. 4.15). Structural studies show that these light-harvesting complexes are extremely accurately arranged within the thylakoid membrane so as to minimize energy losses.

Clearly, the primary absorber may be either a carotenoid or phycobilin or some type of chlorophyll. The condition for it to be a suitable photo-energy transmitter is that it must retain its excited state long enough to transmit it to a neighbour down the energy path.

Now in all light-absorbing molecules the excitation consists in dislodging an electron from a lower-energy orbital to a higher one, apparently in the regions of conjugated double bonds between carbon and nitrogen atoms (Fig. 4.16) where the π–π^* transition can take place. In chlorophylls and phycobilins, a n–π^* transition could also occur as they contain nitrogen and oxygen atoms with occupied nonvalency orbitals. The excitation shifts one of the π-electrons in the, say, pyrrole molecular orbital to an excited singlet state (no change of original spin) from which it can return to the ground state in a number of ways (Fig. 4.17).

1. Radiation-less internal conversion giving off heat, in a number of small packets of energy corresponding to the long wavelengths of thermal emission.
2. Fluorescence. Here the excitation energy is emitted as a quantum of light of slightly longer wavelength than the exciting light as internal conversions preceding fluorescence bring the molecule to a lower vibrational sublevel.
3. Intermolecular energy transfer. This is the basis of propagation of excitation energy in photosynthetic membranes.
4. Trapping of energy by separation of electric charges. This is the process whereby the excitation energy is transferred into knocking out an electron from its molecular orbital into that of a different molecule.

These last two processes will now be considered. As the excited pigment molecule arrives in the vicinity of another molecule whose electron energy oscillations are in resonance with the approaching one, a dipole–dipole

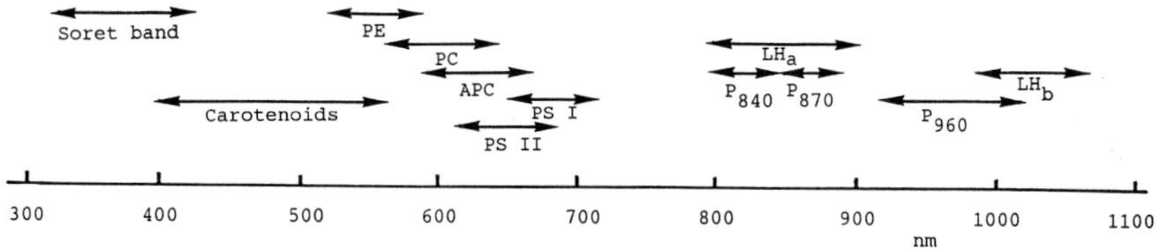

Fig. 4.14. Absorption regions of various photosynthetic pigments. PE, PC and APC stand for cyanobacterial antenna pigments phycoerythrocyanin, phycocyanin and allophycocyanin, LH_a and LH_b are light-harvesting complexes containing chlorophyll *a* and *b*, respectively

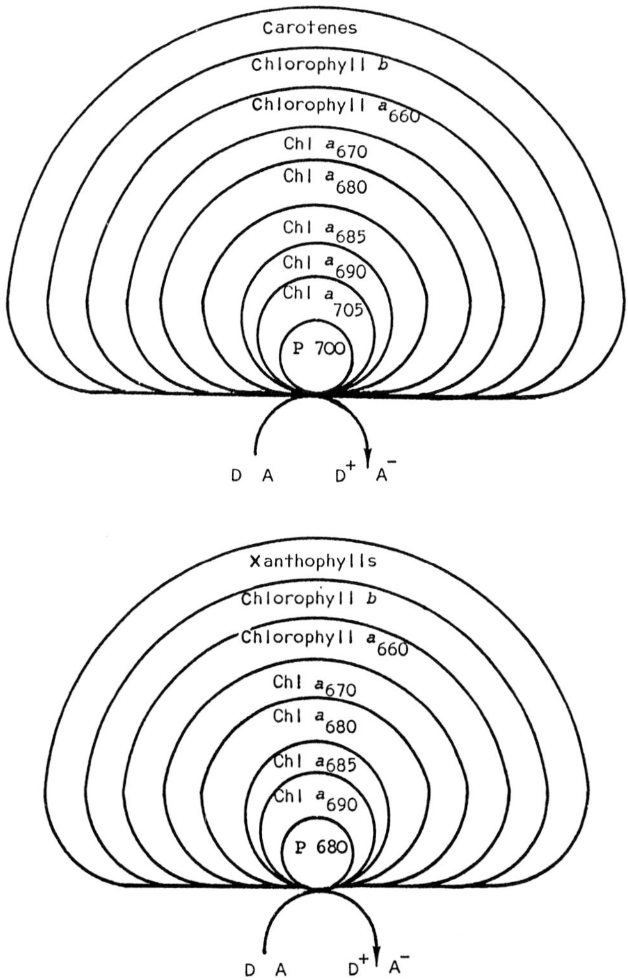

Fig. 4.15. Schematic representation of the light-absorbing pigment organization around photosynthetic systems. The short-wavelength absorbers are seen to lie on the periphery

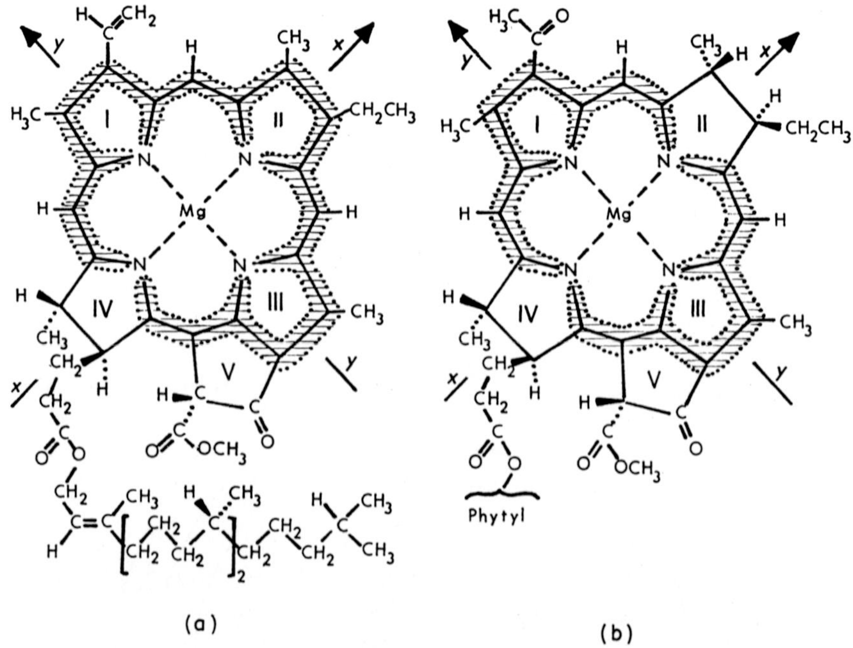

Fig. 4.16. Molecular structure of chlorophyll *a* (a) and bacteriochlorophyll *a* (b) indicating the range of double bonds where π–π^* transitions are likely to take place

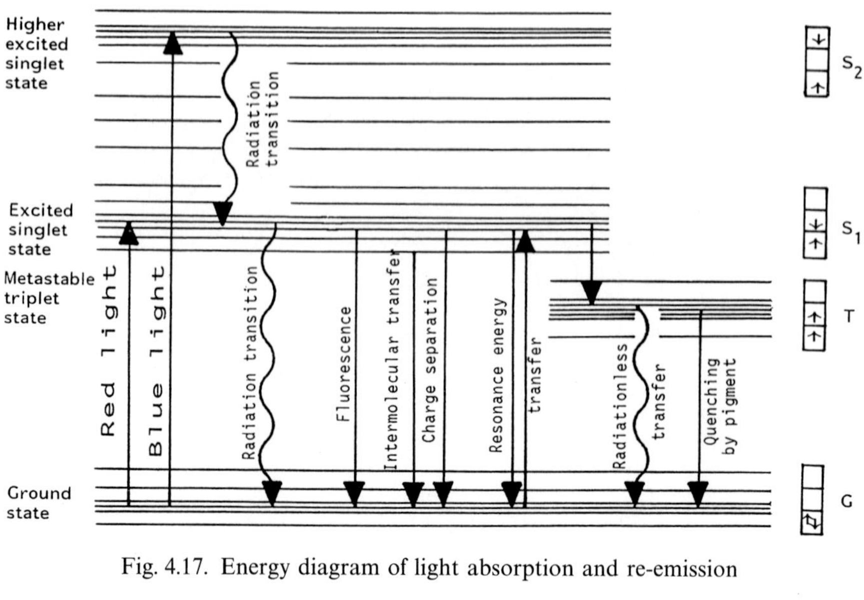

Fig. 4.17. Energy diagram of light absorption and re-emission

resonance interaction between similar vibrational levels may occur so that

$$A^* + B \leftrightarrow A + B^*.$$

This resonance interaction proceeds relatively slowly, with rates of transfer of 10^{11}–$10^{13}\,s^{-1}$ and represents what is called weak coupling in which the electron itself is not transferred from A to B. This is the so-called Förster mechanism of localized excitons which is valid for distances of more than 2 nm.

However, it is now known that many of the pigment molecules are organized in space with high regularity so that they resemble a crystal and their electron clouds overlap. In such a case, an exchange of the excited electron in A* takes place for an excited energy level in B* while the ground-state electron of B shifts to a ground state in A. Such strong coupling interaction is faster, with transfer rates of $10^{15}\,s^{-1}$ (the so-called Dexter mechanism of delocalized excitons which is valid for distances of less than 2 nm). It is now believed that energy transfers between the pigments in a light-harvesting complex occur by strong coupling whereas transfers between the various complexes proceed by weak coupling.

Fluorescence of excited chlorophyll molecules also occurs. To begin with, being a fast process with lifetimes in the case of dissolved chlorophyll of about 5 ns, it 'competes' with the other processes and may represent 30 per cent of the de-excitation of the total excited molecules. This occurs with greater probability as the trapping reaction centres are filled and the incoming excitation energy must be released in ways other than charge separation. Secondly, even after the process of excitation energy transfer has reached the stage of charge separation, it may reverse to regenerate the pigment excited state which then decays with emission of light—the so-called delayed fluorescence.

The exciton thus moves from the light-harvesting antennas to chlorophylls in the reaction centre until a special pair of (bacterio)chlorophylls is reached (the PS I reaction centre perhaps contains a single special molecule). These molecules are mutually placed at a certain angle and are in intimate contact with another (bacterio)chlorophyll and a (bacterio)pheophytin and in close vicinity with two quinones (Fig. 4.18). As the exciton reaches this pair, an electron is lost by the pair and transferred in a sequence through the various members of the complex (Fig. 4.19).

The midpoint potential for the oxidation of the primary donor BChl is about 0.45 V. The reduction of the primary acceptor quinone takes place with a midpoint potential of −0.03 V in *Rhodopseudomonas* (and at −0.5 V in the PS I). Thus, an extremely powerful reductant is formed in the acceptor. The hole remaining after the electron leaves the primary donor bacteriochlorophyll pair is delocalized over the molecular orbitals extending over both halves of the pair. Although both acceptors are ubiquinones with 10 isoprenoid units in the side chain, the quinones are far from being identical. While Q_A can only be reduced to the $QH^{\cdot}$ radical, Q_B can be fully reduced to QH_2. This is apparently due to differences in the apoprotein environment, e.g. Trp is associated very clearly with Q_A and Phe with Q_B. It is not clear what the function of the iron atom

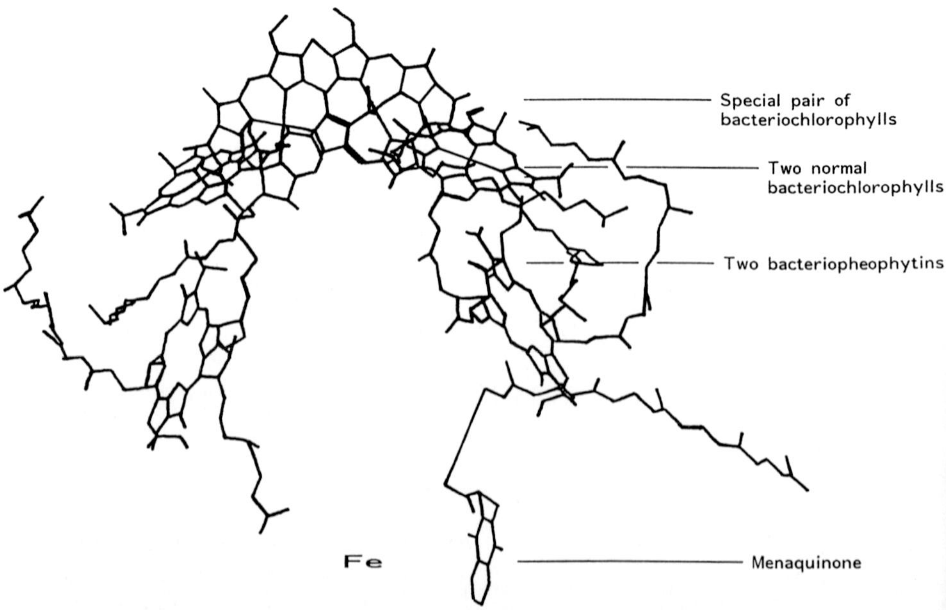

Fig. 4.18. Structural organization of the photosynthetic reaction centre components of *Rhodopseudomonas sphaeroides*

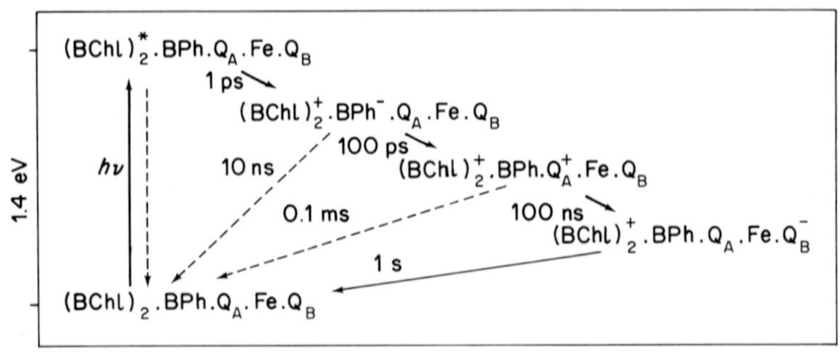

Fig. 4.19. Sequence of charge separation reactions in the reaction centre of *Rhodopseudomonas sphaeroides*

is. It does not undergo oxidation-reduction and can be replaced *in vitro* with Zn, Mn, Co or Ni.

It is seen from Fig. 4.19 that the intermediary complexes can decay in a futile way (but much more slowly) to the ground state of the reaction centre.

A similar sequence apparently proceeds in the reaction centres of the PS II, even if the details are different.

Neutralization of charge in the PS II reaction centre

Let us now go back to the special chlorophyll molecule or chlorophyll pair that was instrumental in the separation of electric charges. While the negative charge has been taken care of by its passage toward NADP reduction, the positive charge on the chlorophyll is rapidly neutralized from a special electron donor.

In green and purple bacteria this is effected by a cyt c_2 molecule which has been reduced by a reverse flow of electrons from either ferredoxin (in green bacteria) or from the ubiquinone pool (purple bacteria). In PS I the natural reductant is plastocyanin which has been reduced by electrons arriving from PS II.

In PS II the situation is more complicated. The electrons arrive here from a complex reaction involving the splitting of water into oxygen and protons plus electrons:

$$2H_2O \rightarrow O_2 + 4H^+ + 4e^-.$$

The system of peripheral polypeptides bordering on the largest integral subunit of the PS II core probably contains four atoms of Mn which are ionized to different levels in the course of the reaction. A significant observation made by measuring oxygen evolution after repeated single-turnover flashes was that four photons must be captured per reaction centre before a molecule of oxygen is produced. However, as the reaction centre contains a single oxidized electron donor per photocycle the electrons used for neutralizing it must arrive one by one. Over the years several models of oxygen evolution at the PS II reaction centre have been proposed, two of which are shown in Fig. 4.20, a well elaborated classical model based on the turnover of two Mn atoms, and a very recent model involving four Mn atoms in generally higher oxidation states.

The transfer of electrons from the Mn-complex to the PS II reaction centre is mediated by a pigment designated Z which is now known to be plastoquinone cation radical with an extremely high redox potential of +1.0 V.

Electron transfer in the photosynthetic membrane

In *green bacteria*, once the [Fe–S] centre has been reduced the electron can either return cyclically via a cytochrome *b* and a cytochrome *c* to compensate the charge in the reaction centre of P 840, or it can proceed toward ferredoxin which then serves as the final reductant of the flavin adenine dinucleotide reductase and NAD^+.

Ferredoxins are a generic name for proteins either relatively small (M_r 6000–15 000) and soluble, such as those in the photosystems, or parts of larger complex enzymes, such as those in mitochondrial respiration, which contain a special iron–sulphur centre of one type or another. These centres can transfer single electrons between suitable redox partners. Most of the [Fe–S] centres have midpoint potentials in the negative range from −60 mV of the simple rubredoxin of obligately anaerobic Clostridia (1 Fe directly linked to 4 Cys residues) to −270 mV of the adrenodoxin of adrenal cortex

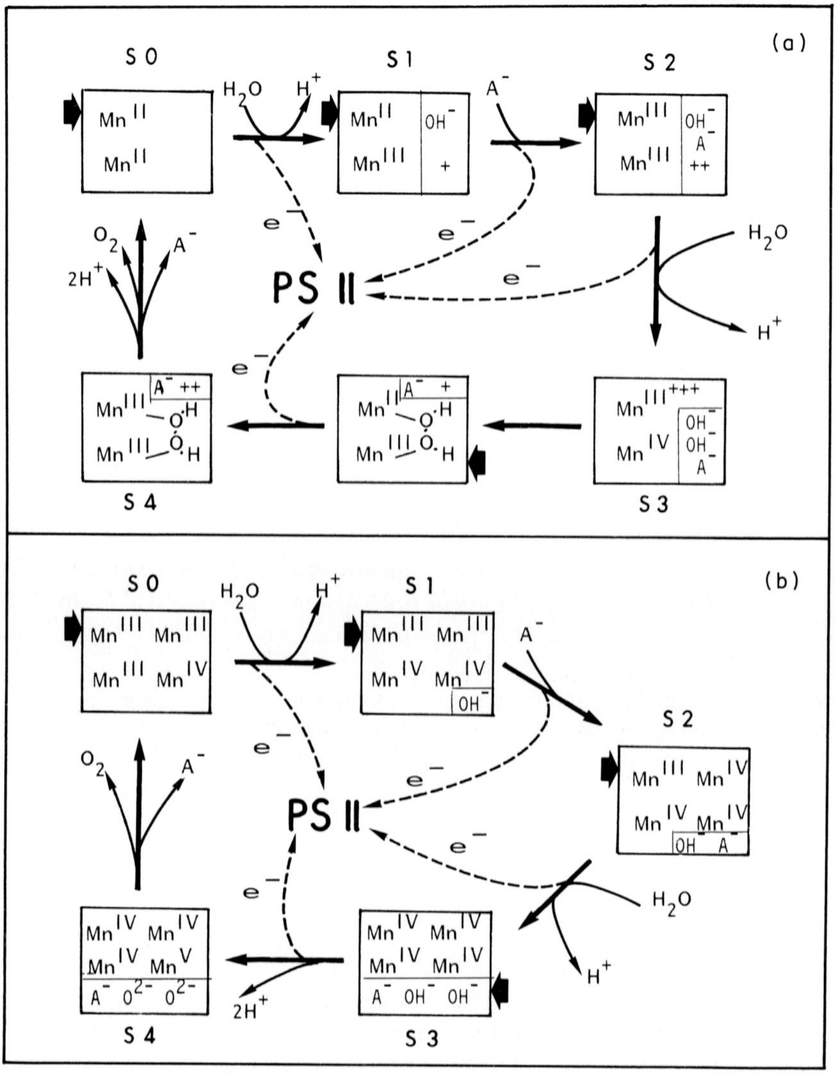

Fig. 4.20. Two possible models of partial reactions taking place in the PS II-associated water-splitting enzyme. (a) A model based on earlier observations, (b) a more recent model. In both cases, the crucial step appears to be that where, between S3 and S4, a peroxide bond is formed between two Mn atoms, the other two not participating in the process. It is believed that S0, S1 and S2 represent a cubane arrangement of the Mn and O atoms (4:4) while S4 is an adamantane with Mn:O = 4:6, the S3 state being transitional. At any rate, two of the four Mn atoms have the unique position, being able to bind the peroxide between them, apparently caused by their special type of coordination with the protein groups. The short heavy arrows indicate steps where photon energy is captured

ferredoxin which donates electrons to cytochrome P-450, to $-490\,\text{mV}$ of the red sulphur bacterial ferredoxin active in photosynthesis. However, the special high-potential iron–sulphur proteins (HiPiP) active in mitochondrial respiration have $E^{\ominus}_{7.0}$ of $+280\,\text{mV}$ and one of the red sulphur bacterium ferredoxins has a midpoint potential of $+360\,\text{mV}$. The usual organization of a [Fe–S] centre is either

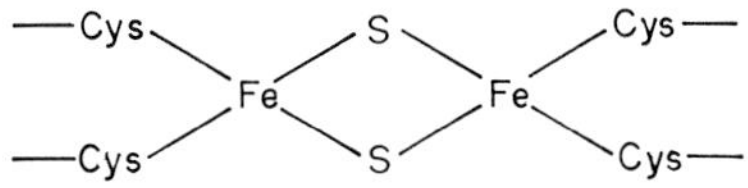

for the $[Fe_2S_2]$ type, or

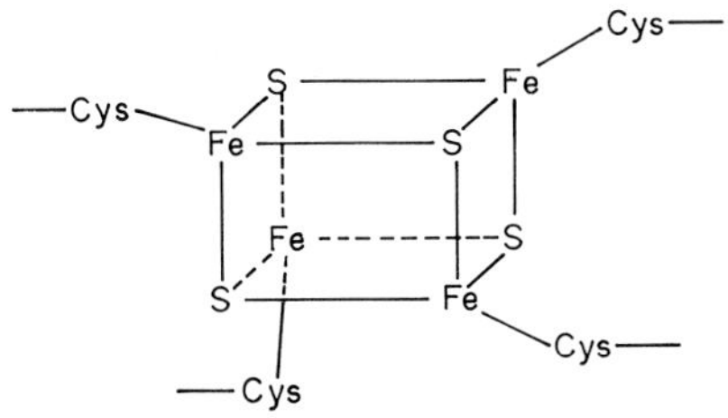

for the $[Fe_4S_4]$ type. A $[Fe_3S_3]$ centre is also known but its structure is probably qualitatively different. The $[Fe_2S_2]$ centres exist as Fe^{3+}–Fe^{3+} and are transiently reduced to Fe^{3+}–Fe^{2+}. The $[Fe_4S_4]$ centres exist as Fe_2^{3+}–Fe_2^{2+} and are transiently reduced to $Fe^{3+}-Fe_3^{2+}$. If they occur in the HiPiP ferredoxins their normal state is Fe_3^{3+}–Fe^{2+} and can be reduced to Fe_2^{3+}–Fe_2^{2+}.

The Fe–S-containing enzymes are fairly ubiquitous and are active not only in photosynthesis and mitochondrial oxidations but also in nitrogen fixation, sulphite reduction, hydroxylation, formation of acetyl phosphate and in CO_2 fixation in the synthesis of 2-oxo acids.

The [Fe–S] centres are apparently synthesized under catalysis of rhodanese (thiosulphate sulphurtransferase, 2.8.1.1).

In green bacteria, on its way from cytochrome *b* to cytochrome *c* the electron gives rise to a translocation of 1–2H^+ ions by an unknown mechanism.

In *purple bacteria*, the electron generated at the reaction centre acquired a formal redox potential of only about -0.22 V and, via special quinones A and B, enters the ubiquinone pool where it takes part in the so-called Q cycle in which, under participation of cytochrome *b*, 2H^+ ions are apparently pumped from the inside of the chromatophores into the cytosol (Fig. 4.21).

It appears that the quinone–quinol transport of H^+ ions is the only true transport of protons across the membrane which thus constitutes the second step in the conversion of incident light energy into an electrochemical potential gradient. However, doubt remains whether the cycle operates in the simple way shown including recycling of electrons. Perhaps a more sophisticated mechanism, such as shown in Fig. 4.29, is correct.

The reduced ubiquinone is then reoxidized by cytochrome b_{50} and cytochrome c_2, and the electron returns to the reaction centre of P 870.

In *oxygenic photosynthetic membranes* the electron emerging from PS II enters the plastoquinone pool, an analogy of the ubiquinone pool of purple bacteria, and, via a Q cycle, manages to translocate one or two protons into

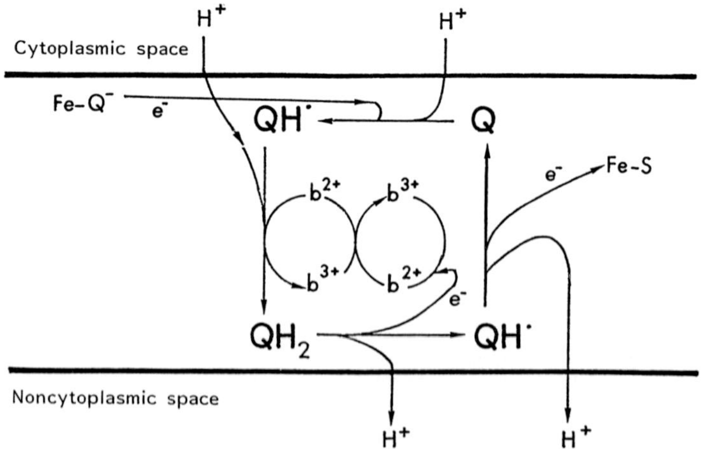

Fig. 4.21. Flow of electrons and protons through a double-loop Q-cycle whereby $2H^+$ are translocated per electron passing through from quinone B to a Fe–S centre in a lower redox-potential component of the electron transfer chain

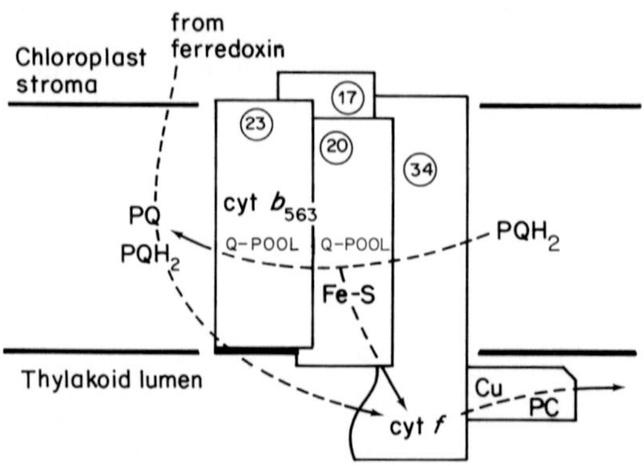

Fig. 4.22. Putative disposition of the cytochrome b_6–f complex in a photosynthetic membrane. The molar masses of the individual polypeptides are circled. Electrons arrive either from the Q-cycle or (in a cyclic system) from ferredoxin and proceed via the cytochromes and plastocyanin to PS I

the lumen of the thylakoids. This process takes place in the cytochrome b_6–f complex (Fig. 4.22). The electron(s) of the reduced plastoquinone are accepted by cytochrome f. Likewise, in cyclic electron flow, the electrons returning from ferredoxin via cytochrome b_{563}, proceed to cytochrome f. Either way, cytochrome f transmits the electron to plastocyanin.

Plastocyanin is a copper-containing protein with a formal redox potential of + 0.37 V and M_r of 23 000. The copper atom is coordinated to N atoms of His-37 and His-38, as well as to the S atom of Met-92. It is covalently bound to the S atom of Cys-84.

The path of the electron ends at PS I where it reduces the positive charge of the charge-transfer complex generated by excitonic energy transfer, in analogy with what the electron from split water does to PS II.

The path of the excited electron of PS I is then qualitatively identical with that in green bacteria, including a [Fe–S] centre, ferredoxin and finally $NADP^+$.

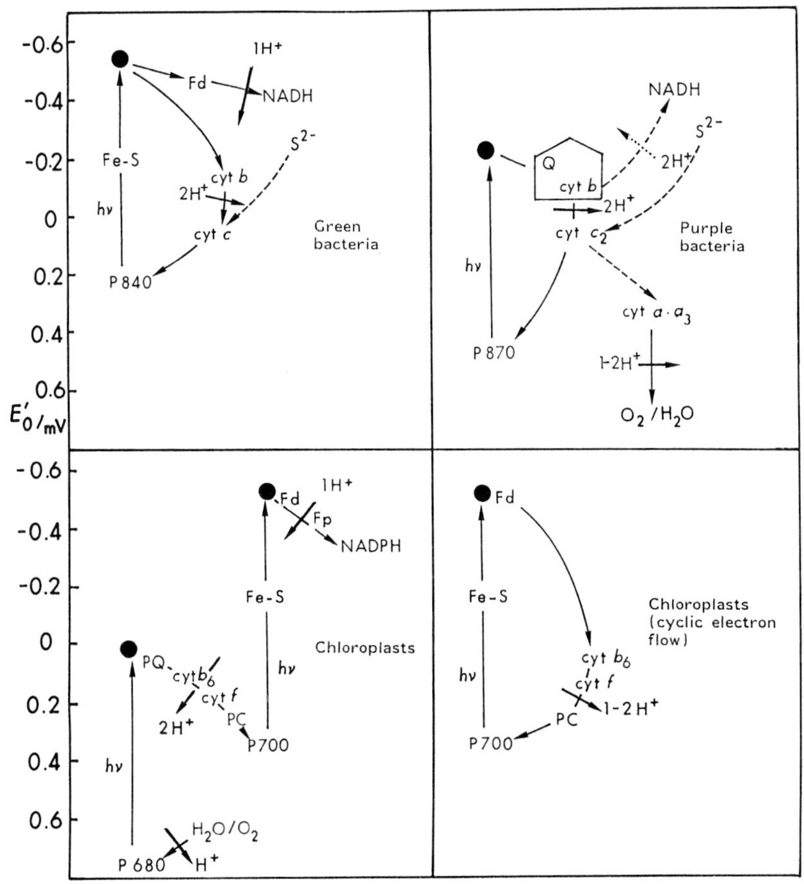

Fig. 4.23. Flow of electrons and protons through photosynthetic systems of different organisms. In photosynthetic bacteria, chemoautotrophic paths (broken lines) can be used in place of photoautotropic ones (full lines). The position of the primary electron donor on the redox potential scale is shown by the heavy dot. All the proton shifts represent uphill movement (a primary active transport), the number expressing an approximate stoichiometry per one electron. The dotted arrow on the path to NADH in purple bacteria is the only case where $2H^+$ move downhill to provide energy for NAD^+ reduction

Sites of proton translocation

The points where energy is transformed in the various centres of photosynthetic systems are depicted summarily in Fig. 4.23. When counting the protons transported across the photosynthetic membranes one should be aware of the fact that not only the ones transported contribute to the established ΔpH. A case in point is the breakdown of water whereby $2H^+$ appear in the thylakoid lumen, another case then the reduction of NADP to NADPH where $1H^+$ disappears from the stroma, hence the total difference will be much more than the two protons actually translocated per electron in the plastoquinone cycle.

Regulation of energy distribution between PS I and PS II

In spite of the fact that the two photosystems are optimally excited at shorter and longer wavelengths, respectively, illumination with nonmonochromatic light at any wavelength of the action range results in comparable rates of photosynthesis. Thus, the two PS centres must be regulated to slow down or speed up their energy-converting process depending on them. This is now known to be achieved by phosphorylation of the apoprotein of the chlorophyll *a*/*b* light-harvesting complex of PS II (Fig. 4.24).

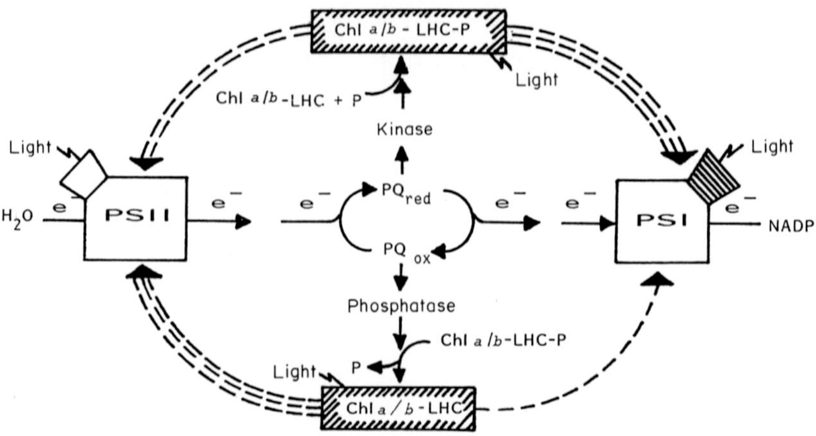

Fig. 4.24. Schematic of the feedback control exerted by electron flow from PS II to PS I to balance the overall rate of photosynthesis. When exposed to a strong electron flow from PS II, the PQ pool, now in a reduced state, activates a phosphokinase which phosphorylates one type of the polypeptides in the LHC (probably one with M_r 25 000). In this form it channels more excitation energy to the PS I reaction centre to keep up with the hard-working PS II. If, on the contrary, PS I is working at full capacity it withdraws electrons from the PQ pool; this is then relatively oxidized and activates a phosphatase which dephosphorylates the LHC peptide. This has the consequence that more excitation energy is channelled into PS II. This differential channelling apparently has its topological counterpart. The phosphorylated LHC peptides (those of the mobile part of the LHC complex) migrate away from the PS II which is located in the appressed region of thylakoids to the unstacked membranes where PS I is known to reside

To summarize, photosynthesis in membranes leads to (1) the generation of an electrochemical gradient of protons; how this is utilized will be discussed in the context of mitochondrial and bacterial oxidation and ATP formation; (2) the generation of reducing power for anabolic reactions—this is a complex of reactions that occur outside membranes, either in the chloroplast stroma or in bacterial cytosol, and they will not be discussed here. Let it only be said that the basic reaction of this complex is one or another of the reactions in which CO_2 is incorporated into an organic molecule and thus creates the increase in mass of green plants which is often used to assess the efficiency of the light-capturing process.

Thus, in the often quoted 'equation of photosynthesis'

$$CO_2 + 2H_2O \rightarrow [CH_2O] + H_2O + O_2$$

it should be noted that all the dioxygen evolved comes from water while all the oxygen of the saccharide unit derives from CO_2.

SYNOPSIS

All light-absorbing systems in cell membranes contain compounds with conjugated double bonds with the possibility of π–π^* transitions. Retinal is found in some archaebacterial systems (bacteriorhodopsin and halorhodopsin) while different types of chlorophyll, as well as carotenoids, capture photons in true photosynthetic membranes.

In the anoxygenic photosynthesis of green and purple bacteria the excitation energy acquired from the photon is used for generating H^+ gradients and for reducing NAD. In oxygenic photosynthesis of cyanobacteria, algae and higher plants H^+-ion gradients are formed, NADP is reduced, and water is cleaved to oxygen, protons and electrons.

The molecules participating in these processes include light-harvesting (antenna) complexes, complex photosynthetic reaction centres, accessory redox systems, quinones, and inorganic ions. They are organized in a highly efficient way in the photosynthetic membranes and make it possible for excitation energy to be used for electric charge separation at a (paired) (bacterio)chlorophyll molecule.

FURTHER READING

Barber, J. Biophysics of photosynthesis, *Rep. Progr. Phys.*, **41**, 1157–99 (1978).

Clayton, R. K. *Photosynthesis. Physical Mechanisms and Chemical Patterns*, Cambridge University Press, Cambridge (1980).

Critchley C. The role of chloride in photosystem II, *Biochim. Biophys. Acta*, **811,** 33–46 (1985).

Danks, S. M., Evans, E. H. and Whittaker, P. A. *Photosynthetic Systems. Structure, Function and Assembly*, John Wiley, Chichester (1983).

Drews, G. Structure and functional organization of light-harvesting complexes and photochemical reaction centres in membranes of phototropic bacteria, *Microbiol. Rev.*, **49,** 59–70 (1985).

Drews, G., and Oelze, J. Photosynthese bei phototrophen Bakterien, *Biologie in unserer Zeit*, **16,** 113–23 (1986).
Foyer, C. H. *Photosynthesis*, John Wiley, Chichester (1984).
Govindjee (ed.). *Bioenergetics of Photosynthesis*, Academic Press, New York (1975).
Miller, K. R. The photosynthetic membrane: prokaryotic and eukaryotic cells, *Endeavour* (New Series), **9,** 175–82 (1985).
Murphy, D. J. The molecular organization of the photosynthetic membranes of higher plants, *Biochim. Biophys. Acta*, **864,** 33–94 (1986).
Prince, R. C. Manganese at the active site of the chloroplast oxygen-evolving complex, *Trends Biochem. Sci.*, **11,** 491–2 (1986).
Sauer, K. A role for manganese in oxygen evolution in photosynthesis, *Acc. Chem. Res.*, **13,** 249–56 (1980).
Siefermann-Harms, D. Carotenoids in photosynthesis. Location in photosynthetic membranes and light-harvesting function, *Biochim. Biophys. Acta*, **811,** 325–55 (1985).
Youvan, D. C., and Daldal, F. (eds). *Microbial Energy Transduction. Genetics, Structure and Function of Membrane Proteins*, Cold Spring Harbor Laboratory (1986).
Zuber, H. Structure of light-harvesting antenna complexes of photosynthetic bacteria, cyanobacteria and red algae, *Trends Biochem. Sci.*, **11**, 414–19 (1986).

4.3 FREE ENERGY OF OXIDATION CONVERTED INTO OTHER FORMS OF ENERGY

4.3.1 General principles

The generation of dioxygen by early oxygenic photosynthetic organisms caused a powerful evolutionary burst by providing a potential source of energy that could be harnessed at the level of biochemical reactions into a useful form to be employed for practically all life processes.

The dehydrogenation reactions used by the primitive fermenting bacteria did apparently produce ATP as it is produced today in the cytosol of many cells by the so-called substrate-level phosphorylation. Likewise, electron transport chains must have evolved before the advent of oxygen and the generation of proton gradients must have led to a membrane-level phosphorylation to produce ATP. However, the redox potential difference between the SH-substrate and NO_3^- is only about 0.7 V while with O_2 it is as much as 1.2 V. Furthermore, oxygen was available in most biotopes and could enter cells by a diffusion mechanism so that access to it was more or less universal.

The evolution of O_2-reducing ability by microorganisms required an alteration of the terminal oxidases such as are found in present-day mitochondria and some aerobic bacteria.

Oxidation of organic substrates to CO_2 and H_2O yields now some 90 per cent of the total nonphotosynthetic free energy obtained and converted to other forms in living cells. Most of this results in the generation of an electrochemical potential gradient of protons (and exceptionally Na^+).

Before proceeding to a detailed description of the oxidation pathways in mitochondria and bacteria let us insert a short general treatise of electron transfer processes.

4.A ELECTRON TRANSFER PROCESSES

The equilibrium of a simple oxidation-reduction reaction

$$\mathrm{Ox}_1 + \mathrm{Red}_2 \rightleftharpoons \mathrm{Red}_1 + \mathrm{Ox}_2 \qquad (4.\mathrm{A}1)$$

is characterized by the equilibrium constant

$$K_\mathrm{a} = \exp(-\Delta G^0/RT) \qquad (4.\mathrm{A}2)$$

where ΔG° is the standard Gibbs energy change of the reaction. To assess quantitatively the oxidative or reductive abilities of each redox couple participating in the reaction, $\mathrm{Ox}_1/\mathrm{Red}_1$ and $\mathrm{Ox}_2/\mathrm{Red}_2$, and, very often, to determine ΔG°, it is convenient to carry out reaction (Eq. (4.A1)) as a cell reaction in a galvanic cell

$$\mathrm{M}|\mathrm{Ox}_2, \mathrm{Red}_2 \| \mathrm{Ox}_1, \mathrm{Red}_1|\mathrm{M} \qquad (4.\mathrm{A}3)$$

where M is an inert electronic conductor (Pt, Au, graphite, in some cases Hg) and the double vertical bar denotes a liquid bridge removing the liquid-junction potential between the two solutions.

When a positive charge passes through the cell from left to right, half-cell reactions

$$\begin{aligned} \mathrm{Red}_2 - ze &\rightleftharpoons \mathrm{Ox}_2 \\ \mathrm{Ox}_1 + ze &\rightleftharpoons \mathrm{Red}_1 \end{aligned} \qquad (4.\mathrm{A}4)$$

proceed at each electrode (z is the charge number of the half-cell reaction).

The electromotive force (EMF) E defined as the difference of electric potentials of identical metallic leads to the right-hand side and to the left-hand side electrode in the schematic representation of the cell (Eq. (4.A3)) is given as the difference of the *electrode potentials* of these electrodes.

(The inner electric potential of a phase should be distinguished from the electrode potential which is the EMF of a cell consisting of the electrode concerned, on the right-hand side of the scheme, and of the standard hydrogen electrode, on the left-hand side of the scheme.)

$$E = E_1 - E_2. \qquad (4.\mathrm{A}5)$$

The electrode potentials (or redox potentials in the present case) are given by the Nernst equation

$$E_1 = E_1^\circ + \frac{RT}{zF}\ln(a_{\mathrm{Ox},1}/a_{\mathrm{red},1}) \qquad (4.\mathrm{A}6)$$

$$E_2 = E_2^\circ + \frac{RT}{zF}\ln(a_{\mathrm{Ox},2}/a_{\mathrm{red},2})$$

where E_1° and E_2° are the standard potentials of the redox systems and the a's are the activities of the participating species. Obviously,

$$\Delta G^\circ = -zF(E_1^\circ - E_2^\circ). \qquad (4.\mathrm{A}7)$$

The larger the E° the stronger oxidant is the oxidized form Ox and, vice versa, for the reduced form Red.

The half-cell reaction only occurs in a limited number of cases according to the simple Eq. (4.A1) while quite often other components of the solution also participate in the redox process (e.g. ligands in the case of metal complexes, ionized forms of the solvent, such as H_3O^+, OH^-, etc.). The latter case is typical of biologically important redox systems where cell reactions occur of the type

$$\mathrm{Ox} + ze + n\mathrm{H}^+ \rightleftharpoons \mathrm{H}_n\mathrm{Red}^{(n-z)}. \qquad (4.\mathrm{A}8)$$

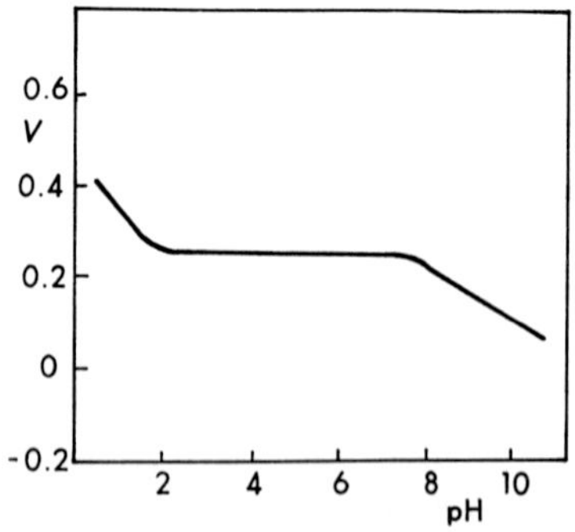

Fig. 4.25. pH dependence of the oxidation-reduction potential of ferri–ferro cytochrome c (in V)

Here, the charge number is usually 1 or 2 and $n \geqslant 0$ often varies depending on the pH of the solution. The change of pH is displayed in a change of ionization of both Ox and Red.

For biological systems a convenient quantity is the formal (conditional) potential $E^{\ominus}$ (sometimes called the midpoint potential), defined for equal overall concentrations of the components of the redox couple and for a definite composition of the solution, appropriate pH, etc. The conditions should be suitably defined at the $E^{\ominus}$ symbol. Figure 4.25 shows the dependence of the formal potential of the Fe^{3+}/Fe^{2+} cytochrome c couple on pH. In subsequent text formal potentials as well as concentrations instead of activities will be used. For a comparison of various redox systems the condition of pH = 7 is usually chosen as it is close to the conditions in biological systems. Table 4.3 lists formal potentials of various biologically important redox systems under these conditions. It should be stressed, however, that these values refer to systems in homogeneous solutions. When the components of a redox system are bound in a membrane the influence of the membrane milieu can change the redox potential quite substantially.

In Eqs (4.A1) and (4.A8) equal stoichiometric coefficients for the oxidized and for the reduced form, equal to 1, were assumed. In the case of dimerization of the oxidized or of the reduced form these stoichiometric numbers become 2 and 1. In other words, the equilibrium between a monomeric and a dimeric reduced form is described as

$$2\,\mathrm{Red} \rightleftharpoons \mathrm{Red}_2 \tag{4.A9}$$

with the equilibrium constant

$$K = c_{\mathrm{red},2}/c_{\mathrm{red}}^2. \tag{4.A10}$$

Equation (4.A6) is modified using concentration terms such as

$$E = E^{\ominus}_{\mathrm{ox,red}} + \frac{RT}{zF}\ln\left(c_{\mathrm{ox}}/c_{\mathrm{red}}\right) \tag{4.A11}$$

where $E^{\ominus}_{\mathrm{ox,red}}$ is the formal electrode potential. Inserting for c_{red} from Eq. (4.A10) into Eq. (4.A11) one obtains

$$E = E^{\ominus}_{\mathrm{ox,red}} + \frac{RT}{zF}\ln\left(c_{\mathrm{ox}}K^{1/2}/c_{\mathrm{red}_2}^{1/2}\right). \tag{4.A12}$$

Table 4.3. Formal oxidation-reduction potentials (midpoint potentials) at pH 7

Redox couple	$E^{\ominus}_{7,0}/V$
O_2^-/O_2	−0.56
Formate/CO_2	−0.42
Soluble ferredoxins	−0.41
H/H^+	−0.41
$NADH_2$/NAD	−0.32
Fe–S centres	−0.3 to +0.28
H_2S/S	−0.27
Ethanol/acetaldehyde	−0.20
Lactate/pyruvate	−0.18
Cytochrome b_6 2+/3+	−0.08
$FMNH_2$/FMN	−0.08
Cytochrome b 2+/3+	−0.07
$FADH_2$/FAD	−0.05
Cytochrome b_T 2+/3+	−0.04
Succinate/fumarate	−0.03
H/H^+ (pH 0)	0.00
Leucomethylene/methylene blue	0.01
Ascorbic/dehydroascorbic acid	0.06
Cytochrome b_K 2+/3+	0.06
Ubiquinone-H_2/ubiquinone	0.10
Plastoquinone-H_2/plastoquinone	0.11
Cytochrome c_1 2+/3+	0.22
Cytochrome c 2+/3+	0.25
Cytochrome a 2+/3+	0.29
H_2O_2/O_2	0.29
Cytochrome c_2 2+/3+	0.34
Cytochrome f 2+/3+	0.36
Ferrocyanide/ferricyanide	0.36
Plastocyanin 1+/2+	0.37
Cytochrome a_3 2+/3+	0.39
NO_2^-/NO_3^-	0.42
P 700, P 870	0.43
Fe^{2+}/Fe^{3+}	0.77
H_2O/O_2	0.82
Mn^{2+}/Mn^{3+}	1.51

In general, the transfer of two or more electrons in one step of a redox reaction is highly improbable kinetically and the reaction proceeds via an intermediate formed by one-electron transfer. This mechanistic view would indeed have no influence on the thermodynamics of the redox reaction but in some cases the intermediate, denoted as I, can be present in an appreciable concentration in the system at equilibrium. The two-step redox reaction proceeds according to

$$\begin{aligned} \mathrm{Ox} + e &\rightleftharpoons \mathrm{I} \\ \mathrm{I} + e &\rightleftharpoons \mathrm{Red.} \end{aligned} \tag{4.A13}$$

These reactions can be combined in a disproportionation reaction

$$2\mathrm{I} \rightleftharpoons \mathrm{Ox} + \mathrm{Red}. \tag{4.A14}$$

Obviously, the concentration of the intermediate is obtained from

$$K_{\text{disprop}} = c_{\text{ox}} c_{\text{red}} / c_{\text{I}}^2. \tag{4.A15}$$

For the quinoid systems which are common components of the electron-transfer chain and of the light phase of photosynthesis the formation of radical intermediates, the semiquinones, is of considerable importance.

Many membrane chemical reactions may be viewed as redox processes but, in fact, these reactions occur by atom or group transfer without genuine electron transfer being involved. On the other hand, there is a limited number of compounds which participate in actual electron-transfer processes at membranes. Among these the enzymes containing porphyrin complexes of iron (in particular, the cytochrome family), nonheme protein complexes of iron, copper and manganese, the already mentioned quinone systems, some adenine and flavin-containing nucleotides and others are included.

With some of the mentioned redox systems the redox potentials can be measured potentiometrically as the equilibrium between the redox couple and the electrons in the electrode is rapidly established. The condition for this is a sufficiently high value of the exchange current density at the electrode. This quantity is connected with the dynamics of the electrode reaction which, at equilibrium, proceeds both as a cathodic and an anodic process. While these processes electrically cancel each other the individual reactions proceed without restriction. The current density corresponding to the anodic reaction at unit concentration ($c = 1\ \text{mol dm}^{-3} = 10^{-3}\ \text{mol cm}^{-3}$) of participating components of the redox couple is equal to the conditional current density. A reasonable condition for a sufficiently rapid establishment of equilibrium at the electrode requires the conditional current density j° to be greater than $10^{-5}\ \text{A m}^{-1}\ \text{mol}^{-1}$.

As examples where potentiometric measurements supply no adequate information on the redox potential of the system under investigation the disulphide/sulphydryl couple and the oxygen/water couple will be considered.

At a platinum electrode the electrode potential in the presence of both cystine and cysteine (or S–S and SH glutathione) does not depend on concentration of the disulphidic form. In a homogeneous solution (and, similarly, in a membrane), however, the redox reaction proceeds rapidly via the free-radical intermediate $RS^{\cdot}$. At a platinum (or gold) electrode this intermediate is adsorbed and can only be oxidized (to cysteic acid RSO_3H). Thus the mechanism of the electrode process does not allow the redox potential to be determined (the formal potential can only be calculated from $\Delta G^{\ominus}$ which is obtained on the basis of thermal data).

One of the fundamental processes in membrane bioenergetics, the reduction of oxygen to water,

$$O_2 + 4e^- + 4H^+ \longrightarrow 2H_2O$$

has a formal potential $E^{\ominus}_{O_2/H_2O}$ (pH = 7) of 0.82 V but this potential has never been measured directly because of the improbable transfer of four electrons in a single step. Even the redox potential of the O_2/H_2O_2 couple is directly obtained only at a mercury electrode in an alkaline medium while the redox potential of the H_2O_2/H_2O couple is not directly measurable at all. On the other hand, in the electron-transfer chain situated in the mitochondrial membrane the oxygen reduction to water proceeds almost reversibly, with little loss of free energy because this process occurs through one-electron steps between the redox components of the chain immobilized in the membrane.

The ability of a substance to be reduced or oxidized depends on its electron affinity or ionization potential in the gas phase, respectively, on the entropy of the redox process in the gas phase and on the difference of the Gibbs energies of interaction of the reduced and of the oxidized form with the medium (solvation, complex formation, etc.) A typical example of the last-named process is the interaction of Fe(II) and Fe(III) ions with the porphyrin cycle of the heme.

In an octahedral arrangement the *d*-orbitals of Fe(II) or Fe(III) belong to two types of symmetry, t_{2g} and e_g, the former having the lowest energy. For each oxidation state two states exist, low-spin Fe(II) and Fe(III) and high-spin Fe(II) and Fe(III), the corresponding electronic configurations being t_{2g}^6 and $t_{2g}^4e_g^2$ for Fe(II) and t_{2g}^5 and $t_{2g}^3e_g^2$ for Fe(III). The effect of the ligand on electron distribution resulting in a change from high- to low-spin Fe(II) increases the formal potential. With porphyrin ligands both spin states for both oxidation states are possible while small changes in the substitution on the cycle as well as effects of axial ligands derived, for example, from the protein molecule, elicit such variations in the spin states that a considerable range of formal potentials can be covered.

The kinetics of electron-transfer reactions is a complex area of physical chemistry. In the simplest case the Born–Oppenheimer or adiabatic approximation is used where the nuclear motion of atoms (forming the molecules of the reactants and of the products) is taken to be independent of the motion of electrons forming the electron shells of these atoms. Even under this assumption the general treatment of the most simple reactions is extremely complicated and the discussion of the elementary step of a reaction is restricted to an analysis of the energy profile showing the dependence of the Gibbs energy of the reacting system on the estimated reaction pathway identified with the reaction coordinate.

In polar media the strong solvation interaction between ions and solvent dipoles, with an energy of the order of $10 - 10^2\,\mathrm{kJ\,mol^{-1}}$ must be considered for electron transfer between the reactants. Similarly, in the case of ions bound in coordination compounds, even larger interaction energies must be taken into account. Thus charge transfer is connected with a marked change in the solvation shell and the activation energy of this transfer depends on the momentary configuration of the inner solvation or coordination shell given by its thermal fluctuations.

In order to provide a rough picture of this process a simple redox reaction

$$\mathrm{A^{m+}} + \mathrm{B^{n+}} \underset{k_b}{\overset{k_f}{\rightleftharpoons}} \mathrm{A^{(m-1)+}} + \mathrm{B^{(n+1)+}} \tag{4.A16}$$

will be considered. The development of the theory of the rate of this reaction (i.e. formulation of a dependence between the rate constants k_f and k_b and the physical parameters of the system) for a general case is a hardly surmountable task. It would be necessary to take into account the vibrational spectrum of the solvation or coordination shell and the quantum-mechanical interactions between the reacting particles. When the problem is restricted to a single energy profile a schematic picture (Fig. 4.26) is obtained in the form of dependence of the potential energy of the reacting system on the reaction coordinate.

If, according to the Marcus theory, the system passes rapidly through the region of intersection of the potential energy curves for the initial state (R) and for the final state (P), there is insufficient time for the electron transfer, so that the system mostly passes through the intercept into the upper part of the R curve; the probability of electron transfer is low, characterizing *nonadiabatic electron transfer*. When the system passes through the intercept region slowly (i.e. with a slow change in the configuration of atomic nuclei), electron transfer almost always takes place. This is characteristic for *adiabatic electron transfer*, corresponding to the Born–Oppenheimer approximation (during their slow motion the atoms cannot 'see' the electron).

The slow process is termed *system reorganization*. According to Marcus, the following contributions are involved.

1. Changes in the energy and entropy of the internal structure of the species as a result of changes in the bond lengths and angles.

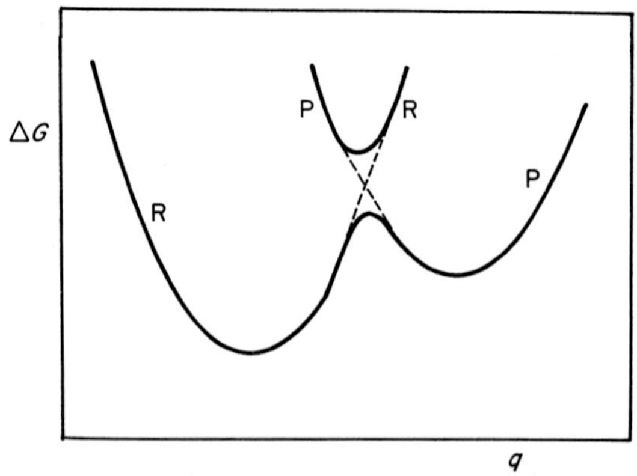

Fig. 4.26. Energy profile of the transition of the reactant R to the product P

2. Changes in the size of the reacting particle.
3. Changes in the vibrational and orientational polarization in the vicinity of the particle.

The applicability of the assumption of an adiabatic process depends on the vibration frequency of the slow system (i.e. of the solvation shell). The criterion is the quantity $kT/\hbar\omega$, with a value at room temperature of 40 THz. As the solvent vibration is characterized by $\omega \ll kT/\hbar$ and the vibration of the electron shell by $\omega \gg kT/\hbar$, the adiabatic character of the process can be assumed. Under these conditions, a further simplification is possible, i.e. the whole problem can be treated classically, where the vibrations of the solvation shell are considered as a set of classical harmonic oscillators.

In a considerably simplified way Marcus deduced the equation for the constant k_f(Eq. (4.A16))

$$k_f = \kappa Z \exp(-\Delta G^{\neq}/kT) \tag{4.A17}$$

where

$$\Delta G^{\neq} = w_r + \frac{\lambda}{4}(1 + \Delta G^{\ominus\prime}/\lambda)^2. \tag{4.A18}$$

In these equations κ describes the nonadiabaticity of the reaction ($\kappa \approx 1$ for an adiabatic reaction, $\kappa \ll 1$ for an nonadiabatic reaction), Z is the bimolecular collision frequency, λ is the reorganization energy and

$$\Delta G^{\ominus\prime} = \Delta G^{\ominus} + w_p - w_r. \tag{4.A19}$$

The quantities w_r and w_p denote the energy of approach and of separation of the reacting particle; $\Delta G^{\ominus}$ is the formal Gibbs energy change of the redox reaction.

The quantum effects are particularly important at low temperatures.

In membrane systems the reaction centres are often rather far apart. Thus, for example, the distance between the heme unit in cyt *c* and cyt *a* of cytochrome *c* oxidase is about 1.65 nm. Under these conditions the effect of distance is taken into account by multiplying the right-hand term in Eq. (4.A18) by the factor $\exp(\alpha r)$ where $\alpha = 0.11$ nm and r is the distance between the reaction centres (nm).

Proton transfers can be treated along similar lines. For a reaction of the type

$$\mathrm{AH + B} \underset{w_r}{\rightleftharpoons} \mathrm{AH\cdots B} \underset{\Delta G^{\neq}}{\rightleftharpoons} \mathrm{A\cdots HB} \underset{w_p}{\rightleftharpoons} \mathrm{A + HB}$$

an analogous equation can be deduced as Eq. (4.A17). The quantum effects are displayed in the coefficient κ.

Several attempts have been made to treat electron transfer in biological membrane systems in the same way as they are analysed in the band theory of solid state. This approach does not seem appropriate as in the membrane systems we have to do with individual molecules of charge-transferring systems while in the solid-state theory highly populated states are considered.

4.3.2 Free energy of oxidation converted into electrochemical potential gradients

The principal oxidizing entities in nature are mitochondrial inner membranes and bacterial membranes.

Terminal oxidation in mitochondria

Components of the electron transfer chain

It is the inner mitochondrial membrane that contains the full complement of enzymes catalyzing the flow of electrons from a hydrogen-rich substrate to oxygen with concomitant generation of a proton electrochemical potential gradient, as well as some auxiliary enzymes, such as NAD–NADPH trans-hydrogenase, and enzymes instrumental in the conversion of the proton gradient into ATP, such as the ATP-synthase and the ADP/ATP carrier plus a number of transport systems for cations and organic acids.

Unlike in the thylakoid membrane no rigid arrangement of the electron carriers has been discovered so that it is questionable whether there is a regular organization of the individual components or not—the nearly integral ratios of their representation might suggest a positive answer. The basic sequence of reactions includes four integral membrane complexes and several soluble factors, both proteins (cyt *c*) and smaller molecules (NAD, FAD, ubiquinone) (Fig. 4.27).

The compositions of and the partial reactions catalyzed by the individual complexes will now be described.

Complex I

The complex as a whole is NADH-ubiquinone reductase and contains 10–12 different subunits with molar masses ranging from 5000 to 75 000. The largest subunit can be identified with the NADH dehydrogenase proper; it carries a binding site for flavinadenine mononucleotide (in some yeast mitochondria it may be flavinadenine dinucleotide) and one of the four nonheme iron groups present ($Fe\text{–}S_1$).

The path of the electron through complex I is

$$\mathrm{NADH} \rightarrow \mathrm{FMN} \rightarrow \mathrm{Fe\text{–}S_1} \rightarrow \mathrm{Fe\text{–}S_4} \rightarrow \mathrm{Fe\text{–}S_3} \rightarrow \mathrm{Fe\text{–}S_2} \rightarrow \mathrm{CoQ}.$$

NADH is the principal electron donor to the system, hence a number of NAD-dehydrogenases supply the electrons as primary sources (3-hydroxy-

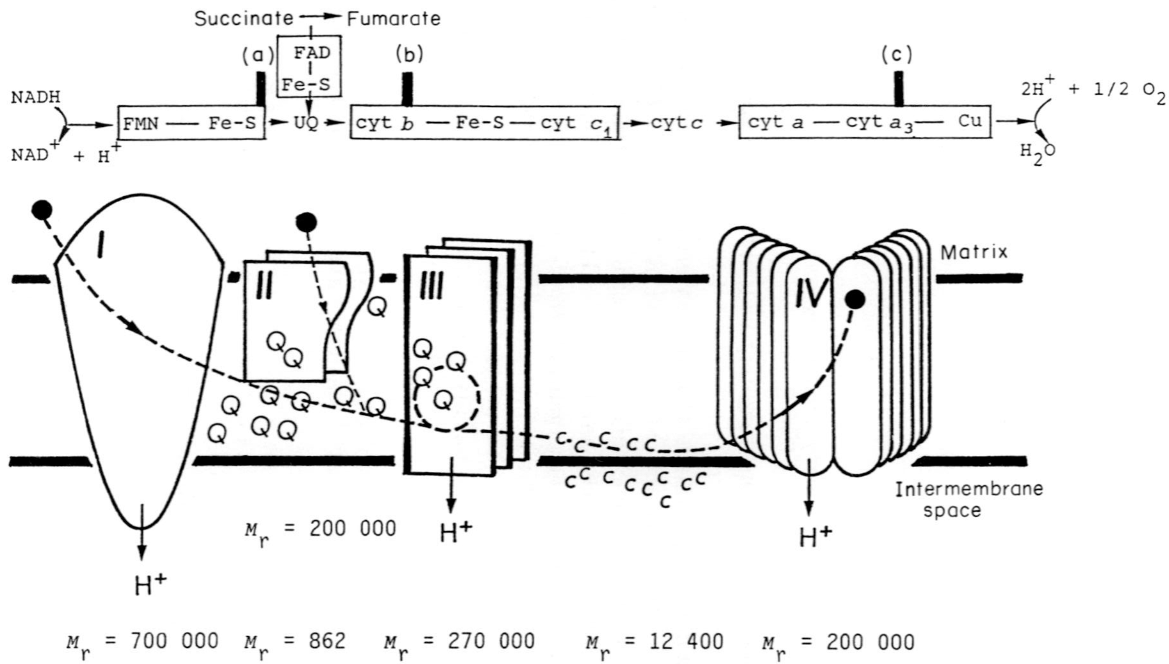

Fig. 4.27. Schematic of electron flow through the redox chain of inner mitochondrial membranes. The upper part shows the individual components of the four complexes (boxed) and the sites of inhibition: (a) rotenone, amobarbital, piericidin; (b) antimycin, mucidin; (c) cyanide, carbon monoxide. The lower part show the individual complexes I–IV as well as ubiquinone (Q) and cytochrome *c* (*c*), together with their relative molecular weights. The approximate stoichiometry of the various redox components related to complex I is indicated by the number of the corresponding geometric forms shown for each component

butyrate, glutamate, isocitrate, malate, pyruvate, 2-oxoglutarate, hydroxyacyl-CoA).

Complex I is the first site where H^+ ions are transported out of the matrix during electron transfer. Either two or three H^+ ions are assumed to cross at the expense of two electrons moving up the respiratory chain. Nothing at all is known about the molecular mechanism of this event.

Complex II

This enzyme, called succinate-ubiquinone reductase is a branch line of feeding electrons into the main chain. It consists of 7–8 nonidentical subunits of M_r ranging from 12 000 to 70 000 and contains FAD attached via a His residue to the apoenzyme, three nonheme iron centres and a molecule of cyt b_{558}. Two of the subunits (70 000 and 27 000) contain all that appears to be necessary for the oxidation reaction, viz. nonheme centres Fe–S_{s-1} and Fe–S_{s-2} and FAD on the 70 000 peptide, and a nonheme centre Fe–S_{s-3} on the 27 000 peptide.

Like with Complex I, some 20 per cent lipid by weight is present in the isolated Complex II.

The electron transport apparently proceeds as follows: succinate → Fe–S_{s-1} → Fe–S_{s-3} → CoQ. The Rieske centre Fe–S_{s-2} and the cytochrome *b* do not fit into this straightforward sequence. The question remains open whether $FADH_2$ which can serve as an independent substrate for this enzyme does not use the other active redox centres. Under normal conditions, the FAD dehydrogenases that can channel electrons into Complex II are also that of glycerol 3-phosphate and of fatty-acyl-CoA's.

No protons are probably transported across the membrane by Complex II.

Ubiquinone (*Coenzyme* Q)

This is a plentiful mobile carrier of charge and hydrogen that links the dehydrogenases and Complex III and can collect stray electrons and protons even at other sites of the electron-transfer chain. It is undoubtedly involved in the proton translocation performed by Complex III. The number of C atoms in the side chain varies: 50 in animal and plant mitochondria (Co Q_{10}); 35, 45 or 50 in algae (Co $Q_{7,9,10}$), 30, 35 or 45 in yeasts (Co $Q_{6,7,9}$).

Complex III

The enzymic activity involved is that of ubiquinol- cyt *c* reductase, the complex containing 11 different subunits (Table 4.4). In bacterial membranes only 4–5 subunits are present.

The path of the electrons is apparently as follows:

$$\tfrac{1}{2}QH_2 \rightarrow \text{cyt}\, b \rightarrow \text{Fe–S}_{\text{III}} \rightarrow \text{cyt}\, c_1 \rightarrow \text{cyt}\, c.$$

Apparently one of the principal functions of Complex III is to translocate protons

Table 4.4. Composition of mitochondrial complex III

Subunit	$M_r(\times 10^{-3})$	Characteristic
I	47.0	Core protein
II	45.0	Core protein
III	42.4	Carries cyt b_{566} and cyt b_{562}
IV	21.6	Iron-sulphur protein
V	28.0	Carries cyt c_1
VI	13.4	Ubiquinone-binding centre
VII	9.5	Binds to I and II
VIII	9.2	Hinge protein associated with cyt c_1
IX	8.0	DCCD-binding ubiquinone centre
X	7.2	Binds to V
XI	6.4	Binds to IV

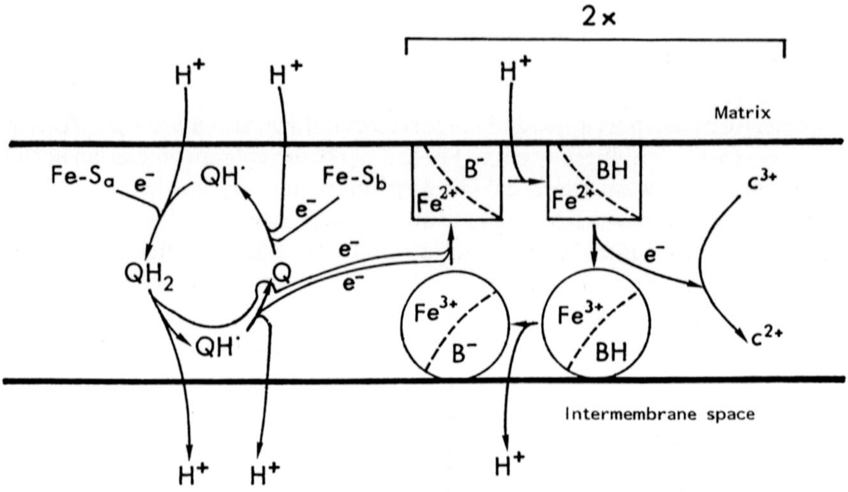

Fig. 4.28. A hypothetical scheme of a simple Q cycle ($2H^+$ transported per two electrons) combined with a vectorial Bohr effect mechanism where a cytochrome *b* is reduced, bringing about a conformational change of a protein channel B which can become protonated from the matrix side and, after reoxidation by cytochrome c_1, returns to the original conformation and releases H^+. As two such centres can be involved a total of $4H^+/2e^-$ can be transported. This version of the Q cycle bears a certain similarity to the revised cycle where different quinone molecules participate in two turnovers of the cycle, with the same overall stoichiometry

out of mitochondria. One can envisage one of the larger subunits to act as a channel across the membrane in which a Q cycle operates. In view of the well established conformational changes of the protein during passage of electrons and of the relatively slow transverse movement of CoQ in the membrane it is attractive to think of this in combination with a vectorial Bohr effect. In this, the conformational change of the protein causes a shift of the p*K* value of at least one functional group of an amino acid residue lining the channel so as to facilitate binding of H^+ from one side, and relaxation of the conformation releases the H^+ ion to the other side. Although direct evidence is still unavailable this conformational change oriented in the membrane (a vectorial Bohr effect) can be envisaged to operate also in the transport of H^+ by Complex I (Fig. 4.28).

Cytochrome c

This cytochrome is a peripheral protein which can move within a certain domain in the intermembrane space of mitochondria and transfer electrons from cytochrome c_1 of Complex III to cytochrome *a* of Complex IV. Cytochrome *c* can be reduced by ascorbate (via phenazine methosulfate or tetramethylphenylene diamine) which thus acts as a substrate for the last segment of the electron-transport chain.

Complex IV

This complex is probably the most thoroughly analyzed part of the mitochondrial electron transport chain, correctly called cytochrome *c* oxidase. It is composed of a number of different subunits, the exact number and molecular weight depending on its source (Table 4.5).

Subunit I carries two hemes cyt *a* and cyt a_3, subunit II carries two Cu atoms (Cu_A and Cu_B) and subunit III probably functions as the H^+ channel. The first three subunits are synthesized in the mitochondria and they have a catalytic activity in both oxygen reduction and H^+ transport, the remaining subunits are synthesized in the cytosol and their function may be regulatory.

Table 4.5. Composition of mitochondrial and bacterial complex IV

	M_r		
Subunit	Bovine heart	Yeast	*Thermus thermophilus*
I	57 000	57 000	52 000
II	20 000	20 000	29 000
III	30 000	30 000	37 000
IVa	17 000	17 000	—
IVb	—	16 000	—
Va	12 500	10 500	—
Vb	10 500	—	—
VIa	9 500	—	—
VIb	10 000	—	—
VIc	8 500	—	—
VIIa	6 500	6 500	—
VIIb	6 000	—	—
VIIc	5 500	—	—
VIII	5 000	—	—

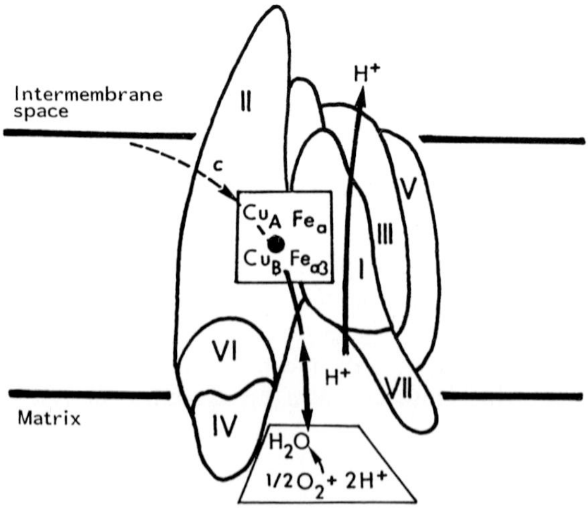

Fig. 4.29. A putative disposition of cytochrome *c* oxidase in the inner mitochondrial membrane. The oxygen reduction centre is boxed

An interesting sequence homology was found between subunit II (the Cu-containing one) and two other copper-containing proteins, viz. azurin of *Pseudomonas fluorescens* and plastocyanin, a component of the electron transport chain of chloroplasts. Particularly interesting is the conservation of two histidines (plus a methionine and cysteine) which are known to bind copper in proteins.

The organization of the enzyme in the membrane is shown in Fig. 4.29, while it is probable that it exists mainly in the form of dimers. The conformation of the reaction centre is still subject to debate. One of the possibilities, together with the path of the electrons, is shown in Fig. 4.30.

Besides the four protons taken up in the reduction of dioxygen (a $2H^+/2e^-$ process) there is an additional channel through cytochrome *c* oxidase, very likely in connection with subunit III, which permits two protons to flow out of the matrix per every two electrons transferred. The subunit has now been defined as an integral protein spanning the membrane seven times and it is assumed to operate by a vectorial Bohr effect mechanism, in analogy with the other two sites at which protons are translocated (Complexes I and III).

In the vectorial Bohr effect it is assumed that at least one of the functional groups involved in H^+ transport changes its pK value during the cycle. However, this need not be the case if certain conditions in a redox system are met.

Let us assume that a two-site reaction centre can exist either as $A^\circ B^+$ or A^+B° (as well A^+B^+ and $A^\circ B^\circ$ but these are assumed to be in rapid equilibrium with the first two). The protein that carries the centre can undergo a conformational change $E_\alpha \rightleftharpoons E_\beta$ such that $E_\alpha A^\circ B^+$ and $E_\beta A^+B^\circ$ forms only exist. Moreover, $E_\alpha A^\circ B^+$ interacts only with

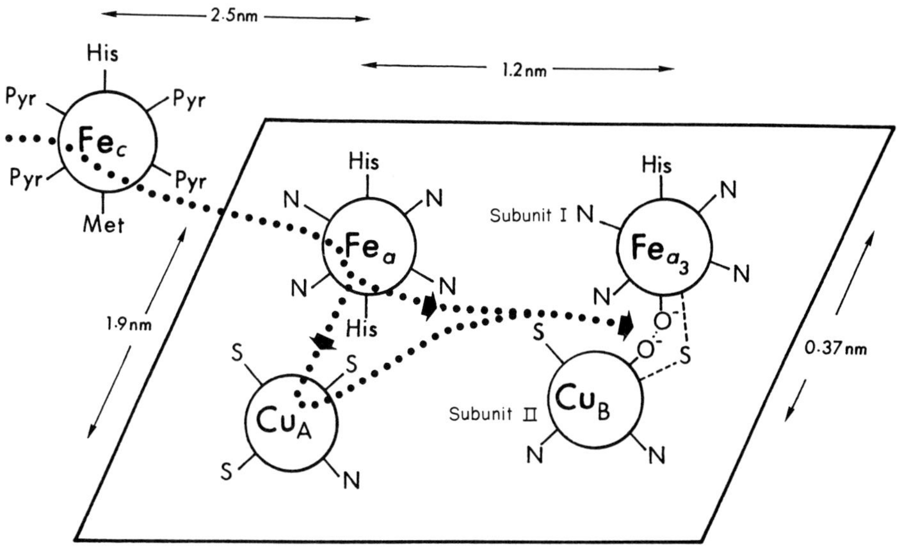

Fig. 4.30. Functional organization of the active centre of cytochrome *c* oxidase. The path of the electron(s) from cytochrome *c* to oxygen is shown by the heavy dotted lines. Both the cyt *a* heme and the Cu_A centre are reduced by (two) cytochrome *c* molecules before they simultaneously reduce the cytochrome a_3 heme and Cu_B. At this stage, a dioxygen molecule interposes itself as a peroxide link between a_3 and Cu_B as shown in the figure. This is preceded by breaking an –S– bond (shown dashed), the S atom then attaching itself to a part of the apoprotein. Two further electrons must arrive via cyt *a* Cu_A to free the oxygen atoms and let them combine with a total of four protons. The exact path of the four protons to the active centre is not known with certainty. Written schematically, the path may be represented as follows

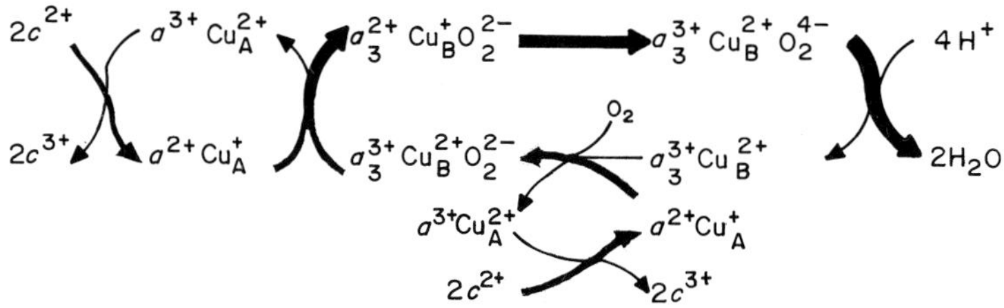

proton at side 1 and $E_\beta A^\circ B^+$ only with proton at side 2. Schematically,

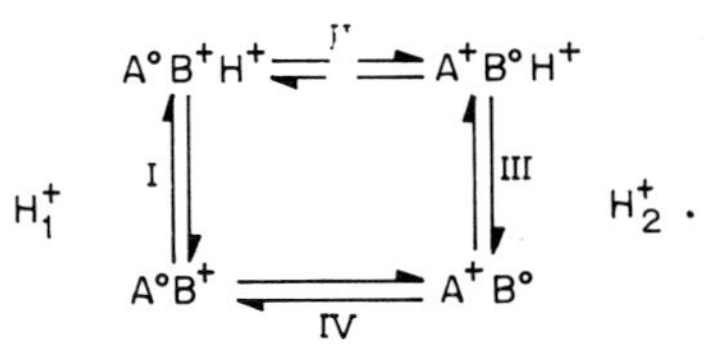

(4.I)

Reactions II and IV are pH-independent but the electron transfer is characterized by a much lower energy barrier for II than for IV. If the difference is substantial (say, 30 kJ mol^{-1}) reaction IV becomes unimportant. Now if the pK of the critical (i.e., potentially proton-translocating group is 5 while $pH_1 = 7$ then $A^\circ B^+ H^+ / A^\circ B^+ = 0.01$.

If a large driving energy is available (such as may be the case in the redox exchange involved) the reaction will proceed to the right but only by picking up a proton at the left and conveying it to the right. If at the target side pH_2 is 6, 90 per cent of the H^+ will be released as $A^+B^\circ H^+/A^+B^\circ = 0.1$. If the left-to-right rate of II is 100 s^{-1} only 10 $H^+ s^{-1}$ will go back so that an effective pH gradient can be established without a change of pK of the critical group. (The turnover number ranges from 40 to 300 H^+ ions per second for different cytochrome *c* oxidases.)

This approach is in fact equivalent to the kinetic approach that leads to thermodynamic results, viz. $J'_{max}K''_T > J''_{max}K'_T$ will result in accumulation because $K_{eq} > 1$ (p. 203).

Summary of reactions

It is now possible to summarize the operation of the electron transfer chain from two aspects, the electron transfer proper and the concomitant translocation of protons.

The path of electrons

The electrons, two of them from NADH to ubiquinone, one only from cytochrome *b* to the terminal oxidation step, move down the ladder of reductants (Fig. 4.31) until they reach O_2. It is worth observing in this context that the movement of the electrons in anoxygenic photosynthetic systems and in the segment between plastoquinone and plastocyanin in oxygenic systems follow a parallel downward path. The difference is in the fact that substrate dehydrogenation producing NADH or an equivalent generates the reducing power in mitochondria whereas absorption of light does that in photosynthetic organisms.

The formal or midpoint potentials shown in Fig. 4.31 reflect the intrinsic reducing power of each of the couples (in fact, they should have been written as succinate/fumarate, cyt c_1^{2+}/cyt c_1^{3+}, etc.) but as the concentrations of various components differ and as they are apparently in a pseudo-equilibrium at various sites, it was possible to determine four levels of equal potential in the electron-transfer chain at which the interacting partners all have the same potential E, regardless of their tabulated $E^\ominus$. These levels are shown by heavy dashed lines in Fig. 4.31.

The paths of protons

The number of H^+ ions transported out of mitochondrial matrix per electron pair wandering from NADH to O_2 is still a matter of controversy. The following table is representative of current thinking.

	$H^+/2e^-$
Complex I	3
Complex III	4
Complex IV	2

The picture would not be complete without considering another proton-translocating enzyme of the inner membrane, the NAD(P)$^+$ transhydrogenase,

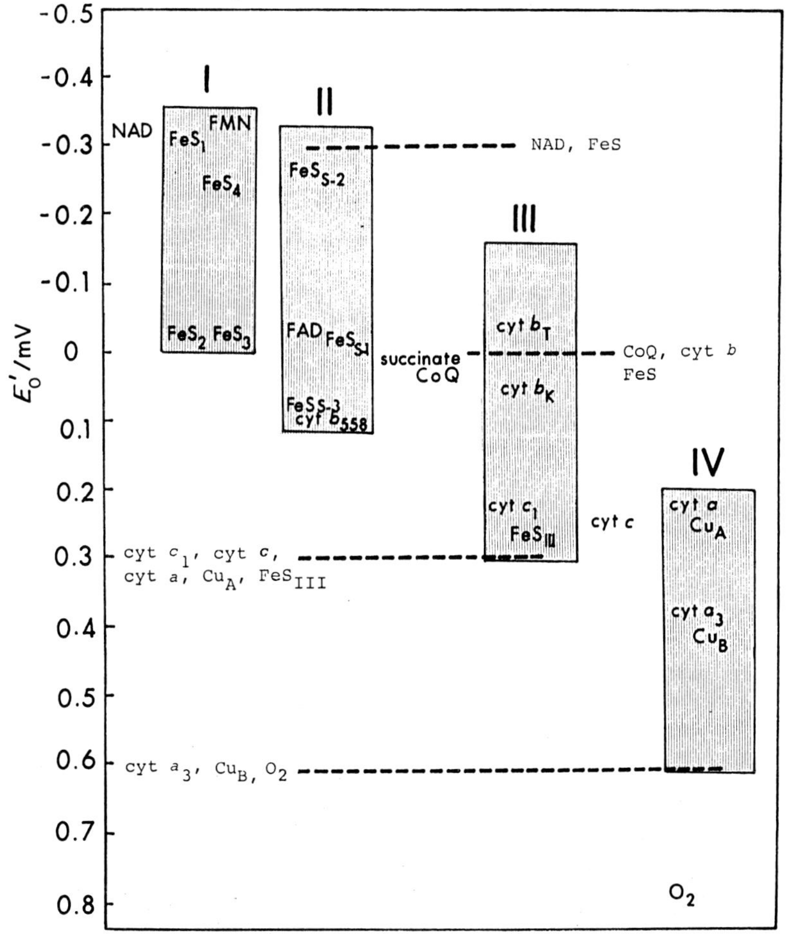

Fig. 4.31. Components of the electron transport chain of mitochondria arranged according to their midpoint potentials (E_0'). Four isopotential levels at which various components apparently occur are shown in heavy dashed lines

catalyzing the following reaction.

$$\text{NADH} + \text{NADP}^+ + 2\text{H}^+_{\text{out}} \rightleftharpoons \text{NAD}^+ + \text{NADPH} + 2\text{H}^+_{\text{in}}$$

Its equilibrium constant is approximately one but either hydrolysis of ATP or respiration-linked generation of a proton gradient cause the equilibrium to be shifted far to the right. The enzyme is now known to be located on the matrix side of the inner mitochondrial membrane and to translocate protons at a ratio of $2H^+/2e^-$ out of mitochondria as NAD^+ is reduced. The protons are not those of NADPH but scalar protons taken up at the matrix side and released in the intermembrane space, possibly again by a vectorial Bohr effect.

The enzyme has a molar mass of 97 000 g/mole and is formed by a single polypeptide.

One more reaction should be mentioned in the context of proton translocations, viz. the synthesis and breakdown of inorganic diphosphate (pyrophosphate). This, the simplest of a series of linear and cyclic polyphosphates, is synthesized in mitochondria and particularly in purple bacteria at rates comparable with those of ATP synthesis. It can be apparently hydrolyzed with concomitant movement of protons out of cells if a suitable diphosphatase (3.6.1.1) is present in the membrane. It thus serves as an alternative to ATP in all its membrane functions. Athough its role in mitochondria is clearly less important than in purple bacteria its very existence should serve as a corrective in quantitative considerations of ATP synthesis vs dissipation.

The protonmotive force

All the movements of protons out of the matrix, no matter how they are achieved, are electrogenic in nature. They carry a charge, which is only incompletely, and with a delay, compensated either by an opposite flow of cations (mainly K^+) or by a parallel movement of anions (mainly Cl^- and HCO_3^-). This gives rise to a membrane potential which ranges from 80 to 200 mV (negative inside the mitochondria). The size of the potential is in mirror relationship to the ΔpH across the membrane in such a way that the total protonmotive force* (the difference in electrochemical potential of H^+ divided by the Faraday constant)

$$\text{pmf} = \Delta\phi - 2.3RT\Delta\text{pH} \tag{4.2}$$

remains nearly constant at about 230 mV.

This value is relatively close to that predicted on the basis of the energy available from oxidation. Oxidation of NADH to O_2 ($\Delta G = \Delta E_m nF$) yields free energy equivalent to $1.14\,\text{V}\cdot 2\cdot 96.5\,\text{kC mol}^{-1} = 220\,\text{kJ mol}^{-1}$. If it were fully converted into the electrochemical potential difference of protons $\Delta\tilde{\mu}_{H^+}$, the value of pmf should be 2.28 V for one H^+ transported but about 250 mV for the assumed nine protons translocated for every pair of electrons.

All the enzyme reactions but the reduction of O_2 are reversible so that a high proton gradient arising, say, from the oxidation of succinate or ascorbate, or hydrolysis of ATP, can reverse the flow of electrons so that NAD is reduced to NADH.

At a different level, the proton gradient can be dissipated by the so-called uncouplers of oxidative phosphorylation, better termed proton conductors. These substances, which include various nitrophenols, salicylanilides, pentachlorophenol, a group of phenylhydrazinomalononitriles (usually called carbonylcyanide phenylhydrazones), dicoumarol, and small or medium-size fatty acids, are distinguished by the fact that they can cross membranes by simple diffusion both in protonated form and as anions, thus C_6Cl_5OH as well as

* The protonmotive force is an analogy to the electromotive force of traditional usage in electrochemistry and, like the latter, does not have dimensions of a force but rather is expressed in volts. Its use has become popular although expressing the energy of a system in terms of the electrochemical potential difference of protons is more appropriate.

$C_6Cl_5O^-$. In this manner, they can efficiently short-circuit the H^+ gradient as well as the membrane potential.

It is perhaps somewhat curious that inner mitochondrial membranes, besides being permeable to the proton conductors, contain a rather specific and high-affinity binding site for them, in contrast with such 'neutral' membranes as those of erythrocytes. Thus, there may be an additional function for the uncouplers to be discovered.

Plant mitochondria

Some differences exist in plant mitochondria and in those of eukaryotic micro-organisms, such as slight shifts in the absorption maxima of the cytochromes, or occurrence of different terminal oxidases that resemble those of bacteria.

Some yeasts can oxidize lactate directly in the inner mitochondrial membrane by an analogue of succinate dehydrogenase; mitochondria of the so-called C_4 plants (with CO_2 fixation different from the Calvin cycle) can oxidize glycine via a NADH-linked dehydrogenase.

The two major distinctions of yeast and plant mitochondria are the following.

1. Two NADH dehydrogenases are operative, one resembling the matrix-oriented enzyme of animal mitochondria, the other oriented to the intermembrane space so that it can use cytosolic NADH directly, without the necessity of shuttling the reducing equivalents from the cytosol to the mitochondria.

2. Starting with the ubiquinone pool, electrons can be channelled into alternative, generally cyanide-insensitive, pathways ending with oxygen but generating no proton gradient. The route is blocked by salicylhydroxamate and its function is a matter of discussion.

Oxidations in bacteria

In full agreement with the endosymbiont theory of mitochondrial origin, most of the enzyme systems of mitochondria (and many additional ones) are found in aerobic bacteria. However, if oxidation is considered from the teleonomic point of view as a process primarily giving rise to a proton electrochemical gradient (and via this gradient to ATP), even anaerobic bacteria should be included in a treatment of bacterial electron transfer chains and this is where our discussion will begin.

Electron transport under anaerobic conditions

The following terminal oxidized acceptors of electrons can serve in different microorganisms in the electron movement through membrane-located oxido-reductases.

1. Nitrate is the 'substrate' of the so-called nitrate respiration. Nitrate reductase A (a Mo-containing enzyme) plus nitrite reductase provide for electron flow coupled with some H^+ ejection in bacteria that can be induced to use nitrate as terminal acceptor of electrons, such as *Paracoccus denitrificans.* If acetate is the reduced substrate, nitrate can be eventually reduced all the way to N_2.

2. Sulphate is the terminal acceptor in the so-called sulphate respiration. With formate or malate as the reduced substrate, sulphate can be reduced in a complicated way, involving the initial formation of adenosine 5′-phospho-

sulphate, via sulphite (and possibly via trithionate and thiosulphate) all the way to sulphide. Cytochromes of the *b* or *c* type are involved. If dihydrogen is the reduced substrate a protonmotive force amounting to 200 mV can be established so that (by analogy with mitochondria) 3 ATP molecules can be synthesized per sulphate reduced, of which at least one is used for the activation of sulphate. Examples of sulphate-reducing bacteria are *Desulfovibrio*, *Desulfobacter* and *Desulfococcus*.

3. Elemental sulphur can serve as the terminal electron acceptor for oxidation of acetate or ethanol in *Desulfuromonas*, being reduced to hydrogen sulphide.

4. An important group of microorganisms, belonging to the archaebacteria, such as *Methanobacterium*, *Methanococcus*, *Methanosarcina*, and *Methanospirillum*, can use CO_2 and reduce it with H_2 to obtain all of their compounds for anabolic pathways of carbon as well as for the production of ATP. The end product of CO_2 reduction is methane.

The reduction proceeds with the participation of some unusual coenzymes, such as mercaptoethane sulphonate, a deazoriboflavine, a pterin derivative, an aminomethylfuran, and a nickel-containing tetrapyrrole where CO can also take part, to a methyl derivative which is then reduced to CH_4, giving rise to a proton gradient. The reduction process is exergonic with $\Delta G = -113\,\text{kJ mol}^{-1}$. While in *Methanosarcina* $\Delta\tilde{\mu}_{H^+}$ is formed during CO_2 reduction and used to synthesize ATP, in *Methanobacterium* ATP is reported to be synthesized directly by the electron chains located on intracytoplasmic membranes.

Some archaebacteria can produce methane by reducing acetate via CO using an unknown reaction sequence with $\Delta G = -32\,\text{kJ mol}^{-1}$ or methanol in two ways: by a disproportionation with water to $CH_3 + CO_2$($\Delta G = -107\,\text{kJ mol}^{-1}$) or by reduction with hydrogen ($\Delta G = -113\,\text{kJ mol}^{-1}$).

5. Some acetogenic bacteria, such as *Clostridium aceticum* or *Acetobacterium woodii*, can reduce CO_2 in a way similar partly to the methanogenic bacteria. It is not clear here where the H^+ extrusion takes place.

6. An interesting case of anaerobic electron transport is that of the fumarate respiration where fumarate is reduced to succinate at the expense of oxidation of such electron donors as glycerol 3-phosphate, NADH, lactate, formate, dihydroorotate and even H_2. It occurs not only in the strictly anaerobic *Bacteroides* but also in facultatively anaerobic *Escherichia*, *Klebsiella*, *Proteus* and *Salmonella*, and even in the aerobic *Propionibacterium*, as well as in parasitic worms that live under practically anaerobic conditions. The reaction in the membrane proceeds according to the outline in Fig. 4.32, from which it follows that a net extrusion of $3H^+$ per $2e^-$ is effected.

The enzyme itself is composed of four subunits (M_r 69 000, 27 000, 15 000 and 13 000). The first two are extrinsic, the last two intrinsic membrane proteins. It carries a FAD coenzyme on subunit I which reacts with a single one (Cys-247) of the 21 cysteine residues present in the peripheral subunits. It is covalently bound to a histidine residue of the apoprotein. Subunit II carries three Fe–S centres with 2, 2 and 4 iron atoms per centre.

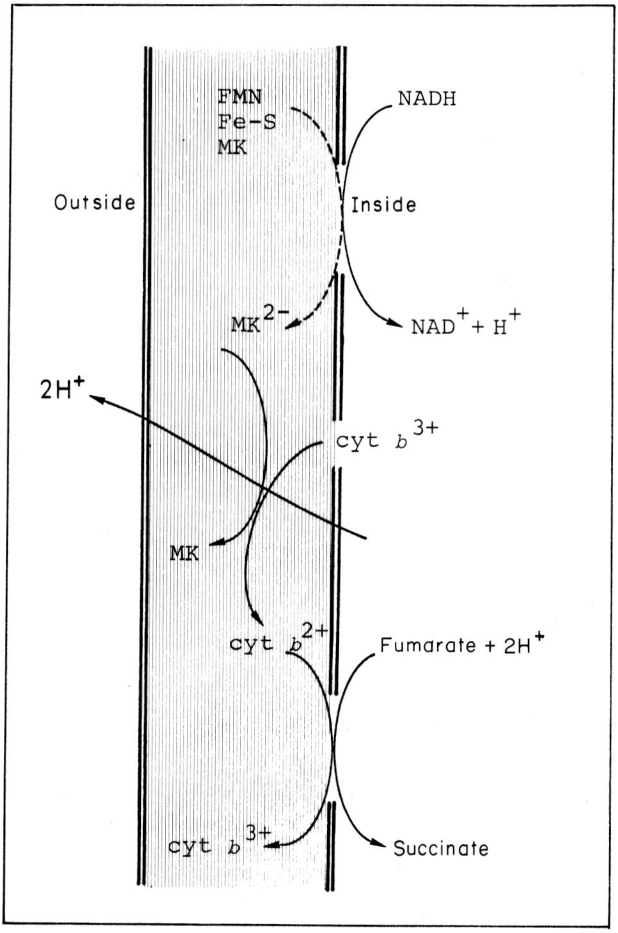

Fig. 4.32. Oxidoreduction processes catalyzed by fumarate reductase. MK—menaquinone. The K_m of fumarate is 0.42 mM, that of succinate 2 mM

It is striking that the apoprotein sequence in the vicinity of these centres is highly conserved in all enzymes containing such centres, ranging from ferredoxin to succinate dehydrogenase of mitochondrial Complex II.

Electron transport under aerobic conditions

A great many bacteria use oxygen as the terminal electron acceptor for oxidative sequences starting with a variety of reduced substrates. In addition to the 'classical' organic substrate D-glucose and other sugars, they include organic acids, sugar alcohols, amino acids and, occasionally, other nitrogen-containing compounds. However, inorganic substrates represent the starting substances of these chains. In contrast to the systems mentioned in the preceding section, oxygen is used here to produce some of the compounds that were used as terminal electron acceptors there.

1. Ammonium ions are oxidized by a sequence including NH_2OH and [NOH] to nitrite by *Nitrosomonas* and *Nitrosococcus*, further from nitrite to nitrate by *Nitrobacter* and *Nitrococcus*.

2. Sulphide, elemental sulphur and thiosulphate are oxidized to sulphate by *Thiobacillus*, *Thiomicrospira* and *Sulfolobus*, via active sulphite intermediates.

3. Various bacteria oxidize metal ions. Thus, *Thiobacillus ferrooxidans*, *Sulfolobus acidocaldarius*, *Gallionella ferruginea* and *Leptothrix ochracea* oxidize preferentially Fe^{2+} to Fe^{3+} but *L. discophorus* manages to oxidize Mn^{2+} to Mn^{4+} and *Stibiobacter senarmontii* Sb^{3+} to Sb^{5+}. The oxidase is a component of the electron transport chain and is probably coupled with a proton-transporting process although it is not easy to see how and why protons could be extruded by a bacterium living at pH of 2, as the *Thiobacillus* and *Sulfolobus* do.

4. A number of bacteria of many groups, such as *Alcaligenes*, *Pseudomonas*, *Aquaspirillum*, *Paracoccus*, *Rhizobium*, *Nocardia*, *Mycobacterium* and some *Bacillus* species, can oxidize molecular hydrogen via a membrane-bound hydrogenase, the process giving rise to NADH which is then oxidized in the usual way by an electron-transport chain. The balance equation here is

$$6H_2 + 2O_2 + CO_2 \rightarrow [CH_2O] + 5H_2O.$$

Some bacteria, e.g. *Pseudomonas carboxidovorans*, can oxidize even carbon monoxide according to

$$14CO + 5O_2 + 2H_2O \rightarrow 2[CH_2O] + 12CO_2.$$

Components of the bacterial electron-transport chain

While the flavins and Fe–S centres of bacteria appear to be identical with those of mitochondria an enormous variety exists among the cytochromes (Table 4.6). Also quinones can occur in several varieties, thus (shown in oxidized form):

Table 4.6. Selection of cytochromes occurring in bacteria

Organism	Cytochromes *b*		Cytochromes *c*		Terminal oxidases
Escherichia coli	b_{556}	b_{562}			*o*
Klebsiella pneumoniae	b_{559}	b_{563}			*o*
Azotobacter vinelandii	b_{560}		c_{551}	c_{555}	o, a_1, d
Beneckea natriegens	b_{557}	b_{562}	c_{547}	c_{554}	o, a_1, d
Micrococcus luteus	b_{557}	b_{562}	c_{549}	c_{552}	a, a_3
Paracoccus denitrificans	b_{556}	b_{562}	c_{546}	c_{550}	o, a, a_3
Mycobacterium phlei	b_{559}	b_{563}	c_{548}	c_{554}	o, a, a_3

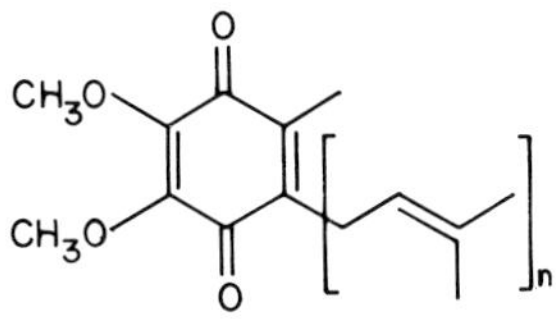

Ubiquinone UQ

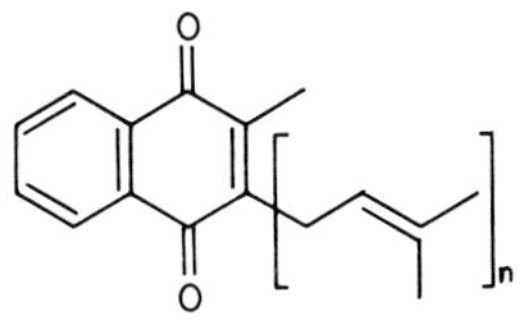

Menaquinone (vitamin K_2) MK

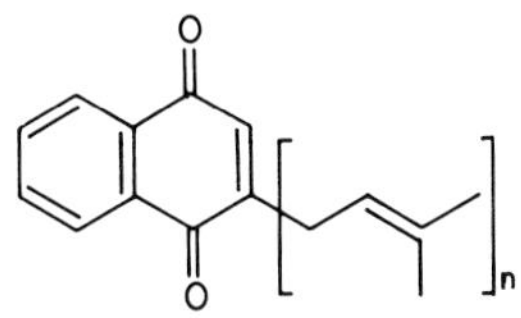

2-Demethylmenaquinone DMK

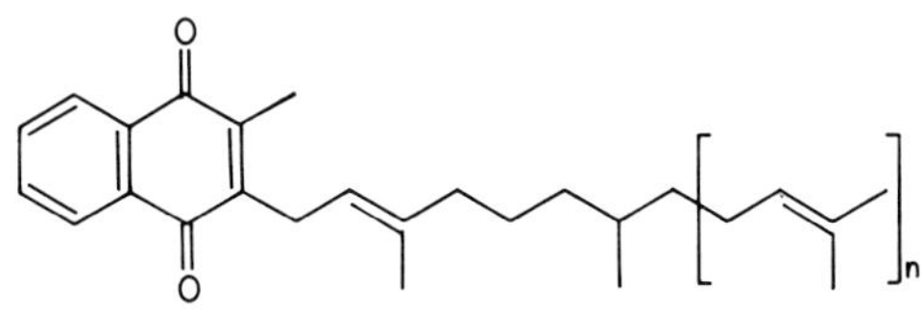

Dihydromenaquinone. MKH_2

Catabolism in 'Na-motive' bacteria

Among the thousands of bacteria several genera were found in which the electrochemical potential gradient formed by catabolic reactions was not that of protons but of sodium ions. Two types of such reactions are now known.

1. The first type is in fact no oxidation but a decarboxylation of organic acids on their way to be oxidized—this is an exergonic reaction with a ΔG of $-30\,\text{kJ}\,\text{mol}^{-1}$. The release of CO_2 is obligatorily coupled to a translocation of probably two Na^+ ions out of cells. So far the following systems have been discovered.

(1) Oxaloacetate → pyruvate	*Klebsiella aerogenes* *Salmonella typhimurium*
(2) Glutaconyl-CoA → crotonyl-CoA	*Acidaminococcus fermentans*, *Peptococcus aerogenes*, *Clostridium symbiosum*
(3) Methylmalonyl-CoA → propionyl-CoA	*Veilonella alcalensis* *Propionigenum modestum*

Reaction (1) is catalyzed by an enzyme of three subunits (M_r 65 000, 34 000, 12 000), reaction (2) is catalyzed by a similar enzyme but with polypeptide chains of M_r 120 000, 30 000 and 30 000. All three enzymes contain biotin bound to the α subunit which is reversibly carboxylated during the transport process while the β subunit in (1) and the γ subunit in (2) appear to act as Na^+ ion channels.

In these organisms apparently Na^+ is the primary driving ion for various energy-requiring translocations.

2. The second type of Na^+-motive reaction is found in the marine alkali-tolerant *Vibrio alginolyticus* and the moderately halophilic *V. costicola* where Na^+ ions are extruded from the cells in the course of oxidation of NADH to menaquinone, generating a membrane potential plus a ΔpNa—this process is completely stopped by cyanide, 2-heptyl-4-hydroxyquinoline-N-oxide and by gramicidin A. The enzyme involved in this extrusion is a NADH-quinone reductase composed of three subunits (M_r 52 000, 46 000 and 32 000) and containing FMN at the α subunit and FAD at the β subunit. Apparently, the α subunit also acts as a Na^+ channel. In full analogy to the H^+-extruding species a membrane potential is generated by the outward movement of Na^+.

The important conclusion one can draw from this is that the Q-loop

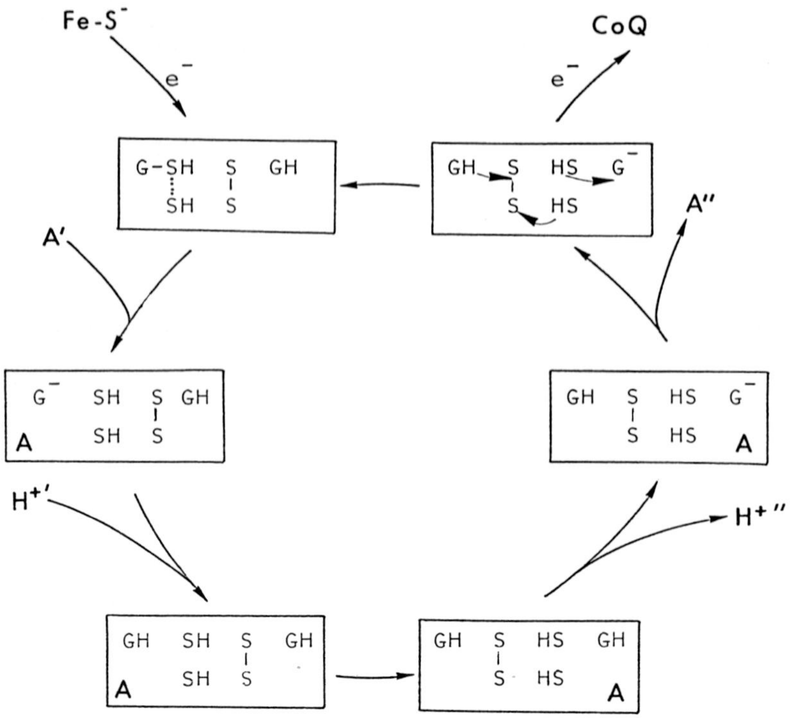

Fig. 4.33. Hypothetical mechanism of proton-driven transport involving reduction of essential S–S bonds in the carrier

mechanisms discussed in connection with H^+ transfer in chloroplasts and mitochondria cannot operate with Na^+. There may be a proton extrusion by the Q cycle even in the alkali-tolerant bacterium but there is no way of attaching two Na^+ ions to a quinone under physiological conditions. Then again it is difficult to envisage a vectorial Bohr effect as feasible mechanism as it also involves association-dissociation of a proton with a functional group in the channel-like path across the membrane. Would it be possible that an ion-specific channel of the type of gramicidin is so gated that it spends more time exposed to the outside than to the inside of a cell? This would be a thermodynamically feasible event to achieve values of $\bar{c}'_{Na}/\bar{c}''_{Na} > 1$.

Redox reactions in secondary active transport

The whole foregoing discussion was meant to convey that the free energy of oxidation can be used to move H^+ out of mitochondria and chloroplasts, and H^+ or Na^+ out of bacteria, with a concomitant generation of a membrane potential, negative inside. Now oxidation and reduction reactions can also energize transports that are apparently of the character of proton symport, not by providing the protons but directly by providing electrons to reduce the transport system's –S–S– bonds and thus alter the conformation of the carrier, with the usual result of changing its affinity for solute (Fig. 4.33). Such systems are known to operate in purple photosynthetic bacteria and in *Escherichia coli* where both the electron-releasing reactions and the transport of solute are located in the same membrane. However, as redox systems (NADH ferricyanide reductase, NADPH ferricyanide reductase, NADH cytochrome b_5 reductase, etc.) are known to occur even in eukaryotic plasma membranes, involvement in solute uptake might be a justification of their existence there.

4.3.3 Free energy of oxidation converted into electromagnetic radiation

Generation of visible light or bioluminescence appears to be the property of a variety of organisms, ranging from deep-sea bacteria, to fungi, radiolarians, jellyfish, marine sponges, mollusks (snails, bivalves, and squid), crustaceans, millipedes, insects, annelids, to some fish species (although much of the luminescence observed there is due to symbiotic bacteria) (Fig. 4.34).

The variety of groups of organisms capable of producing light is almost matched by the variety of molecules and their transformations instrumental in the process. Only some of the proteins involved are associated with membranes and, in most cases, the presence of membranes is not essential for the function of luminescent systems but the several cases where membranes are involved (especially the one involving polynoidin and bacterial luminescence in general) perhaps warrant a more general treatment of the subject within the scope of the chapter on the flow of energy.

All the mechanisms known at present involve an oxidation, either by molecular oxygen directly or by an oxygen-rich compound of peroxy nature, of a cell component that has been given the generic name luciferin. The reaction is in most cases (but not always, e.g. with aequorin) catalyzed by an enzyme of the oxidoreductase class, given the common name luciferase (Table 4.7). If L is

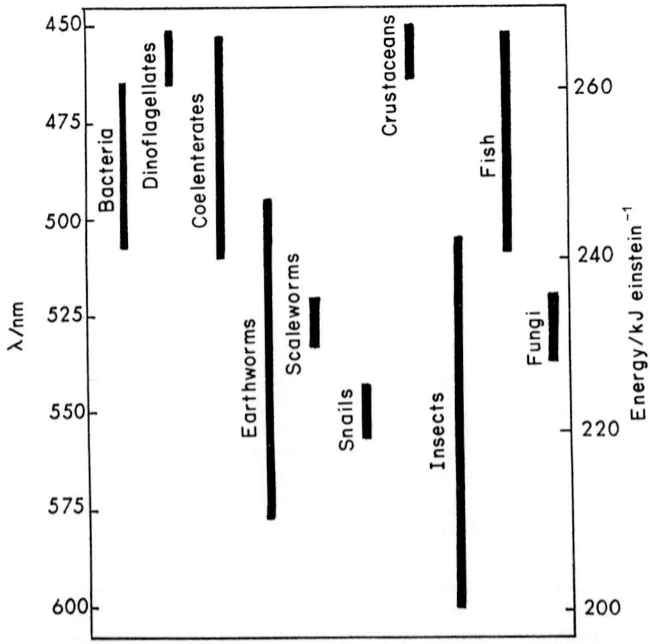

Fig. 4.34. Luminescence maxima (nm) and the corresponding energy of emitted light in kJ per 'mole' of photons. Except for the few fish species (possibly showing secondary bacterial fluorescence) all the animals belong to the Protostomia

luciferin, the reaction may be written in an abbreviated form as follows:

$$L_{red} + [O] \xrightarrow{\text{luciferase}} L^* \rightarrow L_{ox} + h\nu.$$

The excited state of luciferin (or the corresponding chromophore) L* is believed to be due to a singlet oxygen of one of the intermediates directly preceding the light emission which accompanies the return of the oxygen to its ground state. The excited state can be reached in two principal ways: 1. by dissociative electron transfer, involving the formation of radical ions; 2. by opening of highly strained intermediates of dioxetanone type, with concomitant release of CO_2.

Most of the systems would conform with the ring-splitting mechanism which can be exemplified on the *Cypridina* system, with the understanding that *Renilla, Aequorea and Watasemia* would behave identically.

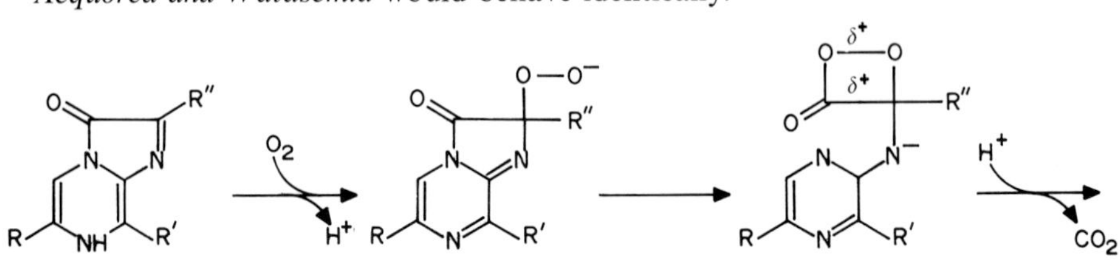

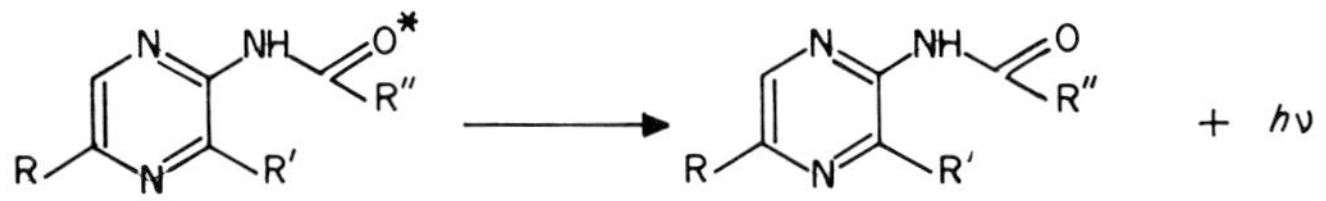

(I)

This mechanism is supported by observations of auto-oxidation of acridine esters which follows the same route.

A basically similar but somewhat more complicated system is the one of the firefly *Photinus* and apparently all other insect species.

The luciferase here catalyzes two different reactions, one involving ATP, the other oxygen (the figures in parentheses under the various intermediates show their enthalpy in kJ/mol).

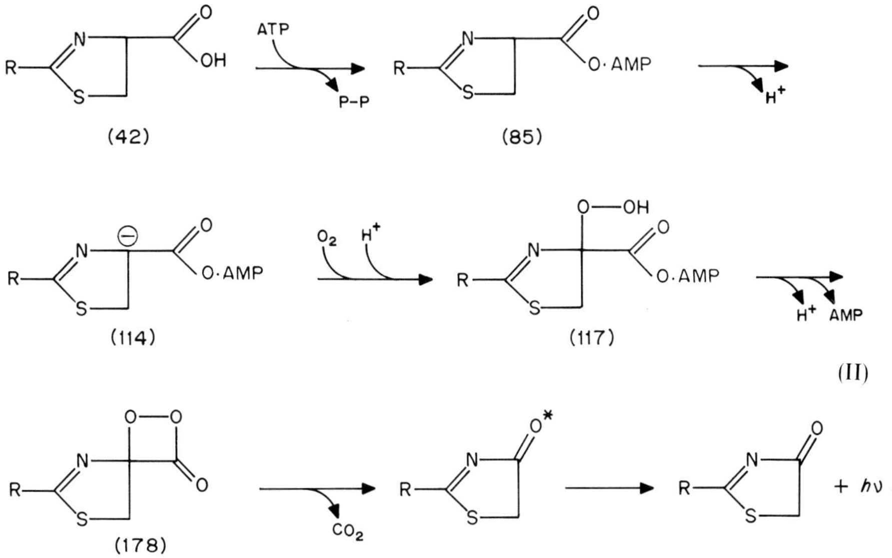

(II)

An alternative sequence of steps, involving an electron transfer mechanism, the so-called CIEEL (for *c*hemically *i*nitiated *e*lectron-*e*xchange *l*uminescence), proceeds from the dioxetanone intermediate as follows:

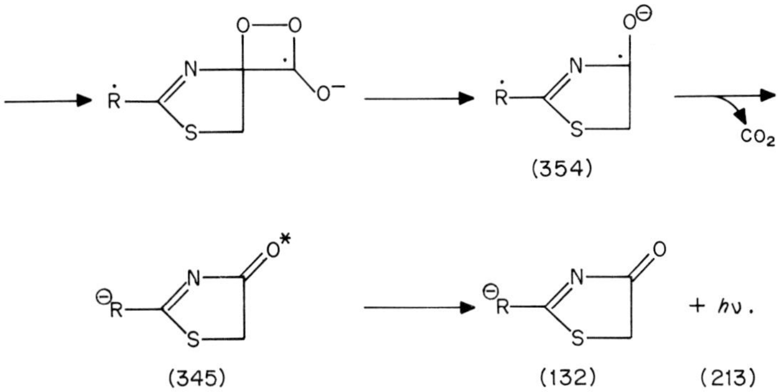

(III)

Table 4.7. Summary of data on bioluminescent systems

Taxonomical class (typical genera)	Luciferin (oxyluciferin, where known)	Luciferase (M_r)	λ_{max}	Additional factors	Note
Bacteriophyceae (*Photobacterium* *Beneckia* *Achromobacter*)	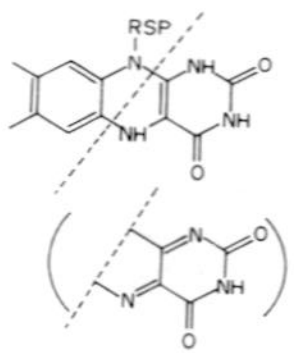	EC 1.4.14.3 alkanal mono-oxygenase, FMN-linked (80 000; two non-identical subunits)	470–505	C_8–C_{16} aldehyde autoinducer	Used for NADP assays
Order Dinoflagellata (*Gonyaulax*)	Unknown chromophore linked to protein (M_r 100 000)	(150 000; a monomer)	470	A binding protein	
Anthozoa (*Renilla*)	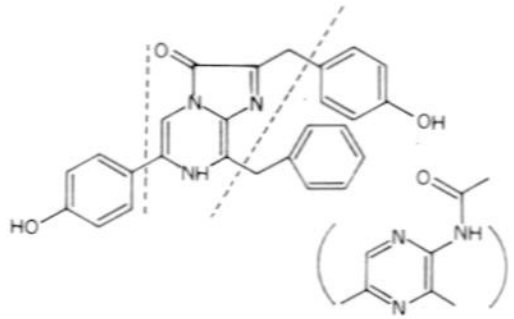	EC 1.13.12.5 Renilla-luciferin 2-monooxygenase	480 with GFP 509	Binding protein (M_r 18 500) binds 2 Ca^{2+} (K_D 0.14 μM) GFP (dimer M_r 54 000)	
Hydrozoa (*Aequorea* *Obelia*)	Same; forms the chromophore of the photoprotein aequorin	Only an oxygenated apoaequorin is known	469	Aequorin (M_r 20 000) binds 3 Ca^{2+}; GFP (M_r 30 000)—its chromophore is a cyclic tripeptide	No luciferase is present. Used for Ca^{2+} and Sr^{2+} assays.
Crustacea (*Cypridina*)	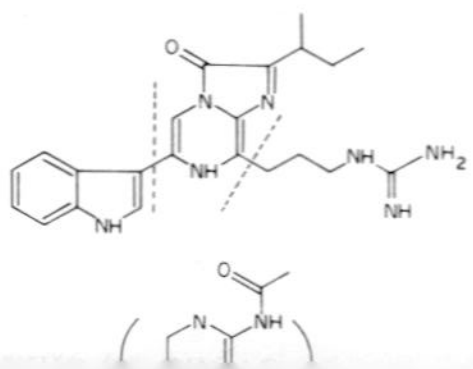	EC 1.13.12.6 Cypridina-luciferin monooxygenase	460	—	Extrinsic fluorescence. Also in fish genera *Neoscopalus* and *Diaphus*

(*Watasemia*)	but sulphated at phenolic hydroxyls	Watasemia-luciferin 2-monooxygenase (possibly identical with EC 1.13.12.5 and/or EC 1.13.12.6)			crustaceans and in fish genera *Apogon* and *Porichthys*
Gastropoda (*Latia*)	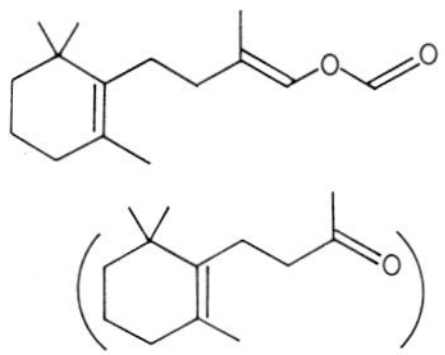	EC 1.14.99.21 Latia-luciferin monooxygenase (demethylating)	535	Purple protein	Extracellular fluorescence
Bivalvia (*Pholas*)	Photoprotein pholasin (M_r 50 000) with no typical chromophore	H_2O_2 serves as oxidant	505	Fe^{2+} stimulates	
Polychaeta (*Harmothoe*, *Acholoe*)	Riboflavin?	—	520	Superoxide radicals, polynoidin with an unknown chromophore	Repetitive luminescence firing
Clitellata (*Diplocardia*, *Diplotrema*, *Octochaetus*, *Spenceriella*)	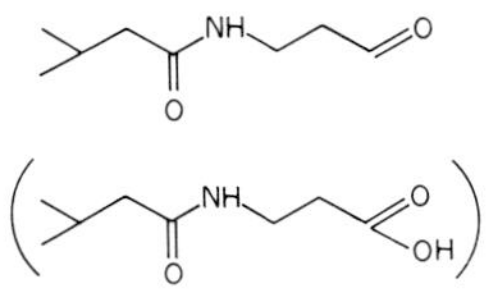	(320 000; subunits of 71 000, 58 000, 14 500) peroxides serve as oxidants	500 545 570 535	Cu^{2+}	Fluorescence of cells exuded from the coelomic cavity
Hexapoda (*Photinus*, *Photuris*, *Phrixothrix*)	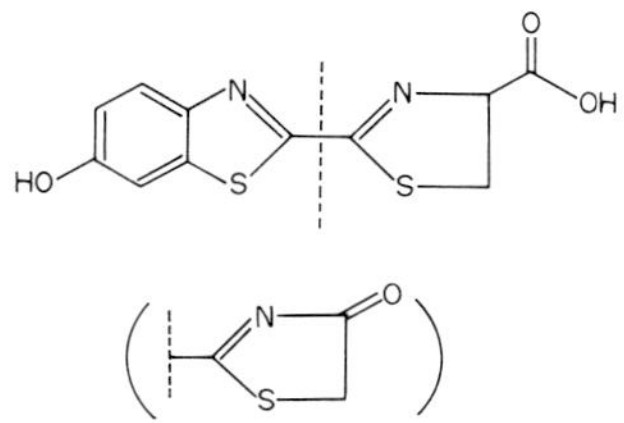	EC 1.13.12.7 Photinus-luciferase 4-monooxygenase; ATP-hydrolyzing (95 000; two non-identical subunits)	575 498 602	ATP, Mg^{2+}	Used for ATP assays

Yet another modification, based on evidence from *in vitro* studies with dimethyldioxetanones, invokes the role of a catalytic luminescence activator, an electron donor D, such as an aromatic hydrocarbon or an amine. Here then

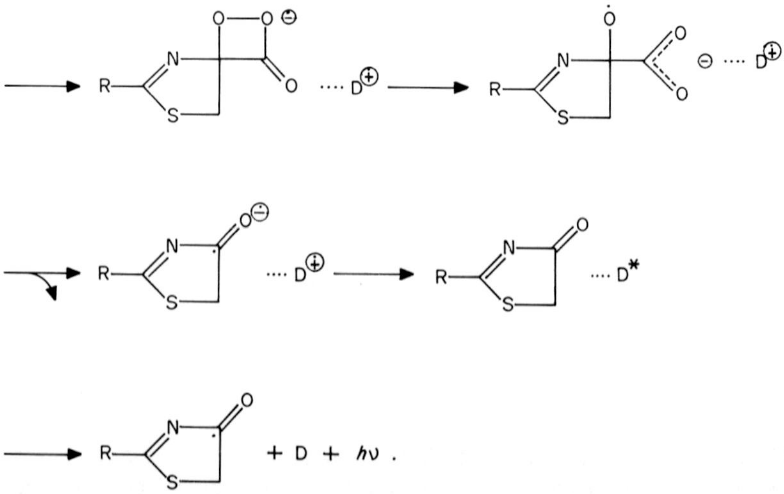

(IV)

While a simple cleavage of the dioxetanone ring (scheme I) is a possible reaction mechanism with most systems, it is not very likely in the case of bacterial luminescence where a dissociative electron transfer with two ion radicals is more probable. The prerequisite for the reaction is the presence of reduced $FMNH_2$ (arising from FMN by reduction with NADPH) which probably acts as a luciferin. The following reaction sequence can be envisaged (up to the last step $FMNH_2$ is bound to the luciferase):

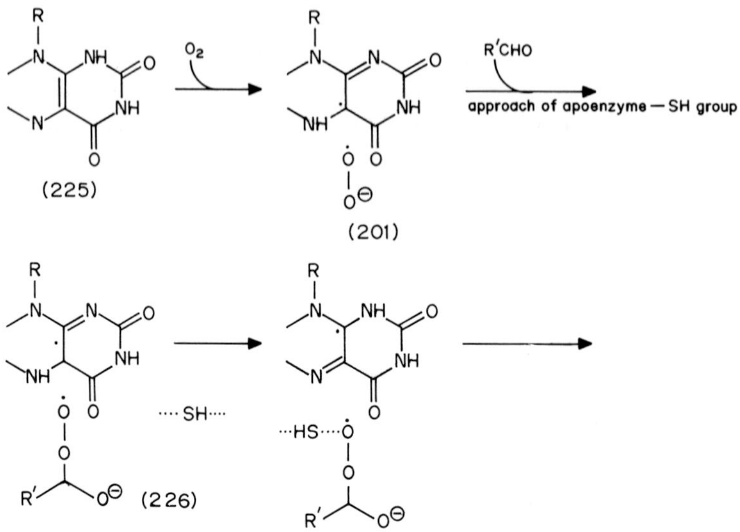

(V)

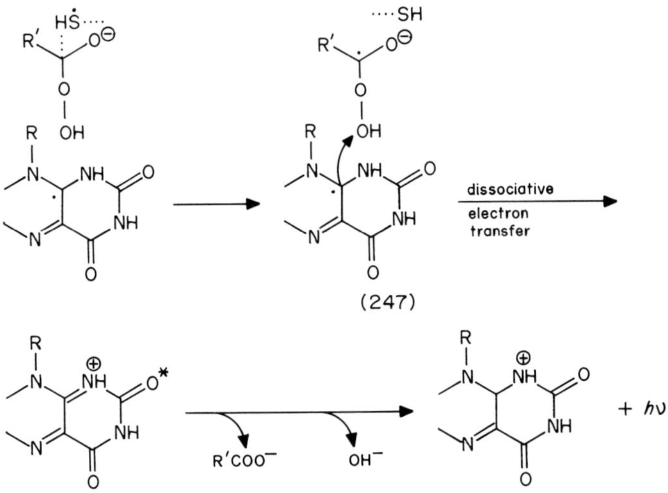

(247)

(368) (117) (247)

It should be observed that the electron transfer mechanisms postulated for the firefly luciferase (scheme III) and that for bacterial luciferase (scheme V) differ in the final upgrading toward the 'energy-rich' state. In luminescent bacteria the dissociative electron transfer lifts the penultimate intermediate directly to the light-emitting species; in the firefly the main upgrading occurs between the dioxetanone and the decarboxylated species which then undergoes 'activation' energetically down-hill to the light-emitting species.

A unique mechanism is that of *Latia* luciferase where a reduced cofactor (A) is required and the whole reaction can be summarized as

$$L_{red} + AH_2 + 2O_2 \rightarrow L_{ox} + CO_2 + HCOOH + A + H_2O + h\nu.$$

The quantum yields of bioluminescent reactions (number of photons emitted per molecule of reacting luciferin) lie between 0.1 and 0.3.

While the above reaction sequences probably account for the light-generating event itself, many adjunct control mechanisms are known that distinguish one organism from another.

1. Bacterial luciferase is an inducible enzyme which is synthesized only in the presence of an auto-inducer, excreted by bacteria into the surrounding medium; the auto-inducer has the formula

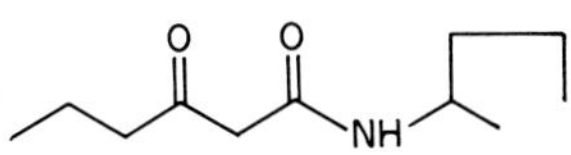

2. Dinoflagellates contain a specific protein that binds luciferin at pH 8 and a luciferase that reacts with luciferin at pH 6. As pH is lowered below 6 in the presence of oxygen, the luciferin is released to react with luciferase. The activity of luciferase changes in a circadian rhythm, being higher at night than during the day.

3. *Renilla* luciferin is normally bound to a binding protein in an inactive form; it is only in the presence of Ca^{2+} that it can react with luciferase (E) and be oxidized. Moreover, under physiological conditions it reacts with a green fluorescent protein (GFP) so that oxygen excites the luciferin first to a L*.E.GFP complex with subsequent intramolecular energy transfer to L.E.GFP*. This is a case also encountered with aequorin which itself acts as a Ca^{2+}-binding protein and then interacts with its specific GFP.

As should be obvious from the diversity of light-emitting mechanisms and of their ethological significance there are several ways by which luminescence can be triggered.

In most bacteria and fungi the light emission appears continuous or, at any rate, may last for minutes or hours, there being apparently a constant supply of luciferin substrate and oxygen, with relatively long lifetimes of the excited molecules. However, it is more likely that the individual cells here flash asynchronously so that a continuous signal is observed.

In some animal species, such as the clam *Pholas* or in the case of extracellular luminescent slime, the light output is macroscopically continuous.

An impulse is usually required to produce luminescence and this is then given off as a flash lasting for fractions of a second. The impulse may have the form of an action potential which can either cause a transient change of pH (in the dinoflagellate scintillons) or a shift of Ca^{2+} ions (in the case of aequorin). The initial signal for the occurrence of the action potential can be mechanical, such as a difference in inertia of flagellum and the cell, optical, such as perception of flashes of partner fireflies, or chemical, such as binding of a triggering compound (say, a pheromone) to a membrane receptor.

A complicated pattern of luminescence firing was described in the case of scale-worm polynoidin where repetitive flashes appear at 40–200 ms duration with 2–20 ms periods in between, for several seconds at a time. The peculiarity of this system lies in the fact that the action potential elicited by acetylcholine which gives rise to the series of flashes must propagate not along the plasma membrane but along the endoplasmic reticulum of the cell with its assorted photosomes and this may involve some extremely rapid ultrastructural changes in the reticulum membranes.

SYNOPSIS

Oxidation of organic substrates represents the principal way of channelling energy into a form that can be used more universally, either as ATP directly or through the intermediate of a proton (and exceptionally Na^{+}) electrochemical potential gradient. The second of these mechanisms is much more efficient and takes place in mitochondria and in bacteria. The principal terminal electron acceptor is oxygen but in many bacteria, and quite exclusively in the anaerobic ones, the terminal acceptor can be sulphate, nitrate, dithionite, sulphur or CO_2.

Although in the majority of cases the oxidative reactions result in the dislocation of H^{+} ions out of the mitochondrion or bacterium, together with the generation of a membrane potential, in some moderately alkalophilic bacteria, Na^+ ions substitute for the H^+ ions. Along the same lines, decarboxylation reactions of oxo-acids in some bacteria drive directly the translocation of Na^+ ions.

An unusual and quantitatively unimportant way of converting the free energy of oxidation is used by various bacteria as well as lower multicellular animals which give off (bio)luminescence.

FURTHER READING

Bassot, J. -M. Fast membrane transformation in a flashing endoplasmic reticulum. In: *Recent Advances in Biological Membrane Studies* (L. Packer, ed.), pp. 259–84, Plenum Press, New York (1985).

Ernster L., and Schatz G. Mitochondria: A historical review, *J. Cell Biol.*, **91**, 227s–55s (1981).

Malmström B. G. Cytochrome *c* oxidase as a proton pump. A transition-state mechanism, *Biochim., Biophys. Acta*, **811**, 1–12 (1985).

Marcus, R. A., Sutin, N. Electron transfers in chemistry and biology, *Biochim. Biophys. Acta*, **811**, 265–322 (1985).

Papa, S., and Lorusso, M. The cytochrome chain of mitochondria: Electron transfer reactions and transmembrane proton translocation. In: *Biomembranes* (R. M. Burton, and F. C. Guerra, eds), pp. 257–90, Plenum Press, New York (1984).

Robillard, G. T., and Konings, W. N. A hypothesis for the role of dithioldisulfide interchange in solute transport and energy-transducing, *Eur. J. Biochem.*, **127**, 597–604 (1982).

Schäfer G., and Klingenberg M. (eds). *Energy Conservation in Biological Membranes*, Springer-Verlag, Berlin (1978).

Schlegel, H. G. *Allgemeine Mikrobiologie*, G. Thieme Verlag, Stuttgart (1985).

Scott, R. A., Mauk, A. G., and Gray, H. B. Experimental approaches to studying biological electron transfer, *J. Chem. Educ.*, **62**, 932–8 (1985).

Skulachev, V. P., and Hinkle, P. C. (eds). *Chemiosmotic Proton Circuits in Biological Membranes*, Addison-Wesley, Amsterdam (1981).

Wikström, M., Krab, M., and Saraste, M. Proton-translocating cytochrome complexes., *Annu. Rev. Biochem.*, **50**, 623–55 (1981).

4.4 ELECTROCHEMICAL POTENTIAL GRADIENT OF IONS CONVERTED INTO OTHER FORMS OF ENERGY

It is common knowledge in bioenergetics that the differences in ion concentrations across membranes expressed as $\Delta\tilde{\mu}_{\text{ion}}$ are a highly convertible currency in cell molecular biology. It was shown in the preceding section how such gradients are formed; let us now examine how they are utilized.

4.4.1 Ion gradients generating gradients of other ions or molecules

This is a field that was discussed in the sections on secondary transport. Let it only be said here that the conversion of energy stored in an ion gradient via a carrier into another ion or nonelectrolyte gradient is the most common way solutes are transported into cells.

One point deserves attention here, viz. whether the primary gradients, especially of H^+, are the most useful form of this convertible currency. It rather appears that it is often expedient to buffer them with a gradient of another, more plentiful ion.

The electrochemical potential gradient of an ion is given by (see p. 204)

$$\Delta\tilde{\mu}_{\mathrm{ion}} = F\,\Delta\phi + RT\ln(c'_{\mathrm{ion}}/c''_{\mathrm{ion}}). \tag{4.3}$$

The electric energy stored in $\Delta\phi$ is equal to

$$E_{\mathrm{el}} = \tfrac{1}{2}SC(\Delta\phi)^2 \tag{4.4}$$

where S is the membrane area of a cell, C its specific electric capacity. If the $\Delta\phi$ is generated by a passive influx of a cation it will be given by its diffusion potential

$$\Delta\phi = (RT/F)\ln(c_{\mathrm{ion}}/c''_{\mathrm{ion}}). \tag{4.5}$$

From this, the electric energy stored in the diffusion potential will be

$$E_{\mathrm{el}} = \frac{SCR^2T^2}{2F^2}\left(\ln\frac{c''_{\mathrm{ion}}}{c'_{\mathrm{ion}}}\right)^2.$$

In contrast, the osmotic energy under the same conditions will be

$$E_{\mathrm{os}} = RTVc''_{\mathrm{ion}}\left(\ln\frac{c''_{\mathrm{ion}}}{c'_{\mathrm{ion}}} - 1 + \frac{c'_{\mathrm{ion}}}{c''_{\mathrm{ion}}}\right) \tag{4.6}$$

as obtained by integration of $RTV\int_{c'}^{c''}(\ln c'_{\mathrm{ion}} - \ln c''_{\mathrm{ion}})\,\mathrm{d}c_{\mathrm{ion}}$ where v is the cell volume.

The ratio of these energies can be illustrated by an example where, say, Na^+ ions flow passively into a yeast cell from a concentration of 0.1 M to 1 mM, the cell having a volume of 250 μm^3 and a surface area of about 200 μm^2, with a capacity of 10 mF/m^{-2}. With these values $E_{\mathrm{os}}/E_{\mathrm{el}} = 385$. Thus the osmotic gradient energy is much greater than that of the electric potential.

If a gradient of H^+ is formed by an outward driven process, such as by respiration or ATP hydrolysis in a microorganism, it is useful for the organism to convert the electric component ($\Delta\phi$) into an osmotic component by allowing, say, K^+ ions to enter, thus creating a ratio of c''_{K^+}/c'_{K^+}. Now, if this leads to an alkalinization of the cell interior, another system may be of help, a Na^+/H^+ antiport which brings in H^+ ions in exchange for Na^+ ions. This then results, with the obvious thermodynamic losses, in a conversion of $\Delta\tilde{\mu}_{H^+}$ into $(RT/F)\ln(c''_{K^+}/c'_{K^+})$ and $(RT/F)\ln(c'_{Na^+}/c''_{Na^+})$.

The energy stored in these gradients is readily available via the ion-transporting systems but, at the same time, forms a buffer in case of a sudden energy demand on the organism.

4.4.2 Ion gradient energy converted into chemical energy

There is one major process that, under physiological conditions, transduces the energy of a cation gradient into the more tangible form of chemical compound where it is stored and from which free energy can be released for various uses. The system that catalyzes the process in most cells is the ATP-synthase or, because of the reversibility of the process, H^+-transporting ATPase or, because of its bipartite composition, the F_1F_0-ATPase, or, especially in older literature, the coupling factor, of the inner mitochondrial, thylakoid and bacterial membrane.

It is probably the most complicated membrane enzyme known, with various identical and nonidentical subunits. It has obviously undergone evolutionary changes, perhaps to increase its efficiency, but in a simple form it apparently functions even in the most primitive bacteria. Its most characteristic feature is its being composed of two sectors, the F_0 sector, highly hydrophobic, residing in the membrane, and the F_1 sector, protruding from the membrane into the space where ATP is formed and from where protons are removed by the adjoint respiratory or photochemical processes. The membrane sector is about 9 nm wide, the protruding head about 12 nm across. The density of ATPases in the energy-transducing membranes may reach $4000/\mu m^2$.

In addition to the two morphologically distinct sectors, the F_1F_0-ATPase contains components that are responsible for the contact of the two parts and confer the sensitivity to oligomycin on the whole molecule (OSCP).

The ATPase shows sensitivity to a number of inhibitors. One group are those that inhibit directly the F_1 sector, such as aurovertin and Dio-9. Another group require for inhibition the complete molecule and include oligomycin, rutamycin, dicyclohexylcarbodiimide, venturicidin, tributyl tin chloride, and various mercurials.

The structural features of the three different basic types are outlined in Fig. 4.35. The disposition of the somewhat anomalous enzymes in some bacteria is not known but it is believed that, e.g. *Lactobacillus casei* contains only six identical subunits of M_r 43 000, and *Micrococcus lysodeikticus* and *Bacillus subtilis* contain only 3α and 3β subunits of M_r about 57 000.

It should be noted that the mitochondrial F_1 sector alone contains an inhibitor designated IF_1 that may have arisen by cleavage of originally larger subunits ε as they now exist in chloroplasts and bacteria where they also play a regulatory role. IF_1 serves to inhibit ATP hydrolysis under conditions of low $\Delta\tilde{\mu}_{H^+}$. If the H^+ gradient is increased IF_1 separates from the ATPase.

The α and β subunits represent the catalytic domain for ATP synthesis (or hydrolysis under conditions favouring that direction of the reaction), the γ subunit appears to hold the assembly together. The β subunits from very different sources (ranging from *Escherichia coli* to human mitochondria) show an 80 per cent homology. The large subunits are organized in a pseudo-six-fold structure (with a superimposed two-fold symmetry) so that the peripheral subunits are α-β-α-β-α-β. The nucleotides are bound at the cleft between each α–β pair. In addition to these three catalytic sites, three different binding sites for ADP/ATP are present.

The role of the contact subunits is particularly expressed in the case of δ and of OSCP, with the F_6 protein assisting in the function.

The protonophoric subunits are instrumental in propelling protons along a

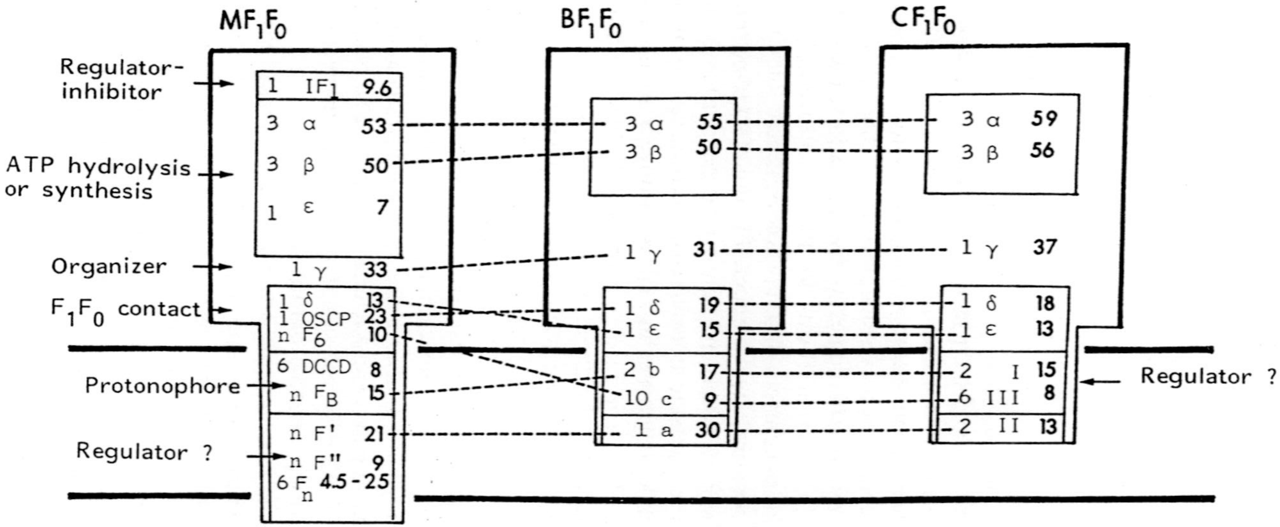

Fig. 4.35. Comparison of three types of F_1F_0 H^+-translocating ATP synthase from beef heart mitochondria (MF_1F_0), from *Escherichia coli* (BF_1F_0) and from spinach chloroplasts (CF_1F_0). The dashed lines indicate sequence homology between the corresponding subunits. The boldface numbers denote relative molecular masses in thousands

channel toward the F_1 sector where ATP is synthesized. Glutamic acid residues on the channel walls belonging to the DCCD-binding protein are apparently involved, as well as contributions by Arg, Tyr, and Cys. The F_0 assembly functions as a channel for H^+ when incorporated into artificial membranes but ceases to do so as soon as the F_1 sector is attached.

The molecular details of ATP synthesis are not definitively established (and they will certainly differ in quantitative aspects between the various ATPase subtypes) but several basic features of the process are now understood.

ADP and inorganic phosphate bind to the α–β interface. The first ADP bound brings about a functional asymmetry of the originally equivalent sites in a negatively cooperative fashion. Further molecules of ADP bind only at higher concentrations. At the first site ATP is synthesized from ADP and P_i and partial equilibrium is reached. At both the first and the second site the affinity changes so as to allow ATP to be released and ADP and P_i to come together to react. In the meantime, a third ADP is bound to the third site and the release and synthesis continues at the second and third sites, respectively. In fact, the catalytic subunits of the enzyme are now believed to rotate during its activity so that the γ subunit acts as a pivot about which the α–β assembly turns by 120 degrees for every ATP synthesized.

The crucial energy-requiring step is apparently the change of conformation of the subunit from which ATP should be released and, although this can be readily visualized as a response to either a redox process or covalent binding, it is still a matter of conjecture how it is brought about. Clearly, no phosphorylated intermediate is formed, in contrast with the transport ATPases to be dealt with in Section 4.5.1.

Electrons acting as carriers of reducing power can be ruled out in this case. There is compelling evidence, from both *in vivo* and *in vitro* studies, that ΔpH and $\Delta\phi$ are the sources of energy for the process and it is also generally believed that three H^+ ions passing through the DCCD-sensitive channel in sequence are required for the synthesis of one molecule of ATP. It would be misleading, however, to connect the three protons with the three α and β subunits. In fact, nine protons would be required for the full rotation of the assembly wherein three molecules of ATP are released. It is fully in the realm of speculation as to how the proton gradient connects with ATP synthesis.

One of the more attractive views is that the ATP-synthase senses the local pH and responds accordingly by a conformational change of the α–β subunits. This is probably achieved by the cluster of 3 Glu (Asp) and 3 Arg (Lys) in the *c* subunit of F_0, from which protons pass through subunit *b* to the β subunit of F_1 where they cause a conformational change leading to release of ATP.

The physiological function of ATP-synthase is to synthesize ATP although it operates perfectly well in reverse if the $c_{ATP}/c_{ADP}\cdot c_P$ ratio is high enough and/or if the $c_{H^+_{in}}/c_{H^+_{out}}$ is sufficiently high. The ATP-splitting activity is in fact what is measured in isolated F_1 segments.

In vivo, the enzyme is obligatorily coupled to the movement of H^+ ions through the F_0 segment, inward passage being connected with ATP synthesis so that

$$ADP + P_i + nH^+_{out} \rightarrow ATP + nH^+_{in}.$$

In an approximation the ATP-synthesizing process will require energy defined by

$$\Delta G = \Delta G^{\ominus} + RT \ln (c_{ATP}/c_{ADP} \cdot c_P) \tag{4.7}$$

$\Delta G^{\ominus}$ is known to be 28.2 kJ/mol (at pH 7, T of 30° C and $c_{Mg^{2+}}$ of 1 m M, with ionic strength at 0.15). In slowly respiring mitochondria one may assume that $c_{ATP}/c_{ADP} = 50$ and that $c_{P_i} = 0.01$ M. From this then $\Delta G = 49.6$ kJ/mol.

Now it was shown before that the free energy of oxidation of NADH to O_2 (a two-electron process is considered here) is 220 kJ/mol—this should suffice for four ATP molecules to be synthesized. The electrochemical potential gradient of protons generated by this oxidation is about 22 kJ/mol. Now, as three protons are required for one molecule of ATP to be synthesized the free energy released from the $\Delta\tilde{\mu}_{H^+}$ would be 66 kJ/mol, more than ample to synthesize ATP.

There is a long-standing controversy about whether the protons used for ATP synthesis are delocalized, i.e. present in the bulk solution, or whether they are local, i.e. in or directly at the membrane.

The classical view of Peter Mitchell involves bulk (delocalized) protons and there is no doubt that in equilibrium, if there ever is one, the c_{ATP}/c_{ADP} ratio measured at static head (no net flow of phosphate equivalents) should be linked to the $c_{H^+_{out}}/c_{H^+_{in}}$ existing between the bulk solutions (then there is also no net flow of H^+ ions) and hence the concentrations at the membrane should be equal to those in the bulk—save for the effects of surface potential.

The critical view of R. J. P. Williams and others considers (probably correctly) the system to be in a local steady state at best so that the concentrations of ions at the membranes are not the same as in the bulk, and argues that the *rates* of ATP production are governed by the concentrations of H^+ ions felt by the system directly in or at the membrane.

On the whole, there is little doubt that protons are involved in ATP synthesis in all membranes where a H^+-ATPase of the F_1F_0 type is present, be it in *Halobacterium* where the proton gradient is generated by capturing light quanta, or in the thylakoid where the energy comes also primarily from light but the form that is directly used is free energy of oxidation, or in bacteria and mitochondria where protons are pumped out at the expense of free energy of oxidation. After all, ATP synthesis could be observed in various reconstituted systems by artificial application of $\Delta\tilde{\mu}_{H^+}$. The intermediary function of the proton gradient between oxidation of H-containing substrate and ATP synthesis is qualitatively supported by the occurrence of at least two phenomena in oxidative phosphorylation.

1. Addition of uncouplers, both to a fully functioning system of mitochondrial oxidative phosphorylation and to one where ATP synthesis is blocked by oligomycin, results in an increased rate of oxidation. This is readily explained

by assuming the rather tightly coupled reaction (grossly simplified)

$$\mathrm{NADH} + n\mathrm{H}^+_{\mathrm{in}} + \mathrm{cyt}\,c^{3+} \rightarrow \mathrm{NAD}^+ + n\mathrm{H}^+_{\mathrm{out}} + \mathrm{cyt}\,c^{2+}.$$

Addition of uncouplers decreases the $c_{\mathrm{H}^+_{\mathrm{out}}}$ and thus permits the reaction to run faster forward. It increases O_2 consumption, although this is in the end an irreversible process, by increasing $c_{\mathrm{cyt}\,c^{2+}}$. Moreover, in addition to the mass-action effect described, uncouplers increase the forward reaction rate by increasing $c_{\mathrm{H}^+_{\mathrm{in}}}$ as they transport H^+ equivalents back to the matrix side.

2. 'Coupled' mitochondria are under so-called respiratory control (Fig. 4.36) which is defined as ADP-stimulated respiration rate (in the terminology of mitochondrial bioenergetics this is a shift from state 4 to state 3). This is obviously accompanied by an increased rate of ATP synthesis. This may again be explained by the fact that ATP synthesis entails movement of H^+ from the outside to the matrix.

The qualitative explanation of the phenomena, however, is not fully complemented by the quantitative aspects particularly as there is no unique

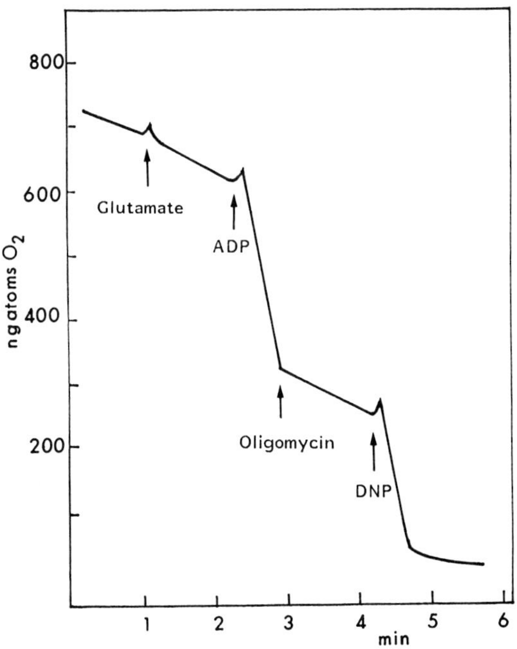

Fig. 4.36. Respiratory activation and inhibition of beef-heart mitochondria. ADP exerts the so-called respiratory control, oligomycin blocks the ATP synthase reaction and thus inhibits respiration while 2,4-dinitrophenol (DNP) short-circuits the H^+ electrochemical potential gradient and thus removes the effect of oligomycin

dependence between the rate of respiration, the $\Delta\tilde{\mu}_{H^+}$ and the rate of phosphorylation. Inasmuch as the lack of straightforward relationship is kinetic in nature, regulatory factors may play a role. Thus, it is now believed that ADP, besides being a substrate for the ATP-synthase, has a controlling role in binding to a noncatalytic site of the ATP-synthase. This is a phenomenon observed in all ATP-synthases and is promoted by a high $\Delta\tilde{\mu}_{H^+}$ which, either directly as in chloroplasts, or in concert with the natural ATPase inhibitor, causes binding of ADP, this bringing about an inactive state of the enzyme.

Some role might be played by proteins called *chargerins* ($M_r = 8000$ and 13 000) isolated in large amounts from mitochondria. As they contain numerous charged amino acids, both cationic and anionic, and are associated with the redox chain (chargerin I) and with the H^+-ATPase (chargerin II) they might transfer charge in a way depending on $\Delta\tilde{\mu}_{H^+}$ generation.

To conclude, it seems reasonable to assume that it is the local protons that are actually used for driving ATP synthesis, with the understanding that there is an overall flux of protons from the matrix bulk to the cytosol bulk as oxidation proceeds, this being reversed if the $\Delta\tilde{\mu}_{H^+}$ is being used up to drive ATP synthesis. This 'compromise' view is supported by measuring the size of actual lateral fluxes of protons along lipid-water interfaces which indicate a rather steep concentration profile between the membrane surface and the bulk liquid.

It was discussed on p. 283 that some organisms contain sodium-motive systems based on decarboxylation of some oxo-acids operating according to

$$A_1 \cdot CO_2 + 2Na^+_{in} \rightarrow A_2 + CO_2 + 2Na^+_{out}.$$

It is perhaps to be expected that these organisms will make use of the $\Delta\tilde{\mu}_{Na^+}$ and convert it into ΔG_{ATP} as other organisms do with $\Delta\tilde{\mu}_{H^+}$. This is indeed so, Na^+-activated ATP-synthases having been found in several bacteria, significantly also in *Vibrio alginolyticus* and in *Propionigenum modestum*, as well as in *Pseudomonas stutzeri*. Little is known about their structure, except that such an enzyme found in *Acholeplasma laidlawii* contains five nonidentical subunits resembling those found in the F_1 sector of the H^+-driven ATP-synthase.

This supports the view that for certain ecological conditions microorganisms switched to a type of Na^+ cycle to replace the more common H^+ cycle. If the molecular mechanisms used by the two ions are identical the conclusion must be made that the proton acts in the various transport systems as a cation rather than by using the kinetic privileges of being a quantum particle.

4.4.3 Conversion of $\Delta\tilde{\mu}_{ion}$ into mechanical energy

One classical case of such conversion is the operation of bacterial flagella. These rather long filaments, often present in dozens on a cell, are used to propel the cell either toward an attractive milieu or away from a repellent one, as will be discussed in the section on chemotaxis.

The flagella are rigid structures made of a single protein kind, flagellin—in contrast with the flagella of eukaryotic organisms that are membrane-covered structures containing sets of microtubules which provide the flagellum with the possibility of a wavy movement energized by ATP as it is split by the dynein ATPase. Hence to move the cell, the bacterial flagella must rotate as a whole. To accomplish this feat (it is in fact the only mechanism in all biology operating on the principle of a wheel) the flagellum is anchored by a hook consisting of nearly 20 different proteins in the bacterial membrane, much like the rotor in the stator of an electric generator (Fig. 4.37). The energy for this rotation is provided by the $\Delta\tilde{\mu}_{H^+}$ (or $\Delta\tilde{\mu}_{Na^+}$ in the sodium-motive bacteria) with no participation of ΔG_{ATP}. In fact it has been calculated that 256 protons must be translocated to move the flagellum around once. As the bacterium may contain 20 flagella and as they rotate at a rate of $100\,s^{-1}$ (moving the bacterium at a speed of 1–10 mm per minute) a flux of about $5 \cdot 10^5$ protons per second would be expected. If one considers that at pH 7 a bacterial cell contains about seven protons it is clear that, irrespective of intracellular buffering, the incoming protons would, within a second, decrease the intracellular pH by several units. Hence, protons must be pumped out at practically the same rate to ensure survival. Conversely, if they are not, the cell cannot rotate its flagellum and move. This example illustrates an important rule that applies to all processes involving the utilization of ion gradients for various purposes. Unless the primary gradient generator is actually working (or unless the given primary gradient is 'buffered' by secondary ones, such as of K^+ or Na^+) no quantitatively significant amount of work can be drawn from the existing $\Delta\tilde{\mu}_{ion}$.

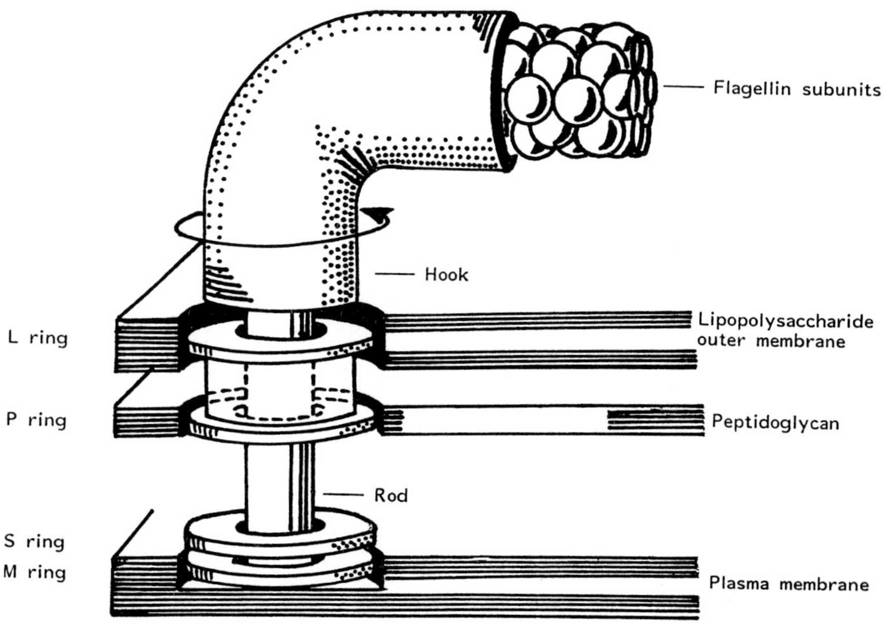

Fig. 4.37. Schematic of a bacterial flagellum

4.4.4 Conversion of $\Delta\tilde{\mu}_{H^+}$ into heat

New-born and hibernating mammals do not possess the normal homoisothermic mechanism wherein catabolic reactions are stimulated to produce heat as a by-product. These animals possess special brown fat tissue which is literally packed with mitochondria. These mitochondria burn fatty acids of this adipose tissue to produce $\Delta\tilde{\mu}_{H^+}$ but not to synthesize ATP. The extruded H^+ ions never manage to generate a gradient as they are efficiently brought back to the mitochondrial matrix.

The mechanism involves a specific protein found nowhere else but akin in many respects to the ADP/ATP carrier, the uncoupling protein or GDP-protein or thermogenin. This is a relatively small protein (M_r 33 000) representing as much as 18 per cent of mitochondrial proteins. It occurs as a dimer of M_r 66 000 which contains probably two functionally distinct channels, one for H^+ (carried down their gradient into mitochondria), the other for halide anions, mainly chloride (carried out of mitochondria). The important feature of the uncoupling protein is that it binds one molecule of GDP per dimer (a half-site reactivity?) with a dissociation constant of about 1 μM, this binding resulting in an inhibition of H^+ transport. Another binding site is revealed with higher concentrations of GDP which can be partly substituted with GTP, ADP or ATP. The principal GDP-binding site lies apparently in the region of the sequence between amino acids 175 and 195.

The regulation of the protein function by GDP is effected by a β-adrenergic pathway, the 'cold' stimulus arriving from a primary sensory receptor, leading both to increased synthesis of the uncoupling protein and to an immediate decrease in GDP level so that the uncoupling protein is more active.

SYNOPSIS

The electrochemical potential gradients of ions, arising during the passage of electrons through sequential oxidation reactions of the components of mitochondrial, thylakoid and bacterial membranes may be considered as a kind of convertible energy currency which is used for different purposes in cells. Some of the ion gradients are utilized to accumulate substances against their concentration gradients in secondary transport. The most important use occurs in all energy-transducing membrane in terms of ATP synthesis from ADP and P_i.

In motile bacteria the ion gradient apparently drives flagellar rotation.

In brown adipose tissue mitochondria the ion gradient is systematically wasted to obtain heat from the more noble kind of energy.

FURTHER READING

Berg, H. C., Manson, M. D., and Couley, M. P. Dynamics and energetics of flagellar rotation in bacteria, *Symp. Soc. Exptl. Biol.*, **35**, 1–31 (1982).

Chernyak, B. V., and Kozlov, I. A. Regulation of H^+-ATPases in oxidative- and photophosphorylation, *Trends Biochem. Sci.*, **11**, 32–5 (1986).

Drachev, A. L., Markin, V. S., and Skulachev, V. P. $\Delta\tilde{\mu}_{H^+}$ buffering by Na^+ and K^+ gradients in bacteria. Model and experimental systems, *Biochim. Biophys. Acta*, **811**, 197–215 (1985).

Ferguson, S. J. Fully delocalized chemiosmotic or localized proton flow pathways in energy coupling. A scrutiny of experimental evidence, *Biochim. Biophys. Acta*, **811**, 47–95 (1985).

Futai, M., and Kanazawa, H. Structure and function of proton–translocating adenosine-triphosphatase (F_0F_1): Biochemical and molecular biological approaches, *Microbiol. Rev.*, **47**, 285–312 (1983).

Glagolev, A. N., Dibrov, P. A., Skulachev, V. P., and Sherman, M. Yu. A new type of energy transduction in biomembranes: Na^+-dependent bacterial motility, *Biologicheskie Membrany* **1**, 27–32 (1984).

Kadenbach, B. Structure and evolution of the 'Atmungsferment' cytochrome *c* oxidase, *Angew. Chemie* (Intern. Ed. Engl.), **22**, 275–83 (1983).

Kopecký, J., Houštěk, J., and Drahota, Z. Molecular aspects of structure-function relationship in mitochondrial H^+-ATPase. In: *Structure and Properties of Biomembranes* (Gh. Benga, ed.), Vol. 2, pp. 77–104, CRC Press, Boca Raton (1985).

Prats, M., Teissié, J., and Tocanne, J.-F. Lateral proton conduction at lipid-water interfaces and its implications for the chemiosomatic-coupling hypothesis, *Nature*, **322**, 756–8 (1986).

4.5 HYDROLYSIS OF CHEMICAL BONDS FOR DRIVING TRANSPORT PROCESSES

Several mechanisms are known in which a bond in an energy-rich compound is hydrolyzed and the free energy of this cleavage is used for driving transport reactions. Most important are the adenosinetriphosphatase reactions, the phosphotransferase reaction and some group translocations not involving phosphate.

4.5.1 Membrane adenosinetriphosphatases

The bond that is cleaved in the transport adenosinetriphosphatase reaction of virtually all membranes is the O–P bond of the terminal phosphoryl group of ATP. This part of the book might equally well be placed in the section on primary active transport as the ATPase-driven fluxes of ions are the most widely represented ones among primary active transports.

Membrane ATPases all appear to be involved in transport of ions but they are of several basic types at the molecular level.

The F_1F_0 type of ATPase was treated in the previous section and was shown to include the mitochondrial H^+-translocating ATPase and the Na^+-translocating ATPase of some 'sodium-motive' bacteria.

The principal group of transport ATPases are those of the E_1E_2 type, which share many properties of quaternary structure as well as amino acid sequence homology (Table 4.8).

The remaining group of ATPases (V type) is rather heterogeneous as it includes molecules of varied size and often with rather unclear function. The only common distinguishing feature between the two groups may rest in the inhibition by

Table 4.8. Properties of transport adenosinetriphosphatases

Transported ion(s) (ion: ATP stoichiometry)	Occurrence Type	Typical K_m(ATP) (mM)	Nucleotide specificity (formation of P-intermediate)	pH_{opt}	M_r (subunits)	Activators	Inhibitors	No effect	Notes
H^+ (2:1 ?)	Yeast, fungal, plant plasma membranes E	0.16–3.5	ATP (+)	6.0	100 000 (single polypeptide)	Mg^{2+}, Mn^{2+}, Co^{2+}, (K^+)	NEM, DES, vanadate, miconazole, DCCD, TPT, Dio 9	Ouabain oligomycin	Activated by substrate oxidation
H^+ (2:1 ?)	Rat liver lysosomes V	—	—	—	—	HCO_3^-, SO_3^{2-}	SITS, DIDS, DCCD, NEM, F^-, N_3^-, oligomycin	—	—
H^+ (2:1 ?)	Yeast, plant vacuoles V	1	ATP (−)	6.3	200 000 or 400 000 (6 subunits)	Cl^-, cAMP, (Dio 9, SO_4^{2-})	NO_3^-, DES, DCCD	Vanadate, N_3^-, oli-gomycin	Transports H^+ into vacuoles
H^+ (1:1 ?)	Chromaffin granules V	0.07	ATP = ITP > GTP > UTP > CTP (−)	7.3	460 000 (5 subunits, catalytic sub-unit 70 000)	—	DCCD, NEM, TMT	Oligomycin, aurovertin reserpine	Transports H^+ into granules
H^+ (?)	*Sulfolobus*, *Thermoplasma* V?	1	dATP, ATP > GTP, ITP > CTP, UTP (−)	6.4 (with SO_3^{2-})	430 000	SO_3^{2-}	NO_3^-, quercetin, erythrosin B	N_3^-, DCCD, ouabain, vanadate, NEM, F^-	— —
Na^+ (?)	*S. faecalis* V	—	— (?)	—	—	Na^+	—	Vanadate	An inducible enzyme
Na^+, K^+ (3:2:1)	Animal plasma membranes E	Depends on Na^+ and K^+	ATP CTP (+)	7.3	270 000–380 000		Ouabain, vanadate	—	—

K^+, H^+ (2:2:1)	Gastric oxyntic cell membranes E	0.07	ATP > CTP > GTP (+)	6.8	100 000	Cl^-	Vanadate, Ca^{2+}, F^- DCCD, PCMB		
K^+ (1:1?)	*Streptococcus faecalis*	0.06	ATP	7.3	78 000	—	Vanadate	—	Activated by phospholipids
Ca^{2+} (2:1)	Sarcoplasmic reticulum E	0.001	ATP > GTP, ITP > CTP (+)	7.5	110 000	Calmodulin	Vanadate, Na^+, La^{3+}, quercetin	Ouabain	Contains 7 phosphate groups per molecule; shows cooperative behaviour
Ca^{2+} (1:1)	Animal plasma membranes E	0.002	ATP > GTP, ITP, CTP (+)	7.4	138 000	Calmodulin	Mersalyl, ethacrynic acid, NEM, La^{3+}	Ouabain oligomycin	—
Ca^{2+} (?)	*Streptococcus* E	0.1	ATP (+)	—	—	K^+	Vanadate	DCCD, FCCP	—
Cl^-	*Amphiuma* or *Aplysia* intestine E	—	ATP (+)	—	—	—	Vanadate, SCN^-	Ouabain oligomycin	—

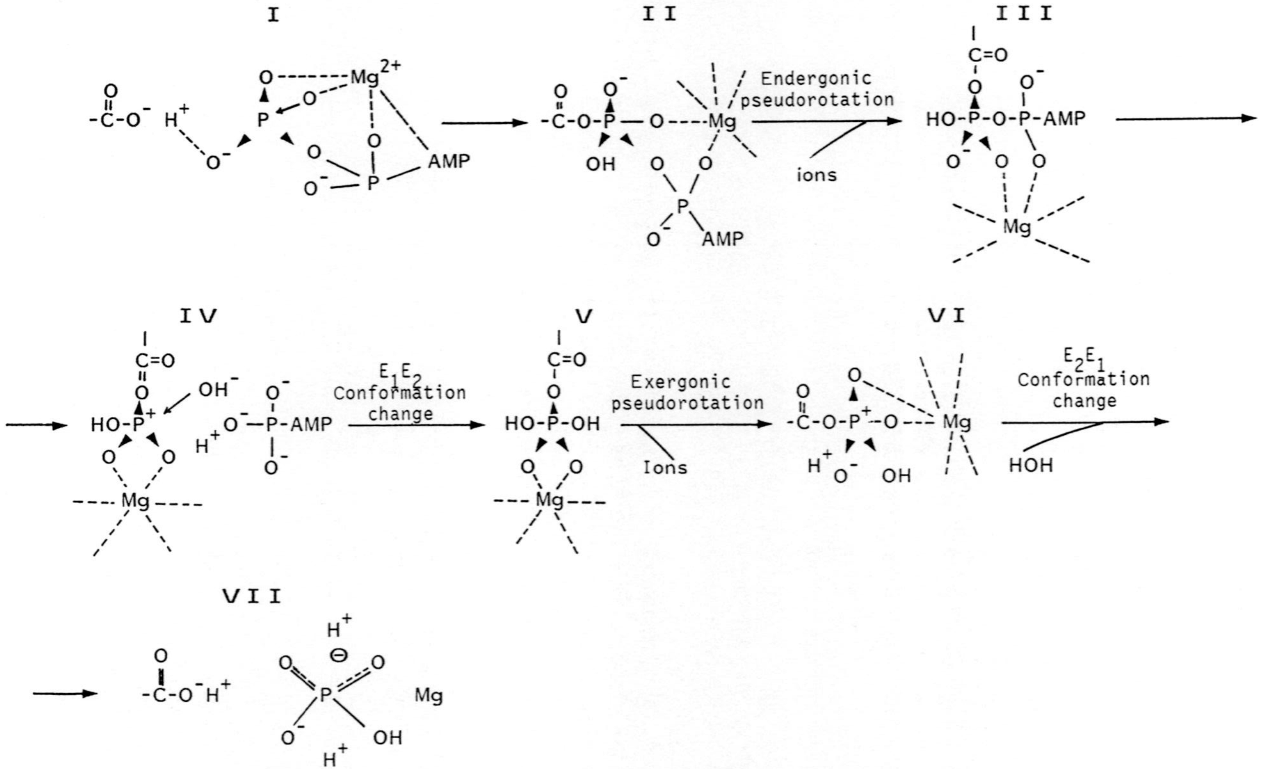

Fig. 4.38. Probable course of ATP splitting in a E_1E_2-type ATPase. I—ATP approaches the β-carboxyl of aspartic acid of the apoenzyme; II—a covalent bond is formed; III—the complex undergoes pseudorotation; IV—water hydrolyzes the γ-phosphoryl bond; V—the enzyme changes its conformation from E_1 to E_2; VI—the phosphate group pseudorotates again; VII—a conformational change back from E_2 to E_1 is followed by hydrolysis of the phosphate group

vanadate of the E_1E_2 and lack of it in the V group. This may be associated with the formation of a phosphorylated intermediate in the E_1E_2 ATPases. In the E_1E_2 type of ATPases the terminal phosphate group of ATP is transferred to a β-carboxyl of an aspartate residue of the enzyme. Only later in the reaction cycle is this phosphoryl group hydrolyzed and released as inorganic phosphate. It is around this phosphorylation site that the amino acid sequence is well conserved among the various ATPases.

The elementary reaction cycle of the ATPases includes binding of the transported ion(s), binding of ATP, release of ADP with formation of the P-intermediate, conformational change of the enzyme which involves a change of exposure in the membrane, release of ion(s), hydrolysis of phosphate, return to original conformation.

At the inframolecular level, it is the dissociation of the ADP and then of the phosphoryl group that are associated with conformational changes in the protein while pseudorotation of the phosphoryl is indirectly associated with binding and unbinding of ions (Fig. 4.38).

Na, K-ATPase

A ubiquitous enzyme in animal plasma membranes, it has now been fully sequenced. It contains subunits α and β (M_r 112 100 and 34 500 + sugar, respectively) which span the membrane several times.

The α subunit contains most of its peptide chain on the cytoplasmic face where it forms $5 \times 5.2\,\text{nm}^2$ knobs, not unlike those seen in the inner mitochondrial membrane that are due to the F_1F_0 ATPase and is the fully catalytic unit of the enzyme.

Its two conformations have a major structural correlate in that the E_1 form can be hydrolyzed by trypsin (i.e., it lies on the surface) at Arg-262 while the E_2 form is hydrolyzed at Arg-438, the intervening segment of molar mass 18 000 being thus involved. The phosphorylated residue is Asp-369 within this segment which also participates in the binding of ouabain, a major inhibitor of Na, K-ATPase, on the external face of the membrane.

Cleavage at Leu-266 stabilizes the E_1 form and sites for cation as well as ATP binding.

The principal function of this enzyme is to pump $2K^+$ into cells and $3Na^+$ out of cells, i.e. by an electrogenic process, at the expense of ATP splitting. The probable reaction cycle is shown in Fig. 4.39.

The role of the β subunit is possible to stabilize the enzyme. It is possible that the whole enzyme forms a dimer in the membrane which is functionally organized in such a way that while one of the molecules transports Na^+, the other transports K^+.

Ca-ATPase

The Ca-ATPase of the sarcoplasmic reticulum has been fully sequenced and is known to contain a single subunit of 1001 amino acids with a peculiar disposition

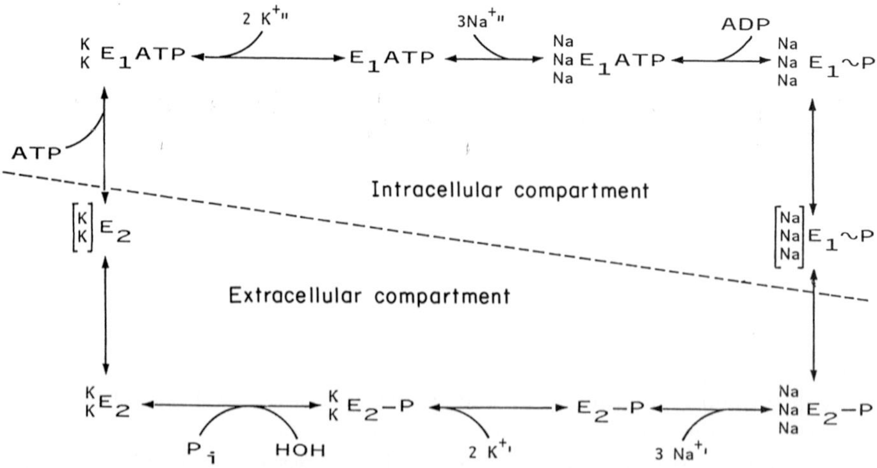

Fig. 4.39. Details of the Na,K-ATPase reaction. The cation sites are exposed on the cytoplasmic surface (″) of the E_1 forms and on the external surface (′) of the E_2 forms. Cations shown in brackets are occluded within the membrane core

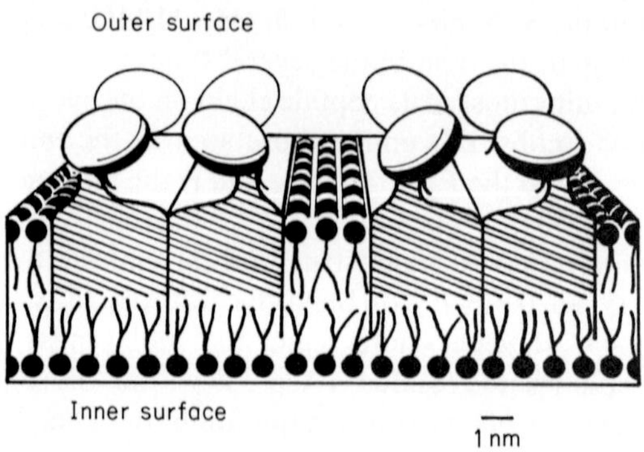

Fig. 4.40. Putative disposition of the Ca-ATPase in the sarcoplasmic reticulum membrane

in the membrane (Fig. 2.19) which has its corollary in the ultramorphology of the sarcoplasmic reticulum membrane (Fig. 4.40).

Its function is to pump calcium ions out of the sarcoplasm into the sarcoplasmic reticulum by binding $2Ca^{2+}$ to one conformational state of the enzyme and releasing them from another conformational state, as shown schematically in Fig. 4.41. The only other cation transported is Sr^{2+} but other divalent cations slightly stimulate ATPase activity.

The Ca-ATPase of erythrocyte and other plasma membranes is highly homologous with the sarcoplasmic reticulum enzyme but has a slightly greater molar mass (138 000) and apparently transports only 1 Ca^{2+} per ATP although this is still a matter of discussion.

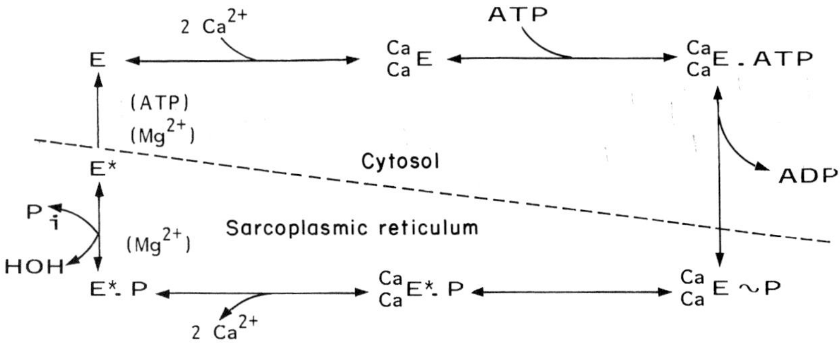

Fig. 4.41. Details of the Ca-ATPase reaction. The conformational changes $E \sim P.Ca_2 \rightleftharpoons {}^*E.P.Ca_2$ and $^*E \rightleftharpoons E$ represent steps where both the exposure of the binding sites and their affinity for Ca^{2+} change

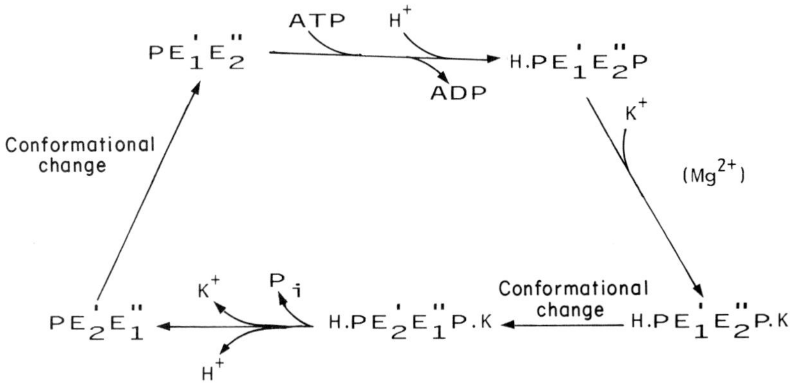

Fig. 4.42. Details of the K,H-ATPase reaction in gastric mucosa. E_1 and E_2 designate two monomers of the enzyme with different affinities for ions and opposite exposure at the membrane faces. (′) at the cytoplasmic, (″) at the luminal face

K, H-ATPase

The substantial acidity of gastric contents arises mainly through the action of a K, H-ATPase of the nonelectrogenic antiport type situated in the oxyntic cell membranes of the stomach lining. Like all E_1E_2 ATPases, it is a single major polypeptide with a phosphorylatable aspartic acid which probably exists as a dimer in the membrane where the two monomers operate in an antiparallel fashion (Fig. 4.42). The monomer exposed to the cytosolic side is always in the phosphorylated state.

H-ATPase

This plasma membrane enzyme appears to be a single polypeptide with a known sequence, spanning several times the membrane, at a M_r of 100 000. There is a clear homology between its sequence and that of the α subunit of the Na, K-ATPase, Ca-ATPase and K, H-ATPase.

H-ATPases isolated from various yeasts, fungi, slime molds and plants are all very similar, showing a clear preference for ATP, a pH optimum at 6.0–6.7 and sensitivity to inhibitors, such as Dio-9, suloctidil, miconazole (all competitive), vanadate, dicyclohexylcarbodiimide, diethylstilbestrol (all noncompetitive) and *p*-hydroxymercuribenzoate (irreversible).

Apparently, $2H^+$ are transported out of cells per 1ATP split. There is good evidence that the rate constants of H^+ are affected by K^+ (but not by Na^+) which suggests that H^+ and K^+ might be transported in an antiport fashion, the H^+/K^+ ratio being greater than unity so that the whole pump would function electrogenically.

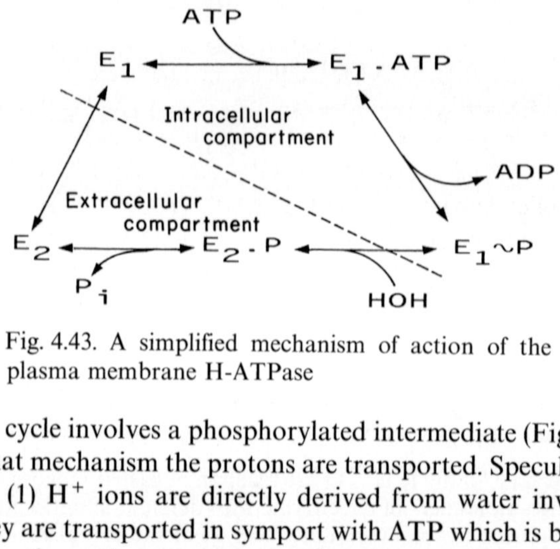

Fig. 4.43. A simplified mechanism of action of the plasma membrane H-ATPase

The reaction cycle involves a phosphorylated intermediate (Fig. 4.43) but it is not clear by what mechanism the protons are transported. Speculations abound in this respect: (1) H^+ ions are directly derived from water involved in ATP splitting; (2) they are transported in symport with ATP which is bound and then dissociates from the active centre near where the β-carboxyl of aspartate is phosphorylated; (3) H^+ ions travel by a membrane Bohr effect mechanism based on conformational changes of the protein; (4) dehydration of the active site may lead to transfer of water to channel sites, accompanied by proton binding and release.

The ATPase can be stimulated (apparently by covalent binding) during active metabolism of glucose and other cytosolic substrates. In this context, a protein kinase is now known to phosphorylate a serine residue of the enzyme.

V-type ATPases

The paradigm of this group of ATPases is the one isolated from yeast tonoplasts (vacuolar membranes). Unlike the F_1F_0 ATPase with its array of multiple subunits, and unlike the single-polypeptide E_1E_2 ATPases, the vacuolar enzyme consists of six subunits (M_r of 75 000, 62 000, 16 000, 14 000, 12 000 and 9000). Its inhibitors dicyclohexylcarbodiimide and N-ethylmaleimide bind to the 9000 subunit, diethylstilbestrol to the 75 000 subunit.

These ATPases do not form a phosphorylated intermediate and are not inhibited by vanadate. They have a pH optimum at 6.5 and in both fungal and plant cells acidify the intravacuolar solution. They show a strong preference for ATP, ADP being completely inactive.

It is not known at present how many protons are translocated per ATP split and the details of this translocation are yet to be elucidated.

The vacuolar ATPase shares the property of being insensitive to vanadate with enzymes found in liver lysosome membranes and in chromaffin granules. Also their apparent molar mass lies between that of the F_1F_0 and of the E_1E_2 ATPases. Whether or not they belong to the same molecular type remains to be established.

4.5.2 The phosphotransferase system

This is the best known of the group translocation systems where the transported solute is chemically changed by a vectorially oriented enzyme. The enzyme relating to phosphoenolpyruvate-dependent sugar:phosphotransferase (PTS) has been found in some twenty bacterial genera comprising cocci, bacilli, spirochetes, purple and green photosynthetic bacteria, and it is probably a ubiquitous enzyme of prokaryotes. Its function is to transport various sugars into cells with concomitant phosphorylation. The sugars so far found to be so transported are D-glucose, D-fructose, D-mannose, L-rhamnose, β-glucosides, lactose, mannitol, glucitol, xylitol, N-acetyl-D-glucosamine, gluconate and dihydroxyacetone, but the spectrum differs in the individual organisms.

The partial reactions that lead to the final translocation involve phosphoenolpyruvate (PEP) as the donor of the phosphoryl group and several enzymes, both membrane-bound and soluble in the cytoplasm (enzyme I, enzyme II, HPr, III), shown in a 'nonstoichiometric' version below.

$$\text{PEP} + \text{EI} \rightleftharpoons \text{P-EI} + \text{pyruvate} \tag{1}$$

$$\text{P-EI} + \text{HPr} \rightleftharpoons \text{P-HPr} + \text{EI} \tag{2}$$

$$\text{P-HPr} + \text{III} \rightleftharpoons \text{P-III} + \text{HPr} \tag{3}$$

$$\text{P-III} + \text{EII} \rightleftharpoons \text{P-EII} + \text{III} \tag{4}$$

$$\text{P-EII} + \text{sugar} \rightleftharpoons \text{P-sugar} + \text{EII} \tag{5}$$

Reactions (1) and (5) both require Mg^{2+} and their equilibrium is heavily displaced to the right.

Enzyme I (M_r 60 000) and HPr (M_r 9500) are constitutive and nonspecific while enzymes II and III are inducible and specific for a given sugar. The phosphorylation of EI proceeds at a histidine residue, giving rise to 3-phosphohistidine. The active center is characterized by a high degree of conservation among various gram-positive and gram-negative bacteria.

HPr is a heat-stable protein which is phosphorylated also at a histidine (His-15) but this time at $N_{(1)}$ of the imidazole ring. Again, the active centre is very well conserved in various HPr proteins. (A fructose-inducible FPr of M_r 8000 also exists.) HPr transmits its phosphoryl group to factor III which is phosphorylated at three sites. These factors are highly specific also in their primary sequence. Their molar masses range from 14 000 ($III^{\beta\text{-glucoside}}$), to 15 000 ($III^{\text{glucitol}}$), to 20 000 ($III^{\text{glucose}}$), to 34 000 ($III^{\text{mannose}}$), to 40 000 ($III^{\text{fructose}}$).

A total of eight enzymes II have been identified so far. They are integral and sugar-specific enzymes of molar masses ranging from 25 000 (EII^{mannose}) to 45 000 (EII^{glucose}) to 65 000 ($EII^{\text{N-acetylglucosamine}}$) to 68 000 ($EII^{\text{mannitol}}$). They apparently occur as dimers that can bind a total of four molecules of III, becoming also phosphorylated during the process. In fact, in the case of the mannitol-specific factor III, it only exists as an insoluble complex with EII.

All these reactions proceed prior to the sugar entering on the scene. The final step is the one where the sugar is phosphorylated via P-E II to a sugar phosphate, the phosphoryl group being in position 6 of glucose but in position 1 in fructose and mannitol.

Besides being a powerful means of sugar transport, the PTS has a regulatory function in that it suppresses the uptake of nonPTS sugars by what is called inducer exclusion. The molecular basis of the exclusion is in the interaction of P-III with adenylate cyclase. The PTS itself is subject to regulation by excess intracellular ATP which phosphorylates Ser-46 in the HPr and prevents it from serving in the PTS.

4.5.3 Group translocation without phosphate

Several other transport systems are known where a substrate at the starting (*cis*) side is different from the product at the target (*trans*) side.

Of the more important ones let us mention the γ-glutamyltranspeptidase which in some vertebrate organs can serve as an accessory system for amino acid transport. An external amino acid interacts with the membrane-bound enzyme which forms a peptide bond between the amino acid NH_2 group and the γ-carboxyl group of glutamate derived from glutathione. The dipeptide emerges in the cell to be then hydrolyzed and the glutathione resynthesized at a considerable energy expense.

Another such system is represented by the carnitine cycle of mitochondrial inner membranes (Fig. 4.44). The two carnitine acyl transferases differ in their affinity for carnitine in such a way that carnitine and acyl carnitine cycle within the membrane between the two enzymes and never escape the membrane. Hence, it is a sort of cryptic group translocation.

SYNOPSIS

Many primary transport processes are driven by the energy released upon hydrolysis of 'energy-rich' compounds. Foremost among them are the transports

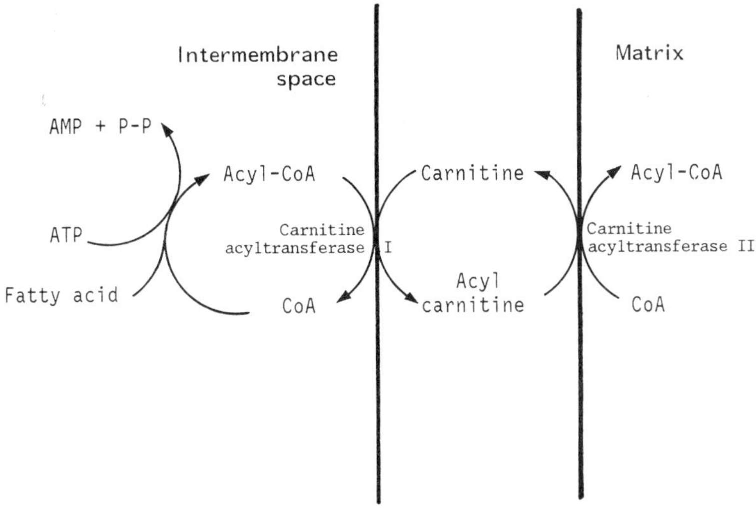

Fig. 4.44. Operation of the carnitine cycle in the inner mitochondrial membrane

of cations and, exceptionally, anions catalyzed by membrane adenosinetriphosphatases. Besides the $F_1F_0H^+$-translocating ATPase which, under physiological conditions, mainly serves the synthesis of ATP, there are two major groups of these enzymes, those that contain a phosphorylated intermediate in their reaction cycle (E_1E_2 ATPases) and those that do not (the vacuolar-type ATPases).

Another transport driven by a hydrolytic splitting of phosphate is the bacterial phosphotransferase where, however, the phosphate appears in the transported species at the target side, the mechanism being called group translocation.

In addition to this, group translocation can be driven by splitting of acyl-CoA or of the glutamyl residue of glutathione under rather special conditions.

FURTHER READING

Carafoli, E., and Scarpa, A. (eds). Transport ATPases, *Ann. N. Y. Acad. Sci.*, **402**, (1982).

Harold, F. M., and Kanikuma, Y. Primary and secondary transport of cations in bacteria, *Ann. N. Y. Acad. Sci.*, **456**, 375–83 (1985).

Lichko, L. P., and Okorokov, L. A. What family of ATPases does vacuolar H^+-ATPase belong to? *FEBS Lett.*, **187**, 349–52 (1985).

Rabon, E., Gunther, R. D., Soumarmon, A., Bassilian, S., Lewin, M., and Sachs, G. Solubilization and reconstitution of the gastric K, H-ATPase, *J. Biol. Chem.*, **260**, 10200–7 (1985).

Villalobo, A., Brown, L., and Roufogalis, B. D. Kinetic properties of the purified Ca^{2+}-translocating ATPase from human erythrocyte plasma membrane, *Biochim. Biophys. Acta*, **854**, 9–20 (1986).

5 Processing of Information in Membranes

The third category of functions performed by biomembranes is characterized by the flow of information. Without any transport of matter across the membranes and with minimum energy involvement, entropy is transmitted to cells via membrane interactions with information-containing molecules or physical signals arriving from the cell exterior.

The relation between entropy and information, as well as the intricacies of distinguishing between the quantity and the quality of information, will be dealt with in the following section.

5.A INFORMATION PRINCIPLES IN BIOLOGY

The principal concept of the information theory which was developed mostly for communication purposes is the amount of information I defined as

$$I = \log_2 (p'/p) \tag{5.A1}$$

where p is the a priori probability of an event and p' its probability after a message concerning the event has been received. Information is expressed in binary units called bits, one bit corresponding to reliable information $p' = 1$ whether an event with an a priori probability of 1/2 occurred or not. Provided that all messages received about the system are reliable

$$I = \log_2 (1/p) = -\log_2 p. \tag{5.A2}$$

From statistical thermodynamics, $1/p$ is equal to thermodynamic probability W, which is the number of ways in which a given system can be realized, i.e., the number of microstates compatible with a certain macrostate. Thus

$$I = \log_2 W \tag{5.A3}$$

and since entropy S is related to the thermodynamic probability W by

$$S = k_B \ln W \tag{5.A6}$$

it is obvious that the amount of information and the entropy of a system are proportional quantities. The entropy of a system corresponds to the information that is necessary for its full description, i.e., for the description of its microstate.

Approximate calculations of the amounts of information stored in macromolecules, cells and organisms are possible with the result that, contrary to a rather wide-spread belief, the living matter does not contain more information or, it is not more ordered than a crystal of the same mass—a bacterial cell is said to contain $4 \cdot 10^{12}$ bits of information, much the same as an equivalent mass of a hydrocarbon. An important difference, however, rests in the fact that the living matter may be likened to an aperiodic crystal, so that the information contained in it is less redundant, is less repetitive than that in a periodic structure and it is rather obvious that it is for this reason more valuable. There is another reason why concepts such as the amount of information or entropy are not suitable to describe specific characteristics of living systems. Such systems should not be considered as statistical systems, but rather as dynamic systems or machines composed of macroscopic parts, from macromolecules to cells, tissues and organs. It is then more important to know how valuable a given piece of information is for the recipient and what problems may be solved after receiving it, provided some 'pool' of knowledge exists prior to the communication of the message. A different measure of information than its amount is then required, which may be suitably called the quality, the sense or the value of information.

The difference between the amount of information and its value is all-important, e.g., an active gene contains the same amount of information as a repressed one, but the information contained in it has no value, at least none while it is repressed.

A formal theory of information that would take into account its quality or value does not yet exist and may be difficult to construct. Thus, it was mentioned already that redundancy decreases the value of information, but, on the other hand, some redundancy is valuable, since it helps to overcome informational noise both during the transmission and reception of a message. Valuable information and nonredundant information are not the same thing. An example from nonliving nature, showing both the reception of a valuable information and its result, is the insertion of a small crystal into a supersaturated solution of the same substance. The system is in a metastable state and thus the information received about how the crystalline lattice should be constructed initiates a spontaneous process, which could possibly occur even without such information, as a result of fluctuations. A crystal of a different substance may contain the same amount of information, but this information is without value in the given case, for it cannot readily initiate crystallization of this particular solution.

The situation in biological trigger systems which exist as systems in metastable state far from equilibrium, is analogous. A spontaneous process is initiated as soon as a useful piece of information stored in some molecule, e.g., hormone, is received. What kind of molecule can pass the useful information has been decided long ago, in the process of evolution, but now the information about it is stored in the system and the molecule is remembered by the receptor in a similar way as a randomly chosen combination of numbers is remembered by a safety lock.

On the whole, the cases of information processing mentioned farther in this chapter are all examples of being able to recognize the arriving message or signal as such. The specificity of this recognition is almost always equivalent to that found in enzyme-substrate systems—occasionally the selectivity of processing the information is even greater than in an enzymic conversion of a substrate into a product as a given information (e.g., hormone) receptor and its associated molecules will determine what effect the received information will have in the cell. In this sense, a relatively small molecule can carry vastly different amounts of information, depending exclusively on the addressee.

The information-carrying signals can be either physical or chemical in nature, thus (1) light, heat, magnetic field, mechanical pressure, electric potential, or (2) chemotactic attractants or repellents, hormones, neurotransmitters and antigens.

It would require writing a sizable volume on cell biology, physiology and immunology to do justice to all the information-transmitting interactions and thus of necessity only examples of each type of interaction will be given.

Likewise, transmission of information not involving membranes (cytosolic enzyme activators, steroid hormones) will be mentioned only when necessary to complement the processes taking place primarily in membranes.

5.1 PHYSICAL SIGNALS

5.1.1 Light as a signal

A flux of photons absorbed by cells can serve three different purposes. The first of these, conversion of radiation energy into ionic gradients and ATP, is the basis of photosynthesis and has been discussed in Section 4.2. The second purpose is to steer a cell toward the optimal level of illuminance—a process observed in various motile unicellular organisms. The third is then to generate a chemical and electric response that is interpreted by the central nervous system as vision.*

The second and third category represent membrane-mediated transfer of information and will be treated separately.

Photomovement

Movement regulated by light in unicellular organisms can be of three types. *Photokinesis* is movement depending on light intensity (irradiance, in Einstein or mole per m^2 per second) regardless of the direction from which the light is coming. A sudden change in irradiance can bring about a *photophobic response*, a sudden but transient change of direction of movement. *Phototaxis* is a movement that is governed by the direction of the incoming light, either positively or negatively. It can proceed either by an actively oriented movement toward or away from the light source, or by a time-averaged movement in the desired direction, consisting, however, of short stretches of movement in a more or less random direction.

The photoresponse in general is associated with the presence of pigments in cell membranes, either those also active in photosynthesis or different. The distinct character of the three photoresponses is seen in their action spectra (Fig. 5.1) in a cyanobacterium.

Not only prokaryotic cyanobacteria, purple bacteria and archaebacteria but also eukaryotic organisms, such as the protozoan *Chlamydomonas*, or even the eyeless mites contain photo-active pigments, generally with retinal as the chromophore-bearing group.

The most carefully investigated phototactic response is that of various *Halobacterium* species where a retinal-containing protein similar to bacteriorhodopsin and halorhodopsin has been found, with absorption peaks at 370 nm and 587 nm, and named sR (for *s*low-cycling or *s*ensory rhodopsin). It is not associated with movement of ions but it undergoes a cycle of intermediate forms (Fig. 5.2). When illuminated with green light at about 587 nm it converts via two intermediates back to the original form—this photocycle produces a

*The pluripotent growth and germination regulator, the red- and far-red-absorbing *phytochrome*, is not an integral membrane component and its action will not be discussed here.

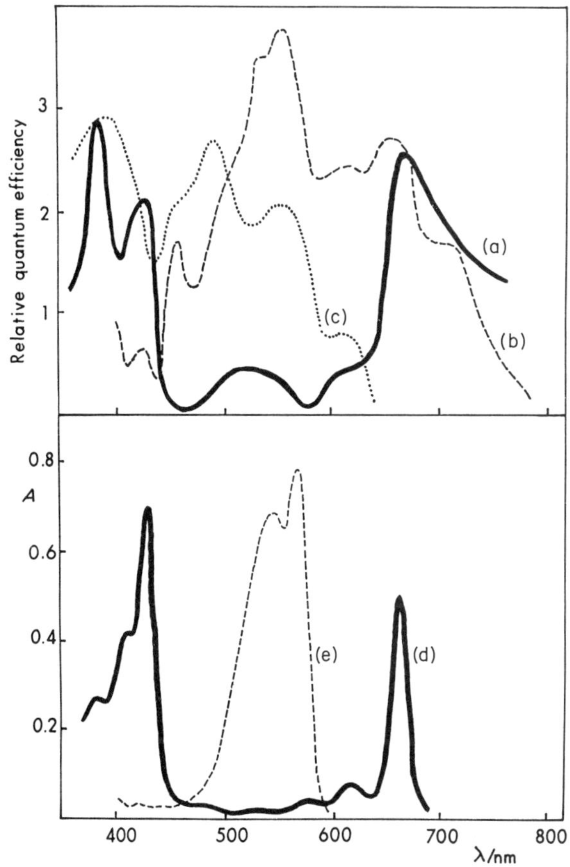

Fig. 5.1. Upper panel: Action spectra of photokinesis (a), photophobic response (b) and phototaxis (c) in the cyanobacterium *Phormidium uncinatum*. Lower panel: Absorbance of chlorophyll *a* (d) and phycoerythrin C (e). Clearly, photokinesis is mediated by chlorophyll *a* as signal receptor while photophobic behaviour is initiated by absorption of light in the phycobilisome. Phototaxis could not be ascribed to any single pigment

positive phototactic response. If, however, it also absorbs light in the blue or ultraviolet range (either simultaneously with or immediately after the green light) the sR_{373} form is converted back to sR_{587} via another form which gives rise to a negative phototactic response. Still, the possibility that actually two distinct pigments P_{560} and P_{370} catalyze the two different responses is not excluded.

The stimulus received by a photosensitive cell is then transduced into movement. This sensory transformation is the least documented part of the process, candidates for the function being (1) a diffusible low-molecular-weight or protein molecule—this might be the case in the halobacteria; (2) membrane electric potential—this might be the mediator in cyanobacterial filaments composed of many cells; (3) the redox state of the membrane—this may be a

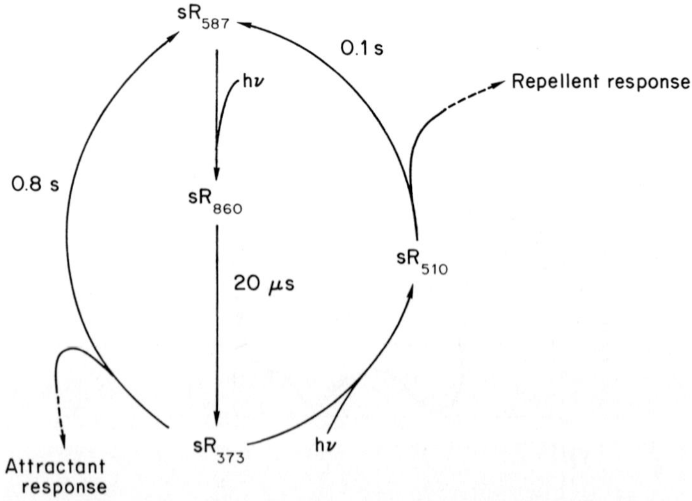

Fig. 5.2. Photocycle of the sensory rhodopsin of halobacteria. Details are in the text

general mechanism; (4) Ca^{2+} ions—these might amplify the membrane potential response. There is, however, evidence that protonmotive force is the universal transducer of the information in photosensitive cells, as well as in chemotaxis as will be shown below.

The mechanism of actual movement is of two types. In the motile bacteria, it involves clockwise and counterclockwise rotation of their flagella, controlled by the polarity of the membrane potential or rather $\Delta\tilde{\mu}_{H^+}$. In the gliding bacteria and cyanobacteria, a vectorially extruded slime bed could propel the filaments in a rocket-like fashion or, more probably, an undulating movement of the cell envelope could move it like a worm.

Visual rhodopsin

In a remarkably conserved evolutionary line, the light-perceiving system of vertebrate as well as invertebrate eyes is always the retinal-containing rhodopsin.

It is an integral protein spanning the membrane seven times like bacteriorhodopsin with a peptide chain of 348 amino acids and with carbohydrate residues at Asn-2 and Asn-15. It is located, in vertebrate cone cells, in multiple invaginations of the plasma membrane, while in vertebrate rod cells it is in an intracellular array of several hundred flattened membranes in the rod outer segment. In invertebrate rods, it is placed in characteristic microvilli.

The chromophoric retinal is in the 11-*cis* form or as 3-dehydroretinal. The maximum absorption of the vertebrate rhodopsin is at 498 nm, with a quantum yield of 0.7.

The absorption of a photon by rhodopsin triggers a complicated sequence of reactions, both photochemical and biochemical in nature.

The first events include the sequence rhodopsin (498 nm, 11-*cis*-H) $\xrightarrow{\text{ps}}$ hypsorhodopsin (430 nm, all-*trans*) $\xrightarrow{\text{ps}}$ bathorhodopsin (543 nm, all-*trans*-H) $\xrightarrow{\text{ns}}$ lumirhodopsin (497 nm, all-*trans*-H) $\xrightarrow{\mu\text{s}}$ metarhodopsin I (478 nm, all-*trans*-H) $\xrightarrow{\text{ns}}$ metarhodopsin II (380 nm, all-*trans*) $\xrightarrow{\text{min}}$ metarhodopsin III (465 nm, all-*trans*-H) $\rightarrow$ all-*trans* retinal + opsin. The conversion of metaR 1 into metaR II entails a major conformational change of the apoprotein opsin.

Probably metaR II, as the most stable intermediate, starts the sequence of biochemical reactions that eventually transmit the information in an adapted form to the nervous system (Fig. 5.3). The figure does not contain data on the amplification character of the system. It appears that metaR II can activate as many as 500 transducin molecules, each of which, via the phosphodiesterase, converts more than 1000 cyclic GMP molecules to linear GMP. The important aspect of the cascade is that cGMP which is present at a relatively high concentration of up to 60 μM, keeps the Na^+ channel in the membrane open. This open form represents the inactive state of the vertebrate rod cell membrane (in invertebrates the inactive state has its Na^+ channels closed) and it requires unbinding of cGMP to close the channels and thus to activate the membrane. There are some 10^5 Na^+ channels per cell but the inactive state has only some 5000 of them actually open. It is these that close down on activation. A single photon can be shown to prevent the inflow of more than 10^6 Na^+ ions. The closing of the channels may (but this is still speculative) enhance the operation of a Ca^{2+}/Na^+ antiport system causing outflow of Ca^{2+} ions into the synaptic cleft across which the rod cell communicates with neurons and the central nervous system. It will be seen later that Ca^{2+} can trigger an electric response of the neuron and thus act as an information carrier between the light-absorbing rod cell and the centripetal cells of the nervous system.

In another theory it is assumed that the hyperpolarization of the membrane attendant on illumination not only closes the Na^{2+} channels but also the voltage-gated Ca^{2+} channels and this then prevents Ca^{2+} from moving in. Ca^{2+} ions are held responsible for the outflow of an inhibitory chemical mediator (see section on neurotransmitters) which thus stops flowing out and the postsynaptic neurons are released from inhibition. Glutamate may play the role of this inhibitory mediator.

5.1.2 Heat as a signal

Thermotaxis is a phenomenon whereby bacteria escape from undesirable temperature. The feeling of hotness or coldness by sensitive bodies in the skin is a sensory input common to all vertebrates and probably many invertebrates.

Although heat is equivalent to the energy of far IR radiation it is not possible to tell whether the membranes of the sensory cell contain absorbers in this spectral

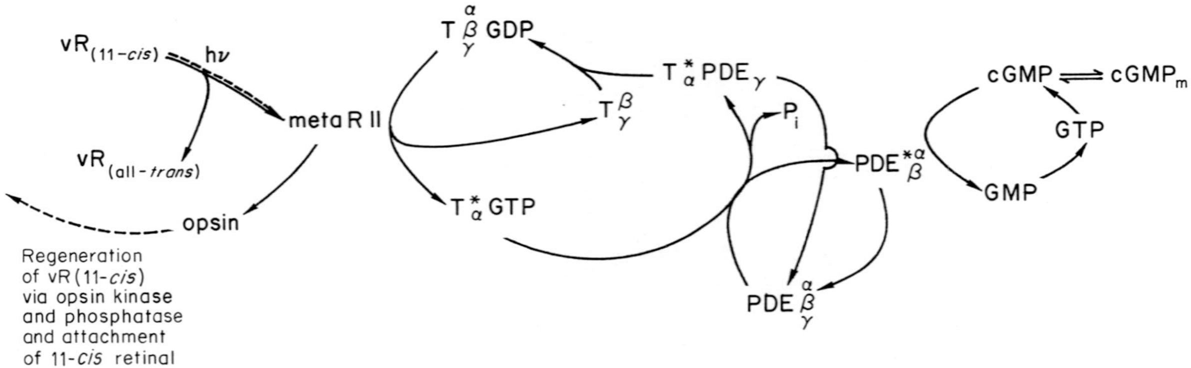

Fig. 5.3. Model of the processing of information of a photon absorbed by the retinal rod outer segment. The active metarhodopsin II interacts with a G-protein analogous to that found in hormone signal transduction, called transducin (T), with subunits of M_r 39 000. 36 000 and 8400, whose α-subunit carries GDP. The interaction breaks the complex and causes an exchange of GDP for GTP. T^*_α (called also lumilamban) now activates phosphodiesterase PDE by complexing its γ-subunit. The activated *PDEαβ now isomerizes cyclic GMP to its linear form 5′–GMP which can be reconverted back to cGMP by energy-requiring reactions

range. It would be tempting to speculate that the state of the membrane lipids could respond to differences in temperature but the graded response of the sensory cells rules out a simple mechanism, such as phase transition.

There is compelling evidence, however, that the temperature sensors are proteins. In *Escherichia coli* the function is performed by the product of the *TSR* gene which is membrane-bound and serves, at the same time, as the chemotactic binding protein for serine, alanine and leucine.

It remains to be seen whether the primary response consists in a reorganization of the receptor protein or in a shift of equilibrium of some enzyme-catalyzed reaction which then brings about the usual train of events, including G proteins, cAMP or cGMP and opening and closing of ionic channels in the cell membrane.

The situation may be different in other cells. In the slime mold *Dictyostelium discoideum* a low-molecular-weight substance is released, called the slug turning factor (STF) which interferes with both phototaxis and thermotaxis. The slug form of the slime mold can compare the concentrations of the STF at both sides of its body and steer its movement toward the light source or optimal temperature where the concentration of STF is lower.

5.1.3 Electric potential as a signal

Many cells in the nervous system are excited by the passage of a depolarization of the membrane, the action potential, that commonly originates in the nerve cell body as a response to a set of chemical and electric signals arriving there from different dendrites. The subsequent opening of sodium channels is an extremely important way of conducting nerve impulses over long distances in the body and it will be discussed in connection with the action of chemical substances on membrane receptors.

However, it can be hardly considered as a signal to which the cell specifically responds—there are no receptors for electric potential or field known, besides a general orientation of dipoles in the membrane at very high electric potential gradients. It is this reorientation that underlies the mechanism of function of voltage-gated channels. It would be a matter of semantics to decide whether this represents a sensing of information. The response to serving as a conductor to electric current in a temporarily closed circuit (after touching a live wire while standing barefooted on a wet floor) is due to a massive depolarization of cell membranes rather than to sensing an external electric field.

5.1.4 Magnetic field as a signal

Response to a magnetic field is a very exceptional phenomenon, limited to some anaerobic motile bacteria which contain grains of magnetite near the membrane base-plates of their flagella. They include bacilli, cocci and spirilla living in fresh-water sediments and at a certain depth in the sea. The magnetic grains are oriented according to the lines of force of the geomagnetic field which dip at a certain angle into the ground. As the bacteria in the northern hemisphere seek the northern magnetic pole (in fact, the south pole) they follow the lines of force down to a certain depth in the sea where they find sufficiently anaerobic conditions. Presumably the ones living in the southern hemisphere are steered to the opposite magnetic pole.

It is not clear what the mechanism of sensing the magnetic field and its orientation might be in migratory birds where this is presumably one of the principal means of their navigation.

5.1.5 Response to mechanical pressure

Taken in a broad sense, two types of membrane-located phenomena can be interpreted as a response to pressure, (1) interaction of cells with and adhesion to a solid support, and (2) registration of sound and of body position by specialized cells of the inner ear.

Cell adhesion

The principal protein involved in adhesion of mammalian cells to an extracellular matrix, such as collagen, as well to nonbiological substrata, (such as glass, polyethylene or polystyrene) is a peripheral protein called fibronectin. It is a V-shaped molecule composed of two peptides (M_r 250 000) connected by two disulphide bridges. Each peptide has a binding site for a cell membrane receptor (M_r 140 000) which, via vinculin, is linked to actin filaments in the cell (Fig. 5.4). It is not clear what binds fibronectin to artificial supports. Ca^{2+} ions are probably involved in the process.

Fibronectin is synthesized in the rough endoplasmic reticulum of many mammalian cells and arrives at the surface within 30–60 min of the stimulus.

Similar peripheral proteins found in other cells are laminin (M_r 2 × 220 000 plus 400 000), chondronectin (M_r 2 × 80 000) and entactin (M_r 158 000). In addition to these, various integral proteins are specifically involved in cell-to-cell adhesion, in particular the CAM-105 protein (for *c*ell-*a*dhesion *m*ediating) of M_r of 150 000 plus 110 000 plus 30 per cent carbohydrate.

There is evidence that all these proteins are synthesized or translocated along the membrane plane upon contact with other cells or with a nonspecific artificially provided surface. There is no known specific receptor that would

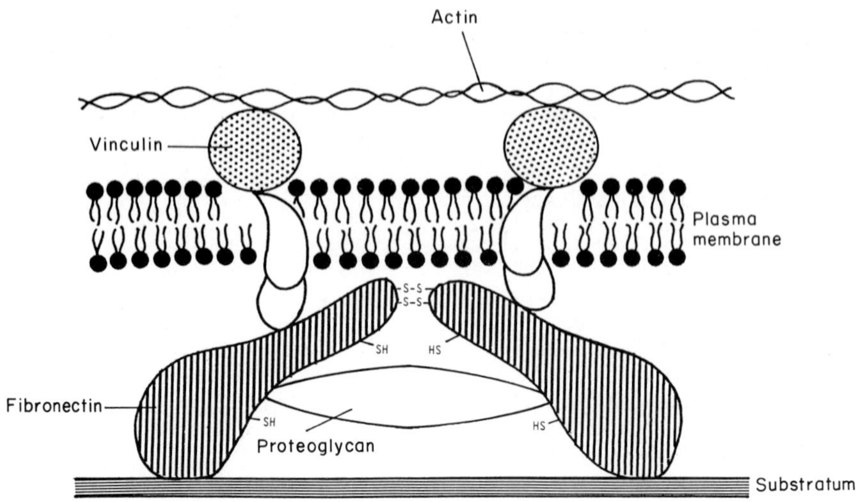

Fig. 5.4. Schematic diagram of fibronectin localization on the plasma membrane

respond to mechanical contact; its role might be played either by the fibronectin receptor itself or by glycoproteins with bulky hydrophilic parts that would somehow sense the mechanical pressure which might be transmitted in some form of intermolecular force.

The pluripotent binding sites of laminin and chondronectin may even catalyze some mechanisms of infection. Thus *Staphylococcus aureus* has surface receptors (about 100 per cell) for laminin which lines blood vessels and epithelial cells. Nonvirulent forms do not possess these receptors.

Inner ear hair cells

The cochlea of the inner ear contains special hair cells, some 25 000 of them in a human ear, that carry on their surface 50–100 projections called *stereocilia*, each on a thin stalk, formed by actin filaments. The cells are covered above the projections by a tectorial membrane. As sound waves propagate along the underlying basilar membrane, the stereocilia are deflected by the tectorial membrane. Because frequency of sound is transformed in the ear into amplitude, different pitch will produce a topologically different response along the tectorial membrane. By being tilted, stereocilia somehow manage to open Ca^{2+} channels, the Ca^{2+} moving in them then causes an outpouring of a chemical transmitter which interacts with the adjacent nerve dendrites and fires an action potential to the central nervous system (Fig. 5.5).

Special hair cells are found in the part of the inner ear where the animal is informed about its position in space. These hair cells contain, in addition to stereocilia, one *kinocilium*, a heavier, microtubule-containing projection which

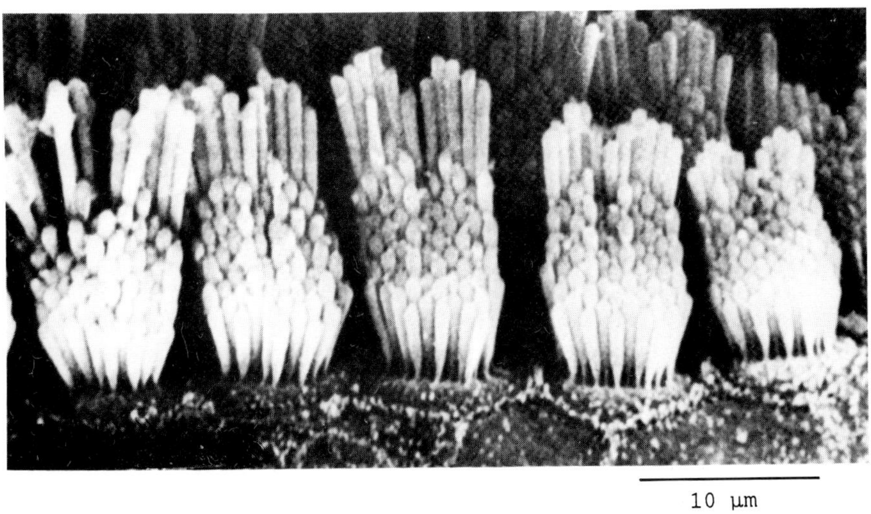

Fig. 5.5. Hair cells of the inner ear

responds to a change in position simply due to its inertia and, like above, tilts the hair cells which then give rise to the same train of events.

SYNOPSIS

Cell membranes contain receptors that respond to various physical signals, among them light, heat, magnetic field, and mechanical pressure. Light mobilizes some photoreceptive motile bacteria and cyanobacteria which seek the optimum level of illuminance. In animals possessing visual sensory organs rhodopsin is activated by light and, through a series of intra- and extracellular reactions, informs the central nervous system about the event. Temperature is sensed by an unknown mechanism by both bacteria and skin receptors of higher organisms while magnetic field elicits a kinetic response only in special magnetobacteria. Mechanical pressure can serve as a signal as cells collide or as they get in contact with a substratum. In special hair cells of the inner ear both the pitch of a sound and body position in space are registered by a mechanical pressure transmission.

FURTHER READING

Abdulaev, N. G., and Artamonov, I. D. Bioorganic chemistry of vision, *Biologicheskie Membrany*, **1**,775–93 (1984).

Blyumenfeld, L. A. Problemy biologicheskoy fiziki, *Nauka*, Moscow (1977).

Findlay, J. B. C., and Pappiu, D. J. C. The opsin family of proteins, *Biochem. J.*, **238**, 625–642 (1986).

Foster, K. W., Saranak, J., Patel, N., Zarilli, G., Okobe, M., Kline, T., and Nakanishi, K. A rhodopsin is the functional photoreceptor for phototaxis in the unicellular eukaryote Chlamydomonas, *Nature*, **311**, 756–9 (1984).

Häder, D.-P. Signal perception and amplification in photomovement of prokaryotes, *Biochim. Biophys. Acta*, **864**, 107–22 (1986).

Hynes, R. Molecular biology of fibronectin, *Annu. Rev. Cell Biol.*, **1**, 67–91 (1985).

Volkenstein, M. V. The amount and value of information in biology, *Foundations of Physics*, **7**, 97–109 (1974).

5.2 CHEMICAL ATTRACTANTS AND REPELLENTS

Motile bacteria respond to the presence of a variety of substances of nutritive character, but also to O_2, H^+ and various toxic compounds, by altering their swimming pattern.

These bacteria move in two different ways. When their flagella rotate counterclockwise (viewed from the cell body) they all move 'in phase' and propel the cell along a smooth path which may last for 1 to 20 s, depending on conditions. When their flagella move clockwise they are out of phase and the cell stops and tumbles before resuming its smooth swimming. It is believed that the rotation of the flagella is driven by the electrochemical potential gradient of H^+ or Na^+ (see Section 4.4.3). It is not known, however, why the motor reverses its direction from time to time—certainly, the proton gradient does not flip back and forth on its own.

The presence of a proton or sodium ion gradient is not the only prerequisite of flagellar rotation. There are proteins that transmit the information to the motor and these proteins are regulated in different ways.

5.2.1 Methylation-dependent chemotaxis

Escherichia coli, Salmonella typhimurium but also bacilli and other prokaryotes contain in their cytoplasmic membrane several different acceptors, four of which have been identified—they are often called MCP, for *m*ethyl-accepting *c*hemotaxis *p*roteins. The proteins bind aspartate, malate and divalent metal ions (Tar), ribose, galactose and glucose (Trg), serine, alanine and leucine (Tsr) and dipeptides (Tap).

While the Tsr protein interacts with its ligands directly, the Trg, Tap and Tar proteins require the presence in the periplasm of so-called binding proteins for some of their respective substrates, these proteins furnishing their ligands also to the appropriate transport systems which represent integral membrane proteins different from Trg and Tar (Fig. 5.6).

Once an MCP binds its ligand it is activated and causes the synthesis of an intracellular mediator which somehow regulates the flagellar motor. It is believed that a *CHE Y* gene product causes it to turn clockwise while the *CHE Z* gene product turns it counterclockwise. Hence, this last-named protein is set loose upon positive activation. A total of at least eight genes regulate this process.

Now the bacterium tends to swim smoothly in its 'normal mode'. If it swims toward the source of positive activation (an attractant), it collects further ligands so that it is maintained in the active ccw mode. After some time it tumbles for reasons that are ill known, and proceeds swimming smoothly, in a random direction. If the direction happens to be away from the attractant it begins to lose its ligands, the MCP is deactivated, the che Z protein is not set free any more and the bacterium tumbles sooner. Averaged over many changes of direction, the time spent in smooth swimming toward the attractant source is thus longer than in the opposite direction so that the bacterium approaches its attractant.

If, instead of an attractant, a repellent that binds to one of the MCPs is present, the MCP would be brought into a state when it sets free the che Y protein and the bacterium would tumble frequently. It would tumble less frequently if it moved away from the repellent because the MCP would be less negatively activated.

However, like in most signal-transducing systems, the chemotactic response is not constant for a longer period of time but rather undergoes an adaptation when it does not respond to the stimulus any more. The adaptation is accomplished by methylation of a glutamate in the protein through the action of a methyltransferase using S-adenosylmethionine as the source of the methyl group. The protein is then brought into an adapted state when it does not set off the release of the che Z protein. When stimulation ceases, i.e. when the ligand dissociates from MCP, the MCP switches to a state resembling that when a repellent is bound and then returns, by demethylation catalyzed by a methyl esterase, to the initial state.

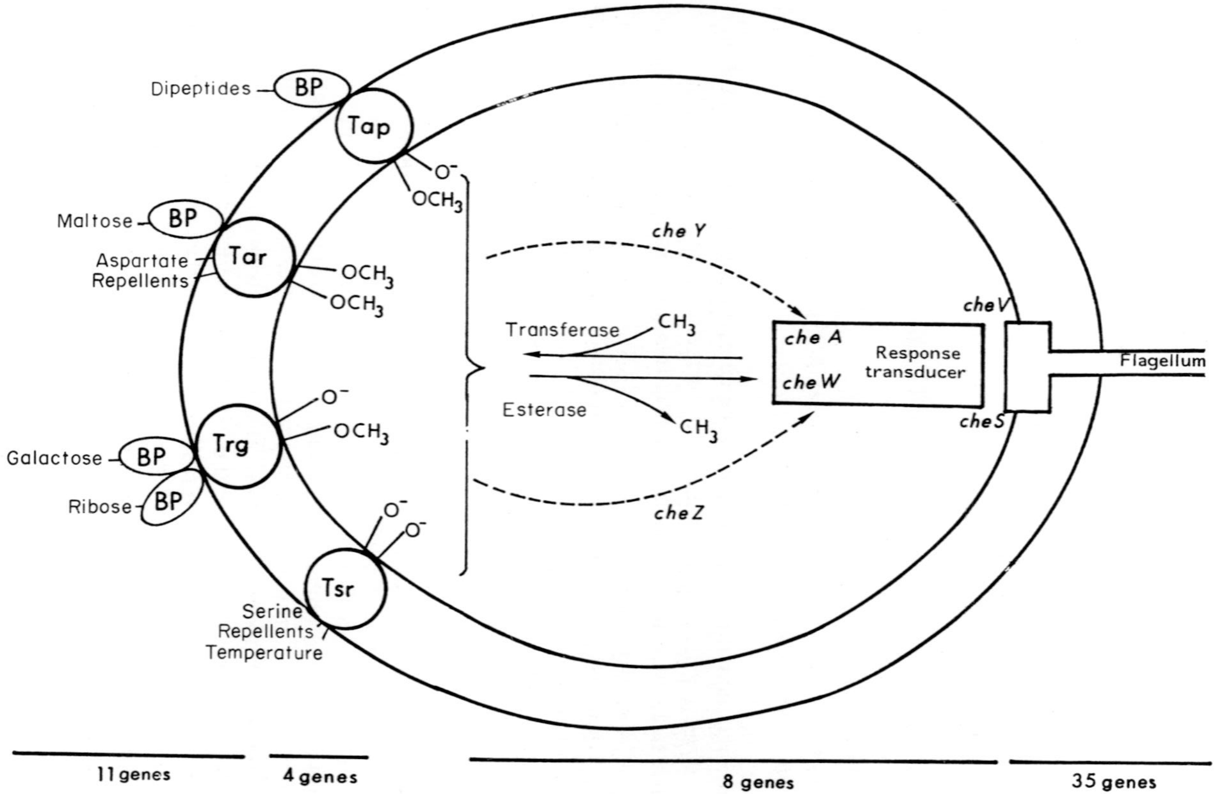

Fig. 5.6. The methylation-dependent chemotaxis in *Escherichia coli*

5.2.2 Methylation-independent chemotaxis

A variety of signals, both positive and negative, interact with bacterial membranes at other sites and elicit a chemotactic response in bacteria, such as *E. coli*, as well as in purple and halophilic bacteria. The sites involve cytochrome oxidase (positive effect of low concentrations of oxygen—this is called *aerotaxis*, and a negative effect of CN^-), nitrate reductase (positive effect of nitrate) and fumarate reductase (positive effect of fumarate). General 'repellents' appear to be valinomycin ($+K^+$), carbonylcyanide chlorophenylhydrazone and other drugs that decrease the protonmotive force across the membrane. The same 'repellent' effect is shown by higher concentrations of oxygen.

It is believed that the sensory response in this case rests in changes of protonmotive force. This would be a straightforward way of influencing the flagellar motor which is probably driven by the electrochemical potential gradient of protons but the situation is more complicated as here, too, an adaptation phenomenon has been observed. This obviously requires an intermediary link, a protein that has been given the name *protometer* and that may undergo phosphorylation or adenylation as the way of adapting it to further stimulus input.

The possibility also exists that not the protonmotive force but simply intracellular pH is the actual signal that is perceived by the protometer.

Enzyme II of the phosphotransferase system specific for mannitol and another enzyme II specfic for glucose also act as chemotactic receptors for their respective ligands. Although their adaptation is also methylation-independent they apparently do not act through the protonmotive force protometer.

Both the methylation-independent and methylation-dependent ways apparently converge at the site of the response transducer associated with the flagellar motor.

Much is to be done in this area particularly with respect to the response to opposing effects of attractants and repellents of the one and of the other type of chemotactic response.

SYNOPSIS

Many bacterial and some eukaryotic species are attracted by nutrients or sources of energy and repelled by toxic agents. The tactic movement toward nutrients and away from repellents is initiated by a membrane-mediated response which is transmitted to intracellular transducers and eventually to the mechanical device capable of moving the cells. The membrane receptors can adapt to continuously arriving signals either by being methylated or by some other chemical reaction, e.g. phosphorylation.

FURTHER READING

Adler, J. The behavior of rats, bacteria and man. In: *Biochemistry of Metabolic Processes* (D. L. F. Lennon, F. W. Stratman, and R. N. Zahlten, eds), pp. 367–83, Elsevier, Amsterdam (1983).

Bonner, J. T. Chemical signals in social amoebae, *Sci. Amer.* **248**, 114–20 (1983).
Taylor, B. L. Role of proton motive force in sensory transduction in bacteria, *Annu. Rev. Microbiol.*, **37**, 551–73 (1983).

5.3 CHEMICAL SIGNALLING IN MULTICELLULAR ORGANISMS

There are two types of chemical signals that are perceived by a multicellular organism (most pronouncedly, a vertebrate), extrinsic and intrinsic. The first are represented by odours and flavours, sensed by olfactory and gustatory receptors, respectively. The second are represented by hormones and a variety of transmitter compounds, foremost among them neurotransmitters.

5.3.1 Extrinsic chemical signals

The organs of smell of higher animals are all characterized by a composite cell layer, the olfactory mucous membrane, which is innervated by olfactory nerve fibres. These fibres end in cell-like dendrites located between supportive sustentacular cells, the dendrites terminating in finger-like cilia. It is in these cilia that olfactory receptors are located, principally glycoproteins of molecular weights in the vicinity of 100 000. Binding of an odoriferous molecule (an odogen) to the receptor causes activation of a stimulating GTP-binding protein (G_s) and brings into a state of alert other G proteins, such as G_i and G_o. The concerted action of these proteins then provides a modulated signal to the adenylate cyclase and, subsequently, to membrane ion channels which open, locally depolarize the membrane and, via an action potential, inform the appropriate parts of the central nervous system. (More on this mechanism will be presented in Section 5.3.2 and 5.3.3.)

Human brain can differentiate between 2000 and 4000 odours, mainly molecules containing 4–20 atoms, the receptors for which are located on the cilia of different dendrites but this does not mean that each of the odours would have its receptors on a special cell. Apparently, different receptors may be present on one dendrite.

Unlike smells, tastes can be easily divided into four categories, sweet, sour, salt and bitter. Receptor cells, terminating with a 'hair' are grouped by 5–18 in taste buds, structures about 60 μm in diameter, that contain also some supporting cells. Taste buds are located in vertebrate tongue epithelium and their primary response to binding a 'tasty' molecule or ion results in Na^+ channel opening and its consequences. However, although some receptor molecules have been isolated (the sweet-sensing receptor of the bovine tongue was the first), it is not clear why substances very different structurally may evoke the same response, e.g. sucrose, saccharin, perillaldehyde aldoxime and aspartylphenylalanine methyl ester.

5.3.2 Intrinsic chemical signals

Chemical compounds circulating in the body fluids of multicellular animals are many and diverse. Their function may be osmotic or nutritional or they may serve as messengers carrying important information from the place of their synthesis to the target cell. These are grouped in the category of hormones and their chemical structure and intracellular effects are extremely varied. Their mechanisms of action, however, follow a few basic patterns and these will represent the subject of the first part of this section. In the second part, we shall describe the mechanisms of action of a special class of compounds, called neurotransmitters which, as the name suggests, carry information between sensory cells, cells of the nervous system and effector cells, such as those of muscles, and often in both directions. It will be seen that some of the substances of this second group are identical with those of the first.

Hormones as first messengers

Although no rigid classification can account for all the signal molecules that exist in a multicellular organism, hormones are usually considered to be products of endocrine gland cells which release them into the bloodstream.

Travelling with blood, the hormones eventually reach their target cells which may be quite distant from the source of the hormone. The hormone secretion itself is controlled by a train of regulatory events that commonly begin in a brain cell which transfers the message to the hypothalamus. In this special region of the brain the message is transformed into producing an agent (a hormone of a kind) that travels through the blood to reach an endocrine gland (the pituitary or hypophysis is the primary target). This then releases another agent (again a hormone) which travels to another endocrine gland which, upon receiving the signal, produces its final hormone. This is the hormone that is destined to reach cells in the body and bring about changes in their metabolism or protein synthesis. Thus, although such hormones are sometimes called the first messengers, they usually represent the fourth link in a chain that starts with a signal in the brain which in its turn, may have crossed several intercellular gaps on its way from the primary sensory cell.

An example of such scenario is the production of growth hormones. An endogenous stimulus of unknown nature arriving to the brain stimulates the hypothalamus to produce a factor which is carried by blood in the pituitary stalk to the anterior lobe of the pituitary. There it initiates the secretion of somatotropin, a polypeptide, which is then carried to liver. This acts now as an endocrine gland to produce another pair of peptides, the somatomedins 1 and 2. These are the final products of the signal chain which are then transported to various organs, bind to cell plasma membranes and initiate growth of the organ.

Three-score such hormones are known from mammals, others occur in other vertebrates, in some invertebrates, and in plants.

An important aspect of hormone action is that the final hormone product may produce different responses in different cells, i.e. the specificity of action in the last instance depends not exclusively on hormone structure but on the receptor with which it interacts.

According to the localization of their receptor, hormones can be divided into those acting at the cell surface and those acting intracellularly. The latter are mainly, if not exclusively, lipid-soluble hormones, first of all steroids, such as progesterone, testosterone, various corticosteroids and aldosterone, but also thyroxine. These interact with receptors associated with the cell nucleus and their effects are mainly oriented toward modification of DNA transcription and stability of messenger RNA so that the macroscopic response may be protracted or even permanent, such as in the case of differentiation of sex organs.

The water-soluble hormones are the ones that bind to surface membranes of cells and initiate a sequence of reactions that have a powerful impact on cell metabolism but, without renewed hormone binding, act only for minutes or hours and no more (Table 5.1).

The first event in the action of a water-soluble hormone on the target cell is its binding to a membrane receptor. The binding is usually governed by a dissociation constant of a magnitude corresponding to the concentration of the hormone in blood (Table 5.2). The binding may occasionally take place at two or more sites which are in cooperative relationship.

Table 5.1. Some hormones acting on the cell surface

Epinephrine (adrenaline) (adrenal medulla)	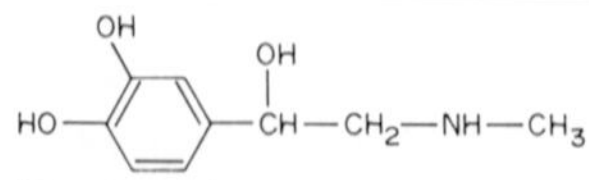 |Increase in pulse rate, contraction of smooth muscles, glycogenolysis; lipid hydrolysis|
Norepinephrine (noradrenaline) (adrenal medulla)	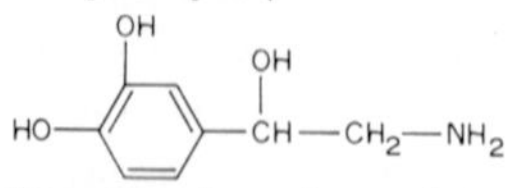 |Contraction of arterioles, fatty acid liberation from arterioles|
Insulin (β cells of pancreas)	A-chain 21 amino acids B-chain 30 amino acids |Stimulation of glucose uptake by cells, stimulation of lipid synthesis; stimulation of protein synthesis and cell proliferation|
Glucagon (α cells of pancreas)	29 amino acids |Glycogenolysis in liver; lipid hydrolysis in fat tissue|
TSH-releasing factor (hypothalamus)	3 amino acids |Stimulation of anterior pituitary to secrete TSH|
LH-releasing factor (hypothalamus)	10 amino acids |Stimulation of anterior pituitary to secrete LH|
Adrenocorticotropic hormone ACTH (anterior pituitary)	39 amino acids |Stimulates lipid hydrolysis, stimulates production of corticosteroids|
Follicle-stimulating hormone FSH (anterior pituitary)	α-chain 92 amino acids β-chain 118 amino acids |Stimulation of growth and estrogen synthesis|

(*Contd. on p. 331*)

Table 5.1. (*Contd.*)

<table>
<tr><td>Luteinizing hormone LH
(anterior pituitary)</td><td>α-chain 92 amino acids
β-chain 115 amino acids
|Stimulation of estrogen and progesterone synthesis|</td></tr>
<tr><td>Thyroid-stimulating hormone
(anterior pituitary)</td><td>α-chain 92 amino acids
β-chain 112 amino acids
|Stimulation of thyroxine release|</td></tr>
<tr><td>Somatotropin
(anterior pituitary)</td><td>191 amino acids
|Stimulation of liver to produce somatomedins|</td></tr>
<tr><td>Somatomedins
(liver)</td><td>Insulin-like peptides
|Stimulation of growth of bone, muscle and other cells|</td></tr>
<tr><td>Somatostatin
(hypothalamus)</td><td>14 amino acids
|Inhibition of somatotropin release|</td></tr>
<tr><td>Secretin
(small intestine)</td><td>27 amino acids
|Secretion of pancreatic digestive enzymes|</td></tr>
<tr><td>Vasopressin
(posterior pituitary)</td><td>9 amino acids
|Increase of water absorption from urine; rise in blood pressure|</td></tr>
<tr><td>Histamine
(mast cells)</td><td>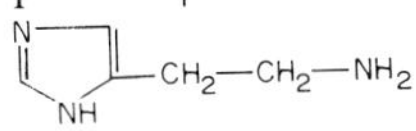
|Dilation of blood vessels|</td></tr>
<tr><td>Prostaglandins
(most body cells)</td><td>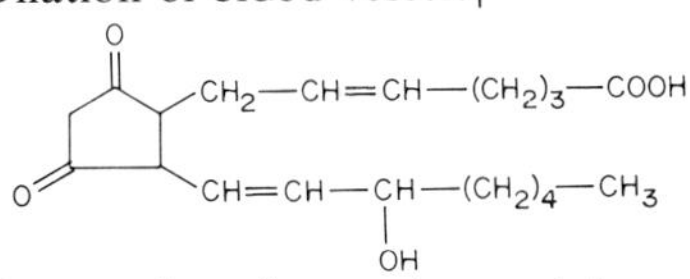
|Contraction of smooth muscle|</td></tr>
</table>

Table 5.2. Dissociation constants of hormones with cell surface receptors

Hormone	Receptor cell	K_D (nM)
Insulin	Adipocyte	1
Glucagon	Adipocyte	4
ACTH	Adipocyte	0.01–10
Angiotensin	Capillary wall	13–55
FSH	Seminal vesicle	1
TSH-releasing factor	Anterior pituitary	11–40
LH-releasing factor	Anterior pituitary	2
Somatotropin	Lymphocyte	0.75
Oxytocin	Frog skin	5
Prolactin	Rabbit mammary gland	0.06
Epinephrine	Frog erythrocyte	5000

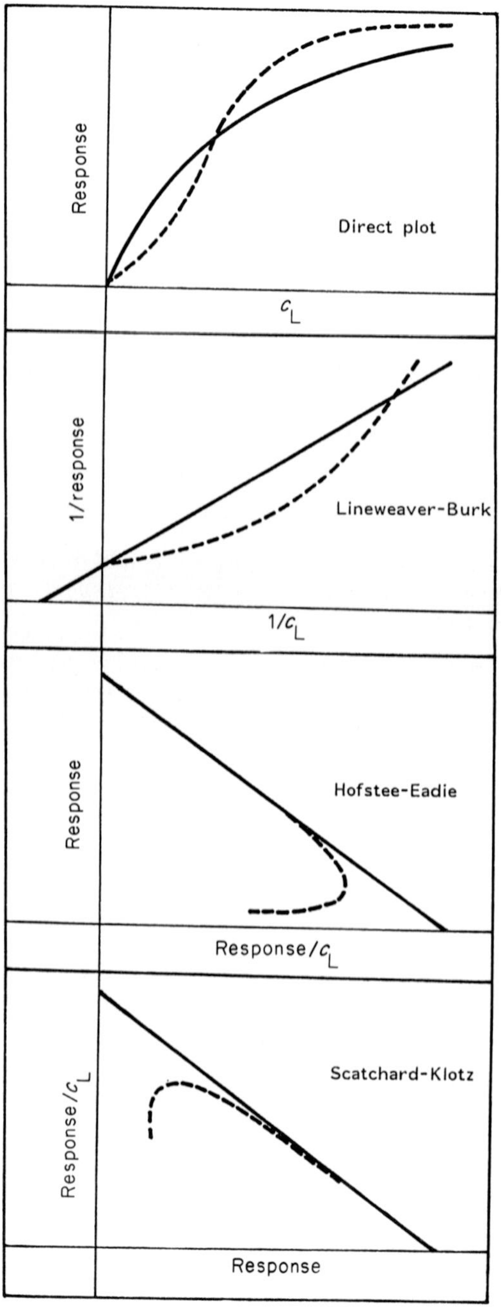

Fig. 5.7. Response of a cell to hormone binding to one (solid lines) or two cooperative (dashed lines) binding sites, plotted in four different ways

The amount of the complex of receptor R with ligand L in a single-site mechanism follows from the dissociation constant

$$c_{RL} = c_R c_L / K_D$$

and, using the conservation equation $c_{R_t} = c_R + c_{RL}$, one obtains an adsorption-type isotherm

$$c_{RL} = c_{R_t} c_L / (c_L + K_D).$$

If the response to the hormone ligand binding is proportional to the concentration of hormone-receptor complex one can write

$$\text{Response} = \alpha c_{R_t} c_L / (c_L + K_D).$$

In a system with two binding sites

$$c_{RL} + c_{RLL} = c_{R_t}(K_{LL} c_L + c_L^2)/(K_L K_{LL} + K_{LL} c_L + c_L^2)$$

where K_L is the dissociation constant of the first binding site, K_{LL} that of the second binding site. If a response to hormone binding requires both sites to be occupied then

$$\text{Response} = \alpha c_{R_t} c_L^2 / (K_{LL} K_L + K_{LL} c_L + c_L^2).$$

Several ways of plotting such responses are shown in Fig. 5.7. While the apparent maximum response is independent of the number of binding sites and possible cooperativity, the cooperative systems are singularly sensitive to changes of ligand concentrations in the range close to the apparent K_m concentration (at which the response is 50 per cent of the maximum).

Several hormone receptors have now been purified, particularly the one for insulin which is composed of two polypeptides with M_r 84 000 and two polypeptides of M_r 70 000, connected by disulphide bridges. Other receptors purified completely are the α_2-adrenergic receptor, the β_2-adrenergic receptor and the epidermal growth factor receptor.

Some of the receptors for hormones show a protein phosphokinase activity, e.g. that for insulin and for the epidermal growth factor. If such receptors are activated by hormone binding they begin to phosphorylate enzymes at the

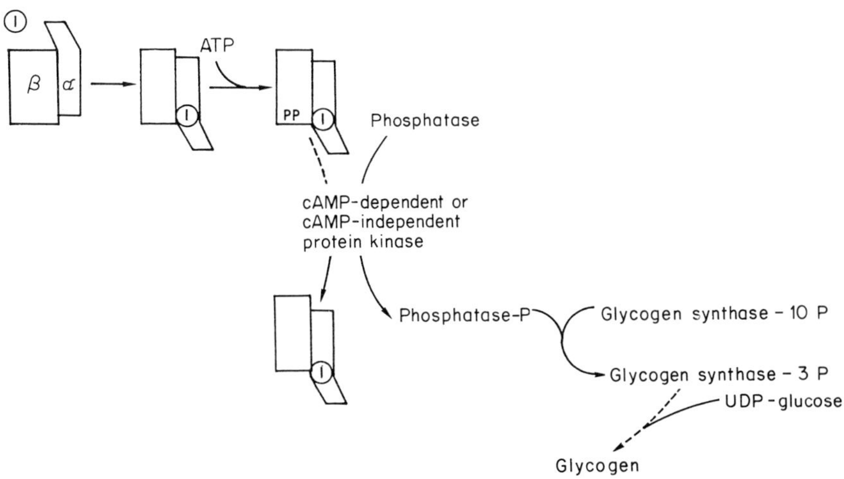

Fig. 5.8. Possible action of insulin (I) in muscle cells

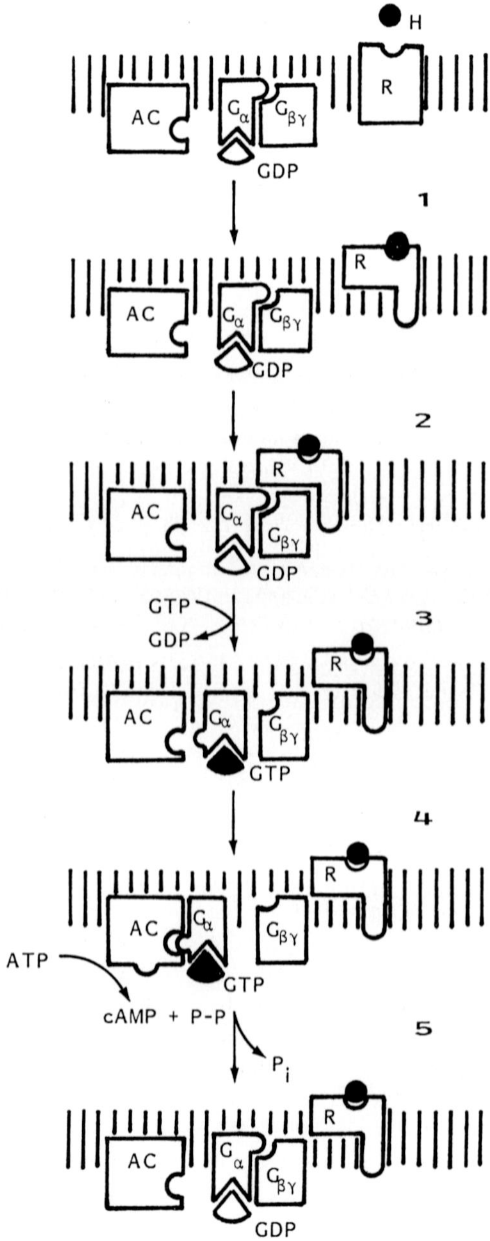

Fig. 5.9. Sequence of steps during hormone activation of adenylate cyclase. H—hormone; R—receptor for the hormone; AC—adenylate cyclase; $G_{\alpha,\beta,\gamma}$—subunits of the GDP-binding protein

expense of ATP. Thus the insulin receptor, in addition to phosphorylating its own β subunit, phosphorylates a phosphatase which becomes active and hydrolytically splits some phosphoryl groups from glycogen synthase which, in its turn, thus becomes active and sets upon synthesizing glycogen (Fig. 5.8).

It must be emphasized, however, that the principal role of insulin appears to be in promoting an exocytotic transfer of glucose carrier proteins to the cell surface which results in enhanced glucose uptake by organ cells and its decreased level in blood.

The second event in the action of a membrane-bound hormone is the activation of membrane transducer, now known to be a guanine-nucleotide-binding protein, or simply a G protein. In the resting state, GDP is bound to the α subunit of this protein. When a nearby receptor binds a hormone it undergoes a conformational change and moves toward the G protein. Interacting with it, it causes a release of GDP and binding of GTP, together with a dissociation of the $\beta\gamma$-subunit complex from the GTP-binding α subunit.

The third momentous event now takes place, viz. binding of the α subunit to membrane-bound adenylate cyclase which is then activated and begins to split ATP to cAMP and diphosphate. However, the GTP-carrying α subunit of the G protein possesses some intrinsic hydrolytic activity and this returns the α subunit into the resting state by hydrolyzing GTP to GDP. This results in a loss of activity of the adenylate cyclase and in restoration of the $\alpha\beta\gamma$-complex of the G protein which is again ready to be stimulated by an active hormone-receptor complex (Fig. 5.9).

Just as was explained with bacterial taxis, the receptors may become adapted to hormonal stimulation and do not respond in the usual way if exposed to high concentrations of the hormone for a prolonged period of time. The adaptation is the result of action of a cAMP-dependent protein kinase which phosphorylates the receptor and inactivates it. This is a typical negative feedback control of a sequence of enzyme-catalyzed reactions.

It is important to note that, on the one hand, several different hormone receptors can interact with one type of G protein and activate adenylate cyclase but, on the other hand, there are different G proteins present in cell membranes, one type mediating cyclase activation, the other cyclase inhibition. Clearly, these proteins are associated with a stimulatory hormone receptor and with an inhibitory hormone receptor, respectively.

The events subsequent to cAMP generation form a cascade of reactions which always involve activation of a protein kinase. The activation consists in binding of four molecules of cAMP to two regulatory subunits of the protein kinase, this leading to a dissociation of two catalytic, active, subunits, from the oligomeric enzymes.

A more or less general rule then applies. The protein kinase phosphorylates enzymes to *activate* those concerned with catabolic, degradative, reactions, but to *inactivate* those involved in anabolic, synthetic, pathways. The best example of this is the turnover of muscle cell glycogen (Fig. 5.10).

Thus cAMP is an important ‘second messenger’ transmitting the information of the hormone ‘first messenger’ to cell metabolic reactions. However, it is not the

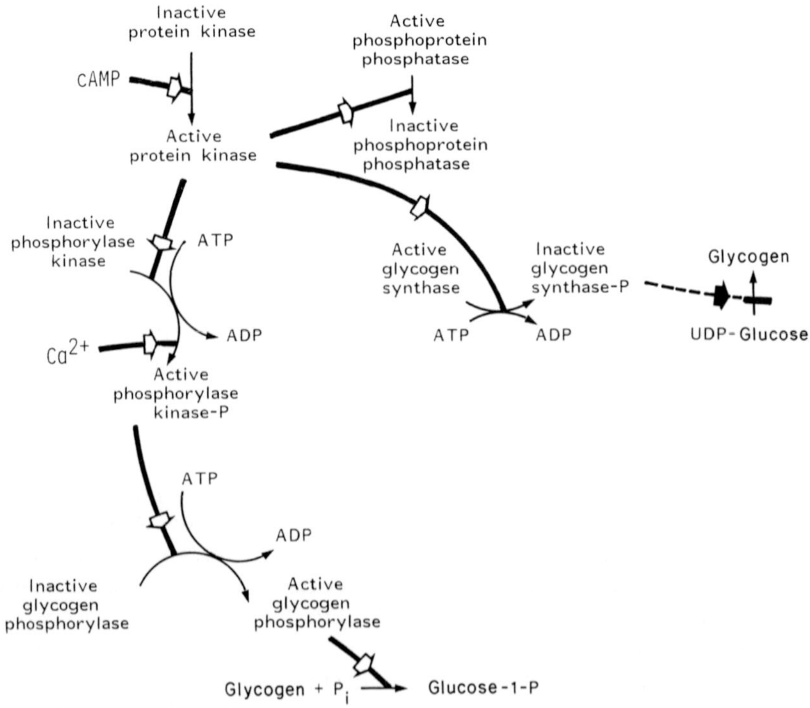

Fig. 5.10 Effect of cAMP on the turnover of glycogen in a skeletal muscle cell. cAMP as well as Ca^{2+} activate enzymes that eventually split glycogen. At the same time, cAMP inactivates the enzyme catalyzing glycogen synthesis

only one. Calcium ions appear as a nearly ubiquitous mediator of enzyme activity. The cytosolic concentration of Ca^{2+} is maintained at a level of 10–100 nM by a variety of stratagems. (1) The plasma membrane ATPase is a powerful electrogenic Ca^{2+}-extruding pump; (2) an antiport system exchanging Na^+ for Ca^{2+} serves the same purpose; (3) there are active pumps or electrophoretically operating translocators sequestering Ca^{2+} in the endoplasmic reticulum and mitochondria, respectively; (4) the cytosol contains now a number of Ca^{2+}-binding proteins, such as calmodulin, calsequestrin, calelectrin, calcimedins, synexins I and II, lipocortin, endonexin, and others, which can bind calcium ions often in a phosphilipid-dependent manner.

Now various external signals, chemical and particularly electric (transmitted through gap junctions) may open gated Ca^{2+} channels either in the plasma membrane or in the sarcoplasmic reticulum and permit an increase of Ca^{2+} in the cytosol. This has a number of consequences, some of them resembling those of increased levels of cAMP.

Taking again the example of glycogen turnover, an increase in cytosolic Ca^{2+} activates glycogen phosphorylase kinase with subsequent rise in glycogen phosphorylase activity and stepped up glycogen breakdown.

While the rush of Ca^{2+} ions into cells through the plasma membrane channels may be a direct response to binding of a chemical signal or to depolarization of the membrane, the release from intracellular stores is now believed to be mediated by another 'second messenger', inositol 1, 4, 5-trisphosphate (IP_3).

Apparently, membranes contain a certain concentration of a phosphatidylinositol 4, 5-bisphosphate which is formed by stepwise phosphorylation of inositol in PI. This compound is cleaved by the hydrolytic activity of hormone-activated, receptor-linked phospholipase C to diacylglycerol and to inositol 1, 4, 5-trisphosphate. The phospholipase C involved in producing IP_3 and diacylglycerol is probably activated by adenylate cyclase which, in its turn, receives the message via a G protein from the hormone receptor. Both of the cleavage products fulfil an information-transmitting role. Diacylglycerol activates membrane-bound protein kinase C which phosphorylates some proteins to activate them (a platelet protein of M_r 40 000, myosin light chain, receptor for epidermal growth factor). The IP_3 binds to the ER membrane and makes it leaky for Ca^{2+}. Whether this is a gating effect or a process involving the multiple negative charge of inositol trisphosphate is unclear.

The protein kinase C was recently identified with the so-called phorbol ester receptor, phorbol being a diterpenic alcohol, with esters substituted at $C_{(12)}$ with an acyl group and esterified at $C_{(13)}$ with acetate or a similar acid. The receptors occur in brain, spleen, lungs, thymus and intestines and, at lower concentrations, in all mammalian tissues.

Phorbol ester binding to the protein kinase C in the presence of Ca^{2+} activates this enzyme just as the binding of diacylglycerol does.

Now it may be seen that at least three substances act as transmitters of the hydrophilic hormone signal from the cell surface to metabolic reactions: cAMP, Ca^{2+} and inositol trisphosphate. They often act in combination and synergistically. In practically all the responses the G proteins play a major role which may comprise cAMP effects (most definitely), Ca^{2+} effects (presumably) and IP_3 effects (indirectly but with certainty).

Information carried by the nerve impulse

The nervous system as is found even in the most primitive multicellular animals (and absent in plants) is a tremendous reservoir of information, different from the information bits stored in hereditary DNA, and immensely varied.

External information flowing through the central nervous system, either to be kept there for future use (short-term and long-term memory) or to be immediately implemented (as a response of the organism to an external stimulus) follows complicated paths of cell-to-cell signal transmission, either afferent (sensory, centripetal) or efferent (motor, centrifugal).

Hundreds of types of nerve cells are known but there are some basic principles that govern the structure of neurons, the cells that actually carry the information (Fig. 5.11). A typical neuron contains a cell body or soma where the nucleus, mitochondria, endoplasmic reticulum and all essential organelles are located. On the afferent side it grows out into tens, hundreds or thousands of thin dendrites that contain regions where they received information from other, more peripheral cells. However, information can also be received directly by the cell body, either

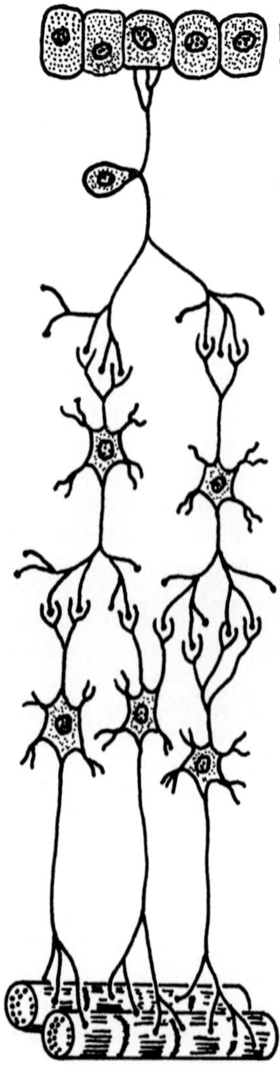

Fig. 5.11. Types of neurons

stimulatory, or inhibitory, and it is there that it is processed, integrated and transformed.

The most significant part of the nerve cell is the axon which sends information out of the cell body to dendrites of other nerve cells or to the effector cells, such as those of muscle or glands. The axon can be a relatively massive part of the neuron, being as wide as 1 mm in cephalopod giant nerve fibre, and as long as 25 m in a large whale. Moreover, for most of its length the motor neuron is covered by convoluted Schwann cells, forming the insulating myelin sheath of the axon.

The passage of information from cell to cell involves the crossing of a synaptic cleft or simply synapse, where the presynaptic membrane releases the information in the form of either a chemical mediator, the neurotransmitter, or of an electric potential change. The first are called chemical synapses, the second electric synapses.

In chemical synapses, the cleft is about 20 nm wide and the neurotransmitter released from the presynaptic membrane in the form of exocytotic vesicles diffuses to receptors on the postsynaptic membrane where the signal is processed in one of several ways. In electric synapses the cleft is bridged by a thousand or more gap junctions through which the passage of current mediated by the dissolved ions and driven by the action potential is instantaneous and the information is carried on in an unchanged form even in the postsynaptic cell.

Although the chemical signal receptor is in the vast majority of cases a protein, lipids apparently can play a role in signal amplification, as they respond to an 'activated' receptor molecule by changes in their fluidity in a highly cooperative fashion.

Neurotransmitters

It is now generally accepted that neurotransmitters synthesized in a sensory or a nerve cell are released into the synaptic cleft by a process of exocytosis, each vesicle containing some 10 000 molecules of the neurotransmitter with three vesicles being emptied during an average impulse.

A great many small molecules serve as neurotransmitters, depending (1) on the type of nervous circuit and (2) on whether the synapse is a stimulatory or an inhibitory one. It will be seen from Table 5.3 that some of the neurotransmitters are identical with hydrophilic hormones, such as epinephrine and its derivatives, and some of the peptides.

Binding of a neurotransmitter to the postsynaptic membrane—this may be a part of the dendrite, patch on the neuron cell body or an area near the presynaptic nerve terminal, brings about several types of response.

1. Immediate increase of ion permeability and hence depolarization of the membrane (e.g., acetylcholine bound to its receptor).

2. Activation of adenylate cyclase which results in increased levels of cAMP, activation of protein kinases and phosphorylation of K^+ channels which are closed (e.g., serotonin in the axon terminal, or norepinephrine on β-receptors in nerve cells).

3. Increase of permeability to chloride ions (e.g., GABA in inhibitory synapses).

There exist several types of receptors for a given neurotransmitter in different cells; thus, the best studied set of acetylcholine receptors at the so-called cholinergic synapses can be divided into two subsets, viz. nicotinic and muscarinic acetylcholine receptors. When nicotine binds to the first subset it has the same effect as acetylcholine, bringing about a rapid and short response (several milliseconds). When muscarine binds to the second subset it brings about a prolonged (several seconds) response which can be either positive or negative, depending on the cell type.

Table 5.3. Molecules believed to act as neurotransmitters

Peptides
Somatostatin (tetradecapeptide)
LH-releasing hormone (decapeptide)
THS-releasing hormone (tripeptide)
Substance P (undecapeptide)
Neurotensin (tridecapeptide)
Angiotensin I (decapeptide)
Angiotensin II (octapeptide)
β-Endorphin (hentriacontapeptide)
Met-enkephalin (pentapeptide)
Leu-enkephalin (pentapeptide)
Other compounds
Acetylcholine
Glycine
Glutamate
Dopamine (2-deoxynorepinephrine)
Norepinephrine
Epinephrine
Octopamine (3′-deoxynorepinephrine)
Serotonin (5′-hydroxytryptamine)
β-Alanine
Histamine
GABA (γ-aminobutyric acid)
Taurine (2-aminoethylsulphonic acid)

The nicotinic acetylcholine receptor is a well characterized protein of M_r 250 000–270 000, containing two α subunits (M_r $2 \times 40\,000$) and one each of $\beta(M_r\,49\,000)$, $\gamma(M_r\,57\,000)$ and $\delta(M_r\,65\,000)$ subunits. The α subunit binds one or two acetylcholine molecules or one α-bungarotoxin (a snake venom) molecule. The subunits are about 40 per cent homologous and each crosses the membrane five times.

As two molecules of the neurotransmitter are bound cooperatively the internal channel (probably formed by the fourth helix of each of the five subunits) opens within 0.1 ms and Na^+ and K^+ ions can move freely (Na^+ somewhat less so) through it down their electrochemical gradients. This depolarizes the membrane from its usual roughly -90 mV to about -15 mV, corresponding to a flow of as many as 30 000 Na^+ or K^+ ions per millisecond.

The depolarization caused by acetylcholine binding can have different effects on different postsynaptic cells. In a muscle cell it will permit a flow of Ca^{2+} into cells through voltage-gated Ca^{2+} channels and cause contraction of muscle fibrils. In a nerve cell it will cause an action potential to form which will propagate along the dendrite toward the cell body and possibly beyond along the axon.

In adrenal chromaffin cells, like in primary sensory cells and nerve endings, the inflow of Ca^{2+} will cause exocytosis of chemical mediators.

Quite generally, the stimulus ceases to function only after acetylcholine has been hydrolyzed by acetylcholinesterase which is located in the synaptic cleft on a trabecular structure formed by collagen.

An interesting phenomenon mediated by acetylcholine is the discharge (through abundant Na^+ channels) of membrane potentials organized in series into a massive electric shock, such as are produced by some species of fish. The electric eel *Electrophorus*, the sting ray *Torpedo*, the tropical African electric catfish *Malapterurus*, the North American stargazer *Astroscopus*, and a number of others possess huge electric organs amounting to as much as 25 per cent of their body weight. The organ is composed of many columns of several thousand cells called electroplaxes that have developed from normal muscle cells. These cells are each separately innervated by cholinergic neurons but only at one surface. An electric stimulus in the form of an action potential arrives at this surface, repolarizes the surface membrane from its usual −90 to +40 mV, whereby a series of alternating potentials arises throughout the whole column. The potentials add up to 100–650 V, depending on the fish species, and are released on closing an external circuit (generally formed by the body of the prey). It is remarkable that in some of the fishes, the discharge can be repeated several hundred times in sequence, each discharge lasting about 5 ms, whereupon, within less than 1 s, the battery must be recharged through return of the cholinergic membrane to a negative potential.

Action potential

Several times before a role of the action potential was mentioned. It is now time to describe the events leading to the generation of an action potential and the way it propagates along the nerve axon.

The resting neuron has a membrane potential of some − 60 to − 90 mV, the potential being maintained at this value by continuous action of the Na,K-ATPase, electrogenically pumping Na^+ ions out of cells and K^+ into cells, and by the limited permeabilities of cations and anions, attempting to compensate the electrogenic flux through the pump. The neuron, particularly along the length of the axon, contains other potentially conducting voltage-dependent ion channels but these are now closed.

The principal one is the sodium channel, a single polypeptide of M_r about 250 000 which is accompanied by smaller, probably regulatory, peptides. It is structured into four homologous segments, each of which contains nearly 300 amino acids that would qualify for membrane-spanning sequences. The channel opens as depolarizing voltage passes by and remains open for an average of 0.7 ms, letting through some 7000 Na^+ ions at a single opening. It is almost equally permeable to Li^+ but much less to K^+ (Table 5.4). The sodium channel is not a particularly abundant protein, compared with various ion pumps and nonelectrolyte carriers. Its density is 30–5000 per μm^2, against 10^4 carriers per μm^2. It is specifically inhibited by tetrodotoxin and saxitonin. In contrast, veratridine and batrachotoxin cause it to remain permanently open. Other types

Table 5.4. Permeability ratios for univalent cations in the nerve sodium and potassium channels

Ion	P_{ion}/P_{Na^+}	P_{ion}/P_{K^+}
Na^+	1.00	0.01
$NH_3^+ \cdot OH$	0.94	0.02
Li^+	0.93	0.02
$H_2N \cdot NH_3^+$	0.59	0.03
Tl^+	0.33	2.3
NH_4^+	0.16	0.13
$CH_3NH_3^+$	0.01	0.02
Rb^+	0.09	0.91

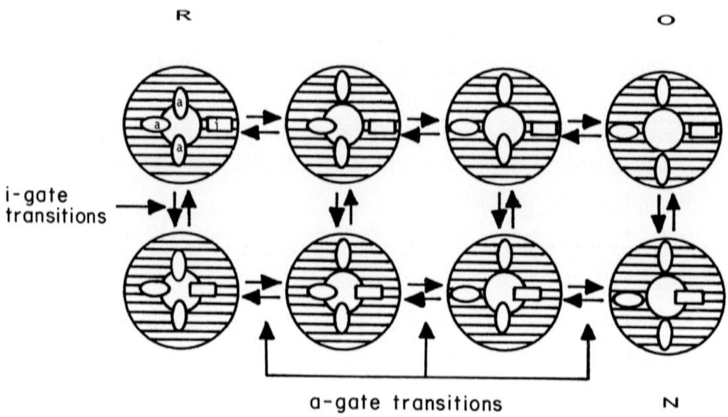

Fig. 5.12. A hypothetical model of the sodium channel. In the resting state R all three activating gates (a) are closed, the inactivating gate (i) is off. The gates open one by one until the open state O is reached. The inactivating gate then shuts down making the channel impermeable to Na^+ ions and, at the same time, inert to inactivation—the refractory state N. The activating gates contain an excess of positive charges so that depolarization of the membrane causes them to align with the newly established direction of membrane potential which is apparently associated with opening of the channel

of inhibitors include scorpion toxins, pyrethroids and ciguatoxin (Fig. 5.12).

Another specific channel is that for K^+ ions which has a three times greater conductivity than the sodium one. It is also voltage-gated, opens as the membrane is depolarized and is selectively blocked by tetraethylammonium ions.

The third voltage-gated channel is that for Ca^{2+} ions which is a highly selective pathway for the entry of Ca^{2+} into cells (from a concentration of 1–10 mM to about 100 nM). It is inhibited by fluoride anions and artificial substances, such as dihydropyridines.

Now the neuron body receives signals of various kinds, some electric, some chemical turned electric as a ligand-gated channel opens to permit ions to flow across the membrane, others still regulating the input of the first two. The integrated sum of them may result in a local depolarization which can be effected by a relatively minute shift of Na^+ ions into the neuron body—perhaps one-millionth of the cytosolic K^+ being replaced with Na^+ is sufficient to depolarize the membrane locally by 20 mV. If such a depolarization is sensed at the axon hillock (the root of the axon) which displays a relatively lower threshold than the rest of the cell an action potential is generated. This is obviously due to the opening of local sodium channels, which permit extracellular sodium to flow in and depolarize the membrane further, creating in fact a repolarized patch on the axon. The extent of this depolarization can be appreciated by using Goldman's constant-field equation for the resting membrane potential (Eq. (3.71) with the difference of surface potentials neglected)

$$\Delta\phi_r = \frac{RT}{F} \ln \frac{P_{Na}c'_{Na^+} + P_K c'_{K^+} + P_{Cl}c''_{Cl^-}}{P_{Na}c''_{Na^+} + P_K c''_{K^+} + P_{Cl}c'_{Cl^-}}. \tag{5.1}$$

If the measured values of ion concentrations are substituted into the equation,

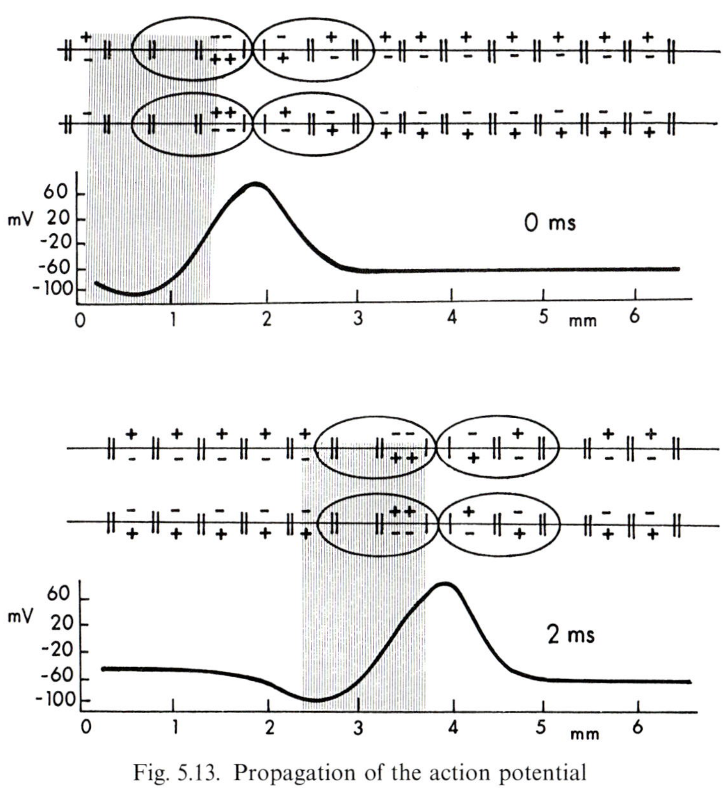

Fig. 5.13. Propagation of the action potential

viz. $c'_{Na^+} = 0.15$ M, $c''_{Na^+} = 0.01$ M, $c'_{K^+} = 0.005$ M, $c'_{K^+} = 0.15$ M, $c'_{Cl^-} = 0.12$ M, $c''_{Cl^-} = 0.005$ M, as well as mobilities, $P_{Na} = 10^{-10}\,\mathrm{m\,s^{-1}}$, $P_K = 10^{-9}\,\mathrm{m\,s^{-1}}$, $P_{Cl} = 10^{-10}\,\mathrm{m\,s^{-1}}$, the resulting potential at 37 °C is -55.3 mV. If the permeability for Na^+ ions is locally increased to $P_{Na} = 10^{-8}\,\mathrm{m\,s^{-1}}$ we have $\Delta\phi_A = +46.7$ mV. Hence, a repolarization results. If, on the other hand, either the permeability to K^+ or Cl^- is increased, a hyperpolarization of the membrane follows (Figs 5.13, 5.14).

The movement and the shape of the action potential are characterized by several factors.

1. As the action potential is triggered by a supraliminal local depolarization it generates so-called loop currents that depolarize the axon farther ahead (Fig. 5.15) by opening further Na^+ channels. These currents pass through the active region and return through the resting part of the plasma membrane. The part of the current that passes outside the nerve axon is defined by

$$i' = \frac{1}{r'}\frac{\partial \phi'}{\partial x} \tag{5.2}$$

where r' is the resistance per unit length in the interstitial fluid, ϕ' is the part of the potential difference outside cells and x is the distance along the axon. Similarly,

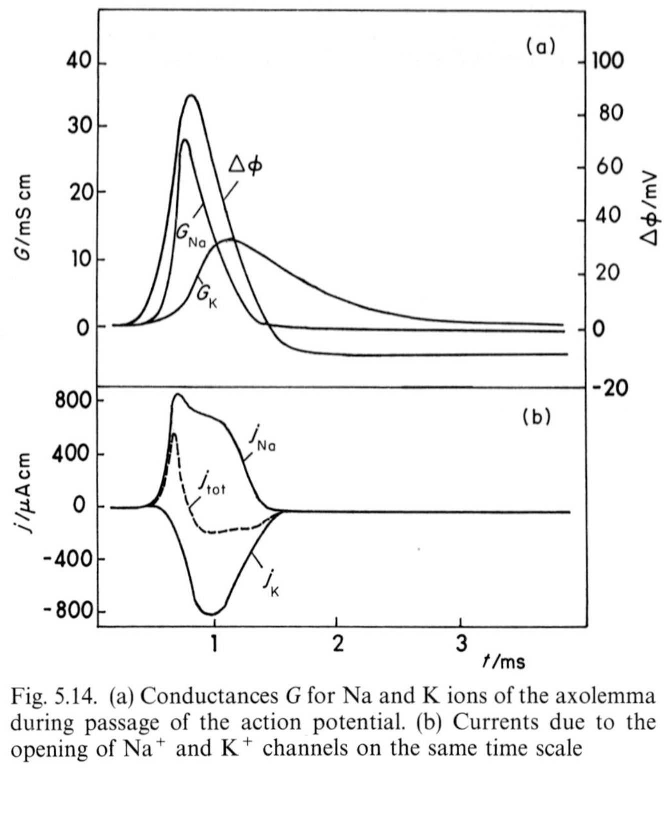

Fig. 5.14. (a) Conductances G for Na and K ions of the axolemma during passage of the action potential. (b) Currents due to the opening of Na^+ and K^+ channels on the same time scale

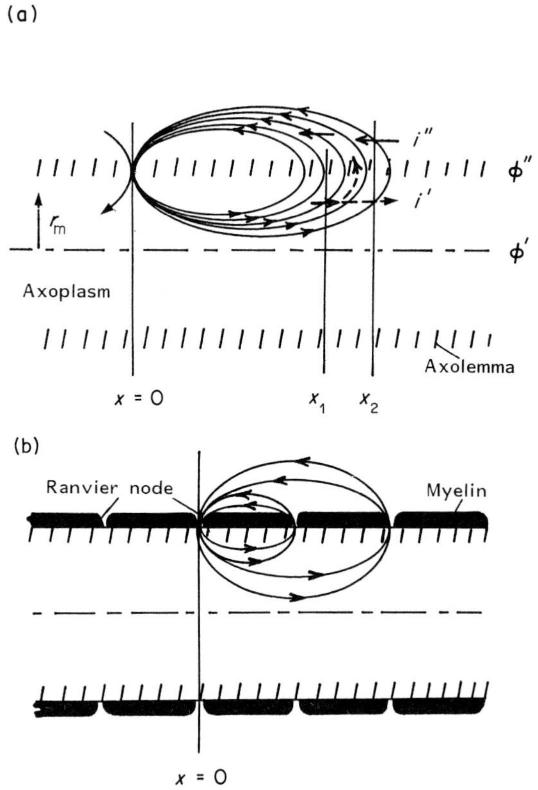

Fig. 5.15. Currents associated with propagating action potential in a bare (unmyelinated) neuron (a) and in a neuron covered with a myelin sheath (b). The momentary position of the action potential peak is at $x = 0$. The symbols in the figure refer to the cable equation developed in the text

for the current inside the cytoplasm (axoplasm)

$$i'' = \frac{1}{r''}\frac{\partial \phi''}{\partial x}. \tag{5.3}$$

Introducing $\phi_m = \phi' - \phi''$ we have

$$\partial \phi_m / \partial x = -r'i' + r''i'' \tag{5.4}$$

and, since the currents form a closed loop

$$i' + i'' = 0 \tag{5.5}$$

so that

$$\partial \phi_m / \partial x = -i'(r' + r''). \tag{5.6}$$

Differentiation of this equation yields

$$\partial^2\phi_m/\partial x^2 = -(r' + r'')(\partial i'/\partial x). \tag{5.7}$$

The current flowing through the membrane is given by

$$-i_m = \partial i'/\partial x = i' + i_c \tag{5.8}$$

where i_c is that part of i_m which flows into the membrane capacity. Using Ohm's law to express the i' we have

$$i' = \Delta\phi_m/r_m \tag{5.9}$$

while

$$i_c = c_m(\partial\phi_m/\partial t) \tag{5.10}$$

where c_m is the capacity of the membrane per unit length. We can now derive the classical cable equation

$$\lambda^2(\partial^2\phi_m/\partial x^2) = \phi_m + \tau(\partial\phi_m/\partial t) \tag{5.11}$$

where

$$\lambda = (r_m/r' + r'') \quad \text{and} \quad \tau = r_m/c_m.$$

Solution of the partial differential equation is complicated but, since under steady-state conditions $\partial\phi_m/\partial t = 0$

$$\phi_m = \lambda^2(\mathrm{d}^2\phi_m/\mathrm{d}t^2). \tag{5.12}$$

Solution of this equation yields

$$\phi_m = C_1\,e^{-x/\lambda} + C_2\,e^{x/\lambda} \tag{5.13}$$

and, since for very large x the ϕ_m must be zero and while for $x = 0$, $\phi_m = \phi''$ the final result is

$$\phi_m = \phi''\,e^{-x/\lambda}. \tag{5.14}$$

The membrane potential change expressed as the action potential would thus spread exponentially and at short distances would disappear. The values of λ (distance at which $\phi_m = \phi''/e$, i.e. 37 per cent) range from 2 to 20 mm so that there would be no way for an action potential initiated at the axon hillock of a long motor neuron to reach its target. Nature has evolved two strategies to overcome this problem.

The first is to envelope the long axons in a myelin sheath which may be up to 500 nm thick and insulates the length of the axon so that no ions can escape from the axon and thus attenuate the signal. The only sites where the axon is naked is at the Ranvier nodes (Fig. 5.16) spaced about 1 mm apart, where a high density of Na^+ channels is found (10 000 per μm^2, compared with 10 per μm^2 under the myelin sheath).

The second strategy to ensure maintenance of the action potential magnitude is the operation of voltage-gated sodium channels which open as the depolarization spreads and allow Na^+ ions to move in, thus keeping the potential spike constant. The action potential at one Ranvier node activates the next one or two Ranvier

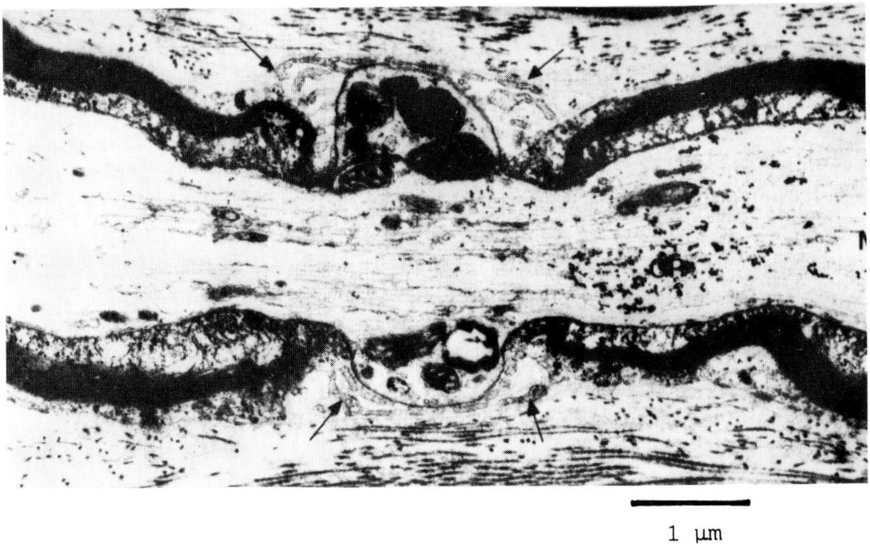

Fig. 5.16. Ultrathin section through a myelinated nerve fibre showing (arrows) the nodes of Ranvier where the myelin sheath is absent

nodes to open their Na^+ channels and thus to 'regenerate' the action potential and keep it moving at a rate of about $100\,m\,s^{-1}$, as compared with $1\,m\,s^{-1}$ in nonmyelinated nerve fibers.

3. The magnitude of the electric impulse transmitted by the action potential is not reflected in its amplitude but repetitive firing of a constant-size potential. The reason for the constant size lies apparently in the fact that at the peak of the action potential Na^+ ions are distributed according to the Na^+ diffusion potential and no net flow of Na^+ ions is possible, this being a passive 'channel' flow where only the existing membrane potential causes an imbalance of Na^+ ion concentrations.

Now let us recapitulate the scenario of transmitting information contained in an external signal to the target where it triggers the final response.

A sensory cell receives a signal in the form of mechanical pressure which locally depolarizes the membrane. The depolarization spreads passively toward the part of the cell where it contains a presynaptic membrane. (The passive spread of depolarization is feasible as we are dealing here with a distance of no more than $10\,\mu m$.) There it activates voltage-dependent Ca^{2+} channels, Ca^{2+} ions rush in and cause immediate fusion of vesicles containing a chemical neurotransmitter with the presynaptic membrane.

The neurotransmitter pours out into the synaptic cleft and binds to receptors in the postsynaptic membrane which may be part of a retrograde axon of a sensory neuron. The receptor is linked with an ion channel or is itself a ligand-gated ion channel so that upon its activation cations, particularly Na^+, rush inside the neuron and trigger an action potential. The action potential may initially move both ways from the postsynaptic membrane but one of the ways is a dead end so that only the action potential that moves toward the neuron body is maintained and propagated. The depolarization it spreads either passively or by opening further Na^+ channels precedes the action potential so that it moves along the axon. The Na^+ channels giving rise to the depolarization remain open for only 0.5–1.0 ms and as they are closing, independent K^+ channels open and cause a transient hyperpolarization of the membrane. Within 10–50 μs, the potential reaches the cell body. The neuron body may receive simultaneously various other signals that have arrived from other sensory cells or other neurons and have all caused slight shifts in membrane potential. The shifts are integrated at the antegrade axon hillock and may trigger another action potential which now travels toward the nerve terminal. There it may be electrically connected to an interneuron postsynaptic membrane so that the potential directly passes through gap junctions connecting the two membranes and travels through the interneuron to its axon terminal. This may involve several interneurons in series until the signal reaches the motor neuron. Near its end plate several synaptic connections may occur which affect the probability that the neuromuscular presynaptic membrane will release its content of neurotransmitter.

The messages of the auxiliary neurons may be electric in nature and cause Ca^{2+} channels to open or they may be chemical and activate adenylate cyclase which eventually makes K^+ channels close down and allows Ca^{2+} to move in electrophoretically. If the overall message is positive, increased intraneuronal Ca^{2+} causes acetylcholine to be released via vesicle fusion. Acetylcholine is bound to the postsynaptic membrane receptors, causes its depolarization which spreads along the muscle cell membrane and opens voltage-gated Ca^{2+} channels. Inflow of Ca^{2+} into the cytosol assists then in muscle contraction.

SYNOPSIS

Various chemical agents are produced and secreted in multicellular animals to carry information either to neighbouring cells or to distant tissues. The principal producers of such compounds are endocrine glands which, in their turn, respond to stimulation by signals arriving mainly from the hypothalamus.

Many of the endocrine gland products, the hormones, are bound to surface receptors in the target cells where they bring about shifts in some membrane components, principally the GTP-binding or G proteins which then stimulate either adenylate cyclase to produce cAMP or cause phosphorylated phosphatidylinositol to be split into diacylglycerol and inositol 1, 4, 5-trisphosphate. Each of these agents is then capable of initiating intracellular reactions which, on the one hand, lead to phosphorylation of enzymes either to activate or deactivate them, and, on the other hand, to permitting Ca^{2+} to enter the cytosol. Calcium ions then exhibit qualitatively similar effects on enzyme activities as cAMP does.

Some of the low-molecular-weight hormones as well as a host of other relatively simple compounds transmit information along the path of a nerve impulse by acting as neurotransmitters. They are released by the end plates of neuronal axons from vesicles and arrive at the postsynaptic membrane to initiate reactions either resembling those triggered by hormones or, usually, to give rise to a local depolarization of the neuron or muscle or gland cell membrane which, in the case of the neuron, propagates along the length of the dendrites and of the axon as an action potential. The action potential is transmitted to further cells either via release of a chemical neurotransmitter or through a gap junction directly.

FURTHER READING

Ashendel, C. L. The phorbol ester receptor: a phospholipid-regulated protein kinase, *Biochim. Biophys. Acta*, **822**, 219–42 (1985).

Barchi, R. L., Tanaka, J. C., and Furman, R. E. Molecular characteristics and functional reconstitution of muscle voltage-sensitive sodium channels, *J. Cell Biochem.*, **26**, 135–46 (1984).

Berridge, M. The molecular basis of communication within the cell, *Sci. Amer.*, **253**, 142–50 (1985).

Carpenter, G., and Cohen, S. Peptide growth factors, *Trends Biochem. Sci.*, **9**, 169–71 (1984).

Changeux, J.-P., Devillers-Thiery, A., and Chemouilli, P. Acetylcholine receptor: an allosteric protein, *Science*, **225**, 1335–45 (1984).

Czech, M. P. (ed.): *Molecular Basis of Insulin Action*, Plenum Press, New York (1985).

Gilman, A. G. G proteins and dual control of adenylate cyclase, *Cell*, **36**, 577–9 (1984).

Hildebrandt, J. D., Sekura, R. D., Codina, J., Iyengar, R., Manclark, C. R., and Birnbaumer, L. Stimulation and inhibition of adenylyl cyclases mediated by distinct regulatory proteins, *Nature*, **302**, 706–9 (1983).

Hille, B. *Ionic Channels in Excitable Membranes*, Sinauer, Sunderland (1984).

Hokin, L. E. Receptors and phosphoinositide-generated second messengers, *Annu. Rev. Biochem.*, **54**, 205–36 (1986).

Levitzki, A. Reconstitution of membrane receptor systems, *Biochim. Biophys. Acta*, **822**, 127–53 (1985).

Manevich, E. M., Tonevitsky, A. G., and Bergelson, L. D. The binding of the B-chain of ricin to Burkitt lymphoma cells. A new approach to ligand-receptor interaction studies, *FEBS Lett.*, **194**, 313–16 (1986).

5.4 SIGNALLING IN THE IMMUNE SYSTEM

A fundamental but highly specialized area where information is transformed into a biochemical response is the immune system of vertebrates, most highly developed in mammals.

There it has two parts, the so-called humoral (or fluid) immunity and the cellular immunity. In the first type of immunity a foreign molecule or particle (an antigen) encounters a *B* lymphocyte (produced in bone marrow) which carries on its surface an immunoglobulin (Ig) of a particular type, such that its binding sites in the variable regions of the molecule fit exactly the antigen, with a dissociation constant of 10^{-8}–10^{-10} M (Fig. 5.17).

The diversity of surface receptors on *B* lymphocytes is astounding, with perhaps 10^{6}–10^{7} different, highly specific, *B* lymphocytes circulating in the body.

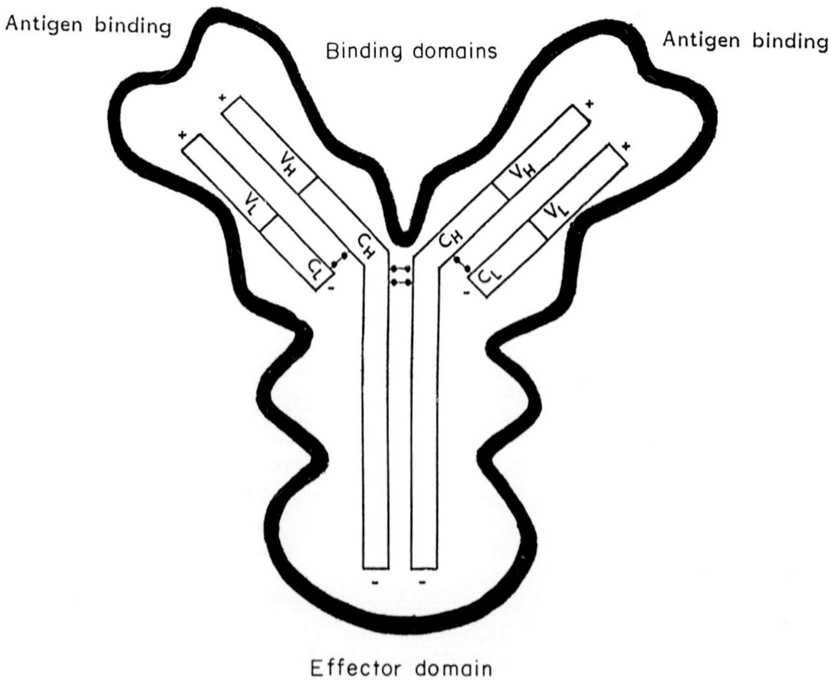

Fig. 5.17. Schematic representation of an antigen-discriminating immunoglobulin (an antibody). The N termini are marked with +, the C termini with −, the dots represent S–S bonds

It is brought about by DNA rearrangement such that antibody genes are not present in the genome as such but occur as gene segments which are brought together during lymphocyte differentiation. Many different variable segments are available to join with any of several constant segments. Imprecision during joining is another factor that leads to an increased variety of the antibodies formed. The number of different variable segments and the different ways of joining the molecule together would allow for the formation of some 10^{11} different immunoglobulin molecules.

The immunoglobulins on the lymphocyte surface are identical with antibodies. To bind properly to the surface antibody the antigen must have a certain minimum size because it is bound at two identical sites of the antibody. Thus, simple organic molecules will not bind, but if they are attached to a protein they will produce a specific response. They thus form the determinant of the large molecule which binds to the antibody. The organic molecule itself is the *hapten* in the sense that it can elicit a cellular response.

Immunoglobulins or antibodies come in four major classes, IgM, IgD, IgG, IgE and IgA, their heavy chains being designated μ, δ, γ, ε and α, respectively. They occur as monomers containing two heavy and two light chains, the heavy chains ranging in M_r from 50 000 to 75 000, the light chains are about 23 000. However, IgA, if secreted in an exocrine way (in colostrum, nasal and intestinal secretions), is a dimer containing, in addition, two peptide chains so that its final M_r is 380 000. IgM can also form a pentamer when not membrane-bound, its M_r being then 970 000. All Ig's contain sugar residues, IgM, IgD and IgE as much as 12 per cent of the total mass. It is above all the IgM monomer that forms the antibody on *B*-cell surfaces although all other Ig classes can be membrane-bound.

Once the *B* cell binds an antigen (Fig. 5.18, *I*) the linked receptors begin to wander along the surface forming patches—apparently a spontaneous process, relatively independent of membrane fluidity. The patches then can migrate toward one pole of the lymphocyte—an energy-requiring process, to form prominent caps (*II*) at this pole which is later engulfed by a type of endocytosis. The antigen is then degraded in the cell and some of its major determinants are exposed (*III*) by a special protein (one of those encoded by the so-called major histocompatibility complex genes). This degraded antigen is then recognized by a helper *T* cell (*IVa*) and the *B* cell is stimulated to divide and eventually to differentiate into a plasma cell capable of producing antibodies, of the type borne on its surface originally (*IVb*).

In the second type of immunity, the cellular one, *T* lymphocytes are instrumental. These carry surface receptors for antigens which, however, are distinct from antibodies and are called *T*-cell receptors. They react with antigens on the surface of other cells. The receptors consist of an α and a β polypeptide, each with a molar mass of 40000, and contain N-linked sugar residues.

Three types of *T* cells are recognized, each with several subdivisions.

Cytotoxic T lymphocytes or *killer T* cells bind to foreign antigen-bearing cells and destroy them by piercing holes in their membranes (*Va*–*IXa*). They are particularly important in combating virus-infected cells and neoplastic (cancer) cells.

Helper T lymphocytes assist other types of *T* cells as well as *B* cells in dealing with an antigen. For instance, a *B* cell by itself will not respond to an antigen with one or two determinant regions but only to multivalent antigens with repeated copies of a determinant. The helper cell recognizes a cell-bound antigen (e.g., processed in a macrophage; *Vb*) and is activated by *interleukin* 1 secreted by macrophages to secrete, in its turn, *interleukin* 2, which stimulates helper and killer lymphocytes to divide (*VIb*, *VIIb*). It also secretes a *B*-cell growth factor which serves as a signal for *B* cells to multiply. There is a further information exchange between the multiplying *B* cells and *T* helper cells. As the number of *B* cells increases the helpers begin to secrete a *B*-cell differentiation factor which causes the *B* cells to differentiate into a plasma cell (*IVb*) and start synthesizing and releasing antibodies. The period between antigen binding and antibody secretion is usually 18–20 h. The secretion then has a capacity of about 2000 molecules of antibodies per cell per second.

The last product of helper *T* cells is γ-*interferon* which (1) activates killer *T* cells (*VIIb*), (2) stimulates further *B* cells to produce antibodies, (3) keeps macrophages at the site of infection.

The third type of *T* lymphocytes, the *suppressor* cells, diminish *B*-cell activity so

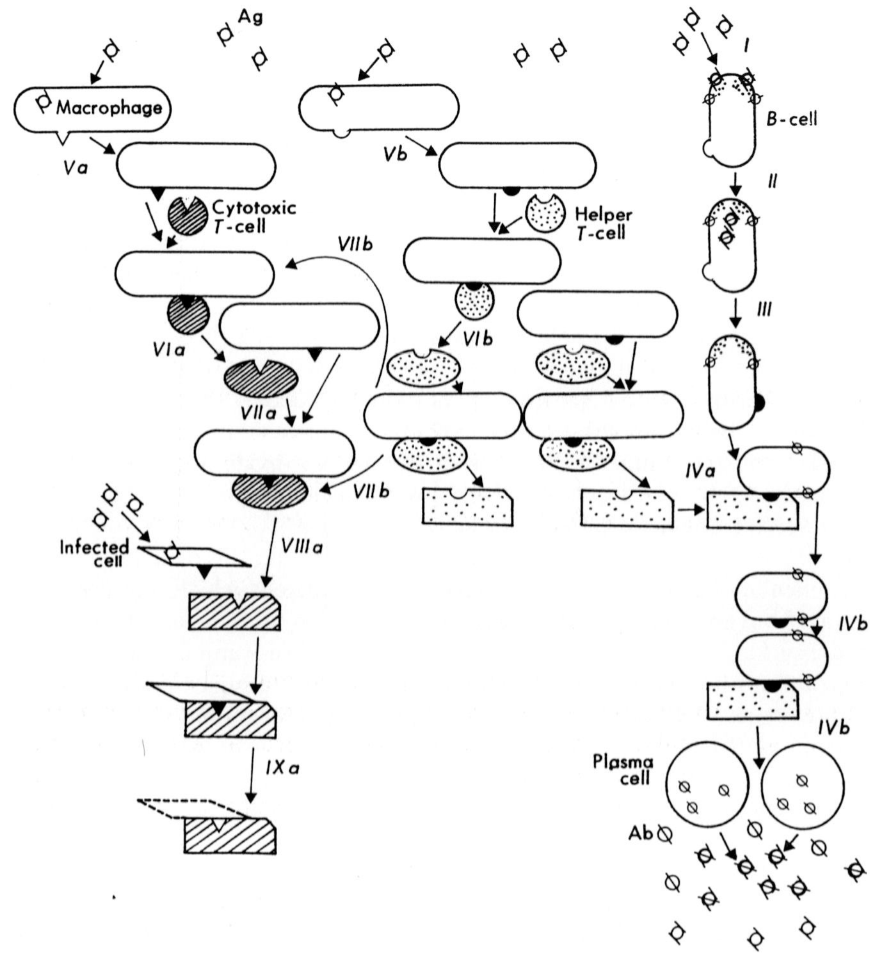

Fig. 5.18. Response of the immune system to an antigen. Description of the individual steps is provided in the text.

that the immune response stops after an infection has been conquered.

Lymphocytes in general and *T* lymphocytes in particular contain a striking variety of receptors and other proteins on their surface, among them infection-site receptors for viruses of several types and for bacterial toxins, primary immunological receptors for antigens and self-antigens, receptors for stimulants, lectins, polyclonal lymphocytes, IgM-bound components of complement, etc. Besides these, special Ca^{2+}-dependent K^+ channels were discovered in lymphocytes, with an elementary conductance of 12–14 pS and open time of 30 ms.

Now in this outline there have been many instances where either a peptide mediator (the interleukins, γ-interferon, etc.), reacted with the cell surface and caused the cell to change its behaviour, or an antigen bound to the cell surface antibody caused the cell to start multiplying and eventually producing antibodies. The binding of these various agents to cells of the immune system brings about a range of changes both in the membrane and in intracellular metabolism, there being no unique response at the molecular level.

Thus, binding of antigen to a *B* lymphocyte is accompanied by (1) methylation of membrane phospholipids which activates serine esterase, (2) fluidization of the membrane which results in relatively smooth movement of antigen-antibody complexes to form patches and subsequently (energized by a nontransporting ATPase) to form caps at one pole of the cell, (3) opening of voltage-gated K^+ channels as well as of Ca^{2+} channels with all its regulatory consequences, (4) effects on adenylate cyclase and guanylate cyclase where cAMP acts as an inhibitor of lymphocyte proliferation and of antibody production while cGMP has a positive effect on these processes, (5) demethylation of genes for parts of the antibody resulting in their activation.

It thus appears that the immune system as a whole as well as in its individual parts, represents a web of information transmission involving all the types discussed in connection with the effects of hormones and neurotransmitters, and much more.

SYNOPSIS

The vertebrate immune system represents an enormously intricate mesh of interacting *B* lymphocytes, *T* lymphocytes, macrophages and a dozen molecules transmitting stimulatory and inhibitory signals once an antigen has been discovered. The *B* lymphocytes are active in humoral immunity where antigens bind to their surface and elicit a sequence of reactions resulting in reproduction of these cells and, eventually, in antibody production and secretion. The *T* lymphocytes are active in cellular immunity and their different types (killer, helper, suppressor) regulate in several ways the activity of other cells of the immune system. Macrophages are cells that phagocytose foreign material and destroy it, exposing, at the same time, its antigenic determinants on their surface to be recognized by helper *T* cells which then initiate and regulate the activity of other parts of the immune system.

The transmission of information arriving with an antigen involves probably all the known transduction systems (phosphoinositides, Ca^{2+}, cAMP, cGMP, as well as covalent modification of DNA).

FURTHER READING

Metzger, H. Antibodies as effector molecules. The Harvey Lectures (series 80). Alan Liss, New York (1986).

Steward, M. W., Steensgaard, J. *Antibody Affinity*, CRC Press, Boca Raton (1983).

Šterzl, J. Evolutionary approach in contemporary studies of antibody formation, *Folia Biol.*, **31**, 432–47 (1985).

Coda

Flow of matter, flow of energy, flow of information. All these processes take place in biomembranes but, clearly, all of them can proceed in the absence of membranes, by different mechanisms and generally with lower efficiency.

Information can be transmitted by direct contact of signal molecules or by direct interaction of physical factors with their effector partners. What is lost is the selectivity and fine control that can be provided by the interposition of a membrane receptor between the source and the consumer of information.

Energy can be transferred by direct coupling of endergonic with exergonic reactions. Absorption of radiation energy and its conversion into separation of electric charges could take place on any other cell structure that would provide suitable organized support for the reaction components. However, its subsequent fate requires a membrane to separate the compartments between which a difference in the electrochemical potential gradient of H^+ or Na^+ ions is established. Likewise, NADH and other reduced substrates can be oxidized by enzymes that need not be organized in a membrane but much of the free energy of the process is wasted unless a membrane separates two distinct compartments where a $\Delta\tilde{\mu}_{H^+}$ or $\Delta\tilde{\mu}_{Na^+}$ can be generated. Hence, a greater efficiency and speed are the consequence of the involvement of a membrane in the process.

Movement of molecules and ions proceeds by diffusion in the aqueous medium where life thrives even without membranes but the nutrients and ions required for its perpetuation would never reach the necessary concentrations near the enzymes or nucleic acids where they are needed if it were not for the membranes that, on the one hand, prevent the nutrients and ions from escaping and, on the other hand, actively bring them into the cell. Transport across membranes is simply the consequence of these needs, not a primary event. Likewise, the transport of ions in energy-transducing membranes is a means of converting energy into a suitable currency, not an end in itself.

In all probability, membranes were here at the beginning of life to keep the cell interior different from the surroundings but, as the lipid bilayer provided fine support for various specific proteins, different functions were gradually invested in membrane-bound proteins, perhaps in the temporal sequence identical with the progress of this volume: transport, utilization of energy, and gain of information.

Apparently, in all life processes everything goes better with membranes.

General References

Basic Principles of Membrane Transport (by S. G. Schultz). Cambridge University Press, Cambridge (1980).
Biomembranes—a series started in 1971 (general editor L. A. Manson, special authors for each volume); continues regularly, at Plenum Press, New York.
Biophysik (by W. Hoppe, W. Lohmann, H. Martel, and H. Ziegler as principal authors), Springer-Verlag, Berlin (1982).
Cell Biology (by N. O. Thorpe), John Wiley, New York (1984).
Cell Membrane Transport (by A. Kotyk and K. Janáček), Plenum Press, New York (1975).
Current Topics in Bioenergetics (edited by R. Sanadi), Vol. 1 (1966)–Vol. 15 (1986); continues regularly under changing editorship, Academic Press, New York.
Current Topics in Membranes and Transport (edited originally by F. Bronner, and A. Kleinzeller), Vol. 1 (1970)–Vol 27 (1987); continues regularly under changing editorship, Academic Press, New York.
Electrical Potentials in Biological Membrane Transport (by E. Heinz), Springer-Verlag, Berlin (1981).
Equations of Membrane Biophysics (by N. Lakshminarayanaiah), Academic Press, Orlando (1984).
Ions, Electrodes and Membranes (by J. Koryta), John Wiley, Chichester (1982).
Membranes and Transport (edited by A. N. Martonosi), Vol. 1, 2, Plenum Press, London (1982).
Membrane Transport (by A. Kotyk, and K. Janáček), Academia, Prague (1977).
Membrane Transport in Biology (edited by G. Giebisch, D. C. Tosteson, and H. H. Ussing), Vol. I–IVB, Springer-Verlag, Berlin (1978).
Membrane Transport Processes (Vol. 1 edited by J. Hoffman; Vol. 2 edited by D. C. Tosteson, Yu. A. Ovchinnikov, and R. Latorre), Raven Press, New York (1977).
Molecular Biology of the Cell (by B. Albers, D. Bray, J. Lewis, M. Raff, K. Roberts, and J. D. Watson as main authors), Garland, New York (1983).
Molecular Cell Biology (by J. Darnell, H. Lodish, and D. Baltimore), Scientific American Books, New York (1986).
Physical Chemistry (by P. W. Atkins), Oxford University Press, Oxford (1986).
Principles and Modes of Biological Transport (by M. H. Friedman), Springer-Verlag, Berlin (1986).
Principles of Electrochemistry (by J. Koryta, and J. Dvořák), John Wiley, Chichester (1987).
Structure and Properties of Cell Membranes (edited by Gh. Benga), Vol. 1–3, CRC Press, Boca Raton (1985).

Index

ADP ribosylation 221
ADP/ATP carrier 33, 84, 86, 114, 130, 214, 269, 302
  lateral diffusion 90
ATP assay 289
ATP hydrolysis 294, 295, 296, 306, 307
ATP synthase 82, 83, 127, 128, 249, 269, 295, 296, 297, 300
  in chloroplasts 37
ATP synthesis 237, 261, 280, 295, 296, 297, 298
Acanthamoeba castellanii 223
Accumulation ratio 187, 202, 203, 205, 206, 207
Acetobacterium woodii 280
Acetogenic bacteria 280
Acetylcholine 340, 348
  as neurotransmitter 339
  in bioluminescence 292
  transport in nerve 198
Acetylcholine receptor 339
  lateral diffusion 90
  muscarinic 339
  nicotinic 339, 340
  rotation 89
Acetylcholinesterase 87, 127, 128
Acetyl-CoA acetyltransferase, in mitochondria 32
N-Acetyl-D-glucosamine, transport by phosphotransferase 311
N-Acetylmuramic acid 41
Acholeplasma 49
  protein-lipid ratio 42
Acholeplasma laidlawii 53, 300
Acholoe 289
Achromobacter 288
Acid phosphatase 87
Acidaminococcus fermentans 283
Acidosome 39
Acid-CoA ligase, in mitochondria 32
Acinar cell 22, 226, 228
Aconitate hydratase, in mitochondria 32
Acrosome 39
Actin 91, 322
  in microfilaments 41
α-Actinin, in microfilaments 41
β-Actinin, in microfilaments 41
Actinogelin 42
Action potential 230, 292, 321, 323, 328, 341, 343, 344, 347, 348
Activated complex 145
Activation-controlled reactions 125
Active transport 197, 207
Activity of a chemical species 120
Acumentin 42
Acyl-CoA dehydrogenase, activation by lipids 126
  in mitochondria 32
Address sequence 108, 115
Adenosine 5′-phosphosulphate 279
Adenosinetriphosphatases (ATPases) 23, 303, 304
S-Adenosylmethionine, in lipid synthesis 106
Adenovirus 230
Adenylate cyclase 221, 312, 328, 334, 335, 339, 353
  activation by lipids 126
Adenylate kinase, in mitochondria 32
Adiabatic closed system 118
Adiabatic electron transfer 267
Adipocyte 331
Adrenal medulla 330
Adrenaline 330
  in chromaffin granules 39
α_2-Adrenergic receptor 333
β_2-Adrenergic receptor 333
Adrenocorticotropic hormone 330, 331
Adrenodoxin 255
Adsorption 98
Adsorption coefficient 98
Aequorea 288
  bioluminescence 286
Aequorin 288, 290

Aerotaxis 327
Affinity of a chemical reaction 199, 202
Agglutinin 222
Agglutinin receptor, lateral diffusion 90
Alamethicin 220
β-Alanine 340
Alanine, transport in bacteria 198
Alcaligenes 282
Alcohol dehydrogenase, in mitochondria 32
Aldehyde dehydrogenase (NAD), in mitochondria 32
Aldolase, in erythrocytes 91
Aldosterone 329
Alkaline phosphatase 87
Alkanal monooxygenase 288
Allophycocyanin 249, 251
Allosteric enzymes 215
Amine oxidase (flavin-containing), in mitochondria 32
Amino acids, transport in avian erythrocytes 198
  transport in bacteria 198
  transport in intestine 198
  transport in kidney tubules 198
  transport in mitochondria 33
  transport in tumour cells 198
  transport in yeasts 198
  transport of in lysosome 25
Aminoacylglycerol 44
Aminoacylglycerolphosphatidic acid 44
γ-Aminobutyric acid 340
  as neurotransmitter 339
Aminomethylfuran, as coenzyme 280
Ammonium ion, oxidation in bacteria 282
Amobarbital 270
Amoeba 20
  plasma membrane 42
Amoeba proteus 22
Amphipathicity 48, 98
Amphiphilicity 48, 98, 143
Amphiuma 304
Amphotericin B 184
Anabaena 243
Anabaena variabilis 182, 183
Anacystis 243
Anacystis nidulans 50
Angiotensin 331, 340
Ankyrin 41, 42, 81, 90, 91
Annelids 40
Annular lipid 93, 94
Anoxygenic photosynthesis 243
Antenna pigments, in photosynthesis 237
Anterior pituitary 330
Anthozoa 288
Antibody 350
Antigen 20, 349, 350, 352
Antigenicity 20
Antimonous ion oxidation in bacteria 282
Antimycin 270
Antiport 24, 130, 207, 309, 310, 319, 336
  in chloroplasts 37
  in mitochondria 33
Aplysia 305
Apocytochrome *b*, synthesis in mitochondria 33
Apogon 289
Aquaspirillum 282
Aqueous channels 162, 163, 181, 182, 185
Aqueous pores 130
Arabinose, transport in bacteria 216
Arachic acid 51
Arachidonic acid 51
Archaebacteria 38, 45, 57, 92, 280, 316
Arrhenius plot 94, 95, 127, 128, 208
Ascorbate, transport in intestine 198
Ascorbate–phenazine methosulphate oxidation in mitochondria 27
Ascorbate–tetramethylphenylene diamine, oxidation in mitochondria 27
Ascorbic acid 265
Asialoglycoprotein 227
Aspartate shuttle, in chloroplasts 36
Aspartylphenylalanine methyl ester, sweet taste 328
Aspergillus niger 221
Astroscopus 341
Asymmetry, of lipids 57, 59
Atmosphere 234
Atractylis gummifera 215
Atractyloside 214
Aurovertin 304
Autolysosome 26, 39
Autophagosome 26
Axolemma 344
Axon 338
Axon hillock 346
Axoplasm 345
Azotobacter vinelandii, cytochromes 282
Azurin 274

B lymphocyte 349, 353
Bacillus 282

Bacillus brevis 218
Bacillus licheniformis 87
  fatty acids 53
Bacillus megaterium 15
Bacillus subtilis 11, 50, 295
Bacteria 12, 14, 20
  bioluminescence 286
Bacteriochlorophyll 39, 49, 243, 244, 247, 252, 253, 254
Bacteriocins 182, 185
Bacteriopheophytin 244, 246, 247, 253, 254
Bacteriophyceae 288
Bacteriorhodopsin 83, 84, 89, 316
  lateral diffusion 90
  rotation 89
Bacteroides 280
Baker's yeast 22
Baltschinski's equation 64
Band 3 protein 84, 86, 91, 213, 214
  lateral diffusion 90
  rotation 89
Bathorhodopsin 319
Behenic acid 51
Bending elasticity 76
Beneckea 288
Beneckea natriegens, cytochromes 282
Binding protein, for dipeptides 326
  for galactose 326
  for maltose 326
  for ribose 326
Binding proteins 216
  in bacterial transport 215
  in chemotaxis 325
Bioenergetics 234
Bioluminescence 40, 285
Bivalvia 289
Blood flukes 13, 14
Blood-group substances 45, 47
Blue-green algae 242, 243
Boltzmann equation 99
Bongkrekic acid 214, 215
Brevin 42
Brownian movement 74
Buckminsterfullerene 224
Bulk flow 174
Bulk lipid 93, 94
Butyryl-CoA dehydrogenase, in mitochondria 32

C4 plants 278
CAM protein 322
cAMP-dependent protein kinase 335
CIEEL, luminescence 287
CURL 225
Calcimedins 336
Calcium channels 230, 319, 323, 336, 340, 342, 347, 348, 353
Calcium ions, as second messenger 336
  in phospholipid bilayers 101
  transport by ATPase 307
  transport in mitochondria 33
Caldesmin 42
Calelectrin 336
Calmodulin 42, 305, 336
  segmental mobility 89
Calsequestrin 336
Calvin cycle 242, 243
Candida utilis 11
Capnophorin 213
Carbon dioxide respiration 280
Carbon monoxide, as respiration inhibitor 270
Carbon monoxide respiration 280
Carbonylcyanide chlorophenylhydrazone, in chemotaxis 327
Carbonylcyanide phenylhydrazones 278
Carboxyatractyloside 214, 215
Carboxypeptidases 110
Cardiolipin 44, 45, 49, 50
Carnitine acetyltransferase, in mitochondria 32
Carnitine acyl transferase 312, 313
Carnitine cycle 33, 312, 313
Carnot thermal cycle 237
β-Carotene 248
Carotenes 49
  synthesis 106
Carotenoids 237, 239, 247, 248, 250, 251
Catalytic number, in membrane transport 208
Catecholamines, transport in chromaffin granules 198
Ca-ATPase 95, 128, 307, 308, 309, 310, 336
  of plasma membranes 308
  of sarcoplasmic reticulum 84, 86, 87, 308
  rotation 89
  with annular lipids 94
Cell adhesion 322
Cell wall 9, 41
α-Cells of pancreas 330
β-Cells of pancreas 330
Cellulose, in cell walls 41
Cephalopoda 288
Ceramide 46
  synthesis 106

Cercidosome 39
Cerebronic acid 51
Cerebroside 45, 46
  synthesis 106
Cerotic acid 51
Chaos chaos 226
Chara corralina 176
Charge separation 238, 250, 252, 254, 255
Chargerins 300
Chemical attractant 324, 325
Chemical potential 120, 131
  of water 169
Chemical repellent 324, 325
  in chemotaxis 327
Chemoautotrophy 259
Chemotaxis, methylation-dependent 325, 326
  methylation-independent 327
Chitin, in cell walls 41
Chlamydomonas 316
Chlorella pyrenoidosa, fatty acids 53
Chlorella vulgaris 198
Chloride anion, transport 241
Chlorobactene 247
Chlorobiaceae 243
Chlorobium 243
Chlorobium limicola 245
Chlorochromatium 243
Chloroflexaceae 243
Chloroflexus 243
Chloroflexus aurantiacus 247
Chloroherpeton 243
p-Chloromercuribenzoate 310
Chlorophyll 2, 49, 237, 239, 241, 243, 244, 248, 250, 251, 252, 253, 260, 317
Chloroplast lamella, protein–lipid ratio 42
Chloroplasts 9, 34, 35, 50, 181, 184, 248, 258, 259, 261, 274
  F_1 F_0 ATPase 296
  protein synthesis 113, 115
Chlorosome 39, 243, 245, 247, 249
Cholera toxin 221
Cholesterol 47, 48, 49, 76, 94, 226
  in phase transition 69
  synthesis 106
Cholic acid 23
Choline dehydrogenase, in mitochondria 32
Chondrite 1
Chondronectin 322, 323
Chromaffin cells 340
Chromaffin granules 39, 198, 304
Chromatiaceae 243
Chromatium 243
Chromatophore 243, 245, 246, 257
Chromosome, in mitochondria 34
Ciguatoxin 342
Circumsporozoite protein 85
Citrate, transport in mitochondria 198
Citrate (*si*)-synthase, in mitochondria 32
Clathrin 25, 26, 41, 224
Claviceps purpurea 26
Clitellata 289
Closed system 117
Clostridium 255
Clostridium aceticum 280
Clostridium symbiosum 283
Coated pits 130, 223, 224
Coated vesicles 24, 41
Cobalamin, transport of in lysosomes 25
Coelenterates 286
  bioluminescence 286
Coenzyme Q 271, 272
Colchicine 224
Colicin A 185
Colicin E_1 185
Colicin E_2 221
Colicin E_3 221
Colicin Ia 185
Colicin Ib 185
Colicin K 185
Colicins 182, 183, 185
Collision theory 122, 124
Compartment, cytoplasmic 8
  noncytoplasmic 8, 9
Compartment hypothesis, of mitochondria origin 30
Competence, of bacterial cells 222
Competitive acceleration 186
Complement 184, 352
Composite channel 189, 195, 197
Compressibility, of membranes 63, 64
Concanavalin A receptor, lateral diffusion 90
Condensing vacuole 228, 229
Conjugate flows 201
Connexin 15, 84, 181
Connexon 15, 17, 181
Convection 135
  forced 135
  natural 135
Convective diffusion 135
Cooperativity, in hormone binding 332
  in transport kinetics 194

Corticosteroids 329
Corticotropin 110
Corynomycolenic acid 52
Cotransport 195
Countertransport 186, 189, 193, 194, 195
Coupling efficiency 201
Coupling factor 295
Coupling of transport 199
Coupling stoichiometry 200
Crustacea 288
Crustaceans, bioluminescence 286
Cubic phase 56, 57
Curie principle 200
Current–voltage characteristic 163, 164, 167
Cyanelles 38
Cyanide, as inhibitor of sodium-motive bacteria 284
  as respiration inhibitor 270
Cyanobacteria 38, 39, 242, 243, 249, 316, 317, 318
Cyanocobalamin, transport in bacteria 183
Cyanophanon 34
Cyclic AMP 321, 334, 335, 336, 337, 339, 353
Cyclic GMP 319, 320, 321, 353
Cyclic electron flow 259
Cycloartenol, synthesis 107
Cypridina 288
  bioluminescence 286
Cypridina-luciferin monooxygenase 288
Cytochalasin B 224
Cytochrome P-450 23, 85, 115, 127
  rotation 89
Cytochrome P-450 reductase 23
Cytochrome *a* 246, 265, 268, 272, 273, 275, 277, 280
  in bacteria 282
  in mitochondrial oxidation 270
Cytochrome a_1, in bacteria 282
Cytochrome a_3 246, 265, 273, 275, 277
  in bacteria 282
  in mitochondrial oxidation 270
  rotation 89
Cytochrome *b* 85, 246, 255, 257, 265, 276, 281
  in mitochondrial oxidation 270
Cytochrome b_5 23, 115
  rotation 89
Cytochrome b_5 reductase 23
  in mitochondria 32
Cytochrome b_{50} 257
Cytochrome b_{558} 271, 277
Cytochrome b_{559} 248
  synthesis in chloroplasts 36
Cytochrome b_{561}, in chromaffin granules 39
Cytochrome b_{562} 272
Cytochrome b_{563} 258
Cytochrome b_{566} 272
Cytochrome b_6 249, 258, 265
  synthesis in chloroplasts 36
Cytochrome b_K 265, 277
Cytochrome b_T 265, 277
Cytochrome *c* 255, 257, 264, 265, 268, 269, 272, 275, 277, 280
  in mitochondrial oxidation 270
  lateral diffusion 90
  lipid dependence 92
  segmental mobility 88
Cytochrome *c* oxidase 83, 84, 127, 128, 268, 273, 275
  activation by lipids 126
  in chemotaxis 327
  in mitochondria 32
  of bovine heart 273
  of *Thermus thermophilus* 273
  of yeast 273
  synthesis in mitochondria 33
  with annular lipids 94
Cytochrome c_1 265, 272, 277
  in mitochondrial oxidation 270
Cytochrome c_2 247, 255, 257, 265
Cytochrome c_{550} 247
Cytochrome *d*, in bacteria 282
Cytochrome *f* 249, 258
  synthesis in chloroplasts 36
Cytochrome *o*, in bacteria 282
Cytochromes *b*, in bacteria 282
Cytochromes *c*, in bacteria 282
Cytokeratins 41
Cytolysis 76
Cytoplasm 22
Cytoskeleton 41, 59, 65, 77, 90, 221
  in protozoan costa 29
Cytoskeleton principle 1

DCCD 304, 305
DCCD-binding centre 272
DCCD-binding protein, of F_0F_1 ATPase 297
DIDS 304
DNA 20
  in immune system 350
  penetration into bacteria 221, 222

DNA transcription, effects of hormones 330
Deazoriboflavin, as coenzyme 280
Debye length 102
Degree of coupling 200
3-Dehydroretinal 318
Delayed fluorescence 253
2-Demethylmenaquinone 283
Dendrites 337, 338
Deoxynucleosides, transport in bacteria 183
2-Deoxy-D-glucose 202
Desmin 41
Desmosome, belt 15
 spot 15, 18
Desmotubule 15, 19
Desulfobacter 280
Desulfococcus 280
Desulfovibrio 280
Desulfuromonas 280
Determinant, in immunity 350
Dexter mechanism, in excitation transfer 252
Diacylglycerol 337
Diacytosis 40
Diaphus 288
Dicoumarol 278
Dictyosome 23
Dictyostelium discoideum 223, 321
Dicyclohexylcarbodiimide 311
 inhibition of H-ATPase 310
Dielectric breakdown 63
Dielectric constant 77
Diethylstilbestrol 304, 311
 inhibition of H-ATPase 310
Differential scanning calorimetry 66
Diffuse electric layer 97, 98, 101, 135
Diffusion 131, 134, 135, 149
Diffusion coefficient 125, 138, 139, 149
 of an unstirred layer 211
Diffusion flux 133, 134
Diffusion layer 133, 135
Diffusion potential 138, 294
Diffusion-controlled reactions 124, 126, 192
Digitonin 184
Diglycerides 43
Dihydrogen, oxidation in bacteria 280, 282
Dihydrogen oxidation 280
Dihydromenaquinone 283
Dihydroorotate, oxidation in bacteria 280
Dihydrosphingosine 46
Dihydrosterculic acid 52
Dihydroxyacetone, transport by phosphotransferase 311
Dinoflagellata 288
Dinoflagellates 286
 bioluminescence 286
Diol esters 49
Dio-9, inhibition of H-ATPase 310
Dipeptides, transport of in lysosomes 25
Diphosphatase 278
Diphosphatidylglycerol 44, 106
 synthesis 105
Diplocardia 289
Diplotrema 289
Disaccharides, transport in yeasts 198
Discocyte 11
Disorder, in membranes 63
Disproportionation reaction 266
Dissipation function 168, 172
Dissociation-field effect 164
Dissociative electron transfer 290
Distribution equilibrium 142
Docking protein 109
Dodecenoyl-CoA isomerase, in mitochondria 32
Dolichol diphosphate 111, 112
Dopamine 340
Dynein 13, 41, 301

E_1E_2 ATPases 85, 295, 303, 305
Earthworms, bioluminescence 286
Ectothiorhodospora 243
Effector cell 338
Efficacy of flow 201
Efficacy of force 201
Einstein's diffusion equation 75
Elaidic acid 51
Elastic energy 77
Electric conductivity 178
Electric discharge 235
Electric double layer 96, 97, 98, 99, 134
Electric field 321
Electric fish 341
Electric potential 115
 as signal 321
Electrochemical potential 121, 135, 151, 154, 261, 278, 283, 293, 294, 298, 299, 324
Electrochemical potential gradient 239, 242, 269
Electrode potential 263
Electrodiffusion 134, 150
Electrofusion 78
Electrogenic pump 159

Electromotive force 263
Electron affinity 266
Electron spin resonance 54, 66
Electron transfer, in photosynthesis 255
Electron-transfer chain 269
Electron-transfer reactions 267
Electro-osmosis 177, 178
Electrophoretic mobility 100
Electrophorus 341
Electrophorus electricus 76
Electroplax 89, 341
Electrostatic interaction 236
α-Eleostearic acid 51
Ellipsosome 39
Elongation factors, synthesis in chloroplasts 36
Endocytosis 221, 230, 351
  adsorptive 223
  constitutive 223
  induced 226
  inducible 223
  receptor-mediated 223, 225, 226, 227
Endonexin 336
Endoplasmic reticulum 7, 15, 19, 40, 108, 109, 111, 226, 292, 336
  rough 6, 9, 22, 23, 110, 229
  smooth 9, 22, 23, 49, 106, 107
Endoproteases 110
β-Endorphin 110, 340
Endosome 39
Endosymbiont theory 30, 181
Endosymbionts 38
Energy, electric 237
  osmotic 237
Enkephalin 110, 340
Enoyl-CoA hydratase, in mitochondria 32
4-Enoyl-CoA reductase, in mitochondria 32
Entactin 322
Enterochelin 183
Enterocyte 11
Enthalpy 119
Entropy 118, 168, 314
Enzyme I, of phosphotransferase 311
Enzyme II, of phosphotransferase 311, 327
Enzyme-catalyzed reactions 125
Epicholesterol 70
Epidermal growth factor receptor 333
  lateral diffusion 90
  rotation 89
Epinephrine 330, 331, 340
Equation of photosynthesis 261
Equilibrium constant 122
Ergosterol 47, 48
Erucic acid 51
Erythrocyte 11, 22, 49, 50, 198, 213
  lipid asymmetry 58
  protein-lipid ratio 42
Erythrosin B 304
Escherichia 280
Escherichia coli 15, 108, 126, 181, 182, 183, 185, 198, 215, 285, 321, 325, 326, 327
  F_1F_0 ATPase 296
  H-ATPase 295
  cytochromes 282
  cytoplasmic membrane 42
  fatty acids 53
  lateral diffusion 75
Estrogen 330, 331
Ethacrynic acid 305
N-Ethylmaleimide 304, 305, 311
Eubacteria 38
Eucytes 38
Excitation energy transfer 250
Exciton 252, 258
  delocalized 253
  localized 252, 253
Exocytosis 130, 221, 226, 228, 230, 339, 340
Extrusome 39
Eyring's equation 123, 138
Eyring's rate theory 163, 188

FAD 265, 269, 277, 284, 296
  in mitochondrial oxidation 270
FAD dehydrogenases 271
F_1F_0 ATPase 113, 216, 298, 300, 303
FMN 265, 277, 284, 290
  in mitochondrial oxidation 270
(Factor) III, of phosphotransferase 312
Faraday's law 138
Fascin 42
Fatty acid synthetase, synthesis in chloroplasts 36
Fatty acids 51, 53, 126, 278
  lateral diffusion 75
  phase transition 67
  synthesis in chloroplasts 38
  transport in mitochondria 33
Ferredoxin 244, 248, 255, 257, 258, 265, 281
Ferredoxin-NADP reductase 244, 248
Ferrichrome 183
Ferricyanide 265
Ferritransferrin 225

Ferrous ion, oxidation in bacteria 282
Fibronectin 322, 323
Fick's first law 131, 132, 138, 139, 144, 168
Fick's second law 132
Filaggrin 41
Filamin 42
Filipin 184
Fimbrin 42
Firefly, bioluminescence 287
First law of thermodynamics 118
First-order reaction 125
Fischerella 243
Flagellin 300, 301
Flagellum 9, 301, 318, 321, 324, 325, 327
  of eukaryotic cells 41
Flavin adenine dinucleotide reductase 255
Flavour, as chemical signal 328
Flexoelectric coefficient 77
Flip-flop movement 5
  of lipids 75, 161
Flow of energy 5, 81
Flow of information 5, 81
Flow of matter 5, 81
Fluidity 63, 65, 66
Fluorescence 237, 250, 252, 253
Fodrin 42
Folch–Lees protein 82
Follicle-stimulating hormone 330, 331
Formal potential 276
Formal (conditional) potential 264
Formate, oxidation in bacteria 280
Forster mechanism, in excitation transfer 253
Fragmin 42
Fret, in chloroplast 36
Fructose, transport by phosphotransferase 311
Fructose bisphosphate aldolase, in chloroplasts 37
Fructosebisphosphatase, in chloroplasts 37
Fumarate hydratase, in mitochondria 32
Fumarate reductase 85, 280, 281
  in chemotaxis 327
Fumarate respiration 280
Fungi 12, 38
  bioluminescence 286
Fusion 20, 223, 230

G protein 221, 320, 321, 328, 334, 335, 337
GDP-binding protein 334
GDP-mannosyl transferase 23
GDP-protein 302
GTP-binding protein 328
Gadolenic acid 51
Galactose, transport in bacteria 198
Galactoside carrier, from *Escherichia coli* 84
Galactosides, transport in bacteria 215
Galactosylglycerides, in chloroplasts 37
Gallionella ferruginea 282
Galvani potential 151
Galvanic cell reaction 263
Gangliosides 45, 46
Gap junction 14, 17, 42, 181, 336
Gastropoda 289
Gated channels 130
Gelsolin 42
Gene-duplication-segregation hypothesis, of mitochondrial origin 30
Genome-invagination hypothesis, of mitochondrial origin 33
Getolenic acid 51
Gibbs energy 236
Gibbs equation 99
Gibbs free energy 119, 121, 143
Gibbs transfer energy 140
Glial filament proteins 41
Gliding bacteria 38, 318
Globoside 46
Gloeocapsa 243
Glucagon 330, 331
Glucans, in cell walls 41
Glucitol, transport by phosphotransferase 311
Gluconate 311
  transport by phosphotransferase 311
Glucose, transport by phosphotransferase 311
Glucose-6-phosphatase 23, 127
  activation by lipids 123
Glucose-6-phosphate, transport in bacteria 198
Glucose-6-phosphate dehydrogenase, in chloroplasts 37
β-Glucosides, transport by phosphotransferase 311
β-Glucuronidase 23
Glucuronosyltransferase, activation by lipids 126
Glutamate, as neurotransmitter 319, 340
  oxidation in mitochondria 271

Glutamic acid, transport in bacteria 198
γ-Glutamyltranspeptidase 130, 312
Glyceraldehyde-3-phosphate dehydrogenase, in erythrocytes 91
Glyceraldehyde-P dehydrogenase (NADP), in chloroplasts 37
Glycerol 3-phosphate, oxidation in bacteria 280
Glycerolipids 43, 51, 105
Glycerol-phosphate dehydrogenase, in mitochondria 32
Glycerophospholipids 107
Glycine, as neurotransmitter 340
  transport in bacteria 198
Glycocalyx 86
Glycogen 333, 335
Glycogen phosphorylase 336
Glycogen synthase 333, 335, 336
Glycogen synthesis 336
Glycogenolysis 330
Glycoglycerolipids 50
Glycolipids 14, 43, 49
Glycophorin 85, 90
  rotation 89
  segmental mobility 89
  with annular lipids 94
Glycoproteins 14, 86, 226, 328
  N-linked 111
  O-linked 111
Glycosidases 25
Glycosome 39
Glycosphingolipids 49
Glycosylation 23
Glyoxylate pathway, in glyoxysomes 28
Glyoxysome 28
Goldman's equation 154, 163, 164, 343
Golgi body (complex, apparatus) 5, 9, 23, 24, 25, 26, 39, 106, 107, 111, 112, 226, 229
Gonyaulax 288
Gouy–Chapman theory 97, 99
Gramicidin A 93, 165, 208, 218, 219, 220, 284
Gramicidin S 220
Granum thylakoid 36
Green bacteria 255, 257, 259
Green filamentous bacteria 38, 242
Green fluorescent protein 290
Green nonsulphur bacteria 243
Green sulphur bacteria 38, 39, 242, 243, 247
Group translocation 130, 211, 312
Growth hormones 329
G-Strophantin 202
Guanylate cyclase 353

H-ATPase 310
  in chromaffin granules 39
  in lysosomes 25
  in microbodies 28
  in vacuole 26
HLA antigens 87
HPr, of phosphotransferase 311, 312
Haemoglobin 2
Hair cells, of inner ear 323
Haldane's relation 203
Half-saturation constant 127, 128, 186, 211
Half-site reactivity 215, 302
Halobacteria 20, 318
Halobacterium 15, 38, 89, 239, 241, 316
  F_1F_0 ATPase 298
Halobacterium halobium 54, 198
Halophilic bacteria, protein-lipid ratio 42
Halorhodopsin 316
Halosulpholipids 49
Hamaker constant 77
Hapten, in immunity 350
Harmothoe 289
Heat, as a signal 319
Heat production, in mitochondria 302
Heliobacterium chlorum 250
Helmholtz free energy 119
Hematoside 46
Hemichrome 91
Hemidesmosome 1, 15
Hepatocyte 50
  lipid asymmetry 58
  plasma membrane 42
2-Heptyl-4-hydroxyquinoline-N-oxide, as inhibitor of sodium-motive bacteria 284
Heterolysosome 26, 39
Heterophagosome 26, 39
Hexagonal phase 56, 57
Hexapoda 289
Hexokinase 127
  lipid dependence 92
High-energy phosphate bond 235, 236
High-potential iron–sulphur proteins 257
Histamine 331, 340
Histidine, transport in bacteria 216
Hodgkin–Horowicz equation 158, 163
Hopanoids 3, 48, 72
Hormone, as chemical signal 328

Hormones 20, 329
H-transporting ATP synthase, in mitochondria 32
Humic acids 221
Hydraulic conductivity 174, 178
Hydrodynamic flow 135
Hydrodynamic layer 135
Hydrogen ion, translocation in mitochondria 271, 278
  transport 240, 257
  transport by ATPase 309
  transport in mitochondria 33
  transport in photosynthesis 259, 260
Hydrogen ion cycle 300
Hydrogenosome 29, 30
Hydrophilicity 140, 164
Hydrophobic effect 143
Hydrophobicity 141, 164
Hydrophobicity index 82
Hydrosphere 234
Hydrostatic pressure 169, 174, 175
  on membranes 64
D-2-Hydroxy-acid dehydrogenase, in mitochondria 32
Hydroxyacyl-CoA, oxidation in mitochondria 271
3-Hydroxyacyl-CoA dehydrogenase, in mitochondria 32
3-Hydroxybutyrate, oxidation in mitochondria 271
3-Hydroxybutyrate dehydrogenase, activation by lipids 126
  in mitochondria 32
β-Hydroxybutyrate dehydrogenase, with annular lipids 94
3-Hydroxybutyryl-CoA epimerase, in mitochondria 32
Hydroxylamine, oxidation in bacteria 282
Hydroxymethylglutaryl-CoA reductase 23
Hydrozoa 288
Hypogaeic acid 51
Hypophysis 329
Hypothalamus 329, 330
Hypsorhodopsin 319

ITIES 97, 140
Immune system 349, 352
Immunity, cellular 349
  humoral 349
Immunoglobulins 87, 349, 350, 352
  segmental mobility 88
Influenza A virus, lipid asymmetry 58
Influenza virus 130, 230
Information, amount 314
  noise 314
  redundancy 314
  value 314
Information theory 314
Infusorians 28
Inner electric potential 263
Inositol 1,4,5-triphosphate as second messenger 337
Insects, bioluminescence 286
Insulin 330, 331, 333, 335
Interfacial lipid 68
Interfacial tension 77
Interferon 87
γ-Interferon 351, 352
Interleukins 351, 352
Intermediate filaments 18, 41
Intermolecular energy transfer 252
Internal energy 118
Internal pressure 171
Interneuron 337
Invertase, from yeast 87
Ion pairing 142
Ionic strength 102
Ionization potential 266
Ion-selective electrode 102
Ion-solvent interaction 142
Iron–sulphur centre 248, 255, 258, 259, 265, 271, 277, 281
  in mitochondria 270
Irreversible thermodynamics 202
Isocitrate, oxidation in mitochondria 271
Isolated system 117
Isolauric acid 51
Isomyristic acid 51
Isopalmitic acid 51
Isopentenyl diphosphate 239
Isoprenoid, synthesis 107
  synthesis in chloroplasts 38
Isuasphaera 2

Joule's heat 238

K,H-ATPase 309, 310
Kephalins 44
Kidney tubules 198
Kinesin 41
Kinks in membrane lipids 62, 161
Kinocilium 323
Klebsiella 280
Klebsiella aerogenes 283

Klebsiella pneumoniae, cytochromes 282
Kraft point 102
Kupffer cell 25
Kynurenine monoxygenase, in mitochondria 32

LDL receptor 85
LH-releasing factor 330, 331
LH-releasing hormone 340
β-Lactamase 113
Lactate, oxidation in bacteria 280
  transport in bacteria 198
L-Lactate dehydrogenase (cytochrome), in mitochondria 32
Lactobacillic acid 52
Lactobacillus casei, ATPase 295
Lactose, transport by phosphotransferase 311
  transport in bacteria 215
Lamellar phase 55, 56, 57, 60, 64, 66, 68, 69, 70, 72, 74, 78, 89, 90, 92
Laminin 85, 322, 323
Langmuir isotherm 98, 102
Lanosterol 48
  synthesis 106
Lateral diffusion 5, 74, 92, 94, 107
Lateral diffusion coefficient 72, 74, 75, 89
Lateral pressure 55, 57
Latia 289, 290
Latia-luciferin monooxygenase 289
Lauric acid 51
Lauroleic acid 51
Leaks in membrane transport 200, 205, 207
Lecithin 44, 55, 65, 232
Leptothrix discophorus 282
Leptothrix ochracea 282
Leukovirus, protein-lipid ratio 42
Level flow, in active transport 201
Leydig cell 23
Light, as information signal 316
Light-harvesting complexes 85, 246, 247, 248, 251, 260
Lignoceric acid 51
Linolenic acid 51
Linolic acid 51
Lipases 25
Lipid bilayer 48, 54, 57, 64, 70, 78, 99, 100, 102, 103, 143, 226, 232
Lipid turnover 107
Lipid-exchange proteins 107, 108
Lipoamino acid 44
Lipocortin 336
Lipopolysaccharide 15, 49, 301
Lipopolysaccharide galactosyltransferase, activation by lipids 126
Lipopolysaccharide glucosyltransferase I, activation by lipids 126
β-Lipoprotein 110
Lipoproteins 87
Liposomes 65
Liquid-junction potential 263
Loculus, in chloroplast 36
Long-chain-fatty acid-CoA ligase, in mitochondria 32
Long-chain-fatty-acid-CoA ligase, activation by lipids 126
Low-density lipoprotein 226
Luciferase 285, 288, 290
Luciferin 285, 286, 288, 290
Lumilamban 320
Luminescence 235, 285
Luminosome 39
Lumirhodopsin 319
Lutein 248
Luteinizing hormone 331
Lymphocytes 20, 352
Lysolecithin 57
Lysophosphatidylcholine 126
Lysophospholipids 44, 57
Lysosomes 9, 25, 39, 111, 223, 226, 304, 311
  in protein turnover 115
  primary 25, 26
  secondary 25, 26

α-Macroglobulin 227
Macromolecules 2
Macrophages 20, 21, 223, 224, 227
Macula adhaerens 15, 18
Magnesium ion, transport in chloroplasts 37
Magnetid field, as signal 321
Magnetite, in bacteria 321
Magnetobacteria 321
Malapterurus 341
Malate, oxidation in mitochondria 271
Malate dehydrogenase, in mitochondria 32
Malate dehydrogenase (decarboxylating), in mitochondria 32
Malate oxidase, activation by lipids 126
Malate shuttle, in chloroplasts 36
Malic enzyme (NADP-linked), in chloroplasts 37
Maltoporin 183

Maltose, transport in bacteria 183, 216
Maltose-binding protein 113
Maltotriose, transport in bacteria 183
Malvalic acid 51
Manganous ion, oxidation in bacteria 282
Mannan, in cell walls 41
Mannitol, transport by phosphotransferase 311
Mannose, transport by phosphotransferase 311
Marcus equation 268
Marcus theory 267
Margaric acid 51
Marine sponges 52
Mass flow 174
Maximum reaction rate 127
Maximum transport rate 186, 211
Mean displacement 134
Mechanical pressure, as signal 322
Mediated diffusion 130
α-Melanocyte-stimulating hormone 110
β-Melanocyte-stimulating hormone 110
γ-Melanocyte-stimulating hormone 110
Melanosome 39
Melibiose, transport in bacteria 198, 215
Melissic acid 51
Melittin 76
  lipid dependence 92
Membrane, bacterial 5
  basilar 323
  cytoplasmic 14
  inner chloroplast 36
  inner mitochondrial 4, 5, 30, 31, 32, 33, 42, 49, 50, 92, 114, 184, 269, 312
  nuclear 6, 7, 20, 21
  outer chloroplast 36
  outer mitochondrial 30, 31, 32, 33, 42, 50, 114, 183
  plasma 8, 9, 11, 12, 13, 14, 17, 19, 20, 224, 226, 230, 247, 301, 304, 305, 322
  postsynaptic 347, 348
  presynaptic 339, 347
  tectorial 323
  unit 12
Membrane anisotropy 199
Membrane biogenesis 104, 226
Membrane capacity 65, 77, 79, 100, 102, 103, 294
Membrane carrier 164, 165, 187, 189, 206, 208, 214, 217, 218
Membrane channel 130, 148, 164, 165, 187, 208, 214, 216, 218
Membrane conductance 344
Membrane conductivity 158
Membrane current 163, 344, 345
Membrane potential 65, 77, 100, 156, 178, 210, 219, 278, 284, 317, 318, 341, 342, 343, 344, 346, 348
Membrane principle 1
Membrane receptors 221
Membrane resistivity 102
Membrane stability 76
Membrane-attack complex 184
Menaquinone 254, 281, 283, 284
Mercaptosulphonate, as coenzyme 280
Mesomorphic states 72
Metarhodopsin 319, 320
Metasequoia glyptostroboides 19
Meteorite 1, 2
Methanobacterium 38, 280
Methanococcus 280
Methanogenic bacteria 280
Methanosarcina 280
Methanospirillum 280
Methyl esterase 326
  in chemotaxis 325
Methylene blue 265
β-Methylgalactoside, transport in bacteria 216
Methyltransferase 326
  in chemotaxis 325
Methyl-accepting proteins (MAP) 325
Mevalonic acid 52
Michaelis constant 128, 186
Michaelis kinetics 125, 188, 191
Miconazole 304
  inhibition of H-ATPase 310
Microbodies 9, 26
  protein synthesis 113
Micrococcus luteus, cytochromes 282
Micrococcus lysodeikticus, ATPase 295
  cytoplasmic membrane 42
Micrococcus radiodurans 15
Micrococcus roseus 50
Microfilaments 41, 221, 224
Microsomes 126
  lateral diffusion 75
Microtubules 9, 41, 224, 323
Microtubule-associated proteins (MAP) 41
Microvillus 16
Microviscosity 63, 64, 92
Midpoint potential 253, 255, 264, 276, 277

of bacterial ferredoxins 257
Mitochondria 6, 7, 9, 28, 30, 38, 126, 181, 198, 237, 266, 269, 336
  F_1F_0 ATPase 296
  fatty acids 53
  in heat production 302
  protein synthesis 113, 115
Mitosis 20
Mobile carrier 189
Mobility 131
  of proteins 88
Molecular activity, in transport 186
Molecular dynamics 148, 163
Monogalactosyldiacylglycerol 44
Monoglycerides 43
Monosaccharides, transport in erythrocytes 193
  transport in intestine 198
  transport in kidney tubules 198
  transport of in lysosomes 25
Montanic acid 51
Montmorillonite 4
Motor neuron 338
mRNA 108, 109
  synthesis in chloroplasts 36
mRNA stability, effects of hormones 329
Mucidin 270
Multisite carriers 194
Muscle contraction 235
Mycobacterium 282
Mycobacterium avium 126
Mycobacterium phlei 52
  cytochromes 282
Mycoplasma 49
Mycoplasma gallisepticum 108
Myelin 82
  protein-lipid ratio 42
Myelin A1 protein 85, 92
  lipid dependence 92
Myelin sheath 13, 20, 338, 345, 346, 347
Myoglobin 3
Myosin, in microfilaments 41
Myristic acid 51
Myristoleic acid 51
Myxovirus, protein–lipid ratio 42

NAD 269, 276, 277, 278, 281, 284, 299
  in mitochondrial oxidation 270
NAD reduction 242, 259
NADH, oxidation in bacteria 280
NADH cytochrome b_5 reductase 285
NADH dehydrogenase, activation by lipids 126
  in mitochondria 32
NADH dehydrogenases 269, 278
NADH ferricyanide reductase 285
NADH transhydrogenase (ubiquinone), in mitochondria 32
NADH-cytochrome *c* reductase 24
NADH-ferrihemoprotein reductase 23
NADH-quinone reductase 284
NADH-ubiquinone reductase 85, 269
NAD-NADPH transhydrogenase 269
NADP 277, 290
  assay 288
NADP reduction 242, 255, 260
NAD(P) transhydrogenase 276
  in mitochondria 32
NADPH ferricyanide reductase 285
NADPH-cytochrome *c* reductase 24
NADP-malate dehydrogenase, in chloroplasts 37
Navier–Stokes equation 136
Na,K-ATPase 84, 127, 128, 307, 308, 310
  with annular lipids 94
Negative pressure 171
Neisseria gonorrhoeae 183, 222
Neolysosome 39
Neoscopalus 288
Neoxanthin 248
Nernst equation 263
Nernst layer 135, 136, 149
Nernst–Donnan equilibrium 121
Nernst–Donnan potential 142, 158
Nernst–Einstein equation 137
Nernst–Planck equation 154, 155, 158
Nerve axon 13
Nerve cell 20
Nerve fibre 338
Nerve impulse 337
Nervonic acid 51
Neuraminidase 85
Neurofilamins 41
Neuron 337, 341, 343, 345, 347, 348
Neurospora 33
Neurospora crassa 83
Neurosporene 247
Neurotensin 340
Neurotransmitters 130, 339, 347, 348
  as chemical signal 328
Nexin 41
Nitrate reductase 279
  in chemotaxis 327

Nitrate respiration 279
Nitrite anion, oxidation in bacteria 282
Nitrite reductase 279
Nitrobacter 282
Nitrococcus 282
Nitrophenols 278
Nitrosococcus 282
Nitrosomonas 282
Nocardia 282
Noise analysis 166
Nonactin 93
Nonadiabatic electron transfer 267
Nonequilibrium thermodynamics 172, 202
Nonspecific permeation 130, 131
Non-Stokesian diffusion 139, 161
Noradrenaline 330
  in chromaffin granules 39
Norepinephrine 330, 339, 340
Nostoc 243
Nuclear magnetic resonance 54, 66, 73, 127
Nuclear pore 9, 21, 221
Nucleases 25
Nucleolus 9
Nucleoside diphosphatase 23
Nucleoside diphosphokinase, in mitochondria 32
Nucleosides, transport in bacteria 183
  transport in lysosomes 25
Nucleoside-triphosphate adenylate kinase, in mitochondria 32
5′-Nucleotidase 23, 24, 87
Nucleus 9, 20, 22
Nystatin 184

OSCP 295
Obelia 288
Octochaetus 289
Octopamine 340
Odour, as chemical signal 328
Ohm's law 168, 346
Oleic acid 51
Oleosome 39
Olfactory nerve fibres 328
Oligoglucan transferase 112
Oligomycin 298, 299, 304
Oligopeptides, transport in bacteria 216
Oligosaccharides 23
Oncorna virus 230
Onsager's law 168, 173, 177, 179, 199
Open system 118
Opsin 319, 320
Opsin kinase 320
Order, in membranes 63, 65
  of lipids 93
Order parameter 65, 66
  molecular 65
Organic acids, transport in chloroplasts 36
  transport in mitochondria 33
Oscillatoria 243
Osmosis 173
Osmotic equilibrium 169
Osmotic pressure 170, 173
  on membranes 64
Ouabain 202, 304, 305, 307
Oxidases, in peroxisomes 27
Oxidation–reduction potential 265
Oxidation–reduction reactions 263
2-Oxoglutarate, oxidation in mitochondria 271
2-Oxoglutarate dehydrogenase complex, in mitochondria 32
3-Oxoacid CoA-transferase, in mitochondria 32
α-Oxoglutarate dehydrogenase, lipid dependence 92
Oxygen reduction, in mitochondria 270
Oxygenic photosynthesis 242, 243, 257
Oxyluciferin 288
Oxyntic cells 304, 309
Oxytocin 331

Palmitic acid 51
Palmitoleic acid 51
Pantothenate, transport in bacteria 198
Paracoccus 30, 282
Paracoccus denitrificans 279
  cytochromes 282
Paramecium 184
Paramyxo virus 230
Partial molal volume 120
Partition coefficient 139, 140, 148
Partition equilibrium 139
Pectins, in cell walls 41
Pelodictyon 243
Pentachlorophenol 278
Peptidoglycan 15, 41, 182, 301
Peptococcus aerogenes 283
Perillaldehyde aldoxime, sweet taste 328
Periplasmic space 182, 216
Permeability coefficient 149
Permeability constant 138, 139, 147, 186

Peroxisomes 27, 28
protein synthesis 113
pH dependence, of transport 209
λ-Phage receptor 113, 183
Phage receptors 182
Phagocytosis 184, 223
Phagosome 39
Phase transition 55, 60, 63, 66, 68, 92
Phaseolus vulgaris 34
Phenylhydrazinomalononitriles 278
Pheophytin 244
Pholas 289, 290
Pholasin 289
Phorbol ester receptor 337
Phormidium uncinatum 317
Phosphatases 25
Phosphate anion, transport in bacteria 216
transport in mitochondria 33
transport in yeasts 198
Phosphatidic acid 43, 45, 57
synthesis 105
Phosphatidylcholine 44, 45, 49, 50, 57, 62, 71, 73
fatty acids 53
lamellar phase 60
lateral diffusion 75
phase transition 67
synthesis 106
Phosphatidylethanolamine 44, 45, 49, 50, 57
fatty acids 53
lamellar phase 60
phase transition 67
synthesis 105, 106
Phosphatidylglycerol 44, 49, 50, 57
lamellar phase 60
phase transition 67
synthesis 105, 106
Phosphatidylinositol 44, 45, 49, 50, 87, 337
fatty acids 53
synthesis 106
Phosphatidylserine 44, 45, 50, 57
fatty acids 53
phase transition 67
synthesis 105, 106
Phosphodiesterase 320
Phosphoenolpyruvate carboxylase, in chloroplasts 37
in mitochondria 32
Phosphofructokinase, in erythrocytes 91
Phosphogluconate dehydrogenase, in chloroplasts 37
Phosphoglycerate kinase, in chloroplasts 37
Phospholipase C 337
Phospholipases 107
Phospholipids 23, 43, 70, 97, 99, 102, 105, 108, 126, 239
in annular region 94
in mitochondria 31
Phosphonolipid 43
Phosphoprotein phosphatase 336
Phosphoribulokinase, in chloroplasts 37
Phosphorylase kinase 336
Phosphorylated intermediate, of ATPase reaction 307, 310
Phosphorylinositol ceramide 46
Phosphotransferase system (PTS) 311, 312, 327
Photinus 289
bioluminescence 287
Photinus-luciferase 4-monooxygenase 289
Photoautotrophy 259
Photobacterium 288
Photobleaching recovery technique 74
Photocycle, of bacteriorhodopsin 239, 240
of halorhodopsin 241
Photokinesis 316, 317
Photomovement 316
Photophobic response 316
Photosomes 40, 292
Photosynthesis 3, 234, 238, 239, 241
regulation 260
Photosynthetic systems 237, 242, 245, 251
synthesis in chloroplasts 36
Photosystem I 242, 248, 249, 251, 253, 255, 258, 260
synthesis in chloroplasts 36
Photosystem II 242, 248, 249, 251, 255, 256, 257, 258, 260
in cyanobacteria 249
synthesis in chloroplasts 36
Phototaxis 316, 317
Photuris 289
Phrixothrix 289
Phthionic acid 52
Phycobilin 250
Phycobiliproteins 249
Phycobilisome 39, 243, 249, 317
Phycocyanin 249, 250, 251
Phycoerythrin 317

Phycoerythrocyanin 249, 251
Phytane 2
Phytochrome 316
Phytol 2
Phytosphingosine 46
Picorna virus 230
Piek's equation 156
Piericidin 270
Piezoelectric effect 241
Pigment Z, in oxygenic photosynthesis 255
Pinocytosis 24, 26
fluid-phase 223
Pinosome 39, 225
Pituitary 329
Pituitary stalk 329
Plasma membrane, lobster nerve 50
rabbit muscle 50
Plasmalemma 15
Plasmalogen 43, 70
Plasmic hypothesis of mitochondrial origin 30
Plasmodesma 9, 15, 19
Plasmodium falciparum 85
Plastid 34
Plastocyanin 244, 248, 255, 258, 265, 274, 276
Plastoquinone 49, 244, 246, 260, 276
Plastoquinone pool 257, 260
Plateau–Gibbs border 101
Platelet, lipid asymmetry 58
Plectin 41
Pleurocapsa 243
Poisson–Boltzmann equation 97
Polar channels 162
Polychaeta 289
Polyene antibiotics, as pore formers 184
Polymorphonuclear leukocytes 223
Polymyxin 220
Polynoidin 285, 292
Polyphosphate, in vacuole 26
Polysaccharides 15
in cell walls 41
Polysomes 9, 113
Pore edge energy 78
Pore formation 78
Pore proteins 183
Porichthys 289
Porin P 183
Porins 84, 130, 181, 182, 183, 184
peptidoglycan-related 182
Porphyrin 2
redox reactions of Fe 267
synthesis in chloroplasts 38
Postsynaptic cell 339
Potassium channels 339, 342, 352
Potassium ion, transport by ATPase 307, 309
transport in chloroplasts 37
transport in mitochondria 33
Potentiometry 266
Pox virus 230
Prandtl layer 100, 135, 149
Precytes 4
Presequence, in protein synthesis 108, 114, 115
Pretransition 70, 71, 72
Pre-Cambrian cells 4
Pre-Cambrian shale 3
Primary active transport 130, 166, 167, 197, 199, 202, 259
Primary electron donor in photosynthesis 247, 248, 253
Primeval cell 3
Pristane 2
Proalbumin 110
Prochloron 243
Prochlorophyta 243
Profilin 42
Progenote 38
Progesterone 329, 331
Proglucagon 110
Proinsulin 110
Prolactin 331
Pro-opiomelanocortin 110
Proparathyroid hormone 110
Propionibacterium 280
Propionigenum modestum 283, 300
Prostaglandins 331
Prostasome 39
Prosthecochloris 243
Prosthecochloris aestuarii 247
Protein kinase 310, 333, 336, 339
Protein kinase C 337
Protein phosphokinase 333
Protein synthesis 108
in mitochondria 33
Protein turnover 115
Proteinases 25
Proteins, extrinsic 81
integral 81
intrinsic 81
peripheral 81
Protein-mediated transport 185, 186
Proteoglycan 322
Proteus 280
Protometer 327
Proton conductors 278

Proton gradient 216, 280
Protonmotive force 278, 280, 327
Protonophore, of F_1F_0 ATPase 296
Protoplast 8, 231
Protozoa 38, 316
Pseudomonas 282
Pseudomonas aeruginosa 183
  cytoplasmic membrane 42
Pseudomonas carboxidovorans 282
Pseudomonas cocovenenans 214
Pseudomonas fluorescens 274
Pseudomonas stutzeri 300
Pseudomurein 239
Pseudorotation, of ATPase active centre 306, 307
Purine bases, transport in yeasts 198
Purple bacteria 255, 257, 259, 316
Purple membrane 239, 241
Purple nonsulphur bacteria 38, 242, 243
Purple sulphur bacteria 38, 242, 243
Pyrethroids 342
Pyrimidine bases, transport in yeasts 198
Pyruvate, oxidation in mitochondria 271
Pyruvate dehydrogenase, lipid dependence 92
Pyruvate dehydrogenase complex, in mitochondria 32
Pyruvate dehydrogenase kinase, in mitochondria 32
Pyruvate dehydrogenase phosphatase, in mitochondria 32
Pyruvate orthophosphate dikinase, in chloroplasts 37
Pyruvate oxidase 127
  activation by lipids 126

Q cycle 257, 284, 285
  in mitochondria 272
Quantum yield, in bacteriorhodopsin 241
  in halorhodopsin 241
Quercetin 304, 305

Radiation energy 20, 238
Ramsaysphaera 2
Ranvier nodes 346, 347
Rapid-equilibrium kinetics 192, 193, 205
Rate theory, of ion transport 152
  of membrane transport 143
Reaction centre, in photosynthesis 245, 246, 254, 255, 257, 265
Reaction enthalpy 236
Reaction kinetics 122
Receptor cell 338
β-Receptors 339
Receptosome 39
Rectangular phase 56, 57
Red algae 249
Redox potential 245, 259
Redox systems 266
Reflection coefficient 175, 176
Renilla 288, 290
  bioluminescence 286
Renilla-luciferin 2-monooxygenase 288
Reserpine 304
Resonance energy transfer 252
Resonance interaction, in excitation transfer 253
Respiratory control, in mitochondria 399
Response, to hormone binding 333
Response transducer, in chemotaxis 327
Retinal 239, 318
Retinal rod outer segment 318, 320
  protein–lipid ratio 42
Reversal potential 167
Reversible electric breakdown 78, 79
Reversible processes 119
L-Rhamnose, transport by phosphotransferase 311
Rhizobium 282
Rhodanese 257
Rhodocyclus 243
Rhodomicrobium 243
Rhodomicrobium vannielii 245
Rhodopin 247
Rhodopseudomonas 243, 253
Rhodopseudomonas sphaeroides 182, 183, 246, 247, 254
Rhodopseudomonas viridis 245, 254
Rhodopsin 83, 318, 319
  lateral diffusion 89
  rotation 89
  with annular lipids 94
Rhodospirillaceae 243
Rhodospirillum 243
Rhodospirillum rubrum 245
Rhodotorula glutinis 198
Ribophorin 109
Ribose, transport in bacteria 216
Ribosephosphate isomerase, in chloroplasts 37
Ribosomes 108, 109, 113, 115
Ribulose-bisphosphate carboxylase 242
  in chloroplasts 37
  synthesis in chloroplasts 36

Ribulose-phosphate-4-epimerase, in chloroplasts 37
Ricinoleic acid 51
Ripoll's equation 159
Rotation 5
of intramolecular bonds 73
of lipid molecules 73
Rotational diffusion 73
Rotational diffusion coefficient 74
Rotenone 270
rRNA, synthesis in chloroplasts 36
synthesis in mitochondria 33
Rubredoxin 255

SITS 304
Saccharin, sweet taste 328
Saccharomyces cerevisiae 6, 50, 192, 198, 228, 231
fatty acids 53
plasma membrane 42
Saddle-splay bending elasticity 78
Saffman–Delbruck equation 89
Salicylanilide 278
Salmonella 280
Salmonella typhimurium 50, 126, 182, 198, 216, 283, 325
Salting-out effect 143
Saponins, as pore formers 184
Sarcina lutea 50
Sarcinic acid 51
Sarcoplasmic reticulum 304, 307
lateral diffusion 75
Saxen's relationship 179
Scaleworms 286
bioluminescence 286
Schistosoma 13
Schistosoma margrebowiei 14
Schlögl's equation 155
Schwann cells 13, 338
Scorpion toxins 342
Second law of thermodynamics 118
Second messenger 335
Secondary active transport 130, 166, 167, 198, 199, 203, 206, 293
Secretin 331
Secretory vesicles 228
Sedoheptulose bisphosphatase, in chloroplasts 37
Semipermeable membranes 170, 171, 173, 174
Semliki forest virus E_2 protein 85
Sensory cell 347
Sensory neuron 337
Sensory rhodopsin 316, 318
Septate junction 14
Serotonin 339, 340
Shear elastic modulus 77
Shear elasticity 76
Short-circuit current 201
Sialic acid 87, 112
transport of in lysosomes 25
Signal amplification 319
Signal recognition particle 109
Signal sequence 108, 109
Signal transmission, afferent 337
centrifugal 337
centripetal 337
efferent 337
motor 337
sensory 337
Simple diffusion 161, 205
Single-channel current 166
Single-file diffusion 162, 219
β-Sitosterol 47, 48
Slime molds 20, 38, 321
Slow-cycling rhodopsin 316
Slug-turning factor 321
Smoluchowski equation 100
Snails, bioluminescence 286
Sodium channels 319, 328, 341, 342, 344, 346, 347
Sodium ion, translocation in bacteria 283
transport by ATPase 307
transport in epithelia 201
transport in mitochondria 33
Sodium ion cycle 300
Sodium-motive bacteria 283, 300, 303
Solvation 140, 142
Solvation Gibbs energy 142
Solvation interactions 267
Solvation shell 267
Solvent-tension theory 171
Soma, of a nerve cell 230
Somatomedins 329, 331
Somatostatin 331, 340
Somatotropin 329, 331
Special chlorophyll 248, 255
Special pair of bacteriochlorophylls 254
Specific channel 183, 187, 221
Specific transport 130
Spectrin 42, 81, 90, 91, 213
lipid dependence 92
Spenceriella 289
Sphacelial cell 26
Spheroidene 247
Spherosome 39
Sphingolipids 43, 45, 46, 47, 50, 51, 105

Sphingomyelin 45, 46, 49, 57
 fatty acids 53
Sphingosine 46
 synthesis 106
Spirillum 15
Spirorchis 13
Spirulina 243
Splay bending elasticity 78
Splitting of water, in photosynthesis 255
Sporopellenin 2
Squid axon 176
Stability constant, of valinomycin complexes 218
Standard potential 263
Staphylococcus aureus 126, 198, 323
Static head, in active transport 201, 205
Stearic acid 51
Sterculic acid 52
Stereocilia 322
Stern equation 99, 102
Steroid hormones 329
Steroids 23
Sterol oxidases 107
Sterols 47, 50, 105, 108, 184
 in mitochondria 31
 in phase transition 69
 synthesis 106
Stibiobacter senarmontii 282
Stigmasterol, synthesis 107
Stochastic processes 129
Stokes–Einstein equation 125, 131, 139
Stokesian diffusion 139
Streaming potentials 178, 179
Streptococcus 15
Streptococcus faecalis 304
Streptomyces fulvissimus 217
Stroma, chloroplast 38
Stroma thylakoid 36
Strong coupling, in excitation transfer 253
Structure-breaking effects, in solvents 143
Structure-forming effect, in solvents 143
Substance P 340
Subtransition 71, 72
Succinate 277
 oxidation in mitochondria 270
Succinate dehydrogenase 128, 278, 281
 activation by lipids 126
 in mitochondria 32
Succinate-ubiquinone reductase 271
Sucrase-isomaltase 85
Sucrose, sweet taste 328
Sugar esters 49
Sulfolobus 38, 45, 282, 304
Sulfolobus acidocaldarius 282
Suloctidil, inhibition of H-ATPase 310
Sulphatases 25
Sulphate, transport in yeasts 198
Sulphate anion, transport in bacteria 216
Sulphate respiration 279
Sulphatide, synthesis 106
Sulphatides 47
Sulphide anion, oxidation in bacteria 282
Sulphite dehydrogenase, in mitochondria 32
Sulpholipids 49
 in chloroplasts 37
 synthesis in chloroplasts 38
Sulphur, oxidation in bacteria 282
Sulphur respiration 280
Superoxide dismutase, in chloroplasts 38
 in mitochondria 32
Surface charge 96, 99
Surface charge density 97
Surface energy 78
Surface potential 45, 49, 97, 128, 210
Surface tension 77, 79
Surfactants 98, 99
Surfactin 220
Symport 130, 195
Synapse 339
Synaptic cleft 230, 338, 347
Synaptic vesicles, protein-lipid ratio 42
Synaptosome 39
Synechococcus 243
Synemin 41
Synexins 336
System reorganization 267

T cell receptor 85
T lymphocyte, helper 351
 killer (cytotoxic) 351
 suppressor 351
T_6 phage receptor 183
TSH-releasing hormone 340
TSH-releasing factor 330, 331
Taste buds 328
Taurine 340
Teichoic acid 15
Telolysosome 26, 39
Temperature dependence, of transport 208
Temperature sensing 321
Teorell's equation 154

Testosterone 329
Tetrahymena 48
  fatty acids 53
Thermal pressure 171
Thermodynamic potential 119
Thermodynamic probability 314
Thermodynamics 117
Thermodynamics of irreversible processes 168
Thermogenin 302
Thermoplasma 45, 304
  protein-lipid ratio 42
Thermoproteus 38
Thermotaxis 319
Thiamin, transport in liver 198
Thiamin diphosphatase 24
Thiobacillus 282
Thiobacillus ferrooxidans 282
Thiocapsa 243
Thiocystis 243
Thiomethylgalactoside, transport in bacteria 198
Thiomicrospora 282
Thiospirillum 243
Thiosulphate anion, oxidation in bacteria 282
Thiosulphate sulphurtransferase 257
Thylakoid 4, 5, 34, 50, 243, 245, 246, 249, 258, 260
  protein synthesis 115
Thyroid-stimulating hormone 331
Throxine 329, 331
Tight junction (*zonula occludens*) 14, 16
Titin 41
Tomatin 184
Tonoplast 26, 310
  protein-lipid ratio 42
Torpedo 89, 341
Transaminase, in chloroplasts 37
Transducin 319, 323
Transduction 221
Transfection 221
Transference number 159
Transferrin 87, 130, 225, 226, 227
Transferrin receptor 85, 87, 225
Transformation, of bacteria 221
Transition state 145
Transition temperature 208
Transition-state theory 122, 124
Transketolase, in chloroplasts 37
Transmission coefficient 143
Trans-gauche transition 62, 88
Tribonema viridis 34
Trichomonads 28
Triglyceride 43
Triglycerides 43, 126
Trimming (of membrane proteins) 23
Triose phosphate shuttle, in chloroplasts 36
Triosephosphate isomerase, in chloroplasts 37
Tritrichomonas foetus 29
tRNA 20
  synthesis in chloroplasts 36
  synthesis in mitochondria 33
  transport from nucleus 221
Tropomyosin, in microfilaments 41
Trypanosomes, number of mitochondria 34
Tubulin 13
α-Tubulin 41
β-Tubulin 41
Tumbling of bacteria 324
Tumour cells 198
Turnover number, in transport 186
Tyrocidin 220

UDP galactose-GlcNAc galactosyl transferase 24
Ubiquinol–cytochrome *c* reductase 271
  in mitochondria 32
Ubiquinone 49, 244, 246, 253, 255, 257, 269, 271, 276, 283
  in mitochondrial oxidation 270
Ubiquinone pool 257, 278
Ubiquitin 115
Ultrafiltration 173
Uncouplers of oxidative phosphorylation 278, 298
Uncoupling protein 302
Undecaprenol kinase, activation by lipids 126
Unidirectional flux 188, 191
Uniport 130
Unstirred layers 149, 150, 176, 210
Ussing's flux ratio 160, 188, 191

V ATPases 303, 310, 311
cis-Vaccenic acid 51
trans-Vaccenic acid 51
Vacuole 9, 14, 26, 304
  in protein turnover 115
  subtegumental 13
Valinomycin 92, 165, 208, 218
  in chemotaxis 327
van der Waals forces 232, 236
Vanadate anion 310
  in ATPase inhibition 304

van't Hoff's formula 170
Vasopressin 331
Vectorial Bohr effect 272, 274, 277, 285, 310
Veilonella alcalensis 283
Vesiculation 130, 221, 222
Vibrio alginolyticus 284, 300
Vibrio costicola 284
Villin 42
Vimentin 41
Vinculin 42, 322
Viruses 49
Viscosity 63
Visual rhodopsin 318, 320
Vitamin B_{12} 185
Vitamin K 49
Vitamin K_2 283
Volume flow 173, 175, 177

Watasemia 289
  bioluminescence 286
Watasemia-luciferin 2-monooxygenase 289
Water splitting, in photosynthesis 256
Water transport 168
Wax esters 49
Weak coupling, in excitation transfer 253
Wien effect 164

Xanthine oxidase 87
Xanthophylls 247, 248, 251
X-ray diffraction 54
Xylans, in cell walls 41
Xylitol, transport by phosphotransferase 311

Yeasts 12, 33
Yersinia pestis 183

Zeroth law of thermodynamics 118
Zeroth-order reaction 125
Zonula adhaerens 15
Zonula occludens 16
Zymogen granules 228, 229